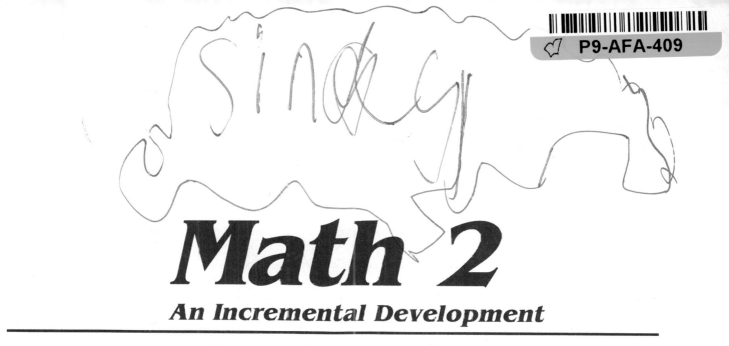

Math 2

An Incremental Development

Home Study Teacher's Edition

Nancy Larson

with

Roseann Paolino
Dee Dee Wescoatt

Saxon Publishers, Inc.

Math 2: An Incremental Development

Home Study Teacher's Edition

Copyright © 1994 by Saxon Publishers, Inc. and Nancy Larson

Printed in the United States of America

ISBN 13: 978-1-56577-015-7
ISBN 10: 1-56577-015-3

Production Supervisor: David Pond
Graphic Artists: Scott Kirby, John Chitwood, Gary Skidmore,
             Tim Maltz, and Chad Threet

13 14 15 0877 17 16 15

4500520775

┌─── *Reaching us via the Internet* ───┐

WWW: http://www.saxonpub.com

E-mail: info@saxonpub.com

Saxon Publishers, Inc.
2450 John Saxon Blvd.
Norman, OK 73071

MATH 2 PROGRAM OVERVIEW

The *Math 2 Home Study Teacher's Edition* contains all the information you will need to teach second grade math concepts to your child. The scripted lessons provide language and techniques that have proven effective for teachers in a variety of settings.

The program is designed so that four lessons are taught each week with the fifth day of the week used for review, if necessary. Each *Math 2* lesson has four components. These components are The Meeting, The Lesson, Class Practice, and Written Practice. Every fifth lesson includes a Written Assessment and every tenth lesson includes an Oral Assessment.

The Meeting occurs at the beginning of each day. During The Meeting your child will use a Meeting Book and a Meeting Strip. A Meeting Strip master is included at the end of the overview. Thirty-three copies of this master will provide enough meeting strips for the year.

Read each day's lesson before teaching it to become familiar with the lesson activity, the materials needed, and the important questions to ask. While teaching The Lesson, you may read the script exactly as it is written. During The Lesson, encourage your child to discuss observations and discoveries and to ask questions. The "doing" part of this program is very important and leads to concept understanding. The Student Workbook includes masters used in lessons, written practice sheets for each lesson (Sides A and B), fact sheets, assessments, and color-coded fact cards. Give your child the appropriate master, written practice sheet, fact sheet, assessment, or fact card page when it is described in a lesson. It is suggested that you keep the workbook and give your child the pages as they are needed.

Each written practice sheet includes a short practice of the new objective and a review of previous concepts. Your child completes Side A with your assistance during the lesson. Side B, which mirrors the problems on Side A, is completed later in the day. This time delay between practice improves retention.

A fact sheet is completed each day. The fact cards and fact sheets provide an opportunity for your child to practice the number facts that have been introduced in previous lessons. All number facts are introduced in groups. Your child is encouraged to use the fact strategies presented in the lessons to find the answers. Automatic fact recall is expected by the end of *Math 3*.

Certain manipulative materials used in the lessons will need to be purchased from an educational supply house. These are pattern blocks, rulers, geoboards, 1" color tiles, tangrams, hundred number chart, demonstration clocks, and balance. A set of six geometric solids and a set of dominoes are optional. Other materials you will need during the year include the following:

Outdoor thermometer

Yardstick

Bathroom scale

Playing cards – 1 deck

Chalkboard or whiteboard

Scrap paper

Construction paper (1 package, multicolored)

3" × 5" cards (about 100)

Crayons

Scissors

Tape

Small plastic bags

Coins (50 pennies, 10 nickels, 30 dimes, and 12 quarters)

TABLE OF CONTENTS

LIST OF MATERIALS

Lesson 1   Math 2 Meeting Book
meeting strip
individual hundred number chart
6 color tiles

Lesson 2   Meeting Book
list of 20 birthdays of family
   members and friends
hundred number chart
1 color tile
Worksheet 2A/2B

Lesson 3   demonstration clock
Worksheet 3A/3B

Lesson 4   20 color tiles (in a bag)
hundred number chart
Worksheet 4A/4B

Lesson 5   Meeting Book
handwriting or scrap paper
addition fact cards — tan
1 envelope, bag, or container to
   store fact cards
Worksheet 5A/5B

Lesson 6   pattern blocks
Fact Sheet AA 1.0
Worksheet 6A/6B

Lesson 7   small plastic animal or other object
pattern blocks (in a basket)
Fact Sheet A 1.2
Worksheet 7A/7B

Lesson 8   10 stuffed animals, dolls, action
   figures, or other toys
color tiles (20 each of 2 colors)
Fact Sheet A 1.2
Worksheet 8A/8B

Lesson 9   pattern blocks
addition fact cards — peach
Fact Sheet A 2.0
Worksheet 9A/9B

Lesson 10   Oral Assessment #1
Written Assessment #1
pattern blocks
Master 2-10 (cut the master in half)
Fact Sheet A 1.2
hundred number chart
handwriting paper
pencil

Lesson 11   10 stuffed animals, dolls, action
   figures, or other toys
basket of pattern blocks
Fact Sheet A 1.2
Worksheet 11A/11B

Lesson 12   demonstration clock
Master 2-12
Fact Sheet A 2.0
Worksheet 12A/12B

Lesson 13   addition fact cards — lavender
Fact Sheet A 3.0
Worksheet 13A/13B

Lesson 14   12 pieces of scrap paper or
   newspaper
crayons — red, blue, yellow, green,
   black, orange, brown, purple
Fact Sheet A 2.0
Worksheet 14A/14B

Lesson 15   Written Assessment #2
Meeting Book
crayons
Fact Sheet A 2.2
Worksheet 15A/15B

Lesson 16   Meeting Book
crayons or markers
demonstration clock
Fact Sheet A 3.0
Worksheet 16A/16B

Lesson 17   Meeting Book
Master 2-17
yellow and green crayons
bag of 20 color tiles
Fact Sheet A 2.2
Worksheet 17A/17B

Lesson 18   Master 2-18
pattern blocks
Fact Sheet A 3.2
Worksheet 18A/18B

Lesson 19   Master 2-18 from Lesson 18
pattern blocks
Fact Sheet A 3.2
Worksheet 19A/19B

Lesson 20   Oral Assessment #2
Written Assessment #3
pattern blocks
Master 2-20
crayons
Fact Sheet A 3.2

Lesson 21   addition fact cards — green
Fact Sheet AA 4.0
Worksheet 21A/21B

Lesson 22   Masters 2-22
plastic bag for shape pieces
Fact Sheet A 3.2
Worksheet 22A/22B

Lesson 23   scrap paper
Fact Sheet AA 4.0
Worksheet 23A/23B

Lesson 24   pattern blocks
construction paper
marker or dark crayon
Master 2-24
scissors
crayons
Fact Sheet A 4.0
Worksheet 24A/24B

Lesson 25   Written Assessment #4
construction paper circle from
    Lesson 24
one 6" construction paper circle (in
    a color different from the circle in
    Lesson 24)
two 6" squares
marker or dark crayon
demonstration clock
Fact Sheet A 4.0
Worksheet 25A/25B

Lesson 26   bag of 20 color tiles
outdoor thermometer
Fact Sheet A 4.0
Worksheet 26A/26B

Lesson 27   Master 2-27
addition fact cards — pink
Fact Sheet A 5.1
Worksheet 27A/27B

Lesson 28   1 cup of 10 dimes
1 cup of 10 pennies
work mat (9" x 12" piece of
    construction paper)
Fact Sheet A 4.2
Worksheet 28A/28B

Lesson 29   Meeting Book
scrap paper
10 pennies
Fact Sheets MA 4.2, A 4.2
Worksheet 29A/29B

Lesson 30   Oral Assessment #3
Written Assessment #5
shape pieces from Lesson 22
Fact Sheet A 5.1

Lesson 31   5 pennies
construction paper mat
Fact Sheet A 5.1
Worksheet 31A/31B

Lesson 32   hundred number chart
10 color tiles
Master 2-32
scissors
small envelope
Fact Sheet A 4.2
Worksheet 32A/32B

Lesson 33   Meeting Book
1 deck of playing cards
addition fact cards — blue
Fact Sheet MA 6.0
Worksheet 33A/33B

Lesson 34   four 8" construction paper circles (1
    each of yellow, blue, red, and
    green)
plastic storage bag (at least 8" x 8")
scissors
a deck of playing cards (tens, jacks,
    queens, and kings removed)
Fact Sheet MA 6.0
Worksheet 34A/34B

Lesson 35   Written Assessment #6
hundred number chart
envelope of hundred number chart
    pieces from Lesson 32
Fact Sheet A 5.1
Worksheet 35A/35B

Lesson 36   7 socks for use as props (3
    matching pairs)
100 color tiles
7 small bags
Fact Sheet A 5.1
Worksheet 36A/36B

Lesson 37   pencil (unsharpened)
color tiles
book
ruler
Master 2-37
Fact Sheet A 5.2
Worksheet 37A/37B

Lesson 38   a deck of playing cards (tens, jacks,
    queens, and kings removed)
Fact Sheet MA 6.0
Worksheet 38A/38B

Lesson 39   3 apples (different types)
3 plates
knife and cutting board
Meeting Book
three 2" construction paper tags
    (See the night before.)
Fact Sheet A 5.2
Worksheet 39A/39B

Lesson 40   Written Assessment #7
Oral Assessment #4
shape pieces from Lesson 22
10 dimes, 10 pennies
Fact Sheet A 5.2

Lesson 41   fraction pieces from Lesson 34
Fact Sheet A 6.2
Worksheet 41A/41B

Lesson 42   addition fact cards — yellow
Fact Sheet A 7.1
Worksheet 42A/42B

Lesson 43   cup of 20 pennies
             cup of 10 dimes
             construction paper work mat
             Fact Sheet A 6.2
             Worksheet 43A/43B

Lesson 44   balance (scale)
             pencil, scissors, marker
             color tiles
             Master 2-44
             Fact Sheet A 7.1
             Worksheet 44A/44B

Lesson 45   Written Assessment #8
             balance
             20 color tiles
             Fact Sheet S 1.2
             Worksheet 45A/45B

Lesson 46   ruler
             yardstick
             tape measure (optional)
             one color tile
             2" x 8" construction paper rectangle
             3" x 9" construction paper rectangle
             Master 2-46
             Fact Sheet A 7.1
             Worksheet 46A/46B

Lesson 47   hundred number chart
             Fact Sheet A 6.2
             Worksheet 47A/47B

Lesson 48   penny, nickel, dime
             cup of 4 dimes and 10 nickels
             Fact Sheet A 7.1
             Worksheet 48A/48B

Lesson 49   9 color tiles
             fact cards
             Fact Sheet S 2.0
             Worksheet 49A/49B

Lesson 50   Written Assessment #9
             Oral Assessment #5
             fraction piece set from Lesson 34
             pattern blocks
             Master 2-50
             Fact Sheet S 3.2

Lesson 51   construction paper
             ten 2" yellow squares
             red and green crayons
             2 yarn or string loops
             Fact Sheet A 7.1
             Worksheet 51A/51B

Lesson 52   yardstick
             poster paint (3 or 4 colors, if
               possible; see lesson)
             white construction paper (4–5
               pieces)
             Fact Sheet A 7.2
             Worksheet 52A/52B

Lesson 53   cups of 10 dimes and 10 pennies
             work mat
             Fact Sheet S 2.0
             Worksheet 53A/53B

Lesson 54   3 small plastic bags (labeled A, B,
               and C) filled with coins as follows:
               Bag A, 4 dimes; Bag B, 23
               pennies; Bag C, 10 nickels
             fifty 3" x 5" cards
             Fact Sheet S 3.2
             Worksheet 54A/54B

Lesson 55   Written Assessment #10
             2 pieces of paper
             2 rulers
             2 color tiles
             Fact Sheet A 7.2
             Worksheet 55A/55B

Lesson 56   2 rulers
             Master 2-56
             Fact Sheet S 3.4
             Worksheet 56A/56B

Lesson 57   1 geoboard and geoband
             Fact Sheet A 7.2
             Worksheet 57A/57B

Lesson 58   cup of 14 pennies
             work mat
             addition fact cards — white
             Fact Sheet A 8.1
             Worksheet 58A/58B

Lesson 59   recipe written on paper
             liquid and dry measuring cups
             measuring spoons
             Fact Sheet A 8.1
             Worksheet 59A/59B

Lesson 60   Written Assessment #11
             Oral Assessment #6
             2 geoboards and geobands
             Master 2-60
             individual clock
             Fact Sheet A 8.1

Lesson 61   Meeting Book
             Fact Sheet S 3.4
             Worksheet 61A/61B

Lesson 62   recipe (from Lesson 59)
             ingredients and supplies for the
               recipe
             Fact Sheet A 8.1
             Worksheet 62A/62B

Lesson 63   demonstration clock
             1 empty egg carton
             Fact Sheet A 8.2
             Worksheet 63A/63B

Lesson 64   50 pennies
             paper cup (optional)
             Fact Sheet S 3.4
             Worksheet 64A/64B

Lesson 65   Written Assessment #12
             fraction pieces (from Lesson 34)
             scrap paper
             small circular object for tracing
             crayons
             Fact Sheet A 8.2
             Worksheet 65A/65B

Lesson 66   8 small cups
             28 dimes and 30 pennies
             Master 2-66
             Fact Sheet A 8.2
             Worksheet 66A/66B

Lesson 67   empty cup
             1 cup of 10 pennies and 1 cup of
                 10 dimes
             Master 2-66 from Lesson 66
             Fact Sheet A 8.2
             Worksheet 67A/67B

Lesson 68   cup of 12 pennies
             work mat
             Fact Sheet S 4.0
             Worksheet 68A/68B

Lesson 69   Masters 2-69 and 2-69A
             crayons
             Fact Sheet A 8.2
             Worksheet 69A/69B

Lesson 70   Written Assessment #13
             Oral Assessment #7
             shape pieces from Lesson 22
             2 geoboards
             5 geobands
             crayon
             Master 2-70
             Teacher's Master 2-70 for the oral
                 assessment
             Fact Sheet S 4.0

Lesson 71   cup of 15 pennies
             cup of 10 dimes
             1 piece each of yellow and white
                 construction paper
             Fact Sheet S 4.0
             Worksheet 71A/71B

Lesson 72   cup of 20 pennies
             cup of 10 dimes
             yellow and white mat from
                 Lesson 71
             Master 2-72
             Fact Sheet A 1-100
             Worksheet 72A/72B

Lesson 73   hundred number chart
             Fact Sheet S 7.0
             Worksheet 73A/73B

Lesson 74   Meeting Book
             ruler
             Master 2-74
             Fact Sheet S 7.0
             Worksheet 74A/74B

Lesson 75   Written Assessment #14
             cup of 20 pennies
             cup of 10 dimes
             yellow and white work mat from
                 Lesson 71
             Fact Sheet S 4.0
             Worksheet 75A/75B

Lesson 76   Master 2-76
             Fact Sheet S 4.4
             Worksheet 76A/76B

Lesson 77   pattern blocks
             scrap paper
             crayon
             pencil
             Master 2-77
             Fact Sheet S 7.0
             Worksheet 77A/77B

Lesson 78   1 large bowl
             2 one-lb packages of beans or 1
                 small box of elbow macaroni
             seven 3" × 5" cards
             Fact Sheet A 1-100
             Worksheet 78A/78B

Lesson 79   bowl of beans from Lesson 78
             4 soup bowls
             scrap paper
             Fact Sheet S 4.4
             Worksheet 79A/79B

Lesson 80   Written Assessment #15
             Oral Assessment #8
             basket of color tiles of mixed colors
             Master 2-80
             geoboard
             geobands
             Fact Sheet S 7.0

Lesson 81   scrap paper
             Fact Sheet S 4.4
             Worksheet 81A/81B

Lesson 82   Master 2-82
             Fact Sheet A 1-100
             Worksheet 82A/82B

Lesson 83   scrap paper
             Fact Sheet S 7.4
             Worksheet 83A/83B

Lesson 84   demonstration clock
             Master 2-84
             Fact Sheet S 7.4
             Worksheet 84A/84B

Lesson 85   Written Assessment #16
             20 small household items or toys
             20 small price tags for the items
             Master 2-85
             Fact Sheet S 7.4
             Worksheet 85A/85B

Lesson 86   5–7 containers of objects (see
            lesson)
            10 small and 4 large paper cups
            piece of paper
            Master 2-86
            Fact Sheet S 7.4
            Worksheet 86A/86B

Lesson 87   demonstration clock
            Fact Sheet S 6.0
            Worksheet 87A/87B

Lesson 88   2 cups of 20 pennies
            Master 2-88
            crayons
            Fact Sheet S 6.0
            Worksheet 88A/88B

Lesson 89   1 small cup of cereal or macaroni
            (less than 100 pieces)
            scrap paper
            multiplication fact cards — green
            Fact Sheet M 10.0
            Worksheet 89A/89B

Lesson 90   Written Assessment #17
            Oral Assessment #9
            1 set of tangrams (or Master 2-
            90A)
            Masters 2-90B and 2-90C
            thermometer
            Fact Sheet S 6.0

Lesson 91   demonstration clock
            Fact Sheet M 10.0
            Worksheet 91A/91B

Lesson 92   Fact Sheet S 5.0
            Worksheet 92A/92B

Lesson 93   advertisements containing $ and ¢
            symbols (from newspapers, etc.)
            scrap paper
            Fact Sheet S 5.0
            Worksheet 93A/93B

Lesson 94   ruler
            tape
            step stool
            Fact Sheet S 6.4
            Worksheet 94A/94B

Lesson 95   Written Assessment #18
            Fact Sheet S 5.0
            Worksheet 95A/95B

Lesson 96   piece of paper
            20 small hard candies (color tiles) in
            a bag
            Master 2-96
            Fact Sheet A 1-100
            Worksheet 96A/96B

Lesson 97   folded paper from Lesson 96
            20 small hard candies (color tiles) in
            a bag
            Master 2-96 from Lesson 96
            Fact Sheet S 5.0
            Worksheet 97A/97B

Lesson 98   Meeting Book
            1 one-dollar bill
            cup of 12 quarters
            Fact Sheet S 6.4
            Worksheet 98A/98B

Lesson 99   multiplication fact cards — beige
            demonstration clock
            Fact Sheet M 10.0
            Worksheet 99A/99B

Lesson 100 Written Assessment #19
            Oral Assessment #10
            20 color tiles
            Master 2-100
            10 dimes and 20 pennies
            scrap paper
            Fact Sheet S 6.4

Lesson 101 Fact Sheet S 8.0
            Worksheet 101A/101B

Lesson 102 10 color tiles
            piece of paper
            3" construction paper square
            Fact Sheet S 8.0
            Worksheet 102A/102B

Lesson 103 set of 6 geometric solids (cone,
            cube, sphere, cylinder,
            rectangular solid, pyramid) or
            objects from your home with
            these shapes
            6 pieces of 1" × 4" construction
            paper
            tape
            Master 2-103
            Fact Sheet S 5.4
            Worksheet 103A/103B

Lesson 104 twenty-five 3" × 5" cards or pieces
            of construction paper
            piece of paper
            Master 2-104
            Fact Sheet A 2-100
            Worksheet 104A/104B

Lesson 105 Written Assessment #20
            inch/centimeter ruler
            2 pieces of $8\frac{1}{2}$" × 11" paper
            Fact Sheet S 8.0
            Worksheet 105A/105B

Lesson 106 elbow macaroni
            small cup
            8 nickels
            multiplication fact cards — yellow
            Fact Sheet M 14.0
            Worksheet 106A/106B

Lesson 107 1 cup of 10 dimes
            1 cup of 20 pennies
            yellow/white mat from Lesson 71
            Fact Sheet M 14.0
            Worksheet 107A/107B

Lesson 108 yellow/white mat from Lesson 71
1 cup of 10 dimes
1 cup of 20 pennies
Fact Sheet S 8.1
Worksheet 108A/108B

Lesson 109 1 cup of 10 dimes
1 cup of 20 pennies
yellow/white mat from Lesson 71
scrap paper
Fact Sheet S 8.1
Worksheet 109A/109B

Lesson 110 Written Assessment #21
Oral Assessment #11
1 set of tangrams
Master 2-110
demonstration clock
individual clock
Fact Sheet S 5.4

Lesson 111 scrap paper
Fact Sheet M 14.0
Worksheet 111A/111B

Lesson 112 Meeting Book
bathroom scale
2 food articles that weigh 1 pound
(butter, sugar, etc.)
10-pound object (e.g., bag of
potatoes)
3 heavy objects that will fit on a
bathroom scale
Fact Sheet S 8.1
Worksheet 112A/112B

Lesson 113 20 cm × 30 cm light color
construction paper
marker
Master 2-113
string
ruler
Fact Sheet A 2-100
Worksheet 113A/113B

Lesson 114 Meeting Book
scrap paper
Fact Sheet S 8.2
Worksheet 114A/114B

Lesson 115 Written Assessment #22
2 geoboards
8 geobands
Fact Sheet S 8.2
Worksheet 115A/115B

Lesson 116 cup of 20 pennies
multiplication fact cards — lavender
Fact Sheet M 13.0
Worksheet 116A/116B

Lesson 117 1 cup of 6 quarters, 1 cup of 5
dimes, 1 cup of 5 nickels, and 1
cup of 10 pennies
Fact Sheet M 13.0
Worksheet 117A/117B

Lesson 118 Meeting Book
11 pink 2" × 3" construction paper
tags
38 yellow 2" × 3" construction
paper tags
Fact Sheet S 5.4
Worksheet 118A/118B

Lesson 119 cup of 20 pennies
Master 2-119
8 pieces of 3" × 4" construction
paper
crayons
Fact Sheet S 8.2
Worksheet 119A/119B

Lesson 120 Written Assessment #23
Oral Assessment #12
2 envelopes
20 tags
writing paper
markers or crayons
Masters 2-120A and 2-120B
tape
10 dimes and 20 pennies
scrap paper
Fact Sheet M 13.0

Lesson 121 20 color tiles (all one color)
Master 2-121
black, red, orange, and yellow
crayons
Fact Sheet A 2-100
Worksheet 121A/121B

Lesson 122 1 rectangular piece of paper
1 geoboard
1 geoband
Master 2-122
Fact Sheet M 14.1
Worksheet 122A/122B

Lesson 123 Master 2-119 from Lesson 119
Fact Sheet S-100
Worksheet 123A/123B

Lesson 124 bag of 30 color tiles
multiplication fact cards — pink
Fact Sheet M 17.0
Worksheet 124A/124B

Lesson 125 Written Assessment #24
1 geoboard
2 geobands
Master 2-125
Fact Sheet M 17.0
Worksheet 125A/125B

Lesson 126 Fact Sheet M 14.1
Worksheet 126A/126B

Lesson 127 Meeting Book
Fact Sheet S-100
Worksheet 127A/127B

Lesson 128 2 geoboards
2 geobands
Fact Sheet M 17.0
Worksheet 128A/128B

Lesson 129 bag of 40 color tiles
multiplication fact cards — blue
Fact Sheet M 15.0
Worksheet 129A/129B

Lesson 130 Written Assessment #25
Oral Assessment #13
set of double six dominoes in a small brown paper bag
Masters 2-130A (optional), 2-130B, and 2-130C
crayons
5 quarters
10 dimes
10 nickels
20 pennies
Fact Sheet M 15.0

Lesson 131 3 store coupons (10¢–50¢ value)
Fact Sheet M 14.1
Worksheet 131A/131B

Lesson 132 bag of 20 color tiles
2 small plates
Fact Sheet S-100
Worksheet 132A/132B

Meeting Strip

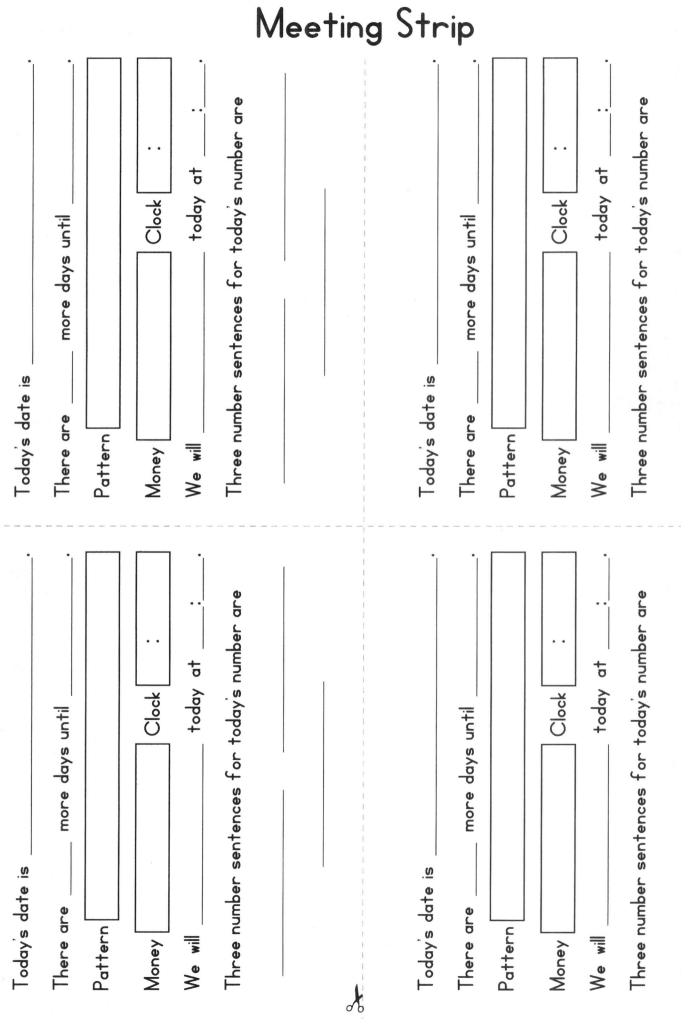

Today's date is _____ . _____ .

There are _____ more days until _____

Pattern

Money | Clock | : |

We will _____ today at _____ : _____ .

Three number sentences for today's number are

_____

_____

Today's date is _____ . _____ .

There are _____ more days until _____

Pattern

Money | Clock | : |

We will _____ today at _____ : _____ .

Three number sentences for today's number are

_____

_____

Today's date is _____ . _____ .

There are _____ more days until _____

Pattern

Money | Clock | : |

We will _____ today at _____ : _____ .

Three number sentences for today's number are

_____

_____

Today's date is _____ . _____ .

There are _____ more days until _____

Pattern

Money | Clock | : |

We will _____ today at _____ : _____ .

Three number sentences for today's number are

_____

_____

L esson 1

reading and identifying numbers to 100
identifying left and right

lesson preparation

materials

Math 2 Meeting Book (This will be used for all lessons.)

meeting strip (This will be used for all lessons. A master is included at the end of the Overview.)

individual hundred number chart

6 color tiles

the night before

• Write the year on each month's calendar in your child's Math 2 Meeting Book.

• Fill in the dates through today's date on this month's calendar. Each day's date will be added during The Meeting. Write Saturday's and Sunday's dates prior to Monday's meeting.

in the morning

• Use 1" color tiles to cover the following numbers on the hundred number chart:

15, 37, 48, 60, 82, 97

THE MEETING

"We will begin each day with a Math Meeting."

"Today you will learn about parts of the Math Meeting."

• Show your child the Math Meeting Book.

"Each morning we will be doing some activities using this Math Meeting Book."

"Let's look through the book."

"What do you see?"

• Spend 2–3 minutes looking through the book with your child.

"Write your name on the first page."

calendar

• Open your child's Meeting Book to this month's calendar.

"What do we call this?" *a calendar*

"Why do we use a calendar?"　to tell us the month, date, year, and day of the week

"We use the calendar to tell us the month, date, year, and day of the week."

- Point to each on the calendar.

"What year is it?"

- Point to the year.

"What month is it?"

- Point to the month.

"Yesterday was the (twenty-eighth of August)."

- Point to the date.

"What do you think today's date is?"　(twenty-ninth of August)

"We write the (twenty-ninth) using the number (twenty-nine)."

"We write it like this."

- Write the date on the calendar.

"What day of the week is it today?"

"How do you know?"

"It's (Thursday) because we wrote today's date under (Thursday)."

- Point to the date and move your finger up to (Thursday).

"Let's read the days of the week together."

- Point as you say the days of the week together.

"Each day we will write today's full date on a meeting strip."

- Show your child the meeting strip.

"We will write the month, the date, and the year."

"What will we write first?"　the month

"What month is it?"

"How do we spell (month)?"

- Write the month on the meeting strip.

"What is the date?"

"Each number is made up of the number symbols from zero to nine."

"We call these digits."

"What digits will we use to write today's date?"

- Write the date on the meeting strip.

"What is the year?"

"What digits will we use to write the year?"

- Record the year on the meeting strip.

  "Let's say the date together."

- Point to the words on the meeting strip as you read, "Today's date is (month, date, and year)."

  "Each morning we will write the date on the calendar, write the full date on a meeting strip, and read the names of the days of the week."

weather graph

"What does the weather outside feel like to you today?"

"Does it feel _____ every day?"

"What else could it be?"   hot, warm, cool, cold

"Each morning we will talk about what the weather outside feels like to you."

"We will use a weather graph to keep track of how the weather feels to you."

"Let's look at this month's weather graph."

- Point to the labels on the weather graph.

  "Let's read the labels on this graph."

- Point to the words "hot," "warm," "cool," and "cold" as you read the labels with your child.

  "What types of clothing do we wear when it's hot?"

  "What kinds of things do we do when it's hot?"

  "What types of clothing do we wear when it's warm?"

  "What kinds of things do we do when it's warm?"

  "What types of clothing do we wear when it's cool?"

  "What kinds of things do we do when it's cool?"

  "What types of clothing do we wear when it's cold?"

  "What kinds of things do we do when it's cold?"

- Discuss with your child how we can use clothing and things we do as clues for describing the weather.

  "How will we know if it is hot or warm?"

  "How will we know if it is cold or cool?"

  "How will we know if it is warm or cool?"

  "Each morning, you will color a box on the weather graph to show what the weather feels like to you that day."

  "You will use a red crayon for hot, yellow for warm, green for cool, and blue for cold."

  "Which color crayon will you use today?"

"What box will you color?"

• Ask your child to color the first box on the graph next to the appropriate word.

counting

"Each day during The Meeting we will practice counting."

• Turn to the hundred number chart in the back of the Meeting Book.

"Let's count from 35 to 50."

• Point to the numbers on the hundred number chart as you count together.

"Let's count backward from 50 to 35."

• Point to the numbers on the hundred number chart as you count together.

• Other information on the meeting strip is not used today.

THE LESSON

Reading and Identifying Numbers to 100

"Today you will learn how to read and identify numbers to 100."

• Show your child the hundred number chart with the covered numbers.

"This is called a hundred number chart."

"Why do you think it is called that?"   there are 100 numbers on the chart

"I covered some numbers on the hundred number chart."

• Point to one of the color tiles.

"What number do you think is under this tile?"

"How do you know?"

"Let's check."

• Remove the tile.

• Repeat with all the tiles.

"Now I will point to a number on the hundred number chart."

• Point to a number.

"What is this number?"

• Repeat with 5–8 numbers.

"Let's read the numbers on the chart together."

"Point to each number as we say it."

• Read the numbers from 1 to 100.

"Look at the numbers in the row that begins with 51."

"Read these numbers."

"What do you notice about the numbers in this row?"

"Did you hear or do you see anything the same?"

"Look at the numbers in the row that begins with 71."

"Read these numbers."

"What do you notice about the numbers in this row?"

"Did you hear or do you see anything the same?"

"If I said the number 64, how would you know where to find it quickly on the hundred number chart?"

"If I said the number 38, how would you know where to find it quickly on the hundred number chart?"

"Point to 29 on the hundred number chart."

"How did you know where to find 29?"

- Repeat with the following numbers: 67, 92, 45, 59, 80, 17, 11, 14

Identifying Left and Right

"Tomorrow we will use the hundred number chart to find a number above, below, to the right, and to the left of a number."

"You will need to know what direction to go when I say 'right' and what direction to go when I say 'left.' "

"Today we will practice right and left."

"I will write the word 'right' on the top right-hand corner of the chalkboard."

"Point to where you think I should write the word 'right.' "

- Write the word "right" on the top right-hand corner of the chalkboard.

"I will write the word 'left' on the top left-hand corner of the chalkboard."

"Point to where you think I should write the word 'left.' "

- Write the word "left" on the top left-hand corner of the chalkboard.

"Stand and face the chalkboard."

"I will give you directions to follow."

"Raise your right hand. Raise your left hand. Hop on your right foot. Touch your left ear. Point to the right. Wave your left hand. Turn to your left. Face the chalkboard. Turn to your right. Face the chalkboard."

Lesson 2

graphing data on a graph
identifying one more and one less than a number

---

lesson preparation

materials

Meeting Book

list of 20 birthdays of family members and friends

hundred number chart

1 color tile

THE MEETING

calendar

- Open your child's Meeting Book to this month's calendar.

  "What year is it?"

- Point to the year.

  "What month is it?"

- Point to the month.

  "Yesterday was the _____th of (month)."

- Point to the date.

  "What do you think today's date is?"

  "We write the _____th using the number _____."

  "We write it like this."

- Write the date on the calendar.

  "What day of the week is it today?"

  "How do you know?"

  "It's _____ because we wrote today's date under _____."

- Point to the date and move your finger up to the day of the week.

  "Let's read the days of the week together."

- Point as you say the days of the week together.

  "Each day we will write today's full date on a meeting strip."

  "We will write the month, the date, and the year."

"What will we write first?" the month

"What month is it?"

"How do we spell (month)?"

- Write the month on the meeting strip.

"What is the date?"

"What digits will we use to write today's date?"

- Write the date on the meeting strip.

"What is the year?"

"What digits will we use to write the year?"

- Record the year on the meeting strip.

"Let's say the date together."

- Point to the words on the meeting strip as you read, "Today's date is (month, date, and year)."

weather graph

"What does the weather outside feel like to you today?"

"Why do you think it feels _____?"

"You will color a box on the graph to show what the weather feels like to you today."

"Which color will you use for _____?"

"What box will you color?"

- Ask your child to color the correct box on the graph. Begin each row on the left. Do not skip boxes.

counting

- Turn to the hundred number chart in the back of the Meeting Book.

"Let's count from 47 to 72 as you point to the numbers on the hundred number chart."

- Count with your child.

"Let's count backward from 72 to 47 as you point to the numbers on the hundred number chart."

- Other information on the meeting strip is not used today.

THE LESSON

Graphing Data on a Graph

"Today you will learn how to show information using a graph."

"I have a list of the birthdays of our friends and family members."

"Let's make a graph so we can see how many people were born in each month."

• Open the Meeting Book to pages 28 and 29.

"We will use this graph to show the names and birth dates of our family members and friends."

"When is your birthday?"

"Where do you think you will write your name on the graph?"

"Write your name in the first box above (name of month)."

"Now write your birth date in the box."

"My birthday is _____."

"Where will you write my name?"

"Write my name and birth date in the box above (name of month)."

• Repeat with the birth dates of 15–20 friends and family members.

"What do you notice about our graph?"

• Encourage your child to offer as many observations as possible.

Identifying One More and One Less Than a Number

"Today you will learn how to find numbers above, below, to the right, and to the left of a number on a hundred number chart."

"You will also learn how to find numbers one more and one less than a number using a hundred number chart."

• Give your child a hundred number chart and a 1" color tile.

"Use the tile to cover the number 24 on your hundred number chart."

"Point to the number above 24."

"What is the number?"   14

"Point to the number below 24."

"What is the number?"   34

"Point to the number to the right of 24."

"What is the number?"   25

"Point to the number to the left of 24."

"What is the number?"   23

• Repeat with 83, 52, 38, 79, and 16.

"Let's read the numbers from 31 to 40."

"Use your finger to point to each number as we say it."

"When we count forward, which way does your finger move?"   right

"Point to 46."

"Point to one more than 46."

"Which way does your finger move?"   right

"What is one more than 46?"   47

"Point to 51."

"Point to one more than 51."

"Which way does your finger move?"   right

"What is one more than 51?"   52

"Let's read the numbers from 80 to 71."

"Use your finger to point to each number as we say it."

"When we count backward, which way does your finger move?"   left

"Point to 69."

"Point to one less than 69."

"Which way does your finger move?"   left

"What is one less than 69?"   68

"Point to 18."

"Point to one less than 18."

"Which way does your finger move?"   left

"What is one less than 18?"   17

• Repeat with other numbers, if necessary.

• Write the following on the chalkboard:

one more than 93

"Read what I wrote on the chalkboard."

"Point to 93 on your hundred number chart."

"Now point to one more than 93."

"What is one more than 93?"   94

• Repeat with each of the following:

one less than 21	one more than 32
one less than 13	one more than 64
one less than 85	

WRITTEN PRACTICE

• Give your child **Worksheet 2A/2B**. (Side A is completed after the lesson and Side B is completed later in the day.)

"We will do Side A together now."

"Write your name and the date at the top of the paper."

"Use the calendar in the Meeting Book to help you spell the month."

"Look at the first problem on Side A."

- Ask your child to read the question or read the question for your child, if necessary.

"How can you find the answer to this question?"

"Use the calendar in the Meeting Book to help you spell (today)."

- Repeat with the other problems.

"Turn your paper over to the other side."

"We will do this side later in the day."

- Complete Side B with your child later in the day.

Name _____ **LESSON 2A**
Date _____ Math 2

1. What day of the week is it today? _____

2. Write the letter **e** to the right of the **n**.
   Write the letter **o** to the left of the **n**.

   o   n   e

3. Use a red crayon to color these numbers on the chart.
   Cross off each number after you color it.

   14, 1, 28, 10, 17, 5, 34, 12, 23
   50, 19, 16, 37, 6, 15, 32, 30, 41

	2	3	4		7	8	9	
11		13		15	16		18		20
21	22		24	25	26	27		29	30
31		33		35	36		38		40
	42	43	44			47	48	49	

4. What number is one more than **28**? _____29_____
   What number is one less than **37**? _____36_____

2-2Wa   Copyright © 1991 by Saxon Publishers, Inc. and Nancy Larson. Reproduction prohibited.

Name _____ **LESSON 2B**
Date _____ Math 2

1. Read these numbers to someone.

   39, 18, 12, 22, 40, 48

2. Write the letter **t** to the left of the **w**.
   Write the letter **o** to the right of the **w**.

   t   w   o

3. Use a red crayon to color these numbers on the chart.
   Cross off each number after you color it.

   22, 14, 37, 8, 23, 34, 42, 17, 48, 6
   12, 44, 27, 46, 4, 32, 47, 7, 24, 2

1		3		5			9	10
11		13		15	16		18	19	20
21				25	26		28	29	30
31		33		35	36		38	39	40
41		43		45				49	50

2-2Wb   Copyright © 1991 by Saxon Publishers, Inc. and Nancy Larson. Reproduction prohibited.

Lesson 3

telling time to the hour

---

lesson preparation

materials
demonstration clock

THE MEETING

calendar

> *"What year is it?"*

- Point to the year.

> *"What month is it?"*

- Point to the month.

> *"Yesterday was the _____th of (month)."*

- Point to the date.

> *"What do you think today's date is?"*

> *"We write the _____th using the number _____."*

> *"We write it like this."*

- Write the date on the calendar.

> *"What day of the week is it today?"*

> *"How do you know?"*

> *"It's _____ because we wrote today's date under _____."*

- Point to the date and move your finger up to the day of the week.

> *"Let's read the days of the week together."*

- Point as you say the days of the week together.

> *"Each day we will write today's full date on a meeting strip."*

> *"We will write the month, the date, and the year."*

> *"What will we write first?"   the month*

> *"What month is it?"*

> *"How do we spell (month)?"*

- Write the month on the meeting strip.

"What is the date?"

"What digits will we use to write the date?"

• Write the date on the meeting strip.

"What is the year?"

"What digits will we use to write the year?"

• Record the year on the meeting strip.

"Let's say the date together."

• Point to the words on the meeting strip as you read, "Today's date is (month, date, and year)."

weather graph

"What does the weather outside feel like to you today?"

"Why do you think it feels _____?"

"You will color a box on the graph to show what the weather feels like to you today."

"What color will you use for _____?"

"What box will you color?"

• Ask your child to color the correct box on the graph. Begin each row on the left. Do not skip boxes.

"What kind of weather have we had most often?"

counting

• Turn to the hundred number chart in the back of the Meeting Book.

"Let's count from 15 to 31 as you point to the numbers on the hundred number chart."

• Count with your child.

"Let's count backward from 31 to 15 as you point to the numbers on the hundred number chart."

graph questions

"Let's look at the birthday graph on pages 28 and 29."

"How many people have a birthday in (month)?"

"(Name of a person)'s birthday is in (month)."

"How many other people have a birthday in that month?"

"How many more birthdays are in _____ than in _____?"

"How do you know?"

"Make up a question about our graph for me to answer."

• Other information on the meeting strip is not used today.

THE LESSON

Telling Time to the Hour

"*Today you will learn how to tell time to the hour.*"

- Show your child the demonstration clock.

- Set the time on the hour (use 6:00 or 7:00).

"*This is the time I got out of bed this morning.*"

"*What time was it?*"

"*How do you know?*"

- Set the time on the hour (use 9:00 or 10:00).

"*This is the time I went to bed last night.*"

"*What time was it?*"

"*How do you know?*"

"*What do you notice about the hands on the clock?*"   one is long and the other is short

"*How do we know that this is (nine) o'clock?*"   the long hand points to the 12 and the short hand points to the (nine)

"*Which hand on the clock tells us the hour?*"   the short hand

"*We call the short hand the hour hand.*"

"*Notice how the hour hand points to the hour, but does not touch the number.*"

"*We have a new hour each time the long hand is pointing to the 12.*"

"*How will we write (nine) o'clock using digital time?*"

- Write the digital time on the chalkboard.

"*When we write the digital time we write the hour first, a colon, and the number of minutes.*"

"*We will write two zeros after the colon because we don't have any extra minutes.*"

- Show 2:00 on the demonstration clock.

"*What time is it?*"

"*How do you know?*"

"*Write the digital time on the chalkboard.*"

- Repeat with 11:00 and 12:00.

- Write "4:00" on the chalkboard.

"*What time is this?*"

"*How will we show this time on the clock?*"

- Ask your child to set the demonstration clock.

  "Where does the hour hand point?"

  "Where does the long hand point?"

- Repeat with 6:00, 1:00, and 12:00.

  "Show three o'clock on the demonstration clock."

- Allow time for your child to do this.

  "Write the digital time on the chalkboard."

- Repeat with 8:00 and 2:00.

WRITTEN PRACTICE

- Give your child **Worksheet 3A/3B**.

  "We will do Side A together now."

  "Write your name and the date at the top of the paper."

  "Use the calendar in the Meeting Book to help you spell the month."

  "Look at the first problem on Side A."

- Ask your child to read the directions for the first problem or read the directions for your child, if necessary.

  "What do you think you will do for this problem?"

- Repeat with the other problems.

  "Turn your paper over to the other side."

  "We will do this side later in the day."

- Complete Side B with your child later in the day.

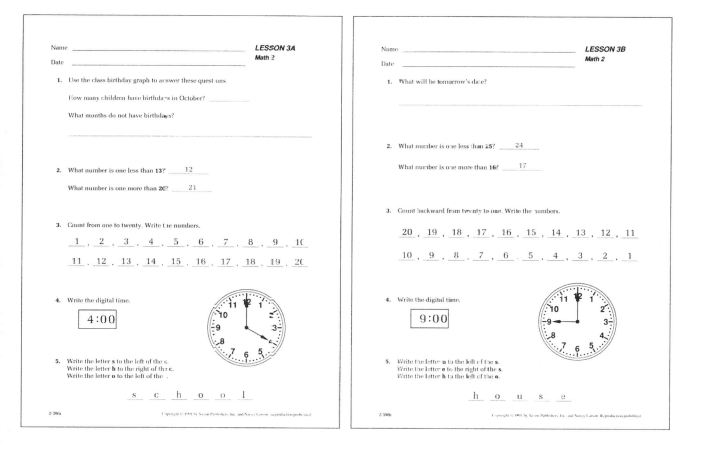

Name _____ LESSON 3A
Date _____ Math 2

1. Use the class birthday graph to answer these questions.

   How many children have birthdays in October? _____

   What months do not have birthdays?

   _____

2. What number is one less than 13? ___12___

   What number is one more than 20? ___21___

3. Count from one to twenty. Write the numbers.

   __1__ , __2__ , __3__ , __4__ , __5__ , __6__ , __7__ , __8__ , __9__ , __10__

   __11__ , __12__ , __13__ , __14__ , __15__ , __16__ , __17__ , __18__ , __19__ , __20__

4. Write the digital time.

   [4:00]

5. Write the letter **s** to the left of the **c**.
   Write the letter **h** to the right of the **c**.
   Write the letter **o** to the left of the **l**.

   __s__ __c__ __h__ __o__ __o__ __l__

2-3Wa

Name _____ LESSON 3B
Date _____ Math 2

1. What will be tomorrow's date?

   _____

2. What number is one less than 25? ___24___

   What number is one more than 16? ___17___

3. Count backward from twenty to one. Write the numbers.

   __20__ , __19__ , __18__ , __17__ , __16__ , __15__ , __14__ , __13__ , __12__ , __11__

   __10__ , __9__ , __8__ , __7__ , __6__ , __5__ , __4__ , __3__ , __2__ , __1__

4. Write the digital time.

   [9:00]

5. Write the letter **u** to the left of the **s**.
   Write the letter **e** to the right of the **s**.
   Write the letter **h** to the left of the **o**.

   __h__ __o__ __u__ __s__ __e__

2-3Wb

Lesson 4

addition facts—doubles to 20

lesson preparation

materials
20 color tiles (in a bag)
hundred number chart

in the morning
• Set the demonstration clock to **7:00**.

THE MEETING

calendar

> *"What year is it?"*

• Point to the year.

> *"What month is it?"*

• Point to the month.

> *"Yesterday was the _____th of (month)."*

• Point to the date.

> *"What do you think today's date is?"*

> *"We write the _____th using the number _____."*

> *"We write it like this."*

• Write the date on the calendar.

> *"What day of the week is it today?"*

> *"How do you know?"*

> *"It's _____ because we wrote today's date under _____."*

• Point to the date and move your finger up to the day of the week.

> *"Let's read the days of the week together."*

• Point as you say the days of the week together.

> *"Let's write today's full date on a meeting strip."*

> *"We will write the month, the date, and the year."*

> *"What will we write first?"*   the month

"What month is it?"

"How do we spell (month)?"

- Write the month on the meeting strip.

"What is the date?"

"What digits will we use to write the date?"

- Write the date on the meeting strip.

"What is the year?"

"What digits will we use to write the year?"

- Record the year on the meeting strip.

"Let's say the date together."

weather graph

"What does the weather outside feel like to you today?"

"Why do you think it feels _____?"

"You will color a box on the graph to show what the weather feels like to you today."

"What color will you use for _____?"

"What box will you color?"

- Ask your child to color the correct box on the graph.

"What kind of weather have we had most often?"

counting

- Turn to the hundred number chart in the back of the Meeting Book.

"Let's count from 64 to 85 as you point to the numbers on the hundred number chart."

- Count with your child.

"Let's count backward from 85 to 64 as you point to the numbers on the hundred number chart."

graph questions

"Let's look at the birthday graph on pages 28 and 29."

"How many people have a birthday in (month)?"

"(Name of a person)'s birthday is in (month)."

"How many other people have a birthday in that month?"

"How many more birthdays are in _____ than in _____?"

"How do you know?"

"Make up a question about our graph for me to answer."

clock

- Point to the demonstration clock.

  "What time is it?"

  "How do you know?"

  "Let's write the digital time on the meeting strip."

- Point to the word "clock" on the meeting strip.

  "Write the digital time in the box next to the word 'clock.' "

- Other information on the meeting strip is not used today.

THE LESSON

Addition Facts—Doubles to 20

"Today you will learn the addition facts called the doubles."

- Write the following on the chalkboard:

  $$1 + 1 = \qquad 2 + 2 = \qquad 3 + 3 = \qquad 4 + 4 = \qquad 5 + 5 =$$
  $$6 + 6 = \qquad 7 + 7 = \qquad 8 + 8 = \qquad 9 + 9 =$$

  "What do you notice about each problem?"   the numbers being added are the same

  "In all of these problems both numbers are the same."

  "When we see two of the same thing, we can say that we are seeing double."

  "That is why we call these problems the doubles."

  "Do you know the answers for any of these problems?"

- Write the suggested answers next to each problem.

  "We will use color tiles to check the answers."

- Give your child a bag of 20 color tiles.

- Point to 1 + 1 on the chalkboard.

  "Let's use the tiles to prove that one plus one is two."

  "Take one tile out of the bag."

  "Now take another tile out of the bag."

  "Count the tiles."

  "Do you have two tiles?"

  "Let's check two plus two."

  "Make a pile of two tiles."

  "Now make another pile of two tiles."

"Are the piles the same height?"

"This is one way we can tell that we have a double."

"How many tiles do you have altogether?" 4

"Let's check three plus three."

"Make a pile of three tiles."

"Now make another pile of three tiles."

"Are the piles the same height?"

"This is one way we can tell that we have a double."

"How many tiles do you have altogether?" 6

• Repeat with all the chalkboard problems.

"Put the tiles back in the bag."

• Write the following on the chalkboard:

$$
\begin{array}{ccccc}
1 & 2 & 3 & 4 & 5 \\
+\,1 & +\,2 & +\,3 & +\,4 & +\,5 \\
\hline
\end{array}
$$

$$
\begin{array}{cccc}
6 & 7 & 8 & 9 \\
+\,6 & +\,7 & +\,8 & +\,9 \\
\hline
\end{array}
$$

"This is another way to write the same problems."

"Let's say the answers together as I write them on the chalkboard."

"Now I will say the problems in random order."

"Let's see how fast you can say the answers."

• Allow your child to refer to the chalkboard answers, if necessary.

"Now we will play Number Fact Jeopardy."

"I will say the answer to a double."

"See if you can tell me the problem."

"Six." three plus three

• Repeat with all combinations.

CLASS PRACTICE

"Now you will use the tiles to practice finding numbers on the hundred number chart."

• Give your child a hundred number chart.

"Find the number 84."

"Put a tile on the number to the right of 84."

"What number did you cover?"   85

"Find the number 38."

"Put a tile on the number to the left of 38."

"What number did you cover?"   37

"Find the number 23."

"Put a tile on the number that is one more than 23."

"What number did you cover?"   24

"Find the number 94."

"Put a tile on the number that is one less than 94."

"What number did you cover?"   93

"Find the number 37."

- Write the following on the chalkboard: 37 + 1

  "What number do you think you will cover for this problem?"   38

- Repeat with 93.

  "Close your eyes."

  "Now I will cover five numbers on the hundred number chart."

- Cover 5 numbers with tiles.

  "Open your eyes and look at the hundred number chart."

  "Try to guess which numbers I covered."

  "Lift the tiles to check your guess."

Written Practice

- Give your child **Worksheet 4A/4B**.

  "We will do Side A together now."

  "Write your name and the date at the top of the paper."

  "Use the calendar in the Meeting Book to help you spell the month."

- Ask your child to read the directions for the first problem or read the directions for your child, if necessary.

  "What do you think you will do for this problem?"

- Repeat with the other problems.

- Complete Side B with your child later in the day.

LESSON 4A

Name _____
Date _____

Math 2

1. What day of the week is it today? _____

   What day of the week was it yesterday? _____

2. Write the letter **h** to the left of the **o**.

   Write the letter **t** to the right of the **■**.

   h   o   t

3. Fill in the missing numbers.

   _1_ , _2_ , _3_ , _4_ , _5_ , _6_ , _7_ , _8_ , _9_ , _10_

   _11_ , _12_ , _13_ , _14_ , _15_ , _16_ , _17_ , _18_ , _19_ , _20_

4. What number is one less than **26**? _25_

   What number is one more than **43**? _44_

5. Use an orange crayon to color these numbers on the chart.

   23, 42, 37, 8, 29, 4, 2   16 + 1   26 + 1   9 + 1   4 + 1
   41, 43, 45, 3, 28, 9, 1   32 + 1   46 + 1   43 + 1   12 + 1   6 + 1

					6			
11	12		14	15	16	18	19	20
21	22		24	25	26			30
31	32		34	35	36	38	39	40
				46	48	49	50	

LESSON 4B

Name _____
Date _____

Math 2

1. What will be tomorrow's date? _____

2. Write the letter **t** to the left of the **o**.

   Write the letter **e** to the right of the **o**.

   t   o   e

3. Fill in the missing numbers.

   _1_ , _2_ , _3_ , _4_ , _5_ , _6_ , _7_ , _8_ , _9_ , _10_

   _11_ , _12_ , _13_ , _14_ , _15_ , _16_ , _17_ , _18_ , _19_ , _20_

4. What number is one more than **76**? _77_

   What number is one less than **52**? _51_

5. Fill in the missing numbers.

1	2	3	4	5	6	7	8	9	10
11	12	13	14	15	16	17	18	19	20
21	22	23	24	25	26	27	28	29	30
31	32	33	34	35	36	37	38	39	40
41	42	43	44	45	46	47	48	49	50

esson 5

counting by 10's to 100
writing numbers to 100

---

lesson preparation

materials
Meeting Book
handwriting or scrap paper
addition fact cards—tan
1 envelope, bag, or container to store fact cards

the night before
• Separate the tan addition fact cards.

in the morning
• Set the demonstration clock to **4:00**.

THE MEETING

calendar

"What year is it?"

• Point to the year.

"What month is it?"

• Point to the month.

"Yesterday was the _____th of (month)."

• Point to the date.

"What do you think today's date is?"

"We write the _____th using the number _____."

"We write it like this."

• Write the date on the calendar.

"What day of the week is it today?"

"How do you know?"

"Let's read the days of the week together."

• Point as you say the days of the week together.

"Let's write today's full date on a meeting strip."

"We will write the month, the date, and the year."

"What will we write first?"   the month

"What month is it?"

"How do we spell (month)?"

• Write the month on the meeting strip.

"What is the date?"

"What digits will we use to write the date?"

• Write the date on the meeting strip.

"What is the year?"

"What digits will we use to write the year?"

• Record the year on the meeting strip.

"Let's say the date together."

weather graph

"What does the weather outside feel like to you today?"

"Why do you think it feels _____?"

"You will color a box on the graph to show what the weather feels like to you today."

"What color will you use for _____?"

"What box will you color?"

• Ask your child to color the correct box on the graph.

"What kind of weather have we had most often?"

counting

"Let's count from 78 to 93."

• Ask your child to point to the numbers on the hundred number chart as you count together.

"Now let's count backward from 93 to 78 as you point to the numbers on the hundred number chart."

graph questions

"Let's look at the birthday graph on pages 28 and 29."

"How many people have a birthday in (month)?"

"(Name of a person)'s birthday is in (month)."

"How many other people have a birthday in that month?"

"How many more birthdays are in _____ than in _____?"

"How do you know?"

"Make up a question about our graph for me to answer."

clock

- Point to the demonstration clock.

"What time is it?"

"How do you know?"

"Write the digital time in the box next to the word 'clock' on the meeting strip."

- Other information on the meeting strip is not used today.

THE LESSON

Counting by 10's to 100

"Let's count by 10's together as I point to the numbers on the hundred number chart."

- Count by 10's with your child as you point to the numbers on the hundred number chart.

"What did you notice about the numbers I pointed to?"   they were in a column

- Open the Meeting Book to page 38.

"Let's count by 10's to 100 as I write the numbers we say on the first counting strip."

- Start at the bottom and work up (similar to the numbering on a thermometer or graph). Write "0" below the 10.

"What do you notice about the numbers on this counting strip?"   they all end in 0

"We will use this counting strip to help us count forward and backward by 10's each morning."

Writing Numbers to 100

"Today you will learn how to write numbers to 100."

"Most numbers are written as they sound."

"For example, what do you hear first when I say 34?"   30

- Write "30" on the chalkboard.

"What do you hear next?"   4

- Write "+ 4" next to the 30 on the chalkboard.

"Instead of writing '30 plus 4,' we write '34.' "

- Write '30 + 4 = 34' on the chalkboard.

- Repeat with 57, 71, and 29.

  "Writing numbers as they sound looks pretty easy."

  "But writing numbers as they sound doesn't always work."

  "Some numbers are tricky."

  "What numbers do you hear for eleven?"   you can't hear any

  "How do we write 'eleven'?"

- Write '11" on the chalkboard.

  "Eleven is a tricky number to write."

  "What other numbers are tricky to write like eleven?"   10 and 12

  "There are some other tricky numbers."

  "What do we hear first in the number fourteen?"   4

- Write "4" on the chalkboard.

  " 'Teen' means 'ten.' "

- Write "teen" on the chalkboard. Cross out one "e."

  "If we take an 'e' out of teen, we have the word 'ten.' "

- Write "+ 10" next to the four.

  "Fourteen is the same as four plus ten, but we write it as 14."

- Write "4 + 10 = 14."

- Repeat with 16, 18, 13, and 15.

  "What was the trick for the teen numbers?"   all of the teen numbers begin with a 1

  "All of the teen numbers begin with a one."

  "Now you will practice writing some numbers."

  "Remember to watch out for the tricky numbers."

  "What are the tricky numbers?"   10, 11, 12, and the teens

- Give your child a pencil and piece of handwriting paper or scrap paper.

- Read the following numbers one at a time. Occasionally ask: "Is this a tricky number?" and "What digits did you write?"

      62, 78, 40, 19, 94, 12, 83, 16, 25, 18

CLASS PRACTICE

addition—doubles fact cards

"Today I will give you fact cards to use to practice the doubles facts."

- Give your child a bag containing the tan addition fact cards.

   "You will use only the addition side now."

- Show your child the 4 + 4 fact card.

   "What is the answer?"

- Show your child the back of the card.

   "Where do you see the answer for four plus four on this card?"

- Repeat with several cards.

   "Now I will show you how you will use these cards to practice the addition facts."

   "First you will check to make sure that all the addition cards are face up."

- Ask your child to check his/her cards.

   "Now you will hold the fact cards in one hand."

   "You will look at the top card and say the answer as quickly as possible."

   "How will you check your answer?"

   "The answer for each problem is the top number on the back of the card."

   "If your answer is correct, put the card on the table."

   "If the answer is incorrect, put the card at the bottom of the pile in your hand."

   "Keep practicing until all the answers are correct."

- Allow time for your child to practice independently.

WRITTEN PRACTICE

- Give your child **Worksheet 5A/5B**.

   "Write your name and the date at the top of Side A."

- Ask your child to read the directions for the first problem or read the directions for your child, if necessary.

   "What do you think you will do for this problem?"

- Repeat with the other problems.

- Complete Side B with your child later in the day.

1. Fill in the missing days of the week.

   Sunday, ___Monday___, Tuesday, ___Wednesday___,

   Thursday, ___Friday___, Saturday

2. Which letter is on the right? __p__   e i p

   Which letter is in the middle? __i__

   Which letter is on the left? __e__

3. Use the birthday graph to answer these questions.

   How many birthdays are in June? _____

   Which month has the most birthdays? _____

4. Count by 10's. Write the numbers

   _10_, _20_, _30_, _40_, _50_, _60_, _70_, _80_, _90_, _100_

5. Fill in the missing numbers.

1	2	3	4	5	6	7	8	9	10
11	12	13	14	15	16	17	18	19	20
21	22	23	24	25	26	27	28	29	30
31	32	33	34	35	36	37	38	39	40
41	42	43	44	45	46	47	48	49	50

2-5Wa

1. Fill in the missing days of the week.

   ___Sunday___, Monday, ___Tuesday___, Wednesday

   ___Thursday___, Friday, ___Saturday___

2. Which letter is on the left? __b__   b y o

   Which letter is on the right? __o__

   Which letter is in the middle? __y__

3. Fill in the missing numbers.

   _1_, _2_, _3_, _4_, _5_, _6_, _7_, _8_, _9_, _10_

   _11_, _12_, _13_, _14_, _15_, _16_, _17_, _18_, _19_, _20_

4. Count backward by 10's. Write the numbers.

   _100_, _90_, _80_, _70_, _60_, _50_, _40_, _30_, _20_, _10_

5. Fill in the missing numbers.

1	2	3	4	5	6	7	8	9	10
11	12	13	14	15	16	17	18	19	20
21	22	23	24	25	26	27	28	29	30
31	32	33	34	35	36	37	38	39	40
41	42	43	44	45	46	47	48	49	50

2-5Wb

Lesson 6

identifying the attributes of pattern blocks

lesson preparation

materials

pattern blocks

Fact Sheet AA 1.0

in the morning

• Set the demonstration clock to **2:00**.

THE MEETING

calendar

"What year is it?"

• Point to the year.

"What month is it?"

• Point to the month.

"Yesterday was the _____th of (month)."

• Point to the date.

"What do you think today's date is?"

"How will we write the _____th?"

• Write the date on the calendar.

"What day of the week is it today?"

"How do you know?"

"Let's read the days of the week together."

"Let's write today's full date on a meeting strip."

"What will we write first?"   the month

"What month is it?"

"How do we spell (month)?"

• Write the month on the meeting strip.

"What is the date?"

"What digits will we use to write the date?"

- Write the date on the meeting strip.

  "What is the year?"

  "What digits will we use to write the year?"

- Record the year on the meeting strip.

  "Let's say the date together."

weather graph

"What does the weather outside feel like to you today?"

"Why do you think it feels _____?"

"What color will you use for _____?"

- Ask your child to color the correct box on the graph.

  "What kind of weather have we had most often?"

counting

"Let's count from 57 to 84."

- Ask your child to point to the numbers on the hundred number chart as you count together.

  "Now let's count backward from 84 to 57 as you point to the numbers on the hundred number chart."

  "Now we will use our counting strip to help us count by 10's to 100."

- Turn to page 38 in the Meeting Book.

  "Let's count by 10's to 100."

- Point to the numbers as you count together.

  "Now let's count backward from 100 by 10's."

graph questions

"Let's look at the birthday graph on pages 28 and 29."

"How many people have a birthday in (month)?"

"(Name of a person)'s birthday is in (month)."

"How many other people have a birthday in that month?"

"How many more birthdays are in _____ than in _____?"

"How do you know?"

"Make up a question about our graph for me to answer."

clock

- Point to the demonstration clock.

  "What time is it?"

"How do you know?"

"Write the digital time in the box next to the word 'clock' on the meeting strip."

- Other information on the meeting strip is not used today.

THE LESSON

Identifying the Attributes of Pattern Blocks

"Today you will learn about pattern blocks."

- Give your child a basket of pattern blocks.

"Put a double handful of pattern blocks on the table."

- Allow time for your child to examine the pattern blocks.

"What do you notice about the pattern blocks?"

- Allow time for your child to offer observations.

"How many different pattern blocks are there?"   6

"We can name the pieces by their color or by their shape."

"What colors are the pattern blocks?"   *yellow, red, blue, green, orange, tan*

"Do you know the names of any of the shapes?"

"The green block is a triangle, the orange block is a square, the yellow block is a hexagon, the red block is a trapezoid, and the blue and tan pieces are parallelograms." (The blue and tan pieces are also rhombuses; a rhombus is a parallelogram with equal sides.)

- Allow your child to refer to the pieces by color or shape. Whenever possible, refer to the pieces by shape or by shape and color to help your child learn the shape names.

"Now you will have a chance to make a design using the pattern blocks."

- Allow 5–7 minutes for creating a design.

"Tell me about your design and the part of your design that you like the best."

- Allow time for your child to describe his/her design.

"Carefully put your pattern blocks in the container."

CLASS PRACTICE

"Let's review the doubles."

"What is an example of a double?"

"Now I will say an example."

"See how quickly you can tell me the answer."

- Say the double facts in random order.

"Today you will practice the doubles facts in a different way."

"You will have a chance to see how many of the doubles facts you can answer in one minute."

"Let's see what one minute feels like."

"When I say 'Go,' close your eyes."

"Open your eyes when I say 'Stop.' "

- Say "Go." After one minute say "Stop."

"Now you will practice the doubles by writing the answers on a fact sheet."

- Give your child **Fact Sheet AA 1.0**.

"There will always be 25 problems on a fact sheet."

"Try to answer at least 15 problems correctly before I say 'Stop.' "

"Work as quickly as you can to get the correct answers."

"Do not skip any problems."

- Fold your child's paper just above the fourth row, if desired. The goal is to finish 15 facts in one minute.

- When your child is ready, say "Go." After exactly one minute, say "Stop." (Unless stated otherwise, all Math 2 fact sheets are timed for one minute.)

"Now you will use a crayon to correct your paper."

"Draw a line after the last problem you answered."

"When you correct your fact sheet, if the answer is correct put a dot next to the answer."

- Demonstrate on the chalkboard.

"If the answer is incorrect, circle the problem and write the correct answer."

- Demonstrate on the chalkboard.

- Read the problems and the answers slowly for only the facts your child answered.

"Count the number of problems you answered correctly before I said stop."

"Write this number at the bottom of your paper."

- Record the score on the **Individual Recording Form**.

"Now finish writing the answers for the rest of the number facts."

• Allow time for your child to do this. Correct these problems with your child.

WRITTEN PRACTICE

• Complete **Worksheet 6A** with your child.

• Complete **Worksheet 6B** with your child later in the day.

Name _____   *LESSON 6A*
                                Math 2
Date _____

1. Fill in the missing days of the week.

   Tuesday, __Wednesday__ , __Thursday__ , Friday

2. Draw a square to the left of the circle.
   Draw a triangle to the right of the circle.

   □   ○   △

3. Use the birthday graph to answer these questions.

   How many birthdays are in February? _____

   Which months have exactly two birthdays?

   _____

4. Count backward by 10's. Write the numbers.

   _100_ , _90_ , _80_ , _70_ , _60_ , _50_ , _40_ , _30_ , _20_ , _10_

5. Show | 2:00 |

2-6Wa   Copyright © 1994 by Saxon Publishers, Inc. and Nancy Larson. Reproduction prohibited.

Name _____   *LESSON 6B*
                                Math 2
Date _____

1. Fill in the missing days of the week.

   __Tuesday__ , Wednesday, Thursday, __Friday__

2. Draw a circle to the right of the square.
   Draw a triangle to the left of the square.

   △   □   ○

3. Fill in the missing numbers.

   _1_ , _2_ , _3_ , _4_ , _5_ , _6_ , _7_ , _8_ , _9_ , _10_

   _11_ , _12_ , _13_ , _14_ , _15_ , _16_ , _17_ , _18_ , _19_ , _20_

4. Count by 10's. Write the numbers.

   _10_ , _20_ , _30_ , _40_ , _50_ , _60_ , _70_ , _80_ , _90_ , _100_

5. Show | 8:00 |

2-6Wb   Copyright © 1994 by Saxon Publishers, Inc. and Nancy Larson. Reproduction prohibited.

Lesson 7

creating and reading a repeating pattern
identifying ordinal position to fifth

lesson preparation

materials

small plastic animal or other object

pattern blocks (in a basket)

Fact Sheet A 1.2

in the morning

• Set the demonstration clock to **10:00**.

THE MEETING

calendar

> *"Yesterday was the _____th of (month)."*

• Point to the date.

> *"What do you think today's date is?"*

> *"How will we write the _____th?"*

• Write the date on the calendar.

> *"What day of the week is it today?"*

> *"How do you know?"*

> *"Let's read the days of the week together."*

> *"Let's write today's full date on a meeting strip."*

> *"What will we write first?"   the month*

> *"What month is it?"*

> *"How do we spell (month)?"*

• Write the month on the meeting strip.

> *"What is the date?"*

> *"What digits will we use to write the date?"*

• Write the date on the meeting strip.

> *"What is the year?"*

"*What digits will we use to write the year?*"

- Record the year on the meeting strip.

"*Let's say the date together.*"

weather graph

"*What does the weather outside feel like to you?*"

"*Why do you think it feels _____?*"

"*What color will you use for _____?*"

- Ask your child to color the correct box on the graph.

"*What kind of weather have we had most often?*"

counting

"*Let's count from 71 to 98.*"

- Ask your child to point to the numbers on the hundred number chart as you count together.

"*Let's count backward from 98 to 71 as you point to the numbers on the hundred number chart.*"

"*Now we will use our counting strip to help us count by 10's to 100.*"

- Turn to page 38 in the Meeting Book.

"*Let's count by 10's to 100.*"

- Point to the numbers as you count together.

"*Now let's count backward from 100 by 10's.*"

graph questions

"*Let's look at our birthday graph.*"

"*How many people have a birthday in (month)?*"

"*(Name of a person)'s birthday is in (month).*"

"*How many other people have a birthday in that month?*"

"*How many more birthdays are in _____ than in _____?*"

"*How do you know?*"

"*Make up a question about our graph for me to answer.*"

clock

- Point to the demonstration clock.

"*What time is it?*"

"*How do you know?*"

"*Write the digital time in the box next to the word 'clock' on the meeting strip.*"

• Other information on the meeting strip is not used today.

THE LESSON

Creating and Reading a Repeating Pattern

"Today you will learn how to use pattern blocks to make and read a repeating pattern."

"You will also learn how to identify ordinal position to fifth."

• Place a small plastic animal on the floor or table.

"We are going to pretend that this is my (cow)."

"I want to build a fence around my (cow) so that she won't wander away."

"I want my fence to have a pattern."

"Watch how I build my fence."

• Use red and green pattern blocks to build a red, green, red, green, etc., fence.

"Let's read the colors together as I build my fence around my (cow)."

• Make sure that when the ends of the fence meet, the pattern continues.

"I have made a repeating pattern."

"We can name this pattern using the colors or the shapes."

"Let's read the pattern together using colors."   red, green, etc.

• Point to each block as you read the pattern.

"Let's read the pattern together using shapes."   trapezoid, triangle, etc.

• Point to each block as you read the pattern.

"We can read the pattern in another way."

"We can use letters."

"We will use a different letter for each pattern block."

"We will call the red trapezoid A and the green triangle B."

"We will read this pattern A, B, A, B, A, B, A, B, etc."

"Let's read the pattern together using letters."

• Point to each block as you read the pattern.

"Now you will have a chance to build a fence around a toy animal."

• Give your child a basket of pattern blocks and the animal.

"Choose two pattern blocks that are different colors."

"Make an AB pattern fence around your (cow)."

- Assist your child, if necessary.
- When your child finishes, continue.

    "Touch each pattern block as you read the colors in your fence."

- Allow time for your child to read his/her fence colors.

    "Touch each pattern block as you read the shapes in your fence."

- Allow time for your child to read his/her fence shapes.

    "Put your pattern blocks in the basket."

Identifying Ordinal Position to Fifth

"Now we will practice ordinal position using the pattern blocks."

"Take one pattern block of each color."

"Put the tan pattern block back in the basket."

"Put the yellow hexagon on the left side of the table."

"This will be the first block."

"Put the triangle second."

"Put the square third."

"Put the red trapezoid fourth."

"Put the blue parallelogram fifth."

"Read the colors of your pattern blocks from first to fifth."

"Now pick up the square and make it fifth."

"Pick up the red trapezoid and make it first."

"Now read the colors of your pattern blocks from first to fifth."   red, yellow, green, blue, orange

"Which pattern block is fourth?"   blue

"Now pick up the triangle and make it fourth."

"Pick up the red trapezoid and make it second."

"Now read the colors of your pattern blocks from first to fifth."   yellow, red, blue, green, orange

"Which pattern block is third?"   blue

"Put your pattern blocks in the basket."

CLASS PRACTICE

"Today you will use your fact cards to practice the addition facts."

"Check your fact cards to make sure that the addition side is face up."

"Read each number fact and say the answer."

"If the answer is correct, put the fact card on the table."

"If the answer is incorrect, put the fact card beneath the pile in your hand."

• Allow time for your child to practice the facts independently.

"Now you will write the answers for the doubles facts on a fact sheet."

"Try to work as quickly as you can to get the correct answers."

"Do not skip any problems."

• Give your child **Fact Sheet A 1.2.** Fold the fact sheet above the fourth row, if desired.

"Try to write all the answers in the first three rows before I say 'Stop.' "

"If you have time, write the answers for the problems in the last two rows."

• Say "Go." After exactly one minute, say "Stop." (Unless stated otherwise, all Math 2 fact sheets are timed for one minute.)

"Now you will use a crayon to correct your paper."

"Draw a line after the last problem you answered."

"When you correct your fact sheet, if the answer is correct put a dot next to the answer."

"If the answer is incorrect, circle the problem and write the correct answer."

• Read the problems and the answers slowly for only the facts your child answered.

"Count the number of problems you answered correctly before I said stop."

"Write this number at the bottom of your paper."

• Record the score on the **Individual Recording Form.**

"Now finish writing the answers for the rest of the number facts."

• Allow time for your child to do this.

• Correct these problems with your child.

WRITTEN PRACTICE

• Complete **Worksheet 7A** with your child.

• Complete **Worksheet 7B** with your child later in the day.

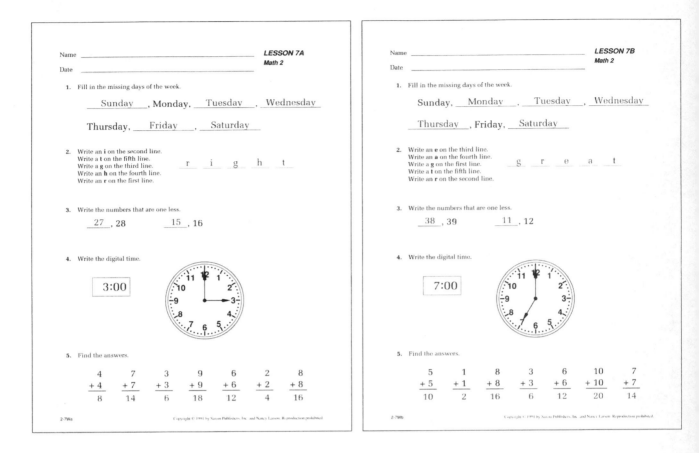

1. Fill in the missing days of the week.

   ___Sunday___ , Monday, ___Tuesday___ , ___Wednesday___

   Thursday, ___Friday___ , ___Saturday___

2. Write an **i** on the second line.
   Write a **t** on the fifth line.
   Write a **g** on the third line.   r i g h t
   Write an **h** on the fourth line.
   Write an **r** on the first line.

3. Write the numbers that are one less.

   ___27___ , 28   ___15___ , 16

4. Write the digital time.

   3:00

5. Find the answers.

4	7	3	9	6	2	8
+4	+7	+3	+9	+6	+2	+8
8	14	6	18	12	4	16

1. Fill in the missing days of the week.

   Sunday, ___Monday___ , ___Tuesday___ , ___Wednesday___

   ___Thursday___ , Friday, ___Saturday___

2. Write an **e** on the third line.
   Write an **a** on the fourth line.
   Write a **g** on the first line.   g r e a t
   Write a **t** on the fifth line.
   Write an **r** on the second line.

3. Write the numbers that are one less.

   ___38___ , 39   ___11___ , 12

4. Write the digital time.

   7:00

5. Find the answers.

5	1	8	3	6	10	7
+5	+1	+8	+3	+6	+10	+7
10	2	16	6	12	20	14

L esson 8

identifying and acting out some, some more stories
comparing numbers to 50

---

lesson preparation _____

materials

10 stuffed animals, dolls, action figures, or other toys

color tiles (20 each of 2 colors)

Fact Sheet A 1.2

in the morning

• Write the following in the pattern box on the meeting strip:

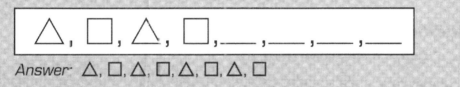

Answer: △, □, △, □, △, □, △, □

THE MEETING

calendar

> *"Yesterday was the _____th of (month)."*

• Point to the date.

> *"What do you think today's date is?"*

> *"How will we write the _____th?"*

• Write the date on the calendar.

> *"What day of the week is it today?"*

> *"How do you know?"*

> *"Let's read the days of the week together."*

> *"Let's write today's full date on a meeting strip."*

> *"What month is it?"*

> *"How do we spell (month)?"*

• Write the month on the meeting strip.

> *"What is the date?"*

"What digits will we use to write the date?"

- Write the date on the meeting strip.

"What is the year?"

"What digits will we use to write the year?"

- Record the year on the meeting strip.

"Let's say the date together."

"Let's read the names of the months together."

- Point as you say the names of the months with your child.

weather graph

"What does the weather outside feel like to you?"

"Why do you think it feels _____?"

"What color will you use for _____?"

- Ask your child to color the correct box on the graph.

"What kind of weather have we had most often?"

counting

"Let's count from 16 to 43."

- Ask your child to point to the numbers on the hundred number chart as you count together.

"Let's count backward from 43 to 16."

"Let's count by 10's to 100."

- Point to the numbers on the counting strip as you count together.

"Now let's count backward from 100 by 10's."

graph questions

"Let's look at our birthday graph."

"How many people have a birthday in (month)?"

"(Name of a person)'s birthday is in (month)."

"How many other people have a birthday in that month?"

"How many more birthdays are in _____ than in _____?"

"How do you know?"

"Make up a question about our graph for me to answer."

patterning

"Each morning you will have a pattern to finish."

- Point to the pattern box on the meeting strip.

"Yesterday we made an AB repeating pattern with our pattern blocks."

"Today's pattern is a repeating pattern."

"Let's read the first part of the pattern together."   triangle, square, ...

"What shape do you think will come next?"   triangle

- Fill in the next shape on the meeting strip.

- Repeat with the rest of the blanks.

"Let's read the shape pattern together."   triangle, square, ...

"Let's read the pattern using letters."   A, B, A, B, A, B, A, B

clock

"Use the demonstration clock to show five o'clock."

"How do you know that this is five o'clock?"

"Write the digital time on the meeting strip."

- Other information on the meeting strip is not used today.

The Lesson

Identifying and Acting Out Some, Some More Stories

"Today you will learn to identify and act out some, some more stories."

"You will also learn how to compare numbers to 50."

"We will pretend that this area of the room (designate an area) will be called the store, the movie theater, the cafeteria, the doctor's office, or another special place in our story."

"Let's try acting out a story."

"We will pretend that your (stuffed animals, dolls, action figures, or other toys) are children."

"Four children went to the store."

- Ask your child to place four (toys) in the designated area.

"Two more children met them there."

- Ask your child to place two more (toys) in the designated area.

"How many children are at the store?"

"What happened in this story?"   some children went to the store and then some more went

"Some children were at the store and then some more came."

"This is a some, some more story."

"Let's try acting out another story."

"Three children went to a party."

• Ask your child to place three (toys) in the designated area.

"Five other children went to the same party."

• Ask your child to place five more (toys) in the designated area.

"What happened in this story?"   some children went to a party and then some more went

"Some children were at the party and then some more came."

"This is a some, some more story."

"Make up another some, some more story for us to act out."

• Act out several of your child's stories. Repeat the above questions.

Comparing Numbers to 50

• Write the following numbers on the chalkboard: 6   6

"Which of these numbers is larger or greater?"   they are both the same

"How do you know?"

"How could you prove it?"

"We will use color tiles to prove that both numbers are the same."

"We will make a pile for each number just like we did when we found the answers for the doubles."

• Give your child 2 bags with 20 color tiles in each bag.

"Make a pile of color tiles for each number."

"Use different color tiles for each pile."

"Put the piles next to each other."

"What do you notice about the piles?"   they are the same height

• Write the following numbers on the chalkboard: 8   12

"Which of these numbers is greater?"

"How can you prove it?"

"Do you think the pile for the greater number will be taller or shorter?"

"Let's use color tiles to prove that 12 is the greater number."

"Make a pile of color tiles for each number."

"Use a different color for each pile."

"Put the piles next to each other."

"What do you notice?"   the pile with 12 is taller

"When we use color tiles, will the taller pile always be the one with more?"

"Why?"   because it contains extra tiles

- Write the following numbers on the chalkboard: 11   7

  "Which of these numbers is smaller?"

  "How can you prove it?"

  "Do you think the pile for the smaller number will be taller or shorter?"

  "Use color tiles to prove that seven is the smaller number."

  "Make a pile of color tiles for each number."

  "Use a different color for each pile."

  "What do you notice?"   the pile with 7 is shorter

  "When we use color tiles, will the shorter pile always be the one with less (or fewer)?"

  "Why?"   because it contains fewer tiles

- Write the following numbers on the chalkboard: 14   9

  "Circle the greater number."

- Repeat with the following combinations: 13, 11; 25, 16; 31, 45; 46, 21; 37, 34. Alternate asking for the greater and the smaller number.

CLASS PRACTICE

"Let's practice the doubles facts."

"Say the answer as I hold up each fact card."

- Hold up each addition fact card as your child says only the answer.

- Repeat several times.

  "Now you will write the answers for the doubles facts on a fact sheet."

  "Try to work as quickly as you can to get the correct answers."

  "Do not skip any problems."

- Give your child **Fact Sheet A 1.2**. Fold the fact sheet above the fourth row, if desired.

  "Try to write all the answers in the first three rows before I say 'Stop.'"

  "If you have time, write the answers for the problems in the last two rows."

- Time your child for one minute.

  "Now you will use a crayon to correct your paper."

  "Draw a line after the last problem you answered."

  "When you correct your fact sheet, if the answer is correct put a dot next to the answer."

  "If the answer is incorrect, circle the problem and write the correct answer."

- Read the problems and the answers slowly for the facts your child answered.

  "Count the number of problems you answered correctly before I said stop."

  "Write this number at the bottom of your paper."

- Record the score on the **Individual Recording Form**.

  "Now finish writing the answers for the rest of the number facts."

- Allow time for your child to do this.

- Correct these problems with your child.

  "A few days ago we practiced writing numbers."

  "Some of the numbers are tricky."

  "What are the tricky numbers?"   *10, 11, 12, teens*

  "Let's practice writing some numbers."

  "Write the numbers on the back of your fact sheet."

- Say the following numbers one at a time: 56, 72, 41, 60, 17, 12, 31, 19

WRITTEN PRACTICE

- Complete **Worksheet 8A** with your child.
- Complete **Worksheet 8B** with your child later in the day.

Name _____   **LESSON 8A**
                                           Math 2
Date _____

1. Use the birthday graph to answer these questions.

    What is the first month of the year? _____ January _____

    How many children have a birthday in that month? _____

2. One of these is my dog.
    Use the clues to find my dog.
    He is not second. Cross out that dog.
    He is not on the left. Cross out that dog.
    Circle my dog.

3. Circle the greater number.

    (24)   17

4. Show   5:00   on the clock.

5. Continue the repeating pattern.

    ○ □ ○ □ ○ □ ○ □ ○ □

6. Fill in the missing numbers.

1	2	3	4	5	6	7	8	9	10
11	12	13	14	15	16	17	18	19	20
21	22	23	24	25	26	27	28	29	30

2-8Wa

Name _____   **LESSON 8B**
                                           Math 2
Date _____

1. What will be tomorrow's date? _____

2. One of these is my grandmother's dog.
    Use the clues to find her dog.
    She is not third. Cross out that dog.
    She is not on the left. Cross out that dog.
    Circle her dog.

3. Circle the greater number.

    38   (43)

4. Show   4:00   on the clock.

5. Continue the repeating pattern.

    △ ○ △ ○ △ ○ △ ○ △ ○

6. Fill in the missing numbers.

1	2	3	4	5	6	7	8	9	10
11	12	13	14	15	16	17	18	19	20
21	22	23	24	25	26	27	28	29	30

2-8Wb

L esson 9

covering pattern blocks with equal size pieces
addition facts—adding one

---

lesson preparation

materials

pattern blocks

addition fact cards—peach

Fact Sheet A 2.0

the night before

• Separate the peach fact cards.

in the morning

• Write the following in the pattern box on the meeting strip:

Answer: ▱, ▢, ▱, ▢, ▱, ▢, ▱, ▢

THE MEETING

calendar

"*Yesterday was the _____th of (month).*"

• Point to the date.

"*What do you think today's date is?*"

"*How will we write the _____th?*"

• Write the date on the calendar.

"*What day of the week is it today?*"

"*Let's read the days of the week together.*"

"*Let's write today's full date on a meeting strip.*"

"*What month is it?*"

"*How do we spell (month)?*"

• Write the month on the meeting strip.

"*What is the date?*"

"What digits will we use to write the date?"

- Write the date on the meeting strip.

"What is the year?"

"What digits will we use to write the year?"

- Record the year on the meeting strip.

"Let's say the date together."

"Let's read the names of the months together."

- Point as you say the names of the months with your child.

weather graph

"What does the weather outside feel like to you?"

"Why do you think it feels _____?"

"What color will you use for _____?"

- Ask your child to color the correct box on the graph.

"What kind of weather have we had most often?"

counting

"Let's count from 26 to 52."

- Ask your child to point to the numbers on the hundred number chart as you count together.

"Now let's count backward from 52 to 26."

"Let's count by 10's to 100."

- Point to the numbers on the counting strip as you count together.

"Now let's count backward from 100 by 10's."

graph questions

"Let's look at our birthday graph."

"How many people have a birthday in (month)?"

"(Name of a person)'s birthday is in (month)."

"How many other people have a birthday in that month?"

"How many more birthdays are in _____ than in _____?"

"How do you know?"

"Make up a question about our graph for me to answer."

patterning

"Each morning you will have a pattern to finish."

- Point to the pattern box on the meeting strip.

"Let's read the first part of the pattern together."   parallelogram, square, …

"This is a repeating pattern."

"What repeats in this pattern?"   parallelogram, square

"What shape do you think will come next?"   parallelogram

- Fill in the next shape on the meeting strip.

- Repeat with the rest of the blanks.

"Let's read the shape pattern together."   parallelogram, square, …

"Let's read the pattern using letters."   A, B, A, B, …

clock

"Use the demonstration clock to show eleven o'clock."

"How do you know that this is eleven o'clock?"

"Write the digital time on the meeting strip."

- Other information on the meeting strip is not used today.

The Lesson

Covering Pattern Blocks with Equal Size Pieces

"Today you will learn how to cover the pattern block pieces with equal size pieces."

"You will also learn the addition facts that are called the adding one facts."

"Today you will use the pattern block pieces in a different way."

"You will try to find which pattern block pieces you can use to cover other pieces."

- Give your child a basket of pattern blocks

"Put a double handful of pattern blocks on the table."

"Hold up the largest pattern block."

"Let's prove that it is the largest by comparing it to each of the other pattern blocks."

"Put the yellow hexagon in front of you on the table."

"Will the red trapezoid cover the yellow hexagon completely?"   no

"Try to cover the yellow hexagon with only red trapezoids."

"Can you do it?"

"How many red trapezoids did you use?"   2

- Repeat with all colors.

  "What color pattern blocks can you use to cover the yellow hexagon if you only use one color at a time?"   red, green, blue

  "Put three yellow hexagons on the table in front of you."

  "Cover the first hexagon using only green triangles."

  "Cover the second hexagon using only blue parallelograms."

  "Cover the third hexagon using only red trapezoids."

- Allow time for your child to do this.

  "How many triangles did you use to cover the hexagon?"   6

  "How many parallelograms did you use to cover the hexagon?"   3

  "How many trapezoids did you use to cover the hexagon?"   2

  "Put the pattern blocks in the basket."

Addition Facts — Adding One

- Write the following on the chalkboard:

0	1	2	3	4	5	6	7	8	9
+ 1	+ 1	+ 1	+ 1	+ 1	+ 1	+ 1	+ 1	+ 1	+ 1

  "What is the same about all of these problems?"

  "We will call these problems the adding one facts."

  "Let's read the problems and say the answers together."

- Write the answers below the problems.

  "What do you notice about the answers?"

  "I'll say the problems in random order."

  "Let's see how fast you can say the answers."

- Allow your child to refer to the chalkboard problems, if necessary.

- Erase the chalkboard answers.

  "Let's play Number Fact Jeopardy."

  "I will tell you the answer."

  "See if you can tell me the adding one problem."

  "Six."   five plus one

- Repeat with all combinations.

- If your child is having difficulty, rewrite the answers on the board.

CLASS PRACTICE

"Today I will give you adding one fact cards."

- Give your child the peach addition fact cards.

  "Use these cards to practice the adding one facts."

- Allow time for your child to practice independently.

  "Put away your fact cards."

  "Now you will write the answer for the adding one facts on a fact sheet."

  "Try to work as quickly as you can to get the correct answers."

  "Do not skip any problems."

- Give your child **Fact Sheet A 2.0**. Fold the fact sheet above the fourth row, if desired.

  "Try to write all the answers in the first three rows before I say 'Stop.' "

  "If you have time, write the answers for the problems in the last two rows."

- Time your child for one minute.

  "Use a crayon to correct your paper."

  "Draw a line after the last problem you answered."

- Read the problems and answers slowly for the facts your child answered.

  "Count the number of problems you answered correctly before I said stop."

  "Write this number at the bottom of your paper."

- Record the score on the **Individual Recording Form**.

  "Now finish writing the answers for the rest of the number facts."

- Allow time for your child to do this.

- Correct these problems with your child.

WRITTEN PRACTICE

- Complete **Worksheet 9A** with your child.

- Complete **Worksheet 9B** with your child later in the day.

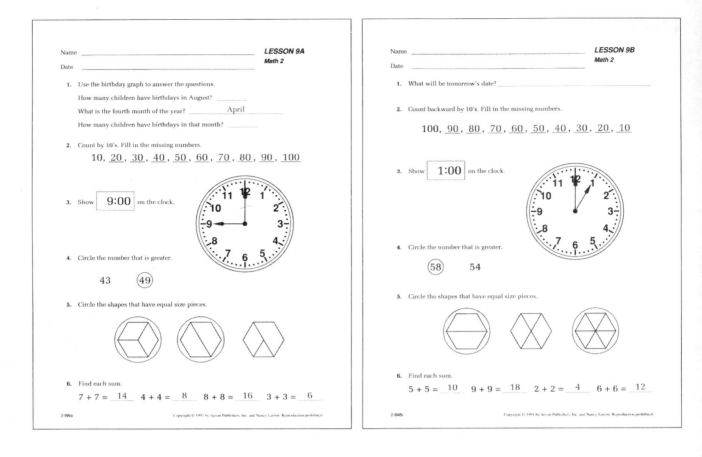

Name _____   **LESSON 9A**
                                 Math 2
Date _____

1. Use the birthday graph to answer the questions.

    How many children have birthdays in August? _____

    What is the fourth month of the year? _____ April _____

    How many children have birthdays in that month? _____

2. Count by 10's. Fill in the missing numbers.

    10, <u>20</u>, <u>30</u>, <u>40</u>, <u>50</u>, <u>60</u>, <u>70</u>, <u>80</u>, <u>90</u>, <u>100</u>

3. Show 9:00 on the clock.

4. Circle the number that is greater.

    43   (49)

5. Circle the shapes that have equal size pieces.

6. Find each sum.

    7 + 7 = <u>14</u>   4 + 4 = <u>8</u>   8 + 8 = <u>16</u>   3 + 3 = <u>6</u>

2-9Wa   Copyright © 1991 by Saxon Publishers, Inc. and Nancy Larson. Reproduction prohibited.

Name _____   **LESSON 9B**
                                 Math 2
Date _____

1. What will be tomorrow's date? _____

2. Count backward by 10's. Fill in the missing numbers.

    100, <u>90</u>, <u>80</u>, <u>70</u>, <u>60</u>, <u>50</u>, <u>40</u>, <u>30</u>, <u>20</u>, <u>10</u>

3. Show 1:00 on the clock.

4. Circle the number that is greater.

    (58)   54

5. Circle the shapes that have equal size pieces.

6. Find each sum.

    5 + 5 = <u>10</u>   9 + 9 = <u>18</u>   2 + 2 = <u>4</u>   6 + 6 = <u>12</u>

2-9Wb   Copyright © 1991 by Saxon Publishers, Inc. and Nancy Larson. Reproduction prohibited.

Lesson 10

covering a design using pattern blocks

THE MEETING

calendar

> *"Yesterday was the _____th of (month)."*

• Point to the date.

> *"What do you think today's date is?"*

> *"How will we write the _____th?"*

• Write the date on the calendar.

> *"What day of the week is it today?"*

> *"How do you know?"*

> *"Let's read the days of the week together."*

> *"Let's write today's full date on a meeting strip."*

> *"What will we write first?"*

"What month is it?"

"How do we spell (month)?"

• Write the month on the meeting strip.

"What is the date?"

"What digits will we use to write the date?"

• Write the date on the meeting strip.

"What is the year?"

"What digits will we use to write the year?"

• Record the year on the meeting strip.

"Let's say the date together."

"Let's read the names of the months together."

• Point as you say the names of the months with your child.

weather graph

"What does the weather outside feel like to you?"

"Why do you think it feels _____?"

"What color will you use for _____?"

• Ask your child to color the correct box on the graph.

"What kind of weather have we had most often?"

counting

"Let's count from 68 to 84."

• Ask your child to point to the numbers on the hundred number chart as you count together.

"Let's count backward from 84 to 68."

"Let's count by 10's to 100."

• Point to the numbers on the counting strip as you count together.

"Now let's count backward from 100 by 10's."

graph questions

• You and your child each ask one question about the birthday graph.

patterning

"Today's pattern is a different repeating pattern."

"Let's read the first part of the pattern together."   *square, triangle, circle, ...*

"What repeats in this pattern?"   *square, triangle, circle*

"What shape do you think will come next?" square

- Fill in the next shape on the meeting strip.

- Repeat with the rest of the blanks.

"Let's read the shape pattern together."

"Let's read the pattern using letters." A, B, C, A, B, C, A, B, C,...

clock

"Use the demonstration clock to show seven o'clock."

"How do you know that this is seven o'clock?"

"Write the digital time on the meeting strip."

- Other information on the meeting strip is not used today.

ASSESSMENT

Written Assessment

- Beginning with this lesson, there will be a written assessment every five lessons. All of the questions on the assessment are based on concepts and skills presented at least five lessons ago. It is expected that your child will master at least 80% of the concepts on the assessment. If your child is having difficulty with a specific concept, reteach the concept the following day.

"Today, I would like to see what you remember from what we have been practicing."

- Give your child **Written Assessment #1**.

- Read the directions for each problem. Allow time for your child to complete that problem before continuing.

- Correct the paper, noting your child's mistakes on the **Individual Recording Form**. Review the errors with your child.

Oral Assessment

- An oral assessment occurs every ten lessons.

- Record your child's response(s) to the oral interview question(s) on the interview sheet.

THE LESSON

Covering a Design Using Pattern Blocks

"Today you will learn how to cover shapes in different ways using pattern blocks."

- Give your child the top half of **Master 2-10.** You will use the bottom half of Master 2-10.

"Use the pattern blocks to cover this shape."

"When you have covered your shape, count how many pattern blocks of each color you used."

"Write the numbers on the lines next to the shape."

"I will cover my shape and fill in the numbers, too, and then we will trade papers."

"See if you can cover my design with the same number of pattern blocks I used."

"I will try to cover your design with the same number of pattern blocks you used."

"I will check your paper when you finish and you can check mine."

CLASS PRACTICE

"Today we will practice the doubles facts together."

- Play Number Fact Jeopardy or review the doubles facts orally.

"Now you will write the answers on a fact sheet."

- Give your child **Fact Sheet A 1.2.**
- Time your child for one minute.
- Read the problems and the answers slowly.

"Count the number of problems you answered correctly before I said stop."

"Write this number at the bottom of your paper."

- Record the score on the **Individual Recording Form.**

"Now finish writing the answers for the rest of the problems."

- Correct these answers with your child.

Teacher _____
Date _____

MATH 2   LESSON 10
Oral Assessment # 1 Recording Form

Materials:
Hundred number chart
handwriting paper
pencil

Students	"Count by 10's to 100."	•Use the hundred number chart. "Point to the number..." "...17." "...to the right of 65." "...to the left of 23."	"Write these numbers." "32, 71, 80, 15, 52, 12, 19."

2-PFn

Copyright © 1991 by Saxon Publishers, Inc. and Nancy Larson. Reproduction prohibited.

Name _____
Date _____

ASSESSMENT 1
LESSON 10
Math 2

1. Use the birthday graph to answer the questions.

   How many children have birthdays in September? _____

   What month has the most birthdays? _____

2. Finish the repeating pattern.

   □ , ○ , □ , ○ , □ , ○ , □ , ○ , □ , ○

3. What number is one more than **34**?   35

   What number is one less than **54**?   53

4. Show 10:00 on the clock.

5. Fill in the missing numbers.

1	2	3	4	5	6	7	8	9	10
11	12	13	14	15	16	17	18	19	20
21	22	23	24	25	26	27	28	29	30

6. Find each sum.

   $2 + 2 =$ 4 $5 + 5 =$ 10 $8 + 8 =$ 16 $3 + 3 =$ 6

   $7 + 7 =$ 14 $4 + 4 =$ 8 $9 + 9 =$ 18 $6 + 6 =$ 12

2-10Aa

Copyright © 1991 by Saxon Publishers, Inc. and Nancy Larson. Reproduction prohibited.

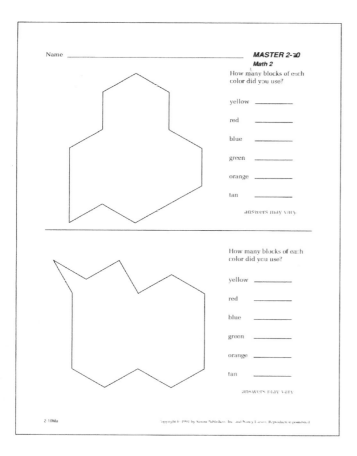

Name _____

MASTER 2-10
Math 2

How many blocks of each color did you use?

yellow _____

red _____

blue _____

green _____

orange _____

tan _____

answers may vary

How many blocks of each color did you use?

yellow _____

red _____

blue _____

green _____

orange _____

tan _____

answers may vary

2-10Ma

Copyright © 1991 by Saxon Publishers, Inc. and Nancy Larson. Reproduction prohibited.

Lesson 11

identifying and acting out some, some went away stories
creating and reading a repeating pattern

---

lesson preparation

materials

10 stuffed animals, dolls, action figures, or other toys

basket of pattern blocks

Fact Sheet A 1.2

in the morning

• Write the following in the pattern box on the meeting strip:

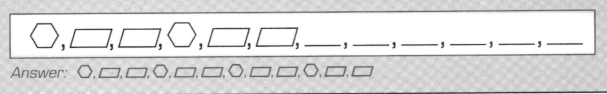

Answer: ⬡,▱,▱,⬡,▱,▱,⬡,▱,▱,⬡,▱,▱

THE MEETING

calendar

> "*Today we will add some new things to The Meeting.*"
>
> "*Let's begin with the calendar.*"
>
> "*Yesterday was the _____ th of (month).*"
>
> "*What do you think today's date is?*"

• Ask your child to write the date on the calendar.

> "*What day of the week is it today?*"
>
> "*Let's read the days of the week together.*"
>
> "*Now we will write today's full date.*"
>
> "*What will we write first?*"   the month
>
> "*What month is it?*"
>
> "*How do we spell (month)?*"

• Write the month on the meeting strip.

> "*What is the date?*"

"What digits will we use to write the date?"

- Write the date on the meeting strip.

  "What is the year?"

  "What digits will we use to write the year?"

- Record the year on the meeting strip.

  "Let's say the date together."

  "Let's read the names of the months together."

- Point as you say the names of the months with your child.

  "What is the first month of the year?"

  "What is the fourth month of the year?"

- Choose an upcoming special event or holiday.

  "I will draw a (star) on the calendar to show when _____ will be."

- Draw a (star) in the appropriate box on the Meeting Book calendar.

  "Let's count together to find the number of days until _____."

- Count the days until the special event will occur. Record the number of days on the meeting strip.

weather graph

"What does the weather outside feel like to you?"

"Why do you think it feels _____?"

"What color will you use for _____?"

- Ask your child to color the correct box on the graph.

  "What kind of weather have we had most often?"

  "How many warm days have we had?"

counting

"Today we will count forward and backward from 51 to 73."

- Ask your child to point to the numbers on the hundred number chart as you count together.

  "Let's count by 10's to 100."

  "If you can count by 10's without looking at the numbers on the counting strip, close your eyes as we count together."

  "Now let's count backward from 100 by 10's."

graph questions

- You and your child each ask one question about the birthday graph.

patterning

"Today's pattern is a different repeating pattern."

"What shapes do you see?"   hexagon, parallelogram

"Let's read the first part of the pattern together."   hexagon, parallelogram, parallelogram, ...

"What repeats in this pattern?"   hexagon, parallelogram, parallelogram

"What shape do you think will come next?"   hexagon

- Fill in the next shape on the meeting strip.
- Repeat with the rest of the blanks.

"Let's read the shape pattern together."   hexagon, parallelogram, ...

"Let's read the pattern using letters."   A, B, B, A, B, B, A, B, B, ...

clock

"Today you will show a time on the clock."

"The time will need to be a time on the hour."

- Ask your child to set the demonstration clock. Make sure the clock is set on the hour.

"What time is it?"

"Write this digital time on the meeting strip."

- Allow time for your child to do this.

"Each day we will record something that will happen during the day on the meeting strip."

- On the meeting strip, write the name of an activity and the time it will occur.

"At what time will _____ occur?"

number of the day

"Each day we will have a number of the day."

"The number of the day will be the date."

- For example, if the date is September 14th, the number of the day will be 14.

"Each day we will write three number sentences that equal the date."

- For example, if 14 is the number of the day, the number sentences could be $1 + 13 = 14$, $16 - 2 = 14$, or $2 \times 7 = 14$.

"Today, we will write three number sentences for _____."

"I will write one number sentence."

- Write a number sentence for the number of the day on the meeting strip.

"What is another number sentence for _____?"

- Record your child's number sentence on the meeting strip.
- Repeat with one more number sentence.
- The money box on the meeting strip is not used today.

THE LESSON

Identifying and Acting Out Some, Some Went Away Stories

"Today you will learn to identify and act out some, some went away stories."

"We will pretend that this area of the room (designate an area) will be called the store, the movie theater, the cafeteria, the doctor's office, or another special place in our story."

"Let's try acting out a story."

"We will pretend that your (stuffed animals, dolls, action figures, or other toys) are children."

"Four children went to the movies."

- Ask your child to place four (toys) in the designated area.

"Three of the children went home early."

- Ask your child to remove three (toys) from the designated area.

"How many children are at the movies now?"   1

"What happened in this story?"   some children went to the movies and then some went home

"Some children were at the movies and then some went away."

"This is a some, some went away problem."

"Let's try acting out another story."

"Eight children jumped in the swimming pool."

- Ask your child to place eight (toys) in the designated area.

"Three children got out of the pool to get a drink of water."

- Ask your child to remove three (toys).

"How many children are in the pool now?"   5

"What happened in this story?"   some children jumped in the swimming pool and then some got out of the pool to get a drink of water

"Some children were in the pool and then some got out."

"This is a some, some went away problem."

"Make up another some, some went away story for us to act out."

- Act out several of your child's stories. Repeat the above questions.

"A few days ago we acted out some, some more stories."

"Make up a some, some more story for us to act out."

• Act out several of your child's stories.

Creating and Reading a Repeating Pattern

"A few days ago we made an AB pattern fence."

• Give your child a basket of pattern blocks.

"Make a straight AB pattern fence using pattern blocks."

• Allow time for your child to make a fence at least eight blocks long.

"Let's read your fence using the names of the colors."

"Let's read your fence using the names of the shapes."

"Let's read your fence using letters."   A, B, A, B, A, B, A, B

"Today you will learn how to make a fence with a different repeating pattern."

"Now I will make a fence that is a little different from your fence."

• Make a fence using the same color pattern blocks that your child used, but with an ABB pattern. (Repeat the second color pattern block.)

"Is my fence the same as your fence?"

"How is it the same?"   same colors

"How is it different?"   there are two of the second color pattern blocks

"Let's read my fence using the names of the colors."

"Let's read my fence using the names of the shapes."

"Let's read my fence using letters."   A, B, B, A, B, B, A, B, B, A, B, B

"Now you will make an ABB repeating pattern fence."

"Choose two pattern blocks that are different colors."

"Make an ABB pattern fence."

• Allow time for your child to make the fence.

"Read the colors in your fence."

• Ask your child to read his/her fence using colors.

"Read the shapes in your fence."

• Ask your child to read his/her fence using shapes.

"Read your fence using letters."

CLASS PRACTICE

• Use the tan fact cards to practice the doubles facts with your child.

"Today you will write the answers for the doubles facts on a fact sheet."

- Give your child **Fact Sheet A 1.2.**
- Time your child for one minute.
- Correct the fact sheet with your child.
- Record the score.
- Allow time for your child to complete the unfinished facts.

WRITTEN PRACTICE

- Complete **Worksheet 11A** with your child.
- Complete **Worksheet 11B** with your child later in the day.

Name _____ **LESSON 11A**
Date _____ **Math 2**

1. One of these is my favorite color. Use the clues to find my favorite color. red (blue) yellow green

   It is not third. Cross out that color.
   It is not on the right. Cross out that color.
   It is not first. Cross out that color.
   Circle my favorite color.

2. Circle the smaller number.

   (28)   34

3. Finish the repeating pattern.

   △, ○, ○, △, ○, ○, △, ○, ○, △, ○, ○

4. What number is one more than 63? 64

   What number is one less than 47? 46

5. Find the number that belongs in the square with the **A.** 25

   Find the number that belongs in the square with the **B.** 46

   Fill in the missing numbers.

1	2	3	4	5	6	7	8	9	10
11	12	13	14	15	16	17	18	19	20
21	22	23	24	A	26	27	28	29	30
31	32	33	34	35	36	37	38	39	40
41	42	43	44	45	B	47	48	49	50

2-11Wa   Copyright © 1991 by Saxon Publishers, Inc. and Nancy Larson. Reproduction prohibited.

Name _____ **LESSON 11B**
Date _____ **Math 2**

1. One of these is my sister's favorite color. Use the clues to find my sister's favorite color. (red) blue yellow green

   It is not third. Cross out that color.
   It is not on the right. Cross out that color.
   It is not second. Cross out that color.
   Circle my sister's favorite color.

2. Circle the smaller number.

   29   (23)

3. Finish the repeating pattern.

   ○, □, □, ○, □, □, ○, □, □, ○, □, □

4. What number is one more than 25? 26

   What number is one less than 73? 72

5. Find the number that belongs in the square with the **A.** 43

   Find the number that belongs in the square with the **B.** 19

   Fill in the missing numbers.

1	2	3	4	5	6	7	8	9	10
11	12	13	14	15	16	17	18	B	20
21	22	23	24	25	26	27	28	29	30
31	32	33	34	35	36	37	38	39	40
41	42	A	44	45	46	47	48	49	50

2-11Wb   Copyright © 1991 by Saxon Publishers, Inc. and Nancy Larson. Reproduction prohibited.

---

esson 12

numbering a clock face
determining elapsed time (one hour)

---

lesson preparation

materials

demonstration clock

Master 2-12

Fact Sheet A 2.0

in the morning

• Write the following in the pattern box on the meeting strip:

Answer: ⬡, ⬡, △, △, ⬡, ⬡, △, △, ⬡, ⬡, △, △

THE MEETING

calendar

> *"Let's begin with the calendar."*
>
> *"Yesterday was the _____th of (month)."*
>
> *"What do you think today's date is?"*

• Ask your child to write the date on the calendar.

> *"What day of the week is it today?"*
>
> *"Let's read the days of the week together."*
>
> *"Now we will write today's full date."*
>
> *"What will we write first?"*   the month
>
> *"What month is it?"*
>
> *"How do we spell (month)?"*

• Write the month on the meeting strip.

> *"What is the date?"*
>
> *"What digits will we use to write the date?"*

• Write the date on the meeting strip.

"What is the year?"

"What digits will we use to write the year?"

- Record the year on the meeting strip.

   "Let's say the date together."

   "Let's read the names of the months together."

- Point as you say the names of the months with your child.

- Choose an upcoming special event or holiday.

   "I drew a (star) on the calendar to show when _____ will be."

   "Let's count together to find the number of days until _____."

- Count the days until the special event will occur. Record the number of days on the meeting strip.

weather graph

"What does the weather outside feel like to you?"

"Why do you think it feels _____?"

"What color will you use for _____?"

- Ask your child to color the correct box on the graph.

   "What kind of weather have we had most often?"

   "How many warm days have we had?"

   "How many more _____ days have we had than _____ days?"

counting

"Today we will count forward and backward from 37 to 59."

- Ask your child to point to the numbers on the hundred number chart as you count together.

   "Let's count by 10's to 100."

   "If you can count by 10's without looking at the numbers on the counting strip, close your eyes as we count together."

   "Now let's count backward from 100 by 10's."

graph questions

- You and your child each ask one question about the birthday graph.

patterning

"Today's pattern is a different repeating pattern."

"What shapes do you see?"   hexagon, triangle

"Let's read the first part of the pattern together."   hexagon, hexagon, triangle, triangle, …

"What happened in this pattern?"

"What shape do you think will come next?"   hexagon

- Fill in the next shape on the meeting strip.
- Repeat with the rest of the blanks.

"Let's read the shape pattern together."

"Let's read the pattern using letters."   A, A, B, B, A, A, B, B, A, A, B, B

clock

"Today you will show a time on the clock."

"Show a time on the hour."

- Ask your child to set the clock. Make sure the clock is set on the hour.

"What time is it?"

"Write this digital time on the meeting strip."

- Allow time for your child to do this.

"Each day we will record something that will happen during the day on the meeting strip."

- On the meeting strip, write the name of an activity and the time it will occur.

"At what time will _____ occur?"

number of the day

"Today's number of the day is the date."

- For example, if the date is September 14th, the number of the day will be 14.

"Each day we will write three number sentences that equal the date."

"I will write one number sentence."

- Write a number sentence for the number of the day on the meeting strip.

"What is another number sentence for _____?"

- Record your child's number sentence on the meeting strip.
- Repeat with one more number sentence.
- The money box on the meeting strip is not used today.

THE LESSON

Numbering a Clock Face
Determining Elapsed Time (One Hour)

"Today you will learn how to number and show the time on a clock face."

"You will also learn how to find the time one hour ago and one hour from now."

- Name the current time to the nearest hour.

"It's almost _____ o'clock now."

- Ask your child to set the demonstration clock to show that time.

"Where does the hour hand point?"

"Where does the long hand point?"

"It's been just about an hour since we _____."

- Refer to something your child did one hour ago.

"What time was it an hour ago?"

"Show that time on the demonstration clock."

- Ask your child to set the demonstration clock.

"Which way does the hour hand move when we find the time one hour ago?"

"We said before that it's almost _____ o'clock now."

- Set the demonstration clock to show that time.

"In one hour from now we will _____."

- Refer to something that will happen one hour from now.

"What time will it be when we _____?"

"Show that time on the demonstration clock."

- Ask your child to set the clock.

"Which way does the hour hand move when we find the time one hour from now?"

"Now let's pretend that it is nine o'clock in the evening."

- Ask your child to set the clock.

"What kinds of things do people do at nine o'clock in the evening?"

"If it is nine o'clock in the evening, one hour from now I will go to bed."

"What time will I go to bed?"

"Show that time on the clock."

- Ask your child to set the clock.

"Which way does the hour hand move when we find the time one hour from now?"

"Now let's pretend that it is seven o'clock in the morning."

- Ask your child to set the clock.

"What kinds of things do people do at seven o'clock in the morning?"

"If it is seven o'clock in the morning now, I got out of bed one hour ago."

"What time did I get out of bed?"

"Show that time on the clock."

• Ask your child to set the clock.

"Which way does the hour hand move when we find one hour ago?"

• Draw a large circle on the chalkboard to represent the clock face.

"What numbers do you see on the clock face?"

"What number is at the top of the clock face?"

• Draw a small line at the top and write "12" beneath it.

"What number is at the bottom of the clock face?"

• Draw a small line at the bottom and write "6" above it.

• Draw a small line where "3" and "9" are located.

"What numbers will I write next to these marks?"

• Write "3" and "9" on the clock.

"When we number a clock face, it is easiest to fill in these numbers first."

"What numbers do I need to put on my clock face now?"

• Fill in all of the other numbers as your child describes their placement.

• Give your child **Master 2-12**.

"Now you will write the numbers on the first blank clock face."

"What number will you write on the clock face first?"   12

"Where will you write the 12?"   at the top

"What number will you write on the clock face next?"   6

"Where will you write the 6?"   at the bottom

• Repeat with 3 and 9.

"Fill in the other numbers on your clock face."

"Fill in the next two clock faces."

• Write "4:00" on the chalkboard.

"What time is this?"

"Point to the clock in the middle of your paper."

"Write this digital time next to the middle clock face."

"Now you will draw the hands on the clock to show four o'clock."

"Where will the long hand point?"

"The long hand goes through the 12 and touches the mark on the edge of the clock."

"Show that on your clock."

"Where will the hour hand point?"

"The hour hand stops just before the number of the hour."

"Show that on your clock."

"If it is four o'clock now, what time was it one hour ago?"   three o'clock

"Point to the clock at the top of the paper."

"Write the digital time for three o'clock next to the clock face at the top of the paper."

"Now you will draw the hands on the clock to show three o'clock."

"Where will the long hand point?"

"The long hand goes through the 12 and touches the mark on the edge of the clock."

"Show that on your clock."

"Where will the hour hand point?"

"The hour hand stops just before the number of the hour."

"Show that on your clock."

"If it is four o'clock now, what time will it be one hour from now?"

"Point to the clock at the bottom of the paper."

"Write the digital time for five o'clock next to the clock face at the bottom of the paper."

"Now you will put the hands on the clock to show five o'clock."

• Repeat steps for putting the hands on the clock.

CLASS PRACTICE

"Use your fact cards to practice the adding one facts."

• Allow time for your child to practice independently.

• Give your child **Fact Sheet A 2.0**.

• Time your child for one minute.

• Correct the fact sheet with your child.

• Record the score.

• Allow time for your child to complete the unfinished facts.

WRITTEN PRACTICE

• Complete **Worksheet 12A** with your child.

• Complete **Worksheet 12B** with your child later in the day.

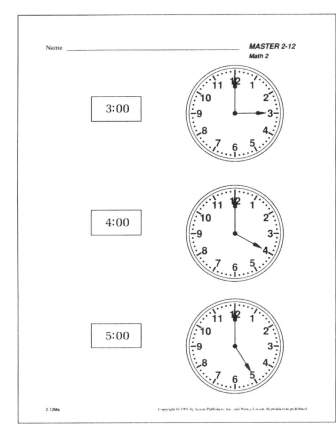

3:00

4:00

5:00

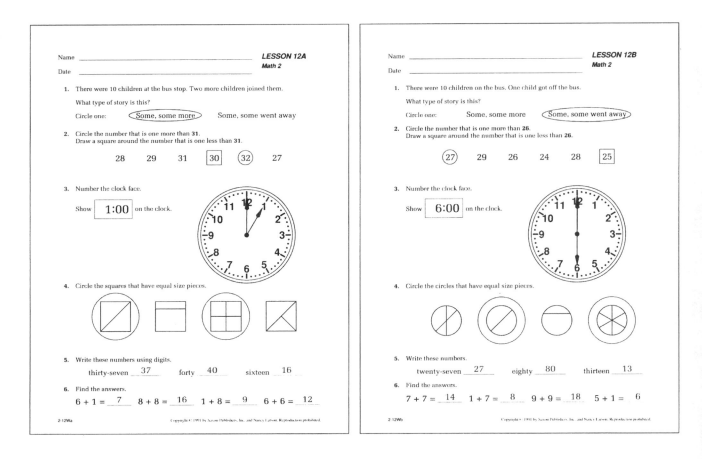

Name _____ **LESSON 12A**
                                         Math 2

Date _____

1. There were 10 children at the bus stop. Two more children joined them.

   What type of story is this?

   Circle one: (Some, some more)   Some, some went away

2. Circle the number that is one more than **31**.
   Draw a square around the number that is one less than **31**.

   28   29   31   [30]   (32)   27

3. Number the clock face.

   Show [1:00] on the clock.

4. Circle the squares that have equal size pieces.

5. Write these numbers using digits.

   thirty-seven __37__   forty __40__   sixteen __16__

6. Find the answers.

   6 + 1 = __7__   8 + 8 = __16__   1 + 8 = __9__   6 + 6 = __12__

Name _____ **LESSON 12B**
                                         Math 2

Date _____

1. There were 10 children on the bus. One child got off the bus.

   What type of story is this?

   Circle one: Some, some more   (Some, some went away)

2. Circle the number that is one more than **26**.
   Draw a square around the number that is one less than **26**.

   (27)   29   26   24   28   [25]

3. Number the clock face.

   Show [6:00] on the clock.

4. Circle the circles that have equal size pieces.

5. Write these numbers.

   twenty-seven __27__   eighty __80__   thirteen __13__

6. Find the answers.

   7 + 7 = __14__   1 + 7 = __8__   9 + 9 = __18__   5 + 1 = __6__

Lesson 13

addition facts — adding zero

lesson preparation

materials

addition fact cards—lavender

Fact Sheet A 3.0

the night before

• Separate the lavender addition fact cards.

in the morning

• Write the following in the pattern box on the meeting strip:

☐ , ☐ , △ , ○ , ☐ , ☐ , △ , ○ , ___ , ___ , ___ , ___

Answer: ☐ , ☐ , △ ○ , ☐ , ☐ , △ , ○ , ☐ , ☐ , △ , ○

THE MEETING

calendar

"Let's begin with the calendar."

"Yesterday was the _____th of (month)."

"What do you think today's date is?"

• Ask your child to write the date on the calendar.

"What day of the week is it today?"

"Let's read the days of the week together."

"Now we will write today's full date."

"What will we write first?" the month

"What month is it?"

"How do we spell (month)?"

• Write the month on the meeting strip.

"What is the date?"

"What digits will we use to write the date?"

• Write the date on the meeting strip.

"What is the year?"

"What digits will we use to write the year?"

- Record the year on the meeting strip.

"Let's say the date together."

"Let's read the names of the months together."

"What is the third month of the year?"

"What is the fifth month of the year?"

- Choose an upcoming special event or holiday.

"I drew a (star) on the calendar to show when _____ will be."

"Let's count together to find the number of days until _____."

- Count the days until the special event will occur. Record the number of days on the meeting strip.

weather graph

"What does the weather outside feel like to you?"

"Why do you think it feels _____?"

"What color will you use for _____?"

- Ask your child to color the correct box on the graph.

"What kind of weather have we had most often?"

"How many cool days have we had?"

"How many more _____ days have we had than _____ days?"

counting

"Today we will count forward and backward from 16 to 42."

- Ask your child to point to the numbers on the hundred number chart as you count together.

"Let's count by 10's to 100."

"If you can count by 10's without looking at the numbers on the counting strip, close your eyes as we count together."

"Now let's count backward from 100 by 10's."

graph questions

- You and your child each ask one question about the birthday graph.

patterning

"Today's pattern is a different repeating pattern."

"What shapes do you see?"   rectangle, triangle, circle

"Let's read the first part of the pattern together."   rectangle, rectangle, triangle, circle, …

"What happened in this pattern?"

"What shape do you think will come next?" rectangle

• Fill in the next shape on the meeting strip.

• Repeat with the rest of the blanks.

"Let's read the shape pattern together." rectangle, rectangle, triangle, circle, ...

"Let's read the pattern using letters." A, A, B, C, A, A, B, C, A, A, B, C

clock

"Show a time on the hour on the demonstration clock."

• Ask your child to set the clock.

"What time is it?"

"Write this digital time on the meeting strip."

"What time was it one hour ago?"

"Each day we will record something that will happen during the day on the meeting strip."

• Write on the meeting strip the name of an activity and the time it will occur.

"At what time will _____ occur?"

number of the day

"Today's number of the day is the date."

"Each day we will write three number sentences that equal the date."

• Write a number sentence for the number of the day on the meeting strip.

"What is another number sentence for _____?"

• Record your child's number sentence on the meeting strip.

• Repeat with one more number sentence.

• The money box on the meeting strip is not used today.

THE LESSON

Addition Facts — Adding Zero

"Today you will learn the addition facts that are called the adding zero facts."

• Write the following problems on the chalkboard:

$$\frac{0}{+0} \quad \frac{1}{+0} \quad \frac{2}{+0} \quad \frac{3}{+0} \quad \frac{4}{+0} \quad \frac{5}{+0} \quad \frac{6}{+0} \quad \frac{7}{+0} \quad \frac{8}{+0} \quad \frac{9}{+0}$$

"What is the same about all of these problems?"

"We will call these problems the adding zero facts."

"Read the problems and tell me the answers."

- Write the answer below each problem.

"What do you notice about the answers?"

"I'll say the problems in random order."

"Let's see how fast you can say the answers."

- Allow your child to refer to the chalkboard problems, if necessary.

"Now I will say the answer to an adding zero problem."

"See if you can tell me the problem."

"Six."   six plus zero

- Repeat with all combinations.

Class Practice

number fact practice

"Today I will give you the adding zero fact cards."

"Add these cards to your doubles and adding one fact cards."

- Give your child the lavender addition fact cards.

"Use your fact cards to practice the adding zero facts."

- Allow time for your child to practice independently.
- Give your child **Fact Sheet A 3.0**.
- Time your child for one minute.
- Correct the fact sheet with your child.
- Record the score.
- Allow time for your child to complete the unfinished facts.

Written Practice

- Complete **Worksheet 13A** with your child.
- Complete **Worksheet 13B** with your child later in the day.

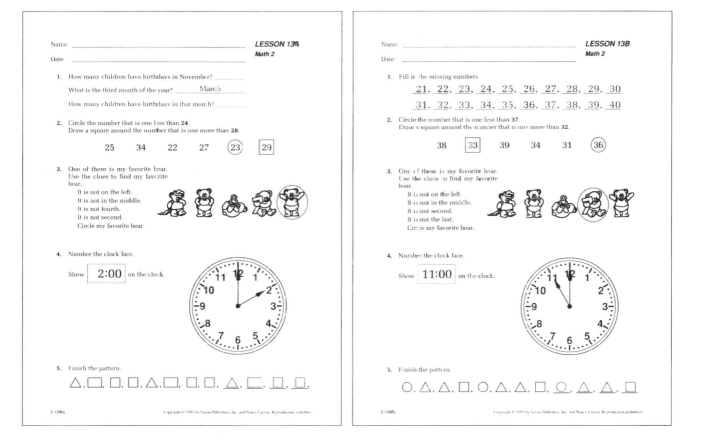

LESSON 13A
Math 2

Name _____

Date _____

1. How many children have birthdays in November? _____

   What is the third month of the year? _____ March

   How many children have birthdays in that month? _____

2. Circle the number that is one less than **24**.
   Draw a square around the number that is one more than **28**.

   25   34   22   27   (23)   [29]

3. One of these is my favorite bear.
   Use the clues to find my favorite bear.
   It is not on the left.
   It is not in the middle.
   It is not fourth.
   It is not second.
   Circle my favorite bear.

4. Number the clock face.

   Show [2:00] on the clock.

5. Finish the pattern.

   △, □, □, □, △, □, □, □, △, □, □, □,

2-13Wa

Copyright © 1991 by Saxon Publishers, Inc. and Nancy Larson. Reproduction prohibited.

LESSON 13B
Math 2

Name _____

Date _____

1. Fill in the missing numbers

   21, 22, 23, 24, 25, 26, 27, 28, 29, 30
   31, 32, 33, 34, 35, 36, 37, 38, 39, 40

2. Circle the number that is one less than **37**.
   Draw a square around the number that is one more than **32**.

   38   [33]   39   34   31   (36)

3. One of these is my favorite bear.
   Use the clues to find my favorite bear.
   It is not on the left.
   It is not in the middle.
   It is not second.
   It is not the last.
   Circle my favorite bear.

4. Number the clock face.

   Show [11:00] on the clock.

5. Finish the pattern.

   ○, △, △, □, ○, △, △, □, ○, △, △, □,

2-13Wb

Copyright © 1991 by Saxon Publishers, Inc. and Nancy Larson. Reproduction prohibited.

Lesson 14

identifying ordinal position to twelfth

---

lesson preparation ——————————————————

materials

12 pieces of scrap paper or newspaper

crayons – red, blue, yellow, green, black, orange, brown, purple

Fact Sheet A 2.0

in the morning

• Write the following in the pattern box on the meeting strip:

THE MEETING

calendar

> *"Let's begin with the calendar."*
>
> *"Yesterday was the _____th of (month)."*
>
> *"What do you think today's date is?"*

• Ask your child to write the date on the calendar.

> *"What day of the week is it today?"*
>
> *"Let's read the days of the week together."*
>
> *"Now we will write today's full date."*
>
> *"What will we write first?"*   the month
>
> *"What month is it?"*
>
> *"How do we spell (month)?"*

• Write the month on the meeting strip.

> *"What is the date?"*
>
> *"What digits will we use to write the date?"*

• Write the date on the meeting strip.

> *"What is the year?"*
>
> *"What digits will we use to write the year?"*

- Record the year on the meeting strip.

  "Let's say the date together."

  "Let's read the names of the months together."

  "What is the _____th month of the year?"

- Choose an upcoming special event or holiday.

  "I drew a (star) on the calendar to show when _____ will be."

  "Let's count together to find the number of days until _____."

- Count the days until the special event will occur. Record the number of days on the meeting strip.

weather graph

  "What does the weather outside feel like to you?"

  "Why do you think it feels _____?"

  "What color will you use for _____?"

- Ask your child to color the correct box on the graph.

  "What kind of weather have we had most often?"

  "How many hot days have we had?"

  "How many more _____ days have we had than _____ days?"

counting

  "Today we will count forward and backward from 51 to 79."

- Ask your child to point to the numbers on the hundred number chart as you count together.

  "Let's count by 10's to 100."

  "If you can count by 10's without looking at the numbers on the counting strip, close your eyes as we count together."

  "Now let's count backward from 100 by 10's."

graph questions

- You and your child each ask one question about the birthday graph.

patterning

  "Today's pattern is a repeating pattern."

  "What shapes do you see?"   parallelogram, hexagon

  "Let's read the first part of the pattern together."   parallelogram, parallelogram, hexagon, hexagon, ...

  "What happened in this pattern?"

  "What shape do you think will come next?"   parallelogram

- Fill in the next shape on the meeting strip.
- Repeat with the rest of the blanks.

> *"Let's read the shape pattern together."* parallelogram, parallelogram, hexagon, hexagon, …
>
> *"Let's read the pattern using letters."* A, A, B, B, A, A, B, B, …

clock

> *"Show a time on the hour on the demonstration clock."*

- Ask your child to set the clock.

> *"What time is it?"*
>
> *"Write this digital time on the meeting strip."*
>
> *"What time was it one hour ago?"*

- On the meeting strip, write the name of an activity and the time it will occur.

> *"At what time will _____ occur?"*

number of the day

> *"Today's number of the day is the date."*
>
> *"Each day we will write three number sentences that equal the date."*

- Write a number sentence for the number of the day on the meeting strip.

> *"What is another number sentence for _____?"*

- Record your child's number sentence on the meeting strip.
- Repeat with one more number sentence.
- The money box on the meeting strip is not used today.

The Lesson

Identifying Ordinal Position to Twelfth

> *"Today you will learn the names that tell us the place of something when there are twelve things in a row."*
>
> *"Where are some places where people stand in line?"* fast food restaurants, banks, post office, supermarket, cafeteria, etc.
>
> *"Today we will pretend that we are at a bank."*
>
> *"This table will be the teller's counter."*

- Put 12 pieces of paper in a row on the floor beginning at the table.

> *"These 12 pieces of paper show where people stand in line."*
>
> *"Where will the first person in line stand?"*
>
> *"I will stand on the first paper."*

"You stand on the second paper."

• Allow time for your child to do this.

"Now stand on the next paper."

"In what position are you now?"   third

"Now stand on the sixth paper."

• Repeat with the eighth, ninth, seventh, and tenth.

• Stand on the seventh paper.

"I am standing on the seventh paper."

"Stand in front of me."

"What paper are you standing on?"   sixth

• Stand on the twelfth paper.

"Now I am standing on the twelfth paper."

"Stand in front of me."

"What paper are you standing on?"   eleventh

• Repeat with other positions, if desired.

"Now you will use crayons to practice ordinal position."

• Give your child 8 crayons (one each of red, blue, yellow, green, black, orange, brown, and purple).

"I will tell you how to line up the crayons on the table."

"You will start at the left side of the table."

• Ask your child to identify the left side of the table.

"Put the red crayon first."

"Put the green crayon second."

"Put the yellow crayon third."

"Put the brown crayon fourth."

"Put the orange crayon fifth."

"Put the black crayon sixth."

"Put the blue crayon seventh."

"Put the purple crayon eighth."

"What color is the fourth crayon?"   brown

"What color is the seventh crayon?"   blue

"What color is the second crayon?"   green

"What color is the eighth crayon?"   purple

"Put the purple crayon second."

"Hold up the crayon that is seventh now."   black

"Put this crayon in the box."

"Hold up the crayon that is in the middle now."   yellow

"Put this crayon in the box."

"Hold up the crayon that is third now."   green

"Put this crayon in the box."

- Repeat until all the crayons are returned to the box.

CLASS PRACTICE

number fact practice

- Use the peach fact cards to practice the adding one facts with your child.
- Give your child **Fact Sheet A 2.0.**
- Time your child for one minute.
- Correct the fact sheet with your child.
- Record the score.
- Allow time for your child to complete the unfinished facts.

WRITTEN PRACTICE

- Complete **Worksheet 14A** with your child.
- Complete **Worksheet 14B** with your child later in the day.

Name _____   **LESSON 14A**
                                     Math 2
Date _____

1. Crystal's dog had seven puppies. She gave four of the puppies to friends.

   What type of story is this?

   Circle one: Some, some more   (Some, some went away)

2. Circle the fifth letter.
   Circle the twelfth letter.
   Circle the eighth letter.   O (S) P A (E) W A (V) O R (E) N
   Circle the eleventh letter.
   Circle the second letter.

3. Circle the smallest number.
   Put an X on the largest number.   2̸7̸   (15)   24

4. Finish these repeating patterns.

   □, △, □, △, □, △, □, △, □

   ○, △, △, ○, △, △, ○, △, △, ○, △, △

5. Find the answers.

   4 + 0 = __4__   8 + 8 = __16__   3 + 1 = __4__
   6 + 6 = __12__   0 + 7 = __7__   8 + 0 = __8__
   1 + 9 = __10__   6 + 1 = __7__   7 + 7 = __14__

2-14Wa   Copyright © 1991 by Saxon Publishers, Inc. and Nancy Larson. Reproduction prohibited.

Name _____   **LESSON 14B**
                                     Math 2
Date _____

1. Carol had 17 pennies. She found 2 more pennies on the sidewalk.

   What type of story is this?

   Circle one: (Some, some more)   Some, some went away

2. Circle the fifth letter.
   Circle the twelfth letter.
   Circle the second letter.   S (E) N T (I) (G) A C (H) R E (T)
   Circle the ninth letter.
   Circle the sixth letter.

3. Circle the smallest number.
   Put an X on the largest number.   39   (34)   4̸7̸

4. Finish these repeating patterns.

   △, □, △, □, △, □, △, □, △

   □, □, ○, □, □, ○, □, □, ○, □, □

5. Find the answers.

   5 + 5 = __10__   8 + 1 = __9__   9 + 0 = __9__
   4 + 1 = __5__   0 + 5 = __5__   4 + 4 = __8__
   0 + 6 = __6__   9 + 9 = __18__   1 + 5 = __6__

2-14Wb   Copyright © 1991 by Saxon Publishers, Inc. and Nancy Larson. Reproduction prohibited.

Lesson 15

identifying weekdays and days of the weekend

---

lesson preparation

materials

Written Assessment #2

Meeting Book

crayons

Fact Sheet A 2.2

in the morning

• Write the following in the pattern box on the meeting strip:

▱, ◯, ◯, ▯, ▱, ◯, ◯, ▯, ____, ____, ____, ____

Answer: ▱, ◯, ◯, ▯, ▱, ◯, ◯ ▯, ▱, ◯, ◯, ▯

THE MEETING

calendar

"*Let's begin with the calendar.*"

"*Yesterday was the _____th of (month).*"

"*What do you think today's date is?*"

• Ask your child to write the date on the calendar.

"*What day of the week is it today?*"

"*Let's read the days of the week together.*"

"*Now you will write today's full date on the meeting strip.*"

"*What month is it?*"

"*How do we spell (month)?*"

• Your child writes the month on the meeting strip.

"*What is the date?*"

"*What digits will you use to write the date?*"

• Your child writes the date on the meeting strip.

"*What is the year?*"

• Your child records the year on the meeting strip.

"Let's say the date together."

"Let's read the names of the months together."

"What is the _____th month of the year?"

"What month of the year is _____?"

"What is the month before _____?"

"What is the month after _____?"

• Choose an upcoming special event or holiday.

"Let's count together to find the number of days until (special event)."

• Record the number of days on the meeting strip.

weather graph

• Ask your child to describe how the weather feels to him/her.

• Ask your child to color the correct box on the graph.

"What kind of weather have we had most often?"

"How many _____ days have we had?"

"How many more _____ days have we had than _____ days?"

counting

"Today we will count from 54 to 82."

• Write "54" and "82" on a piece of paper.

"Let's count from 54 to 82 as I write the numbers on a piece of paper."

• Write the numbers as you count together.

"Let's count backward from 82 to 54."

• Count by 10's to 100 and backward from 100 by 10's.

graph questions

• You and your child each ask one question about the birthday graph.

patterning

"Today's pattern is a repeating pattern."

"What shapes do you see?"   *parallelogram, circle, rectangle*

"Let's read the first part of the pattern together."   *parallelogram, circle, circle, rectangle, ...*

"What happened in this pattern?"

"What shape do you think will come next?"   *parallelogram*

• Fill in the next shape on the meeting strip.

- Repeat with the rest of the blanks.

   "Let's read the shape pattern together." *parallelogram, circle, circle, rectangle, ...*

   "Let's read the pattern using letters." *A, B, B, C, A, B, B, C, ...*

clock

- Ask your child to set the clock on the hour.

   "What time is it?"

   "Write the digital time on the meeting strip."

   "What time was it one hour ago?"

- On the meeting strip, write the name of an activity and the time it will occur.

   "At what time will _____ occur?"

number of the day

   "Today you will write three number sentences for the date."

- Record the three number sentences on the meeting strip.

- The money box on the meeting strip is not used today.

ASSESSMENT

Written Assessment

   "Today, I would like to see what you remember from what we have been practicing."

- Give your child **Written Assessment #2**.

- Read the directions for each problem. Allow time for your child to complete that problem before continuing.

- Correct the paper, noting your child's mistakes on the **Individual Recording Form**. Review the errors with your child.

THE LESSON

Identifying Weekdays and Days of the Weekend

   "Today you will learn how to identify the weekdays and the days of the weekend."

   "Each morning we have been reading the names of the days of the week."

- Open the Meeting Book to this month's calendar.

"Monday, Tuesday, Wednesday, Thursday, and Friday are called the weekdays."

"Let's read the names of the weekdays together."

"Saturday and Sunday are the days that make up the weekend."

"Some people say that the first day of the new week is Monday."

"On our calendar the first day of the new week is Sunday."

"We will use our calendar when we talk about what day of the week it is."

"We will call Sunday the first day of the week."

"What will be the last day of the week?"   Saturday

"What will be the second day of the week?"   Monday

"What will be the fifth day of the week?"   Thursday

"Many people do special things on different days of the week."

"Many people also have a favorite day of the week."

"Some people's favorite day of the week is the day of the week when they do something special."

"My favorite day of the week is _____."

"That is because I _____ on that day."

"What is your favorite day of the week?"

"Why?"

"Let's make a graph to show our favorite day of the week and the favorite day of people we know."

• Open the Meeting Book to pages 30 and 31.

"You said that your favorite day of the week is _____."

"Find the abbreviation for _____ on this graph."

"Point to the box above _____."

"Draw a picture of why this is your favorite day."

"Write your name below the picture."

• Allow time for your child to do this.

"I said that my favorite day of the week is _____ because I _____."

"Find the abbreviation for _____ on this graph."

"Point to the box above _____."

"Draw a picture of why this is my favorite day."

"Write my name below the picture."

• Allow time for your child to do this.

"Tomorrow we will find the favorite day of the week of some of our family members and friends."

"What day of the week do you think _____ will choose?"

- Name several family members, friends, or neighbors.

CLASS PRACTICE

- Use the tan and peach fact cards to practice the doubles and adding one facts with your child.

- Give your child **Fact Sheet A 2.2.**

- Time your child for one minute.

- Correct the fact sheet with your child.

- Record the score.

- Allow time for your child to complete the unfinished facts.

WRITTEN PRACTICE

- Complete **Worksheet 15A** with your child.

- Complete **Worksheet 15B** with your child later in the day.

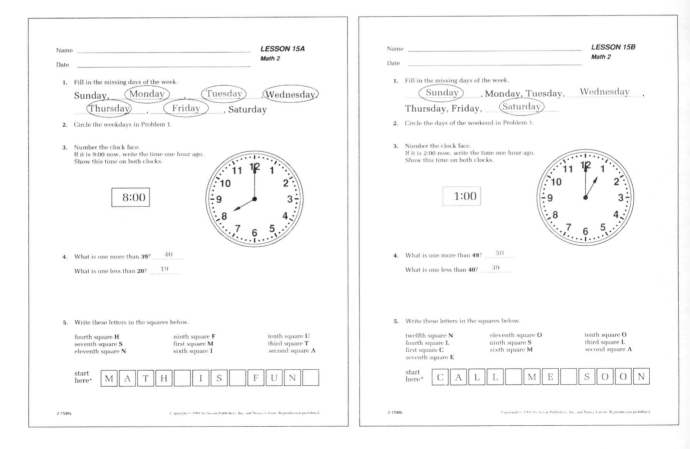

Name _____ ASSESSMENT 2

Date _____ **LESSON 15**

Math 2

1. Write a **K** in the fifth square.
   Write an **A** in the middle square.
   Write a **B** in the first square.
   Write a **C** in the fourth square.
   Write an **L** in the second square.

   B L A C K

2. Circle the smaller number.   46 ④②

3. Count by 10's. Fill in the missing numbers.

   _10_, _20_, _30_, _40_, _50_, _60_, _70_, _80_, _90_, _100_

4. Fill in the missing numbers on the chart.

   What number belongs in the square with the **A**? _27_

   What number belongs in the square with the **B**? _35_

   What number belongs in the square with the **C**? _19_

1	2	3	4	5	6	7	8	9	10
11	12	13	14	15	16	17	18	C	20
21	22	23	24	25	26	A	28	29	30
31	32	33	34	B	36	37	38	39	40

5. Draw a square to the right of the triangle.
   Draw a circle to the left of the triangle.

   ○   △   □

6. Add.

   8 + 1 = _9_   3 + 1 = _4_   6 + 1 = _7_   7 + 1 = _8_

2-15Aa   Copyright © 1991 by Saxon Publishers, Inc. and Nancy Larson. Reproduction prohibited.

Name _____ **LESSON 15A**

Date _____ Math 2

1. Fill in the missing days of the week.

   Sunday, (Monday) , (Tuesday) , (Wednesday) , (Thursday) , (Friday) , Saturday

2. Circle the weekdays in Problem 1.

3. Number the clock face.
   If it is 9:00 now, write the time one hour ago.
   Show this time on both clocks.

   8:00

4. What is one more than **39**? _40_

   What is one less than **20**? _19_

5. Write these letters in the squares below.

   fourth square **H**   ninth square **F**   tenth square **U**
   seventh square **S**   first square **M**   third square **T**
   eleventh square **N**   sixth square **I**   second square **A**

   start here▸ M A T H   I S   F U N

2-15Wa   Copyright © 1991 by Saxon Publishers, Inc. and Nancy Larson. Reproduction prohibited.

Name _____ **LESSON 15B**

Date _____ Math 2

1. Fill in the missing days of the week.

   (Sunday) , Monday, Tuesday, Wednesday , Thursday, Friday, (Saturday)

2. Circle the days of the weekend in Problem 1.

3. Number the clock face.
   If it is 2:00 now, write the time one hour ago.
   Show this time on both clocks.

   1:00

4. What is one more than **49**? _50_

   What is one less than **40**? _39_

5. Write these letters in the squares below.

   twelfth square **N**   eleventh square **O**   tenth square **O**
   fourth square **L**   ninth square **S**   third square **L**
   first square **C**   sixth square **M**   second square **A**
   seventh square **E**

   start here▸ C A L L   M E   S O O N

2-15Wb   Copyright © 1991 by Saxon Publishers, Inc. and Nancy Larson. Reproduction prohibited.

L esson 16

creating and reading a pictograph

THE MEETING

calendar

• Ask your child to write the date on the calendar.

"Now you will write today's full date on the meeting strip."

"What month is it?"

"How do we spell (month)?"

• Your child writes the month on the meeting strip.

"What is the date?"

"What digits will you use to write the date?"

• Your child writes the date on the meeting strip.

"What is the year?"

• Your child records the year on the meeting strip.

"Let's say the date together."

• Recite the days of the week together once a week.

"What are the weekdays?"

"What are the days of the weekend?"

"What is the _____th day of the week?"

"What day of the week is _____?"

"What day of the week was it yesterday?"

"Let's read the names of the months together."

"What is the _____th month of the year?"

"What month of the year is _____?"

"What is the month before _____?"

"What is the month after _____?"

- Choose an upcoming special event or holiday.

"Let's count together to find the number of days until (special event)."

- Record the number of days on the meeting strip.

weather graph

- Ask your child to describe how the weather feels to him/her.

- Ask your child to color the correct box on the graph.

"What kind of weather have we had most often?"

"How many _____ days have we had?"

"How many more _____ days have we had than _____ days?"

counting

"Today we will count from 13 to 35."

- Write "13" and "35" on a piece of paper.

"Today you will write the numbers we say as we count from 13 to 35."

- Your child writes the numbers as you count together.

"Let's count backward from 35 to 13."

- Count by 10's to 100 and backward from 100 by 10's.

graph questions

- You and your child each ask one question about the birthday graph.

patterning

"Today's pattern is a different repeating pattern."

"Let's read the first part of the pattern together."

"What happened in the pattern?"

"What shape do you think will come next?"

- Fill in the next shape on the meeting strip.

- Repeat with the rest of the blanks.

"Let's read the shape pattern together."

"Let's read the pattern using letters."   A, B, C, C, A, B, C, C, ...

clock

- Ask your child to set the clock on the hour.

  "What time is it?"

  "Write the digital time on the meeting strip."

  "What time was it one hour ago?"

- On the meeting strip, write the name of an activity and the time it will occur.

  "At what time will _____ occur?"

number of the day

"Today you will write three number sentences for the date."

- Record the three number sentences on the meeting strip.

- The money box on the meeting strip is not used today.

THE LESSON

Creating and Reading a Pictograph

"Today you will learn how to create and read a pictograph."

"Yesterday, you drew a picture of your favorite day of the week and my favorite day of the week on a graph."

"Today you will complete the graph to show the favorite day of the week of our friends, relatives, and neighbors."

"Whose favorite day of the week would you like to show on the graph?"

- List 8–10 names on the chalkboard.

  "Let's call these people to find their favorite day of the week and why it is their favorite."

- Assist your child as he/she calls each person on the list. Record each person's favorite day of the week and reason why on the chalkboard.

- Open the Meeting Book to pages 30 and 31.

  "Draw a picture and write each person's name in a box on the graph to show his/her favorite day of the week."

- When all the pictures have been drawn on the graph, ask the following questions:

  "What day did most people choose as their favorite day?"

  "How many more people chose _____ than _____?"

- This graph will be used for graph questions during The Meeting.

Class Practice

"Today you will use a clock to show the time one hour ago and the time one hour from now."

- Give your child a demonstration clock.

  "Let's pretend that it is seven o'clock now."

  "Show seven o'clock on your clock."

  "What time will it be one hour from now?"   *8:00*

  "Show that time on your clock."

  "Let's pretend that it is two o'clock now."

  "Show two o'clock on your clock."

  "What time was it one hour ago?"

  "Show that time on your clock."

- Repeat, using 1:00, 11:00, 6:00, and 3:00. Alternate using one hour ago and one hour from now.

number fact practice

- Use the peach fact cards to practice the doubles facts with your child.
- Give your child **Fact Sheet A 3.0.**
- Time your child for one minute.
- Correct the fact sheet with your child.
- Record the score.
- Allow time for your child to complete the unfinished facts.

Written Practice

- Complete **Worksheet 16A** with your child.
- Complete **Worksheet 16B** with your child later in the day.

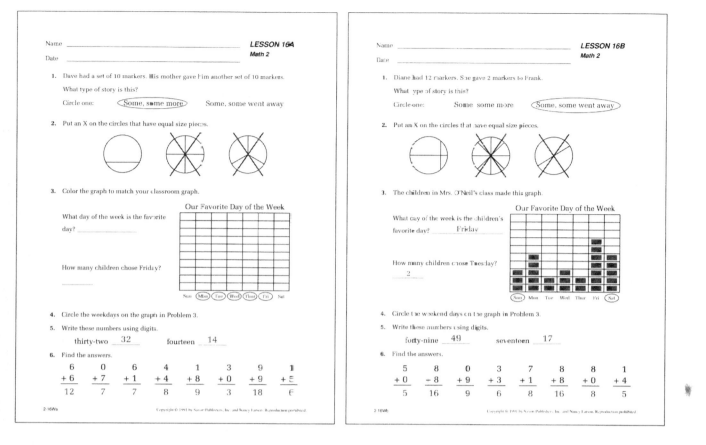

LESSON 16A
Math 2

Name _____

Date _____

1. Dave had a set of 10 markers. His mother gave him another set of 10 markers.

   What type of story is this?

   Circle one:　(Some, some more)　　Some, some went away

2. Put an X on the circles that have equal size pieces.

3. Color the graph to match your classroom graph.

   Our Favorite Day of the Week

   What day of the week is the favorite day? _____

   How many children chose Friday?

   _____

   Sun (Mon) (Tue) (Wed) (Thur) (Fri) Sat

4. Circle the weekdays on the graph in Problem 3.

5. Write these numbers using digits.

   thirty-two ___32___　　fourteen ___14___

6. Find the answers.

6 +6 12	0 +7 7	6 +1 7	4 +4 8	1 +8 9	3 +0 3	9 +9 18	1 +5 6

2-16Wa

LESSON 16B
Math 2

Name _____

Date _____

1. Diane had 12 markers. She gave 2 markers to Frank.

   What type of story is this?

   Circle one:　　Some, some more　　(Some, some went away)

2. Put an X on the circles that have equal size pieces.

3. The children in Mrs. O'Neil's class made this graph.

   Our Favorite Day of the Week

   What day of the week is the children's favorite day? ___Friday___

   How many children chose Tuesday?

   ___2___

   (Sun) Mon Tue Wed Thur Fri (Sat)

4. Circle the weekend days on the graph in Problem 3.

5. Write these numbers using digits.

   forty-nine ___49___　　seventeen ___17___

6. Find the answers.

5 +0 5	8 −8 16	0 +9 9	3 +3 6	7 +1 8	8 +8 16	8 +0 8	1 +4 5

2-16Wb

Lesson 17

identifying odd and even numbers

THE MEETING

calendar

• Ask your child to write the date on the calendar.

"Now you will write today's full date on the meeting strip."

"What month is it?"

"How do we spell (month)?"

• Your child writes the month on the meeting strip.

"What is the date?"

"What digits will you use to write the date?"

• Your child writes the date on the meeting strip.

"What is the year?"

• Your child records the year on the meeting strip.

"Let's say the date together."

• Recite the days of the week together once a week.

"What are the weekdays?"

"What are the days of the weekend?"

"What is the _____ th day of the week?"

"What day of the week is _____?"

"What day of the week was it yesterday?"

"Let's read the names of the months together."

"What is the _____ th month of the year?"

"What month of the year is _____?"

"What is the month before _____?"

"What is the month after _____?"

- Choose an upcoming special event or holiday.

"Let's count together to find the number of days until (special event)."

- Record the number of days on the meeting strip.

weather graph

- Ask your child to describe how the weather feels to him/her.

- Ask your child to color the correct box on the graph.

"What kind of weather have we had most often?"

"How many _____ days have we had?"

"How many more _____ days have we had than _____ days?"

counting

"Today we will count from 82 to 111."

- Write "82" and "111" on a piece of paper.

"Write the numbers we say as we count from 82 to 111."

- Your child writes the numbers as you count together.

"Let's count backward from 111 to 82."

- Count by 10's to 100 and backward from 100 by 10's.

graph questions

"Let's look at our 'Favorite Day of the Week Graph.'"

"How many people chose _____ as their favorite day of the week?"

"(Name of person)'s favorite day is _____."

"How many other people chose _____ as their favorite day?"

"How many more people chose _____ than _____?"

"How do you know?"

patterning

"Today's pattern is a different repeating pattern."

"Let's read the first part of the pattern together."

"What do you notice about the pattern?"

"What shape do you think will come next?"

- Fill in the next shape on the meeting strip.
- Repeat with the rest of the blanks.

"Let's read the shape pattern together."

clock

- Ask your child to set the clock on the hour.

"What time is it?"

"Write the digital time on the meeting strip."

"What time was it one hour ago?"

- On the meeting strip, write the name of an activity and the time it will occur.

"At what time will _____ occur?"

number of the day

"Today you will write three number sentences for the date."

- Record the three number sentences on the meeting strip.
- The money box on the meeting strip is not used today.

THE LESSON

Identifying Odd and Even Numbers

"Each morning we have had a repeating pattern on our pattern strip."

"Today you will make a repeating pattern using colors."

"You will use yellow and green crayons to make the pattern."

"You will make the color pattern on a hundred number chart."

- Give your child **Master 2-17** and a yellow and green crayon.

"Lightly color the first number yellow."

"What number did you color yellow?"   1

"Lightly color the second number green."

"What number did you color green?"   2

"Color the next number yellow."

"What number did you color yellow?" 3

"Color the next number green."

"What number did you color green?" 4

• Repeat until your child finishes coloring the 12.

"Now you will continue coloring one number yellow and the next number green until you reach 100."

• When your child finishes coloring, continue.

"What happened when you colored a yellow, green pattern?"

"Point to the green numbers as we read them together."

• Read the numbers together.

"These numbers are called the even numbers."

"I will write the even numbers to 20 on a counting strip in the Meeting Book as you read them."

• Open the Meeting Book to page 38. Begin at the bottom. Leave room for "0" below the "2." The strip should look like the following:

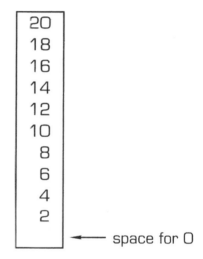

"We will use this counting strip to help us count forward and backward by 2's each morning."

"When we are saying these numbers, we are saying the even numbers."

• Give your child a bag of 20 color tiles.

"Let's choose an even number from the first two rows on the hundred number chart."

"What number could we choose?" 2, 4, 6, 8, 10, 12, 14, 16, 18, 20

• Ask your child to choose a number.

"Take (that number) tiles from the bag."

"Divide the tiles into two piles."

"Try to make both piles the same."

"What happened?" they are the same height

"Let's try this again using a different even number."

"What number could we choose?"

• Ask you child to choose a number.

"Take (that number) tiles from the bag."

"Divide the tiles in two piles."

"Try to make both piles the same."

"What happened?"

• Repeat one or two more times.

"An even number of tiles can be divided equally into two piles."

"Some people say that both piles are level or even."

"Put the tiles in the bag."

"Point to the yellow numbers as we read them together."

• Read the numbers together.

"These numbers are called the odd numbers."

"I will write the odd numbers from 1 to 19 on a counting strip in the Meeting Book as you read them."

• Open the Meeting Book to page 38. Begin at the bottom. The strip should look like the following:

| 19 |
| 17 |
| 15 |
| 13 |
| 11 |
| 9 |
| 7 |
| 5 |
| 3 |
| 1 |

"We will use this counting strip to help us say the odd numbers each morning."

"Now let's choose an odd number from the first two rows on the hundred number chart."

"What number could we choose?"   1, 3, 5, 7, 9, 11, 13, 15, 17, 19

• Ask your child to choose a number.

"Take (that number) tiles from the bag."

"Divide the tiles into two piles."

"Try to make both piles the same."

"What happened?"   there is one extra tile

"Let's try this again using a different odd number."

"What number could we choose?"

"Take (that number) tiles from the bag."

"Divide the tiles into two piles."

"Try to make both piles the same."

"What happened?"   there is one extra tile

• Repeat one or two more times.

"An odd number of tiles cannot be divided equally into two piles."

"Put the tiles in the bag."

CLASS PRACTICE

- Use the tan and peach fact cards to practice the doubles and adding one facts with your child.
- Give your child **Fact Sheet A 2.2.**
- Time your child for one minute.
- Correct the fact sheet with your child.
- Record the score.
- Allow time for your child to complete the unfinished facts.

WRITTEN PRACTICE

- Complete **Worksheet 17A** with your child.
- Complete **Worksheet 17B** with your child later in the day.

Name _____ **MASTER 2-17**
                                        Math 2

1	2	3	4	5	6	7	8	9	10
11	12	13	14	15	16	17	18	19	20
21	22	23	24	25	26	27	28	29	30
31	32	33	34	35	36	37	38	39	40
41	42	43	44	45	46	47	48	49	50
51	52	53	54	55	56	57	58	59	60
61	62	63	64	65	66	67	68	69	70
71	72	73	74	75	76	77	78	79	80
81	82	83	84	85	86	87	88	89	90
91	92	93	94	95	96	97	98	99	100

Parents: This is a hundred number chart which we will use to examine number
patterns. We will use this often during Math 2. Use this at home to help
with homework assignments.

Odd numbers will be colored yellow. Even numbers will be colored green.

2-17Ma   Copyright © 1991 by Saxon Publishers, Inc. and Nancy Larson. Reproduction prohibited.

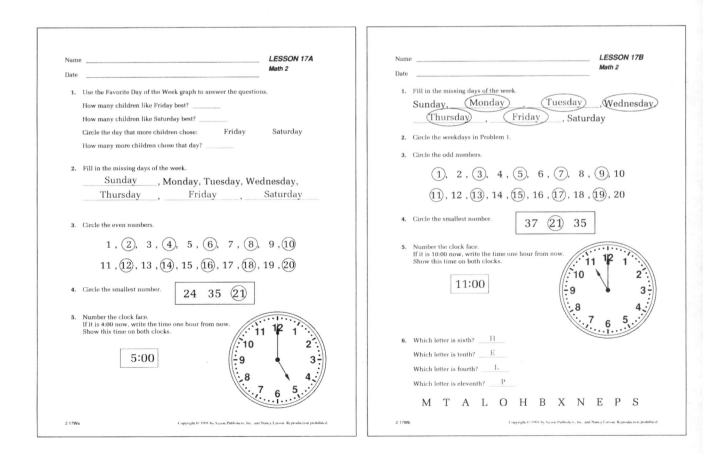

Name _____ **LESSON 17A**
                                        Math 2
Date _____

1. Use the Favorite Day of the Week graph to answer the questions.

   How many children like Friday best? _____

   How many children like Saturday best? _____

   Circle the day that more children chose: Friday   Saturday

   How many more children chose that day? _____

2. Fill in the missing days of the week.

   __Sunday__ , Monday, Tuesday, Wednesday,

   __Thursday__ , __Friday__ , __Saturday__

3. Circle the even numbers.

   1, ②, 3, ④, 5, ⑥, 7, ⑧, 9, ⑩

   11, ⑫, 13, ⑭, 15, ⑯, 17, ⑱, 19, ⑳

4. Circle the smallest number.   24   35   ㉑

5. Number the clock face.
   If it is 4:00 now, write the time one hour from now.
   Show this time on both clocks.

   5:00

2-17Wa   Copyright © 1991 by Saxon Publishers, Inc. and Nancy Larson. Reproduction prohibited.

Name _____ **LESSON 17B**
                                        Math 2
Date _____

1. Fill in the missing days of the week.

   Sunday, ⟨Monday⟩ , ⟨Tuesday⟩ , ⟨Wednesday⟩
   ⟨Thursday⟩ , ⟨Friday⟩ , Saturday

2. Circle the weekdays in Problem 1.

3. Circle the odd numbers.

   ①, 2, ③, 4, ⑤, 6, ⑦, 8, ⑨, 10

   ⑪, 12, ⑬, 14, ⑮, 16, ⑰, 18, ⑲, 20

4. Circle the smallest number.   37   ㉑   35

5. Number the clock face.
   If it is 10:00 now, write the time one hour from now.
   Show this time on both clocks.

   11:00

6. Which letter is sixth? __H__

   Which letter is tenth? __E__

   Which letter is fourth? __L__

   Which letter is eleventh? __P__

   M T A L O H B X N E P S

2-17Wb   Copyright © 1991 by Saxon Publishers, Inc. and Nancy Larson. Reproduction prohibited.

L esson 18

identifying common geometric shapes

materials

Master 2-18

pattern blocks

Fact Sheet A 3.2

in the morning

• Write the following in the pattern box on the meeting strip:

□, ○, ○, △, ▱, □, ○, ○, △, ▱, □, ___, ___, ___, ___

Answer: □, ○, ○, △, ▱ □ ○, ○, △, ▱ □ ○, ○, △, ▱

THE MEETING

calendar

   • Ask your child to write the date on the calendar.

   "Now you will write today's full date on the meeting strip."

   "What month is it?"

   "How do we spell (month)?"

   • Your child writes the month on the meeting strip.

   "What is the date?"

   "What digits will you use to write the date?"

   • Your child writes the date on the meeting strip.

   "What is the year?"

   • Your child records the year on the meeting strip.

   "Let's say the date together."

   • Recite the days of the week together once a week.

   "What are the weekdays?"

   "What are the days of the weekend?"

   "What is the _____th day of the week?"

"What day of the week is _____?"

"What day of the week was it yesterday?"

"Let's read the names of the months together."

"What is the _____th month of the year?"

"What month of the year is _____?"

"What is the month before _____?"

"What is the month after _____?"

- Choose an upcoming special event or holiday.

  "Let's count together to find the number of days until (special event)."

- Record the number of days on the meeting strip.

weather graph

- Ask your child to describe how the weather feels to him/her.

- Ask your child to color the correct box on the graph.

  "What kind of weather have we had most often?"

  "How many _____ days have we had?"

  "How many more _____ days have we had than _____ days?"

counting

"Today we will count from 103 to 122."

- Write "103" and "122" on a piece of paper.

  "Write the numbers we say as we count from 103 to 122."

- Your child writes the numbers as you count together.

  "Let's count backward from 122 to 103."

- Count by 10's to 100.

- Count backward from 100 by 10's.

  "Let's count by 2's to 20."

  "Do you remember what we call these numbers?"   even numbers

- Point to the numbers on the Meeting Book counting strip as you count together.

  "Let's count backward from 20 by 2's."

- Point to the numbers as you count together.

  "Now let's say the odd numbers together."

- Point to the numbers on the Meeting Book counting strip as you count together.

  "Now let's say the odd numbers backward from 19 to 1."

- Point to the numbers as you count together.

graph questions

"Let's look at our 'Favorite Day of the Week Graph.' "

"How many people chose _____ as their favorite day of the week?"

"(Name of person)'s favorite day is _____."

"How many other people chose _____ as their favorite day?"

"How many more people chose _____ than _____?"

"How do you know?"

patterning

"Today's pattern is a different repeating pattern."

"Let's read the first part of the pattern together."

"What do you notice about the pattern?"

"What shape do you think will come next?"

- Fill in the next shape on the meeting strip.

- Repeat with the rest of the blanks.

"Let's read the shape pattern together."

clock

- Ask your child to set the clock on the hour.

"What time is it?"

"Write the digital time on the meeting strip."

"What time was it one hour ago?"

- On the meeting strip, write the name of an activity and the time it will occur.

"At what time will _____ occur?"

number of the day

"Today you will write three number sentences for the date."

- Record the three number sentences on the meeting strip.

- The money box on the meeting strip is not used today.

THE LESSON

Identifying Common Geometric Shapes

"Today you will learn how to identify common geometric shapes."

"You will also learn how to cover the yellow pattern block using the other pattern blocks."

- Draw the following shapes on the chalkboard:

 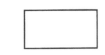

  "What is the name of each of these shapes?"   square, circle, triangle, rectangle

- Write the names under the shapes.

  "Now you will try to find things in our house that have these shapes."

- Give your child **Master 2-18**.

  "When you find something that has one of these shapes, draw a picture of what you found."

  "You will have five minutes to find as many shapes as possible."

- Allow time for your child to do this.

  "What did you find with the shape of a triangle?"

- Repeat with the other shapes.

  "There are some other shapes we have used."

- Draw the following shapes on the chalkboard:

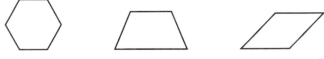

  "Where have you seen these shapes before?"   pattern blocks, pattern of the day

  "Do you know the name of one of these shapes?"   hexagon, trapezoid, parallelogram

- Write the names under the shapes.

- Put a yellow, red, blue, orange, green, and tan pattern block on the table.

  "What shape is the red pattern block?"   trapezoid

  "What shape is the green pattern block?"   triangle

  "What shape is the blue pattern block?"   parallelogram

  "What shape is the yellow pattern block?"   hexagon

  "What pattern block is the largest?"   yellow hexagon

  "Now you will cover the four hexagons on Master 2-18 using only one color pattern block at a time."

- Give your child a basket of pattern blocks.

  "Cover each hexagon on the paper using only one color pattern block."

- When your child finishes, continue.

  "How many yellow pieces did you use to cover the hexagon?"   1

"How many trapezoids did you use to cover the hexagon?"   2

"Are the two trapezoids the same size?"

"How can you be sure that they are the same size?"

"If the two pieces are the same size, we can say that each piece is one half of the hexagon."

"How many parallelograms did you use to cover the hexagon?"   3

"Are the three parallelograms the same size?"

"How can you be sure that they are the same size?"

"If the three pieces are the same size, we can say that each piece is one third of the hexagon."

"How many green triangles did you use to cover the hexagon?"   6

"Are the six triangles the same size?"

"How can you be sure that they are the same size?"

"If the six pieces are the same size, we can say that each piece is one sixth of the hexagon."

"We will cover the hexagons and trace the pattern blocks tomorrow."

- Save **Master 2-18** for use in Lesson 19.

CLASS PRACTICE

number fact practice

"Use your fact cards to practice the addition facts."

- Allow time for your child to do this.
- Give your child **Fact Sheet A 3.2.**
- Time your child for one minute.
- Correct the fact sheet with your child.
- Record the score.
- Allow time for your child to complete the unfinished facts.

WRITTEN PRACTICE

- Complete **Worksheet 18A** with your child.
- Complete **Worksheet 18B** with your child later in the day.

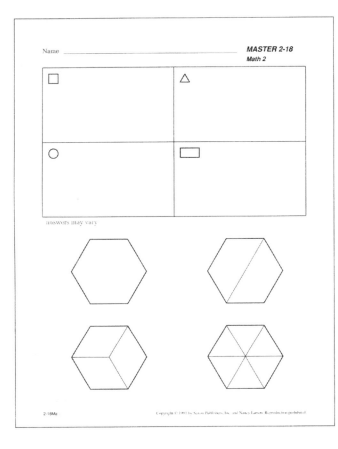

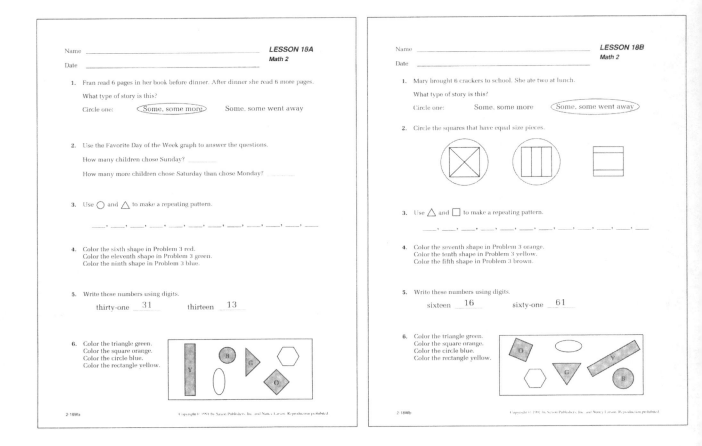

Name _____ LESSON 18A
Date _____ Math 2

1. Fran read 6 pages in her book before dinner. After dinner she read 6 more pages.

    What type of story is this?

    Circle one: (Some, some more)   Some, some went away

2. Use the Favorite Day of the Week graph to answer the questions.

    How many children chose Sunday? _____

    How many more children chose Saturday than chose Monday? _____

3. Use ◯ and △ to make a repeating pattern.

    ___, ___, ___, ___, ___, ___, ___, ___, ___, ___, ___, ___

4. Color the sixth shape in Problem 3 red.
    Color the eleventh shape in Problem 3 green.
    Color the ninth shape in Problem 3 blue.

5. Write these numbers using digits.

    thirty-one ___31___   thirteen ___13___

6. Color the triangle green.
    Color the square orange.
    Color the circle blue.
    Color the rectangle yellow.

2-18Wa   Copyright © 1994 by Saxon Publishers, Inc. and Nancy Larson. Reproduction prohibited.

Name _____ LESSON 18B
Date _____ Math 2

1. Mary brought 6 crackers to school. She ate two at lunch.

    What type of story is this?

    Circle one: Some, some more   (Some, some went away)

2. Circle the squares that have equal size pieces.

3. Use △ and □ to make a repeating pattern.

    ___, ___, ___, ___, ___, ___, ___, ___, ___, ___, ___, ___

4. Color the seventh shape in Problem 3 orange.
    Color the tenth shape in Problem 3 yellow.
    Color the fifth shape in Problem 3 brown.

5. Write these numbers using digits.

    sixteen ___16___   sixty-one ___61___

6. Color the triangle green.
    Color the square orange.
    Color the circle blue.
    Color the rectangle yellow.

2-18Wb   Copyright © 1994 by Saxon Publishers, Inc. and Nancy Larson. Reproduction prohibited.

Lesson 19

identifying fractional parts of a whole

---

lesson preparation

materials

Master 2-18 from Lesson 18

pattern blocks

Fact Sheet A 3.2

in the morning

• Write the following in the pattern box on the meeting strip:

Answer:

THE MEETING

calendar

• Ask your child to write the date on the calendar.

"Now you will write today's full date on the meeting strip."

"What month is it?"

"How do we spell (month)?"

• Your child writes the month on the meeting strip.

"What is the date?"

"What digits will you use to write the date?"

• Your child writes the date on the meeting strip.

"What is the year?"

• Your child records the year on the meeting strip.

"Let's say the date together."

"What are the weekdays?"

"What are the days of the weekend?"

"What is the _____th day of the week?"

"What day of the week is _____?"

"What day of the week was it yesterday?"

"Let's read the names of the months together."

"What is the _____th month of the year?"

"What month of the year is _____?"

"What is the month before _____?"

"What is the month after _____?"

- Choose an upcoming special event or holiday.

"Let's count together to find the number of days until (special event)."

- Record the number of days on the meeting strip.

weather graph

- Ask your child to describe how the weather feels to him/her.

- Ask your child to color the correct box on the graph.

"What kind of weather have we had most often?"

"How many _____ days have we had?"

"How many more _____ days have we had than _____ days?"

counting

"Today we will count from 173 to 197."

- Write "173" and "197" on a piece of paper.

"Write the numbers we say as we count from 173 to 197."

- Your child writes the numbers as you count together.

"Let's count backward from 197 to 173."

- Count by 10's to 100.

- Count backward from 100 by 10's.

"Let's count by 2's to 20."

"Do you remember what we call these numbers?"   even numbers

- Point to the numbers on the counting strip as you count together.

"Now let's count backward from 20 by 2's."

- Point to the numbers as you count together.

"Now let's say the odd numbers together."

- Point to the numbers on the counting strip as you count together.

"Now let's say the odd numbers backward from 19 to 1."

- Point to the numbers as you count together.

graph questions

"Let's look at our 'Favorite Day of the Week Graph.' "

"How many people chose _____ as their favorite day of the week?"

"(Name of person)'s favorite day is _____."

"How many other people chose _____ as their favorite day?"

"How many more people chose _____ than _____?"

"How do you know?"

patterning

"Today's pattern is a repeating pattern."

"Let's read the first part of the pattern together."

"What do you notice about the pattern?"

"What do you think will come next?"

- Fill in the next shape on the meeting strip.

- Repeat with the rest of the blanks.

"Let's read the pattern together."

clock

- Ask your child to set the clock on the hour.

"What time is it?"

"Write the digital time on the meeting strip."

"What time was it one hour ago?"

- On the meeting strip, write the name of an activity and the time it will occur.

"At what time will _____ occur?"

number of the day

"Today you will write three number sentences for the date."

- Record the three number sentences on the meeting strip.

- The money box on the meeting strip is not used today.

THE LESSON

Identifying Fractional Parts of a Whole

"Today you will learn how to identify fractional parts of a whole."

"Yesterday you covered the yellow pattern block using only one shape pattern block at a time."

"What pattern blocks did you use to cover the yellow hexagon exactly?" trapezoid, parallelogram, triangle

"You will do that again today."

- Give your child **Master 2-18** from Lesson 18 and a basket of pattern blocks.

  "Cover each hexagon on the paper using only one color pattern block."

- Allow time for your child to cover each hexagon.

  "Now we will work together to trace the pattern blocks you used."

  "You will take away one piece at a time and trace along the edges of the pieces that are left."

  "I will hold the pattern blocks as you trace along the edges."

- Allow time for your child to trace the pattern blocks.

  "How many trapezoids did you use to cover the hexagon?"   2

  "Are the two trapezoids the same size?"

  "How can you be sure that they are the same size?"

  "If the two pieces are the same size, then we can say that each piece is one half of the hexagon."

  "How many parallelograms did you use to cover the hexagon?"   3

  "Are the three parallelograms the same size?"

  "How can you be sure that they are the same size?"

  "If the three pieces are the same size, we can say that each piece is one third of the hexagon."

  "How many triangles did you use to cover the hexagon?"   6

  "Are the six triangles the same size?"

  "How can you be sure that they are the same size?"

  "If the six pieces are the same size, we can say that each piece is one sixth of the hexagon."

- Draw and label the following on the chalkboard:

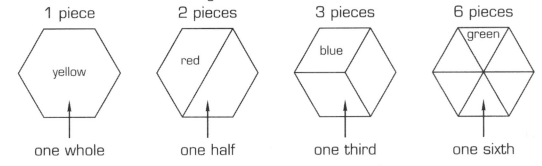

1 piece	2 pieces	3 pieces	6 pieces
yellow	red	blue	green
one whole	one half	one third	one sixth

- Hold up the red trapezoid.

  "What fractional part of the hexagon is this piece?"   one half

- Repeat with the blue parallelogram and the green triangle.

  "Hold up the pattern block that is one third of the hexagon."

  "What is the name of this shape?"   parallelogram

  "Hold up the pattern block that is one sixth of the hexagon."

"What is the name of this shape?" triangle

"Hold up the pattern block that is one half of the hexagon."

"What is the name of this shape?" trapezoid

CLASS PRACTICE

number fact practice

- Use the fact cards to practice the addition facts with your child.
- Give your child **Fact Sheet A 3.2**.
- Time your child for one minute.
- Correct the fact sheet with your child.
- Record the score.
- Allow time for your child to complete the unfinished facts.

WRITTEN PRACTICE

- Complete **Worksheet 19A** with your child.
- Complete **Worksheet 19B** with your child later in the day.

Name _____ LESSON 19A
Date _____ Math 2

1. Fill in the missing days of the week.
   Sunday, _Monday_ , _Tuesday_ , _Wednesday_ ,
   Thursday, _Friday_ _Saturday_

2. Match each name with the correct piece.
   one half
   one third
   one sixth

3. Write five odd numbers.
   ___ , ___ , ___ , ___ , ___

4. Circle the number that is one less than **28**.
   Put a square around the number that is one more than **17**.
   [18]   16   30   29   19   (27)

5. What time does the clock show?
   5:00
   What time was it one hour ago?
   4:00

2-19Wa   Copyright © 1991 by Saxon Publishers, Inc. and Nancy Larson. Reproduction prohibited.

Name _____ LESSON 19B
Date _____ Math 2

1. Fill in the missing numbers on this piece of a hundred number chart.

41	42	43	44	45	46	47	48	49	50
51	52	53	54	55	56	57	58	59	60
61	62	63	64	65	66	67	68	69	70
71	72	73	74	75	76	77	78	79	80

   What number is to the right of **56**? _57_
   What number is to the left of **56**? _55_
   What number is above **56**? _46_
   What number is below **56**? _66_

2. Write five even numbers.
   ___ , ___ , ___ , ___ , ___

3. Circle the number that is one less than **42**.
   Put a square around the number that is one more than **23**.
   (41)   21   [24]   43   44   22

4. What time does the clock show?
   7:00
   What time was it one hour ago?
   6:00

2-19Wb   Copyright © 1991 by Saxon Publishers, Inc. and Nancy Larson. Reproduction prohibited.

esson 20

creating a color pattern

---

lesson preparation ─────────────────────────────────

materials

Oral Assessment #2

Written Assessment #3

pattern blocks

Master 2-20

crayons

Fact Sheet A 3.2

in the morning

• Write the following in the pattern box on the meeting strip:

| ⊖, ⊖, ⊖, ⊖, ⊖, ⊖, ___, ___, ___, ___ |

Answer: ⊖, ⊖, ⊖, ⊖, ⊖, ⊖, ⊖, ⊖, ⊖, ⊖

THE MEETING

calendar

- • Ask your child to write the date on the calendar.

    "Now you will write today's date on the meeting strip."

    "What month is it?"

    "How do we spell (month)?"

- • Your child writes the month on the meeting strip.

    "What is the date?"

    "What digits will you use to write the date?"

- • Your child writes the date on the meeting strip.

    "What is the year?"

- • Your child records the year on the meeting strip.

    "Let's say the date together."

    "What are the weekdays?"

"What are the days of the weekend?"

"What is the _____th day of the week?"

"What day of the week is _____?"

"What day of the week was it yesterday?"

"Let's read the names of the months together."

"What is the _____th month of the year?"

"What month of the year is _____?"

"What is the month before _____?"

"What is the month after _____?"

- Choose an upcoming special event or holiday.

"Let's count together to find the number of days until (special event)."

- Record the number of days on the meeting strip.

weather graph

- Ask your child to describe how the weather feels to him/her.

- Ask your child to color the correct box on the graph.

"What kind of weather have we had most often?"

"How many _____ days have we had?"

"How many more _____ days have we had than _____ days?"

counting

"Today we will count from 146 to 165."

- Write "146" and "165" on a piece of paper.

"Write the numbers we say as we count from 146 to 165."

- Your child writes the numbers as you count together.

"Let's count backward from 165 to 146."

- Count by 10's to 100 and backward from 100 by 10's.

"Let's count by 2's to 20."

"Do you remember what we call these numbers?" even numbers

- Point to the numbers on the counting strip as you count together.

"Now let's count backward from 20 by 2's."

- Point to the numbers as you count together.

"Now let's say the odd numbers together."

- Point to the numbers on the counting strip as you count together.

"Now let's say the odd numbers backward from 19 to 1."

- Point to the numbers as you count together.

graph questions

"Let's look at our 'Favorite Day of the Week Graph.' "

"How many people chose _____ as their favorite day of the week?"

"(Name of person)'s favorite day is _____."

"How many other people chose _____ as their favorite day?"

"How many more people chose _____ than _____?"

"How do you know?"

patterning

"Today's pattern is a repeating pattern."

"Let's read the first part of the pattern together."

"What do you notice about the pattern?"

"What do you think will come next?"

- Fill in the next shape on the meeting strip.

- Repeat with the rest of the blanks.

"Let's read the pattern together."

clock

- Ask your child to set the clock on the hour.

"What time is it?"

"Write the digital time on the meeting strip."

"What time was it one hour ago?"

- On the meeting strip, write the name of an activity and the time it will occur.

"At what time will _____ occur?"

number of the day

"Today you will write three number sentences for the date."

- Record the three number sentences on the meeting strip.

- The money box on the meeting strip is not used today.

ASSESSMENT

Written Assessment

- All of the questions on the assessment are based on concepts and skills presented at least five lessons ago. It is expected that your child will master at least 80% of the concepts on the assessment. If your child is having difficulty with a specific concept, reteach the concept the following day.

"Today I would like to see what you remember from what we have been practicing."

- Give your child **Written Assessment #3**.

- Read the directions for each problem. Allow time for your child to complete that problem before continuing.

- Correct the paper, noting your child's mistakes on the **Individual Recording Form**. Review the errors with your child.

Oral Assessment

- Record your child's response(s) to the oral interview questions on the interview sheet.

THE LESSON

Creating a Color Pattern

"You have made patterns with the pattern blocks and you have colored patterns on the hundred number chart."

"Today you will learn how to make a color pattern using your name."

- Give your child **Master 2-20**.

"Start in the first box at the upper left-hand corner and write one letter of your name in each box."

- Assist your child as he/she begins the pattern.

"When you get to the last letter of your name, start again from the beginning."

"If you come to the end of the line, go to the next line to finish writing your name."

"Write your name over and over until you fill all of the boxes on the grid."

B	O	B	B	O	B	B	O	B	B
O	B	B	O	B	B	O	B	B	O
B	B	O	B	B	O	B	B	O	B
B	O	B	B	O	B	B	O	B	B
O	B	B	O	B	B	O	B	B	O

Name = Bob
B = green
O = yellow

"Now you will use your crayons to color each letter a different color."

"Whenever you see the same letter, you will color it the same color."

"All of the _____'s will be one color, all of the _____'s will be another color, and so on."

- Assist your child as he/she begins the color pattern.

- Allow time for your child to complete the color pattern.

  "What do you notice about your pattern?"

- Allow time for your child to offer observations.

- Repeat with the next two grids.

CLASS PRACTICE

number fact practice

- Allow time for your child to practice the addition facts independently.

- Give your child **Fact Sheet A 3.2.**

- Time your child for one minute.

- Correct the fact sheet with your child.

- Record the score.

- Allow time for your child to complete the unfinished facts.

Teacher _____

Date _____

MATH 2   LESSON 20

Oral Assessment # 2 Recording Form

Materials:
Pattern blocks

Students	A. *"Name the days of the week."* B. *"What are the weekdays?"* C. *"What are the days of the weekend?"*	•Place 12 pattern blocks in a row. *"Point to the first block."* *"Point to the fourth, ...eighth, ...eleventh."*

2-PFo

Name _____

Date _____

ASSESSMENT 3
LESSON 20
Math 2

1. John has 8 pennies. Susan gave him 2 pennies. How many pennies does John have now?

   What type of story is this?

   Circle one: (some, some more)   some, some went away

   Four children were playing. One child went home. How many children are playing now?

   What type of story is this?

   Circle one: some, some more   (some, some went away)

2. Finish the repeating pattern.

   △ ○ ○ □ △ ○ ○ □ △ ○ ○ □ △

3. Write these numbers using digits.

   Fourteen ___14___   thirty-five ___35___

   sixty-one ___61___   seventy ___70___

4. Circle the shapes that have equal size pieces.

5. What is one more than **49**? ___50___

   What is one less than **20**? ___19___

6. Add.

   6 + 1 = __7__   5 + 0 = __5__   8 + 8 = __16__   0 + 7 = __7__

   7 + 7 = __14__   1 + 8 = __9__   3 + 3 = __6__   9 + 9 = __18__

2-20Aa

Name _____

MASTER 2-20
Math 2

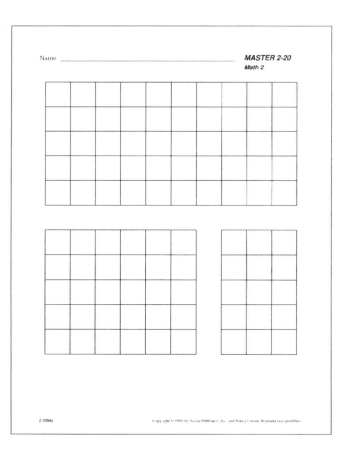

2-20Ma

Lesson 21

addition facts—adding two

lesson preparation

materials

addition fact cards — green

Fact Sheet AA 4.0

the night before

• Separate the green addition fact cards.

in the morning

• Write the following in the pattern box on the meeting strip:

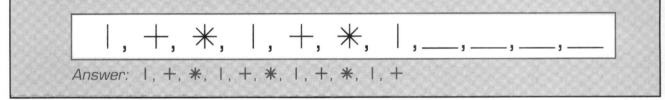

Answer: I, +, *, I, +, *, I, +, *, I, +

THE MEETING

• If necessary, refer to Lesson 20 for the questioning sequence.

calendar

• Ask your child to write the date on the calendar and meeting strip.

• Ask your child the following:

days of the week, weekdays, days of the weekend

months of the year, _____th month, month before, month after

"What day of the week was it four days ago?"

"What day of the week will it be seven days from today?"

• Record on the meeting strip a special event and the number of days until it occurs.

weather graph

• Ask your child to graph the weather.

• Ask questions about the graph.

counting

- Count forward and backward by 1's from 98 to 121 as your child writes these numbers.

- Count by 10's to 100 and backward from 100 by 10's.

- Say the even numbers to 20 and backward from 20.

- Say the odd numbers to 19 and backward from 19.

graph questions

- You and your child each ask a question about any of the graphs.

patterning

- Ask your child to do the following:

  identify the pattern as a repeating pattern

  identify the shapes to complete the pattern

  read the pattern

clock

- Ask your child to set the clock on the hour.

- Ask your child to write the digital time on the meeting strip.

  "What time was it one hour ago?"

  "What time will it be one hour from now?"

- Record on the meeting strip the time an activity will occur.

number of the day

- Write three number sentences for the number of the day on the meeting strip.

- The money box on the meeting strip is not used today.

THE LESSON

Addition Facts — Adding Two

- Write the following on the chalkboard:

$$\begin{array}{r} 0 \\ +\ 2 \\ \hline \end{array}$$

  "What is the answer for this problem?"

- Write the answer on the chalkboard.

  "I will write the even numbers next to the zero."

• Write the following on the chalkboard:

$$
\begin{array}{ccccc}
0 & 2 & 4 & 6 & 8 \\
\underline{+\,2} & & & & \\
2 & & & &
\end{array}
$$

"Now we will add two to each of these numbers."

• Write the following on the chalkboard:

$$
\begin{array}{ccccc}
0 & 2 & 4 & 6 & 8 \\
\underline{+\,2} & \underline{+\,2} & \underline{+\,2} & \underline{+\,2} & \underline{+\,2} \\
2 & & & &
\end{array}
$$

"Let's find each of these answers."

"What is two plus two?"

"What is four plus two?"

• Repeat with the remaining problems.

• Record the answers on the chalkboard.

"What did you notice about the answers when we added two to an even number?" the answer is the next even number

"Whenever we add two to an even number, the answer is the next even number."

• Do not erase these problems.

• Write the following on the chalkboard:

$$
\begin{array}{c}
1 \\
\underline{+\,2}
\end{array}
$$

"What is the answer for this problem?"

"I will write the odd numbers next to the one."

$$
\begin{array}{ccccc}
1 & 3 & 5 & 7 & 9 \\
\underline{+\,2} & & & & \\
3 & & & &
\end{array}
$$

"Now I will add two to each of these numbers."

• Write the following on the chalkboard:

$$
\begin{array}{ccccc}
1 & 3 & 5 & 7 & 9 \\
\underline{+\,2} & \underline{+\,2} & \underline{+\,2} & \underline{+\,2} & \underline{+\,2} \\
3 & & & &
\end{array}
$$

"Let's find each of these answers."

"What is three plus two?"

"What is five plus two?"

• Repeat with the remaining problems.

• Record the answers on the chalkboard.

"What did you notice about the answers when we added two to an odd number?" the answer is the next odd number

"Whenever we add two to an odd number, the answer is the next odd number."

"I'll say the problems in random order."

"See how fast you can say the answers."

- Allow your child to refer to the chalkboard problems, if necessary.
- Erase the chalkboard answers only.

"Now I will say the answer to an adding two problem."

"See if you can tell me the problem."

"Six."   *four plus two*

- Repeat with all combinations.
- If your child is having difficulty, rewrite the answers on the board.

CLASS PRACTICE

number fact practice

- Give your child the green addition fact cards.
- Allow your child to practice independently.
- Give your child **Fact Sheet AA 4.0.**
- Time your child for one minute.
- Correct the fact sheet with your child.
- Record the score.
- Allow time for your child to complete the unfinished facts.

writing numbers to 100

"Now you will practice writing numbers."

"Write each number I say on the back of your fact sheet."

"Remember to watch out for the tricky numbers."

"What are the tricky numbers?"   *10, 11, 12, and the teens*

- Read the following numbers:

    42, 38, 71, 60, 19, 94, 11, 83, 17, 75, 15

WRITTEN PRACTICE

- Complete **Worksheet 21A** with your child.
- Complete **Worksheet 21B** with your child later in the day.

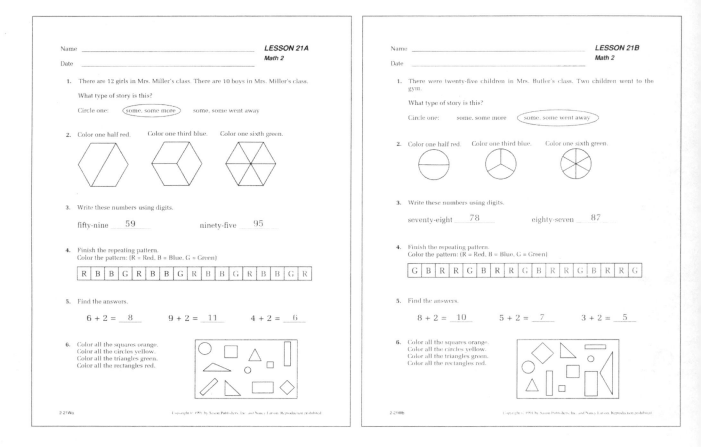

Lesson 22

identifying and sorting common geometric shapes by attribute

---

lesson preparation

materials

Masters 2-22

plastic bag for shape pieces

Fact Sheet A 3.2

the night before

• Cut out the shapes on Masters 2-22.

in the morning

• Write the following in the pattern box on the meeting strip:

△, ◯, △, ▢, △, ◯, △, ▢, ___, ___, ___, ___

Answer: △, ◯, △, ▢, △, ◯, △, ▢, △, ◯, △, ▢

THE MEETING

• If necessary, refer to Lesson 20 for the questioning sequence.

calendar

• Ask your child to write the date on the calendar and meeting strip.

• Ask your child the following:

    days of the week, weekdays, days of the weekend

    months of the year, _____th month, month before, month after

"What day of the week was it seven days ago?"

"What day of the week will it be three days from today?"

• Record on the meeting strip a special event and the number of days until it occurs.

weather graph

• Ask your child to graph the weather.

• Ask questions about the graph.

counting

- Count forward and backward by 1's from 93 to 114 as your child writes these numbers.
- Count by 10's to 100 and backward from 100 by 10's.
- Say the even numbers to 20 and backward from 20.
- Say the odd numbers to 19 and backward from 19.

graph questions

- You and your child each ask a question about any of the graphs.

patterning

- Ask your child to do the following:

  identify the pattern as a repeating pattern

  identify the shapes to complete the pattern

  read the pattern

clock

- Ask your child to set the clock on the hour.
- Ask your child to write the digital time on the meeting strip.

  "What time was it one hour ago?"

  "What time will it be one hour from now?"

- Record on the meeting strip the time an activity will occur.

number of the day

- Write three number sentences for the number of the day on the meeting strip.
- The money box on the meeting strip is not used today.

THE LESSON

Identifying and Sorting Common Geometric Shapes by Attribute

  "Today you will learn how to name geometric shape pieces by attribute."

- Give your child the bag of shape pieces from **Masters 2-22**.

  "What do you notice about the pieces I gave you?"

- Allow time for your child to offer observations.

  "What colors do you have?"   red, green, yellow, blue

"What shapes do you have?" squares, rectangles, circles, triangles

"How many green squares do you have?" 2

"Are the green squares exactly the same?" no

"How are they different?" size

"How many yellow triangles do you have?"

"Are the yellow triangles exactly the same?" no

"How are they different?" size

"Put together the pieces that are alike in some way."

- When your child finishes sorting, continue.

  "How did you sort the pieces?"

- Ask your child to identify how he/she sorted the pieces.

  "Now sort your pieces in a different way."

- Ask your child to describe how he/she sorted the pieces this time.

- When your child finishes, continue.

  "How did you sort the pieces?"

  "What are the different ways we could sort the pieces?" color, size, shape

- Hold up the small yellow square.

  "Describe this piece for me."

- If your child does not describe all of the attributes, ask the following question:

  "What else can you tell us about this piece?"

  "This piece has three attributes."

  "It is small, it is yellow, and it is square."

- Repeat with several more pieces.

  "Each piece has three attributes: size, color, and shape."

  "Find the piece that is a small blue triangle."

- Repeat with several more pieces.

  "Now we will play a guessing game."

  "Mix up your pieces."

  "Close your eyes."

  "I will take a piece."

- Remove a piece.

  "Open your eyes."

  "Tell me the missing piece."

- Ask your child to name the missing piece using all three attributes.
- Take turns removing and identifying the missing piece.

"Put the pieces in the bag so we can use them again."

CLASS PRACTICE

number fact practice

- Use the fact cards to practice the doubles, adding one, and adding zero facts with your child.
- Give your child **Fact Sheet A 3.2.**
- Time your child for one minute.
- Correct the fact sheet with your child.
- Record the score.
- Allow time for your child to complete the unfinished facts.

WRITTEN PRACTICE

- Complete **Worksheet 22A** with your child.
- Complete **Worksheet 22B** with your child later in the day.

MASTER 2-22
Math 2

2-22Ma   Copyright © 1991 by Saxon Publishers, Inc. and Nancy Larson. Reproduction prohibited.

Name _____   **LESSON 22A**
Date _____   Math 2

1. Use the Favorite Day of the Week graph to answer the questions.

   How many children chose a weekday as their favorite day? _____

   How many children chose a day of the weekend as their favorite day? _____

2. Write the even numbers from 2 to 20.

   __2_ , __4_ , __6_ , __8_ , _10_ , _12_ , _14_ , _16_ , _18_ , _20_

   Write the odd numbers from 1 to 19.

   __1_ , __3_ , __5_ , __7_ , __9_ , _11_ , _13_ , _15_ , _17_ , _19_

3. Circle the greatest number.   47   43   ⑭9

4. How are these shapes the same?   ▭   ▭

5. Number the clock face.

   If it is 12:00 now, write the digital time one hour ago.

   | 11:00 |

   Show this time on the clock face

2-22Wa   Copyright © 1991 by Saxon Publishers, Inc. and Nancy Larson. Reproduction prohibited.

Name _____   **LESSON 22B**
Date _____   Math 2

1. Fill in the missing numbers on this piece of a hundred number chart.

61	62	63	64	65	66	67	68	69	70
71	72	73	74	75	76	77	78	79	80
81	82	83	84	85	86	87	88	89	90

   What number is to the right of 74? _75_   What number is to the left of 74? _73_

   What number is above 74? _64_   What number is below 74? _84_

2. Write the even numbers from 20 to 2.

   _20_ , _18_ , _16_ , _14_ , _12_ , _10_ , __8_ , __6_ , __4_ , __2_

   Write the odd numbers from 19 to 1.

   _19_ , _17_ , _15_ , _13_ , _11_ , __9_ , __7_ , __5_ , __3_ , __1_

3. Circle the greatest number.   ㊱   34   33

4. How are these shapes different?   ▭   ▭

5. Number the clock face.

   If it is 6:00 now, write the digital time one hour ago.

   | 5:00 |

   Show this time on the clock face.

2-22Wa   Copyright © 1991 by Saxon Publishers, Inc. and Nancy Larson. Reproduction prohibited.

esson 23

drawing pictures and writing number sentences for some, some more, and some, some went away stories

---

lesson preparation

materials

scrap paper

Fact Sheet AA 4.0

in the morning

• Write the following in the pattern box on the meeting strip:

○, △, ▱, ▭, ▭, ○, △, ▱, ▭, ▭, ___, ___, ___, ___, ___

Answer: ○, △, ▱, ▭, ▭, ○, △, ▱, ▭, ▭, ○, △, ▱, ▭, ▭

THE MEETING

• If necessary, refer to Lesson 20 for the questioning sequence.

calendar

• Ask your child to write the date on the calendar and meeting strip.

• Ask your child the following:

> days of the week, weekdays, days of the weekend
>
> months of the year, _____th month, month before, month after

> **"What day of the week was it six days ago?"**

> **"What day of the week will it be four days from today?"**

• Record on the meeting strip a special event and the number of days until it occurs.

weather graph

• Ask your child to graph the weather.

• Ask questions about the graph.

counting

- Count forward and backward by ´'s from 128 to 153 as your child writes these numbers.
- Count by 10's to 100 and backward from 100 by 10's.
- Say the even numbers to 20 and backward from 20.
- Say the odd numbers to 19 and backward from 19.

graph questions

- You and your child each ask a question about any of the graphs.

patterning

- Ask your child to do the following

    identify the pattern as a repeating pattern

    identify the shapes to complete the pattern

    read the pattern

clock

- Ask your child to set the clock on the hour.
- Ask your child to write the digital time on the meeting strip.

    "What time was it one hour ago?"

    "What time will it be one hour from now?"

- Record on the meeting strip the time an activity will occur.

number of the day

- Write three number sentences for the number of the day on the meeting strip.
- The money box on the meeting strip is not used today.

THE LESSON

Drawing Pictures and Writing Number Sentences for Some, Some More, and Some, Some Went Away Stories

"Today you will learn how to draw pictures and write number sentences for some, some more stories and some, some went away stories."

"We acted out some, some more stories and some, some went away stories before."

"Sometimes we will have a some, some more or a some, some went away story that will be difficult to act out."

"Instead of acting out the story, we will draw a picture to show what happened in the story."

"I will tell you a story."

"Darleen went fishing with her sister. She caught two fish in the morning. In the afternoon, she caught three fish."

"What happened in this story?"   Darleen caught 2 fish in the morning and 3 fish in the afternoon

"What kind of story is this?"   some, some more

"Let's draw a picture to show what happened."

"What will we draw first?"   2 fish

- Draw a picture of two fish on the chalkboard.

"What will we draw next?"   three more fish

- Draw a picture of three more fish.

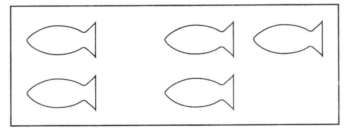

"What number sentence will we write to show what happened in this story?"

- Write "2 fish + 3 fish = 5 fish" on the chalkboard.

"How many fish did Darleen catch that day?"

"The answer to my question is five fish."

- Write "5 fish" on the chalkboard.

"When we answer a story problem, we will write our answer with a label."

"The label tells us whether we have five fish or five dogs or five hamburgers or five flowers."

"I will tell you another story."

"There were two birds in the bird house."

"Their eggs hatched and they had four baby birds."

"What happened in this story?"

"What kind of story is this?"   some, some more

"Let's draw a picture to show what happened."

"What will we draw first?"   2 birds

- Draw the following on the chalkboard:

"What will we draw next?"   4 more birds

- Draw the following on the chalkboard:

"What number sentence will we write to show what happened in this story?"

- Write "2 birds + 4 birds = 6 birds" on the chalkboard.

"How many birds are in the bird house now?"   6 birds

"When we answer a story problem, we will write our answer with a label."

"The label tells us whether we have six birds or six dogs or six hamburgers or six flowers."

"I will tell you another story."

"There were six ice cream bars in the package."

"Felicia and her friend each ate one ice cream bar."

"What happened in this story?"

"What kind of story is this?"   some, some went away

"Let's draw a picture to show what happened in this story."

"What will we draw first?"

- Draw the following on the chalkboard:

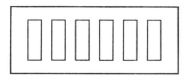

"What happened next?"

"I will show this by crossing out the ice cream bars they ate."

"How many ice cream bars did they eat?"   2

- Draw the following on the chalkboard:

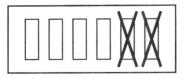

"What number sentence will we write to show what happened in this story?"

- Write "6 bars – 2 bars = 4 bars" on the chalkboard.

"How many ice cream bars are in the package now?"

"When we answer a story problem, we will write our answer with a label."

"I will tell you another story."

"Cecil had eight dimes in his pocket."

"He spent three dimes for a pencil."

"What happened in this story?"

"What type of story is this?"   some, some went away

"Let's draw a picture to show what happened in this story."

"What will we draw first?"

- Draw the following on the chalkboard:

"What happened next?"

"How will I show that?"

"I will show that by crossing out the dimes he spent."

- Draw the following on the chalkboard:

"What number sentence will we write to show what happened in this story?"

- Write "8 dimes – 3 dimes = 5 dimes" on the chalkboard.

"How many dimes are in his pocket now?"   5 dimes

"When we answer a story problem, we will write our answer with a label."

"Now you will practice drawing pictures and writing number sentences for some, some more, and some, some went away stories."

- Give your child a piece of scrap paper.
- Write the following on the chalkboard:

    Cheryl had 5 markers. She gave 3 markers to Steven.

"Read this story."

"What happened first in the story?"   Cheryl had 5 markers

"What happened next in the story?"   she gave 3 markers to Steven

"What type of story is it?"   some, some went away

"Draw a picture to show what happened first in the story."

- Allow time for your child to draw the picture.
- Draw the following on the chalkboard:

"How will you show what happened next in the story?"   cross out 3 markers

- Cross out three markers on the chalkboard picture.

"What number sentence will we write to show what happened in this story?"

- Write "5 markers – 3 markers = 2 markers" on the chalkboard.

"How many markers does Cheryl have now?"   2 markers

"Let's try another story."

- Write the following on the chalkboard:

    Joan put 4 papers on the table. Kara put 3 papers on the table.

"Read this story."

"What happened first in the story?"

"What happened next in the story?"

"What type of story is this?"   some, some more

"Draw a picture to show what happened first in the story."

- Allow time for your child to draw the picture.
- Draw the following on the chalkboard:

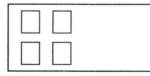

"How will we show what happened next?" draw 3 more pieces of paper

- Draw three more pieces of paper on the chalkboard picture.

"What number sentence will we write to show what happened in our story?"

- Write "4 papers + 3 papers = 7 papers" on the chalkboard.

"How many pieces of paper are on the table now?" 7 pieces of paper

"Make up a some, some more or a some, some went away story that you can draw a picture for."

- Ask your child to tell a simple story.

"What type of story is this?"

"Draw a picture and write a number sentence for this story."

- Allow time for your child to draw the picture and write the number sentence.

"What number sentence did you write?"

"What is the answer?"

- Repeat with several stories.

CLASS PRACTICE

number fact practice

- Practice the adding two facts together.
- Give your child **Fact Sheet AA 4.0.**
- Time your child for one minute.
- Correct the fact sheet with your child.
- Record the score.
- Allow time for your child to complete the unfinished facts.

WRITTEN PRACTICE

- Complete **Worksheet 23A** with your child.
- Complete **Worksheet 23B** with your child later in the day.

Name _____   *LESSON 23A*
Date _____   **Math 2**

Draw a picture and write a number sentence for each story. Write the answer with a label.

1. Susan has two new pencils and three old pencils. How many pencils does she have?

   [ ]

   Number sentence _____ 2 pencils + 3 pencils = 5 pencils _____
   Answer _____ 5 pencils _____

2. Steven had 10 markers. He gave 3 markers to his friend. How many markers does he have now?

   [ ]

   Number sentence _____ 10 markers – 3 markers = 7 markers _____
   Answer _____ 7 markers _____

3. Write the names of the weekdays.

   _____ Monday,   Tuesday,   Wednesday _____
   _____ Thursday,   Friday, _____

4. Match each name with the correct piece.

   one half •
   one sixth •
   one third •

Name _____   *LESSON 23B*
Date _____   **Math 2**

Draw a picture and write a number sentence for each story. Write the answer with a label.

1. Sharon's mom bought four blueberry muffins and three apple muffins at the store. How many muffins did she buy?

   [ ]

   Number sentence _____ 4 muffins + 3 muffins = 7 muffins _____
   Answer _____ 7 muffins _____

2. There were 12 eggs in the carton. Scott's family ate 5 eggs for breakfast. How many eggs are left in the carton now?

   [ ]

   Number sentence _____ 12 eggs – 5 eggs = 7 eggs _____
   Answer _____ 7 eggs _____

3. Circle the names of the days of the weekend.

   Sunday, Monday, Tuesday, Wednesday, Thursday, Friday, Saturday

4. Ask the members of your family what their favorite days of the week are. Write each person's name under the day they chose in Problem 3.

5. Find something at home that has the shape of a rectangle and something that has the shape of a triangle. Draw pictures of what you found.

rectangle	triangle

131

esson 24

dividing a shape in half
shading one half of a shape

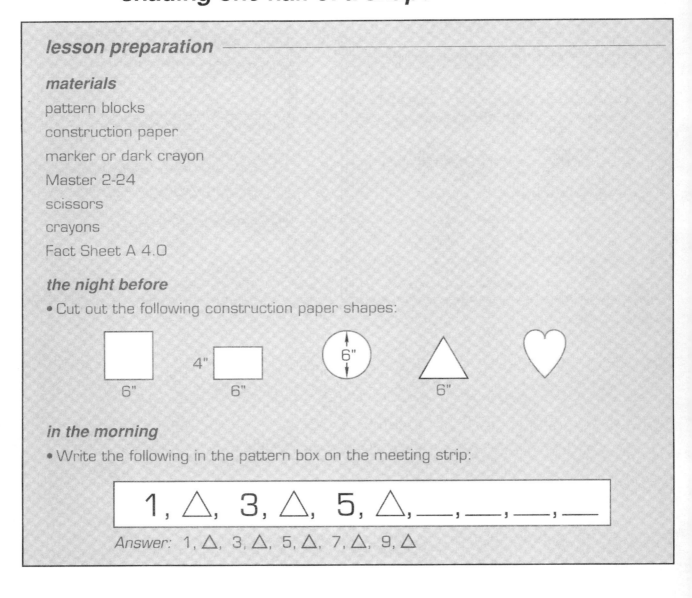

lesson preparation

materials

pattern blocks

construction paper

marker or dark crayon

Master 2-24

scissors

crayons

Fact Sheet A 4.0

the night before

• Cut out the following construction paper shapes:

in the morning

• Write the following in the pattern box on the meeting strip:

1, △, 3, △, 5, △,___, ___, ___, ___

Answer: 1, △, 3, △, 5, △, 7, △, 9, △

THE MEETING

• If necessary, refer to Lesson 20 for the questioning sequence.

calendar

• Ask your child to write the date on the calendar and meeting strip.

• Ask your child the following:

days of the week, weekdays, days of the weekend

months of the year, _____th month, month before, month after

"What day of the week was it seven days ago?"

"What day of the week will it be six days from today?"

- Record on the meeting strip a special event and the number of days until it occurs.

weather graph

- Ask your child to graph the weather.
- Ask questions about the graph.

counting

- Count forward and backward by 1's from 188 to 202 as your child writes these numbers.
- Count by 10's to 100 and backward from 100 by 10's.
- Say the even numbers to 20 and backward from 20.
- Say the odd numbers to 19 and backward from 19.

graph questions

- You and your child each ask a question about any of the graphs.

patterning

"What do you notice about today's pattern?"

"This is a repeating and a continuing pattern."

"The numbers continue and the shapes repeat."

- Ask your child to do the following:

  identify the numbers and shapes to complete the pattern

  read the pattern

clock

- Ask your child to set the clock on the hour.
- Ask your child to write the digital time on the meeting strip.

  "What time was it one hour ago?"

  "What time will it be one hour from now?"

- Record on the meeting strip the time an activity will occur.

number of the day

- Write three number sentences for the number of the day on the meeting strip.
- The money box on the meeting strip is not used today.

THE LESSON

Dividing a Shape in Half
Shading One Half of a Shape

"Today you will learn how to divide and shade half of a shape."

- Show your child the yellow, red, green, and blue pattern blocks.

"What pattern block will cover one half of the yellow hexagon?"

"How do you know?"

"How could you prove it?"

- Ask your child to show that two red trapezoids exactly cover the yellow hexagon.

- Show your child the construction paper shapes one at a time.

"What shape is this?"

- Repeat with each shape.

- Draw each construction paper shape on the chalkboard.

"Let's divide each shape in half so that we have two pieces that are equal in size."

"How can we divide the square in half?"

- Fold the square in half in the following way:

"Do we have two equal pieces?"

"How do you know?"

"I will draw a line along the fold to show the line that divides the square in half."

- Draw a line along the fold with a marker or a dark crayon.

"Draw a line on the chalkboard square to show the line that divides it in half."

"Use the chalk to shade one half of the square."

- Repeat with each shape.

- Give your child **Master 2-24**, scissors, and crayons.

"Carefully cut out each shape."

"Fold the shape in half."

"Draw a line along the fold."

"Color one half of each shape."

• Save the construction paper circle for use in Lesson 25.

CLASS PRACTICE

number fact practice

• Write the following on the chalkboard:

$$
\begin{array}{cc}
2 & 7 \\
+\,7 & +\,2
\end{array}
$$

"What do you notice about these number facts?"

"How are they different?"

"How are they the same?"

• Practice the adding two facts together.

• Give your child **Fact Sheet A 4.0.**

• Time your child for one minute.

• Correct the fact sheet with your child.

• Record the score.

• Allow time for your child to complete the unfinished facts.

WRITTEN PRACTICE

• Complete **Worksheet 24A** with your child.

• Complete **Worksheet 24B** with your child later in the day.

Name _____

MASTER 2-24
Math 2

Name _____

Date _____

LESSON 24A
Math 2

Draw a picture and write a number sentence for each story. Write the answer with a label.

1. There were 4 plants in Room 2. The children brought in 3 more. How many plants are in the room now?

    Number sentence _____ 4 plants + 3 plants = 7 plants _____

    Answer ____ 7 plants ____

2. Melanie bought five lunch tickets. She used two. How many tickets does she have now?

    Number sentence _____ 5 tickets − 2 tickets = 3 tickets _____

    Answer ____ 3 tickets ____

3. Number the clock face.

    If it is 5:00 now, write the digital time one hour from now.

    | 6:00 |

    Show this time on the clock face.

4. Draw a line to divide each shape in half. Shade one half.
    answers may vary

Name _____

Date _____

LESSON 24B
Math 2

Draw a picture and write a number sentence for each story. Write the answer with a label.

1. There were 7 fish in Room 5's fish tank. Mrs. Weber put 2 more fish in the tank. How many fish are in the tank now?

    Number sentence _____ 7 fish + 2 fish = 9 fish _____

    Answer ____ 9 fish ____

2. There were five birds at the bird feeder. Three flew away. How many birds are at the feeder now?

    Number sentence _____ 5 birds − 3 birds = 2 birds _____

    Answer ____ 2 birds ____

3. Number the clock face.

    If it is 6:00 now, write the digital time one hour from now.

    | 7:00 |

    Show this time on the clock face.

4. Draw a line to divide each shape in half. Shade one half.

L esson 25

dividing a square in half two ways
telling time to the half hour

---

lesson preparation

materials

Written Assessment #4

construction paper circle from Lesson 24

one 6" construction paper circle (in a color different from the circle in Lesson 24)

two 6" squares

marker or dark crayon

demonstration clock

Fact Sheet A 4.0

in the morning

Note: This lesson may take two days to complete.

• Write the following in the pattern box on the meeting strip:

$$0, \square, 2, \square, 4, \square, 6, \square, \underline{\quad}, \underline{\quad}, \underline{\quad}, \underline{\quad}$$

Answer: 0, $\square$, 2, $\square$, 4, $\square$, 6, $\square$, 8, $\square$, 10, $\square$

THE MEETING

• If necessary, refer to Lesson 20 for the questioning sequence.

calendar

• Ask your child to write the date on the calendar and meeting strip.

• Ask your child the following:

    days of the week, weekdays, days of the weekend

    day of the week _____ days ago, day of the week _____ days from now

    months of the year, _____th month, month before, month after

"What is the fifth day of the week?"

• Record on the meeting strip a special event and the number of days until it occurs.

weather graph

- Ask your child to graph the weather.

- Ask questions about the graph.

counting

- Count forward and backward by 1's from 193 to 213 as your child writes these numbers.
- Count by 10's to 100 and backward from 100 by 10's.
- Say the even numbers to 20 and backward from 20.
- Say the odd numbers to 19 and backward from 19.

graph questions

- You and your child each ask a question about any of the graphs.

patterning

"What do you notice about today's pattern?"

"This is a repeating and a continuing pattern."

"The shapes repeat and the numbers continue."

- Ask your child to do the following:

    identify the numbers and shapes to complete the pattern

    read the pattern

clock

- Ask your child to set the clock on the hour.

- Ask your child to write the digital time on the meeting strip.

    "What time was it one hour ago?"

    "What time will it be one hour from now?"

- Record on the meeting strip the time an activity will occur.

number of the day

- Write three number sentences for the number of the day on the meeting strip.
- The money box on the meeting strip is not used today.

ASSESSMENT

Written Assessment

"Today I would like to see what you remember from what we have been practicing."

- Give your child **Written Assessment #4**.

- Read the directions for each problem. Allow time for your child to complete that problem before continuing.

- Correct the paper, noting your child's mistakes on the **Individual Recording Form**. Review the errors with your child.

THE LESSON

Dividing a Square in Half Two Ways

"Today you will learn how to divide a square in half two different ways."

"You will also learn how to tell time to the half hour."

"Yesterday you learned one way to divide a square in half."

- Give your child a construction paper square.

"Fold this square in half."

- Allow time for your child to fold the square.

"Are the two halves the same size?"

"How do you know?"

"What shape is each half?"   rectangle

- Draw a square on the chalkboard.

"Draw a line to divide this square in half so that you have two pieces that are equal in size."

- Ask your child to draw a line to divide the chalkboard square in half.

"How many halves are in one whole square?"   2

"Let's find another way to divide a square in half."

- Give your child another paper square.

"Try to fold this square in half a different way."

- Allow time for your child to do this.

"Did you find a different way to fold a square in half?"

"How did you do it?"

"Are the two halves the same size?"

"How do you know?"

"What shape is each half?"   triangle

"Draw a line along the fold to show the line that divides the square in half."

- Ask your child to draw a line along the fold with a marker or a dark crayon.

- Draw another square on the chalkboard.

"Draw a line on this chalkboard square to show the line that divides the chalkboard square in half."

"How do we know that each piece is one half of the square?"   the pieces are the same size

"Now use the chalk to shade one half of the square."

Telling Time to the Half Hour

"Yesterday we folded a paper circle in half."

- Show your child the folded construction paper circle placed over another construction paper circle of a contrasting color.

"We can turn the folded circle around in any direction and this line will still divide the other circle in half."

- Demonstrate by turning the folded circle and showing how the line dividing the circle in half moves around the circle.

- Draw several circles on the chalkboard.

"When we divide a circle in half, we can draw any line that goes through the center of the circle."

- Demonstrate, using the chalkboard circles.

"The center of the circle is just like the center of a clock."

"When we have a new hour on a clock, where does the long hand point?"   to the 12

"Where does the hour hand point?"   to the hour

"When we find half past an hour, both of the hands on the clock will move."

"The long hand will move halfway around the clock."

- Hold the folded paper circle against the face of a demonstration clock.

"When it is a new hour, the long hand points to the 12."

"When it is half past the hour, the long hand has gone halfway around the circle."

"What number is halfway around the circle?"   6

- Point to the number 6 at the bottom of the half circle.

- Show your child the demonstration clock set to 8:00.

"What time does the clock show?"

"What time will it be one hour from now?"

- Show how the minute hand moves around the circle and the hour hand moves to nine.

- Reset the clock to 8:00.

"In one hour the long hand moves around the circle and back to the 12."

"In one half hour the long hand moves halfway around the circle."

"What number is halfway around the circle?" 6

"The hour hand moves halfway between the eight and the nine."

"We put the hour hand in the middle, halfway between the eight and the nine."

- Demonstrate on the clock.

"Show one o'clock on the clock."

"Show half past one."

"Where is the long hand pointing?" 6

"Where is the hour hand pointing?" *halfway between the 1 and 2*

- Repeat with three o'clock, half past three, ten o'clock, half past ten, one o'clock, half past one, twelve o'clock, and half past twelve.

"How will we show half past two on the clock?"

- Ask your child to set the clock.

"How do we know that this is half past two?"

"When it is half past two, the hour hand is in the middle between the two and the three and the long hand is halfway around the clock."

"We write half past two using digital time like this."

- Write the digital time on the chalkboard.

"When it is half past the hour, we write the hour first, a colon, and 30."

- It is not necessary to explain at this time that 30 minutes is half of 60 minutes. This will be discussed in a future lesson.

- Repeat for half past seven and half past eleven.

Class Practice

number fact practice

- Practice the adding two facts together.

- Give your child **Fact Sheet A 4.0**.

- Time your child for one minute.

- Correct the fact sheet with your child.

- Record the score.

- Allow time for your child to complete the unfinished facts.

WRITTEN PRACTICE

- Complete **Worksheet 25A** with your child.

- Complete **Worksheet 25B** with your child later in the day.

Name _____ **ASSESSMENT 4**
Date _____ **LESSON 25**
                                        Math 2

1. Use the Favorite Day of the Week graph to answer these questions.

    What day of the week did the most children choose? _____

    How many children chose the first day of the week as their favorite day? _____

    How many more children chose Saturday than chose Wednesday? _____

2. Write the names of the weekdays. You may use the calendar to help you spell the names of the days.

    _____ Monday,   Tuesday,   Wednesday, _____

    _____ Thursday,   Friday _____

3. Write these letters in the squares below.

    sixth square E   tenth square I   eighth square S   ninth square M
    first square S   third square E   twelfth square E   eleventh square L
    fifth square M   second square E

S	E	E	M	E	S	M	I	L	E

4. Circle the even numbers.

    17 (6) (20) 5 (14) (8)

5. Number the clock face. If it is two o'clock now, write the time one hour from now. Show this time on both clocks.

    [3:00]

6. Color the triangle green.
    Color the square orange.
    Color the circle yellow.
    Color the rectangle blue.

2-25Aa

Name _____ **LESSON 25A**
Date _____ Math 2

Draw a picture and write a number sentence for this story. Write the answer with a label.

1. Four girls were playing at the playground. Stephanie joined them. How many girls are playing now?

2. Circle the even numbers.

    (14) 7 9 (8) (4) (18) (6) 3

    Number sentence _____ 4 girls + 1 girl = 5 girls _____

    Answer _____ 5 girls _____

3. Write the names for the first four months.

    _____ January _____ February _____ March _____ April _____

4. Show two different ways to divide a square in half.

5. Number the clock face.

    Show half past three.

2-25Wa

Name _____ **LESSON 25B**
Date _____ Math 2

Draw a picture and write a number sentence for this story. Write the answer with a label.

1. At the first bus stop, four children got on the bus. At the next stop, three children got on the bus. How many children are on the bus now?

    Number sentence _____ 4 children + 3 children = 7 children _____

    Answer _____ 7 children _____

2. Circle the odd numbers.

    14 (7) (9) 8 4 18 6 (3)

3. What month comes just before April? _____ March _____

    What month comes just after January? _____ February _____

4. Divide each shape in half. Shade one half.   answers may vary

5. Number the clock face.

    Show half past eight.

2-25Wb

143

Lesson 26

reading a thermometer to the nearest ten degrees
addition facts—doubles plus one

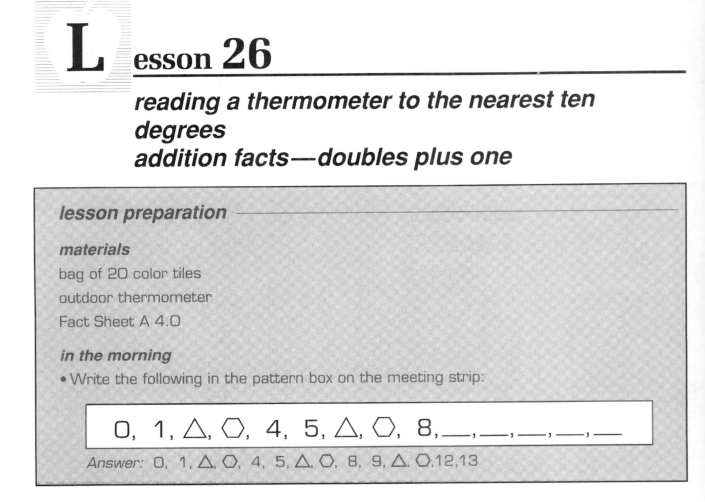

lesson preparation

materials

bag of 20 color tiles
outdoor thermometer
Fact Sheet A 4.0

in the morning

• Write the following in the pattern box on the meeting strip:

0, 1, △, ◯, 4, 5, △, ◯, 8,___,___,___,___,___

Answer: 0, 1, △, ◯, 4, 5, △, ◯, 8, 9, △, ◯,12,13

THE MEETING

calendar

• Ask your child to write the date on the calendar and meeting strip.

• Ask your child the following:

days of the week, weekdays, days of the weekend

day of the week _____ days ago, day of the week _____ days from now

months of the year, _____th month, month before, month after

"What is the second day of the week?"

• Record on the meeting strip a special event and the number of days until it occurs.

weather graph

• Ask your child to graph the weather.

• Ask questions about the graph.

counting

- Count forward and backward by 1's from 147 to 175 as your child writes these numbers.
- Count by 10's to 200 and backward from 200 by 10's.
- Say the even numbers to 20 and backward from 20.
- Say the odd numbers to 19 and backward from 19.

graph questions

- You and your child each ask a question about any of the graphs.

patterning

"What do you notice about today's pattern?"

"This is a repeating and a continuing pattern."

"The shapes repeat and the numbers continue."

- Ask your child to do the following:

    identify the numbers and shapes to complete the pattern

    read the pattern

clock

"Beginning today, you may set the clock on the half hour or hour."

- Ask the following:

    time shown on the clock

    time one hour ago

    time one hour from now

- Ask your child to write the digital time on the meeting strip.
- Record on the meeting strip the time an activity will occur.

number of the day

- Write three number sentences for the number of the day on the meeting strip.
- The money box on the meeting strip is not used today.

THE LESSON

Reading a Thermometer to the Nearest Ten Degrees

"Today you will learn how to read a thermometer to the nearest ten degrees."

"Each morning you have been coloring your weather graph to show how the weather feels to you."

"People also use numbers to describe how the weather feels to them."

"A number that tells us how hot or cold it is is called the temperature."

"Do you know what we call the instrument we use to measure the temperature?"   thermometer

- Give your child an outdoor thermometer.

- Allow time for your child to examine the thermometer.

"What do you notice about the thermometer?"

- Allow time for your child to offer observations.

"A thermometer is an instrument that we use to measure temperature."

"There is a glass tube on the thermometer that contains a silver liquid called 'mercury' or a red liquid called 'alcohol.' "

"When we handle a thermometer, it is very important to handle it carefully so the glass tube does not break."

"Thermometers have numbers on them that we read to tell us the temperature."

"A thermometer is usually marked in 10's."

"What numbers do you think will be on our thermometer?"

"Each day we have been counting by 10's."

"Let's count by 10's to 100."

- Draw a large model of your thermometer on the chalkboard. Number only the 10's on the model.

"Each morning you will read our thermometer."

"You will look at the thermometer to see which ten the mercury in the tube is closest to."

"Let's practice that."

- Shade the current height of the mercury on the chalkboard thermometer. (Your child will determine the closest ten by visual inspection, not by a rule.)

"The mercury in our thermometer looks like this now."

"Which ten is the temperature closest to?"

"How do you know?"

- Draw another model of the thermometer on the chalkboard. Shade the thermometer to show about 29°.

"About what temperature is it on this thermometer?"

"How do you know?"

"We write the temperature like this."

- Write 30° on the chalkboard.
- Shade the thermometer to show 43°.

  "About what temperature is shown on the thermometer?"   40°

  "We write the temperature like this."

- Write 40° on the chalkboard.
- Repeat with 78°.

  "Each morning you will read the thermometer and write the temperature to the nearest ten degrees in the box you colored on the weather graph."

Addition Facts—Doubles Plus One

"Today you will learn how to make a doubles plus one fact."

"We will use the doubles facts to help us do that."

"What are the doubles facts?"

- List the doubles on the chalkboard as your child names them.

$$\begin{array}{ccccccccc} 1 & 2 & 3 & 4 & 5 & 6 & 7 & 8 & 9 \\ +1 & +2 & +3 & +4 & +5 & +6 & +7 & +8 & +9 \end{array}$$

"Let's show the doubles using tiles."

- Give your child a bag of 20 color tiles.

  "How will we show seven plus seven using the color tiles?"

  "Make two piles of seven tiles."

  "Now we have an example of a doubles fact."

  "How do you think we will make this into a doubles plus one fact?"   add one to one pile

  "Add one tile to one of your piles."

  "How many tiles are in each pile now?"

  "Seven plus eight is a doubles plus one fact."

  "We started with a double and added one more to one of the numbers."

  "Make a pile of four tiles."

  "Now make another pile of four tiles."

  "What doubles problem did we show?" $4 + 4 = 8$

  "Add one tile to one of your piles."

  "How many tiles are in each pile now?"

  "Four plus five is a doubles plus one fact."

  "We started with a double and added one more to one of the numbers."

  "Make a pile of eight tiles."

  "Now make another pile of eight tiles."

"What doubles problem did we show?" 8 + 8 = 16

"How will we make this a doubles plus one problem?"

"Add one tile to one of your piles."

"How many tiles are in each pile now?"

"Eight plus nine is a doubles plus one fact."

"We started with a double and added one more to one of the numbers."

"How can we make another doubles plus one problem?"

- Allow time for your child to use the tiles to make the doubles plus one problems.

"Put your tiles in the bag."

"Let's list the doubles and the doubles plus one facts on the chalkboard."

"If we start with one plus one, what will we write for the doubles plus one problem?" 1 + 2

- Write the problems next to each other on the chalkboard. Write the next problems in a list below the first two chalkboard problems.

"If we start with two plus two, what will we write for the doubles plus one problem?" 2 + 3

"If we start with three plus three, what will we write for the doubles plus one problem?" 3 + 4

- Repeat with 4 + 4, 5 + 5, 6 + 6, 7 + 7, and 8 + 8.

"What do you notice about the numbers in a doubles plus one problem?"

- Allow time for your child to offer observations.

"The numbers in a doubles plus one problem will always be next to each other on the number line."

"What are two numbers that are next to each other on the number line?"

"We will use only numbers less than ten."

- Repeat until all combinations are named.

"These are the doubles plus one problems."

"Tomorrow you will practice finding the answers for doubles plus one problems."

CLASS PRACTICE

number fact practice

"Practice the adding two facts using the green addition fact cards."

- Allow time for your child to do this.

- Give your child **Fact Sheet A 4.0**.
- Time your child for one minute.
- Correct the fact sheet with your child.
- Record the score.
- Allow time for your child to complete the unfinished facts.

WRITTEN PRACTICE

- Complete **Worksheet 26A** with your child.
- Complete **Worksheet 26B** with your child later in the day.

Name _____ LESSON 26A
Date _____ Math 2

Draw a picture and write a number sentence for this story.

1. There are four green lunch boxes and three yellow lunch boxes on the shelf in Mr. Taylor's room. There are two green lunch boxes on the floor. How many green lunch boxes are there altogether?

[G G G G Y Y Y G G]

Number sentence __4 green boxes + 2 green boxes = 6 green boxes__

Answer ____6 green boxes____

2. Write the 10's on the thermometer. Which number on the thermometer is the temperature closest to? __50__ °F

3. Finish the repeating pattern. Color the pattern. (R = Red, G = Green, Y = Yellow)

[R G Y G R G Y G R G Y G R G Y G]

4. Divide the squares in half two different ways.

Color one half of each square red. answers may vary

5. Change these doubles examples into doubles plus one examples.

```
  4    4      7    7      5    5
+ 4  + 5    + 7  + 8    + 5  + 6
  8    9     14   15     10   11
```

Name _____ LESSON 26B
Date _____ Math 2

Draw a picture and write a number sentence for this story.

1. Maureen put four cartons of grape juice and three cartons of apple juice on the table. There are two cartons of apple juice in the refrigerator. How many cartons of apple juice does Maureen have?

[G G G G A A A A A]

Number sentence __3 cartons + 2 cartons = 5 cartons of apple juice__

Answer ____5 cartons of apple juice____

2. Write the 10's on the thermometer. Which number on the thermometer is the temperature closest to? __70__ °F

3. Finish the repeating pattern. Color the pattern. (R = Red, G = Green, Y = Yellow)

[R R Y G G R R Y G G R R Y G G]

4. Divide the squares in half two different ways.

Color one half of each square red. answers may vary

5. Change these doubles examples into doubles plus one examples.

```
  3    3      8    8      6    6
+ 3  + 4    + 8  + 9    + 6  + 7
  6    7     16   17     12   13
```

Lesson 27

addition facts—doubles plus one

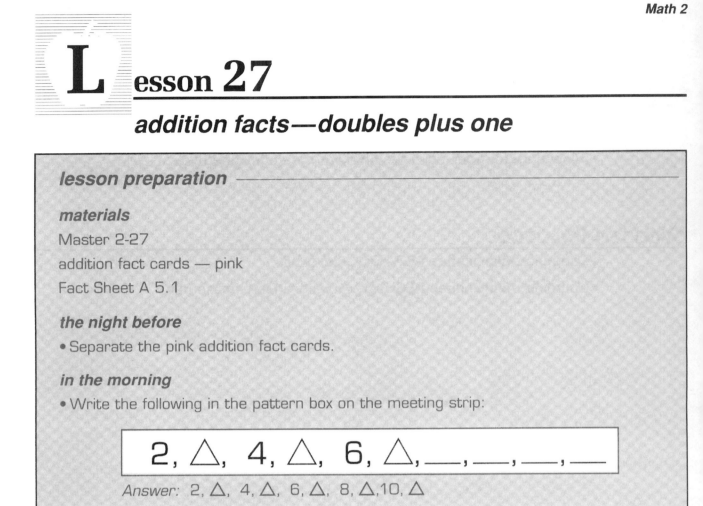

lesson preparation

materials

Master 2-27

addition fact cards — pink

Fact Sheet A 5.1

the night before

• Separate the pink addition fact cards.

in the morning

• Write the following in the pattern box on the meeting strip:

2, △, 4, △, 6, △, ___, ___, ___, ___

Answer: 2, △, 4, △, 6, △, 8, △,10, △

THE MEETING

calendar

• Ask your child to write the date on the calendar and meeting strip.

• Ask your child the following:

days of the week, weekdays, days of the weekend

day of the week _____ days ago, day of the week _____ days from now

months of the year, _____th month, month before, month after

"What is the seventh day of the week?"

• Record on the meeting strip a special event and the number of days until it occurs.

weather graph

• Ask your child to color the graph and write the temperature to the nearest ten degrees in the box he/she colored.

• Ask questions about the graph.

counting

- Count forward and backward by 1's from 253 to 276 as your child writes these numbers.

- Count by 10's to 200 and backward from 200 by 10's.

- Say the even numbers to 20 and backward from 20.

- Say the odd numbers to 19 and backward from 19.

graph questions

- You and your child each ask a question about any of the graphs.

patterning

- Ask your child to do the following:

  identify the pattern (repeating, continuing, or both)

  identify the numbers and shapes to complete the pattern

  read the pattern

clock

- Ask your child to set the clock on the half hour or hour.

- Ask the following:

  time shown on the clock

  time one hour ago

  time one hour from now

- Ask your child to write the digital time on the meeting strip.

- Record on the meeting strip the time an activity will occur.

number of the day

- Write three number sentences for the number of the day on the meeting strip.

- The money box on the meeting strip is not used today.

THE LESSON

Addition Facts — Doubles Plus One

"Today you will learn how to find the answers for the doubles plus one facts."

"Yesterday we used tiles to make doubles plus one facts."

"How did we do that?"

"What is an example of a doubles plus one fact?"

- Write the example on the chalkboard.

"This is a doubles plus one fact because we started with a double and we added one more."

"What is another doubles plus one fact?"

- Repeat until all the doubles plus one facts are listed on the chalkboard. List the doubles plus one facts in the following order:

0 + 1 =	5 + 6 =
1 + 2 =	6 + 7 =
2 + 3 =	7 + 8 =
3 + 4 =	8 + 9 =
4 + 5 =	

"Let's look at these pairs of numbers."

"What do you notice about the numbers in a doubles plus one problem?"   they are next to each other on the number line; one number is one more than the other

"A doubles plus one problem will always have numbers that are next to each other on the number line."

"There is a trick for adding numbers that are next to each other on the number line."

"We find the smaller number, double it, and add one."

"Let's try that."

- Point to the 7 + 8 combination.

"Which is the smaller number?"

- Circle the 7.

"If seven and seven is fourteen, then seven plus eight is how many?"   15

"Why?"   because 8 is one more than 7

- Point to the 5 + 6 combination.

"Which is the smaller number?"

- Circle the 5.

"If five and five is ten, then five plus six is how many?"   11

"Why?"   because 6 is one more than 5

- Write   7   on the chalkboard.
      + 6

"Is this a doubles plus one fact?"

"How do you know?"   one number is one more than the other

"Which number will we double?"   6

"Why?"   it is the smaller number

- Circle the 6.

"How will we find the answer for this doubles plus one problem?" double six and add one

"If six and six is twelve, what is seven plus six?" 13

"Now you will have a chance to identify and find the answers for some doubles plus one facts."

- Give your child **Master 2-27**.

"Look at the problems in the first box."

"Which of these problems is a doubles plus one problem?" 5 + 4

"How do you know?" the numbers are next to each other on the number line; one number is one more than the other

"Circle the doubles plus one problem."

"Look at the problems in the second box."

"Circle the doubles plus one problem."

"Which problem did you circle?" 7 + 8

- Repeat with the third box.

"What do you notice about the problems at the bottom of the paper?" they are doubles and doubles plus one facts

"Write the answers for all the doubles facts."

- Allow time for your child to do this.

"Read the answers for the doubles facts in the first row."

- Repeat with each row.

"All of the other problems are doubles plus one facts."

"How will you find the answers for the doubles plus one facts?"

"We double the smaller number and add one."

"Circle the smaller number in the first doubles plus one problem."

"Which number did you circle?" 5

"How will you find the answer for five plus six?" double five and add one

"Double five and add one."

"What is the answer?" 11

"Five plus five is ten, so five plus six is eleven."

"Circle the smaller number in the next doubles plus one problem."

"Which number did you circle?" 7

"How will you find the answer for eight plus seven?" double seven and add one

"Double seven and add one."

"What is the answer?"   15

"Seven plus seven is fourteen, so eight plus seven is fifteen."

"Circle the smaller number in the next doubles plus one problem."

"Which number did you circle?"   4

"How will you find the answer for five plus four?"   double four and add one

"Double four and add one."

"What is the answer?"   9

"Four plus four is eight, so five plus four is nine."

- Repeat with the rest of the doubles plus one problems.

"Now I will give you a set of fact cards for the doubles plus one facts."

- Give your child the pink addition fact cards.

"Match the fact cards that have the same answers."

- Allow time for your child to do this.

"What do you notice about the numbers on the fact cards you matched?"   they switch places; sometimes the smaller number is at the top, sometimes at the bottom

"Which number will you double to find the answer?"   the smaller number

"Practice saying the answers to yourself."

"Remember to double the smaller number and add one."

"Turn over each card to check the answer."

- Allow 3–5 minutes for practice.

CLASS PRACTICE

number fact practice

- Give your child **Fact Sheet A 5.1**.

"What facts do you see on this fact sheet?"   doubles plus one

"How will you find the answers?"   double the smaller number and add one

- Time your child for one minute.
- Correct the fact sheet with your child.
- Record the score.
- Allow time for your child to complete the unfinished facts.

WRITTEN PRACTICE

- Complete **Worksheet 27A** with your child.
- Complete **Worksheet 27B** with your child later in the day.

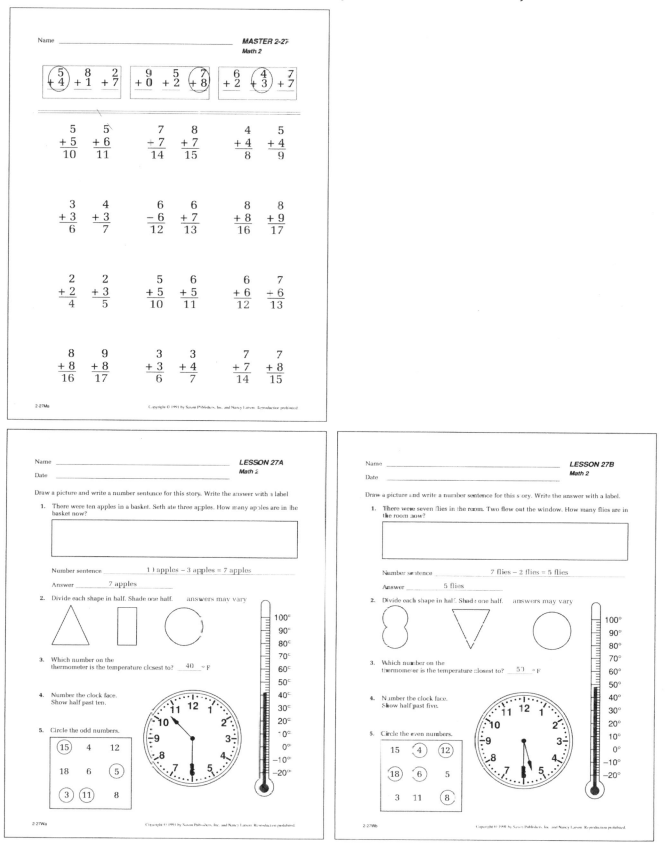

Lesson 28

counting dimes and pennies

---

lesson preparation

materials

1 cup of 10 dimes

1 cup of 10 pennies

work mat (9" × 12" piece of construction paper)

Fact Sheet A 4.2

in the morning

• Write the following in the pattern box on the meeting strip:

Answer: ■,□□,□■,■□,■□,□■,■□,□□,■□,■□,□□

THE MEETING

calendar

• Ask your child to write the date on the calendar and meeting strip.

• Ask your child the following:

  days of the week, weekdays, days of the weekend

  day of the week _____ days ago, day of the week _____ days from now

  months of the year, _____th month, month before, month after

  "What is the fourth day of the week?"

• Record on the meeting strip a special event and the number of days until it occurs.

weather graph

• Ask your child to color the graph and write the temperature to the nearest ten degrees in the box he/she colored.

• Ask questions about the graph.

156

counting

- Count forward and backward by 1's from 232 to 254 as your child writes these numbers.
- Count by 10's to 200 and backward from 200 by 10's.
- Say the even numbers to 20 and backward from 20.
- Say the odd numbers to 19 and backward from 19.

graph questions

- You and your child each ask a question about any of the graphs.

patterning

- Ask your child to do the following:

  identify the pattern (repeating, continuing, or both)

  identify the shapes to complete the pattern

  read the pattern

clock

- Ask your child to set the clock on the half hour or hour.
- Ask the following:

  time shown on the clock

  time one hour ago

  time one hour from now

- Ask your child to write the digital time on the meeting strip.
- Record on the meeting strip the time an activity will occur.

number of the day

- Write three number sentences for the number of the day on the meeting strip.
- The money box on the meeting strip is not used today.

THE LESSON

Counting Dimes and Pennies

"Today you will learn how to count dimes and pennies."

"If you could have a dime or a penny, which would you rather have?"

"Why?"

- Give your child a cup of 10 dimes, a cup of 10 pennies, and a work mat (9" × 12" piece of construction paper).

"Put the pennies on the mat."

"We say that one penny is worth one cent."

- Write "1¢" on the chalkboard.

"How many pennies do you have?"   10

"Each dime is worth the same as the ten pennies on your mat."

"Put all the dimes on your mat."

"How many dimes do you have?"   10

"When we count dimes we count by 10's."

"Let's count by 10's to find out how much all the dimes are worth."

- Count the dimes.

"Push the dimes and pennies off your mat."

- Write "40¢" on the chalkboard.

"Show 40¢ using your dimes."

"How many dimes did you use?"   4 dimes

- Repeat with 20¢, 80¢, 50¢, and 90¢.

- Write "36¢" on the chalkboard.

"How much money is this?"

"You will use as many dimes as possible to make 36¢."

"How many dimes can you use?"   3

"Put three dimes on your mat."

"Let's count the three dimes."   10, 20, 30

"Now add pennies until you reach 36¢."

- Count with your child as he/she adds one penny at a time.

31, 32, 33, 34, 35, 36

"How many pennies did you use?"   6

- Record on the chalkboard: 36¢ = 3 dimes + 6 pennies

"When we count dimes and pennies, we will count the dimes first."

- Repeat with 52¢, 75¢, 14¢, and 67¢. Continue with additional amounts, if desired.

"Each morning I will write an amount of money on the meeting strip."

"You will put coins in a coin cup to show that amount."

"You will use only dimes and pennies."

- Collect the cups of coins and the work mat.

CLASS PRACTICE

number fact practice

- Practice the addition facts with your child.
- Give your child **Fact Sheet A 4.2**.
- Time your child for one minute.
- Correct the fact sheet with your child.
- Record the score.
- Allow time for your child to complete the unfinished facts.

WRITTEN PRACTICE

- Complete **Worksheet 28A** with your child.
- Complete **Worksheet 28B** with your child later in the day.

Name _____   **LESSON 28A**
   Math 2
Date _____

Draw a picture and write a number sentence for this story. Write the answer with a label.

1. Lee and Gail are having a party. Six friends are coming to the party. How many children will be at the party?

Number sentence ___ 2 children + 6 children = 8 children

Answer ___ 8 children

2. Count by 10's. Write the numbers.

10 , 20 , 30 , 40 , 50 , 60 , 70 , 80 , 90 , 100

3. How much money is this? ___ 34¢

4. Which number on the thermometer is the temperature closest to? ___ 80 ___ °F

5. How are these shapes the same? [ ] [ ]

6. Finish the pattern.

△ , □ , ○ , △ , □ , ○ , △ , □ , ○ , △ , □ , ○

2-28Wa   Copyright © 1991 by Saxon Publishers, Inc. and Nancy Larson. Reproduction prohibited.

Thermometer scale: 100° 90° 80° 70° 60° 50° 40° 30° 20° 10° 0° −10° −20°

Name _____   **LESSON 28B**
   Math 2
Date _____

Draw a picture and write a number sentence for this story. Write the answer with a label.

1. Marsha and Susan went to the movies with three friends. How many children went to the movies together?

Number sentence ___ 2 children + 3 children = 5 children

Answer ___ 5 children

2. Count backward by 10's. Write the numbers.

100 , 90 , 80 , 70 , 60 , 50 , 40 , 30 , 20 , 10

3. How much money is this? ___ 25¢

4. Which number on the thermometer is the temperature closest to? ___ 20 ___ °F

5. How are these shapes different? [ ] [ ]

6. What time (to the nearest half hour) do you usually wake up on a school day?
answers may vary

[:]

2-28Wb   Copyright © 1991 by Saxon Publishers, Inc. and Nancy Larson. Reproduction prohibited.

Thermometer scale: 100° 90° 80° 70° 60° 50° 40° 30° 20° 10° 0° −10° −20°

Lesson 29

creating and reading a bar graph
identifying missing addends

lesson preparation

materials

Meeting Book
scrap paper
10 pennies
Fact Sheets MA 4.2, A 4.2

in the morning

• Write the following in the pattern box on the meeting strip:

$$10, 20, \bigcirc, \triangle, 30, 40, \bigcirc, \triangle, \underline{\quad}, \underline{\quad}, \underline{\quad}, \underline{\quad}, \underline{\quad}$$

Answer: 10, 20, ○, △, 30, 40, ○, △, 50, 60, ○, △, 70

• Write ⬚31¢⬚ on the meeting strip. Provide a cup of 10 dimes and a cup of 10 pennies.

THE MEETING

calendar

• Ask your child to write the date on the calendar and meeting strip.

• Ask your child the following:

days of the week, weekdays, days of the weekend

day of the week _____ days ago, day of the week _____ days from now

months of the year, _____th month, month before, month after

"What is the first day of the week?"

• Record on the meeting strip a special event and the number of days until it occurs.

weather graph

• Ask your child to color the graph and write the temperature to the nearest ten degrees in the box he/she colored.

• Ask questions about the graph.

counting

- Count forward and backward by 1's from 195 to 230 as your child writes these numbers.

- Count by 10's to 200 and backward from 200 by 10's.

- Say the even numbers to 20 and backward from 20.

- Say the odd numbers to 19 and backward from 19.

graph questions

- You and your child each ask a question about any of the graphs.

patterning

- Ask your child to do the following:

    identify the pattern (repeating, continuing, or both)

    identify the numbers and shapes to complete the pattern

    read the pattern

money

"Each morning I will write an amount of money in the box on the meeting strip."

"You will put coins in the coin cup to show this amount of money."

"Today you will put coins in the coin cup to show 31¢."

"For now, we will use only dimes and pennies for our coin cup."

"Show 31¢ using the fewest number of pennies."

- Ask your child to put the coins in the cup

"What coins did you use?"

"Let's count to check to see if this is 31¢."

- Count the money with your child.

clock

- Ask your child to set the clock on the half hour or hour.

- Ask the following:

    time shown on the clock

    time one hour ago

    time one hour from now

- Ask your child to write the digital time on the meeting strip.

- Record on the meeting strip the time an activity will occur.

number of the day

- Write three number sentences for the number of the day on the meeting strip.

THE LESSON

Creating and Reading a Bar Graph

"Today you will learn how to read a bar graph."

"You will also learn how to find missing addends."

"You have made graphs that show our birthdays and our favorite days of the week and the birthdays and favorite days of the week of our friends."

"Today you will make another graph."

"You will make a graph that shows the times we wake up on weekdays."

"You will include the wake-up times of our friends, relatives, and neighbors."

"What are the weekdays?"

"What do we usually do on weekdays?"

"Yesterday you wrote on side B of your practice sheet the time you usually wake up in the morning."

- Open the Meeting Book to page 32.

"Write your initials in the box above the time you wake up each weekday morning."

"I usually wake up at _____ on weekdays."

"Write my initials on the graph."

"Let's ask some other people what time they wake up on weekdays so we can write their initials on the graph."

"Who would you like to ask?"

- List 8–10 names on the chalkboard.

- Assist your child as he/she calls each person on the list. Record each person's wake-up time on the chalkboard.

"Write each person's initials in a box on the graph to show the time he/she wakes up on weekdays."

- When all the wake-up times are graphed, continue.

"What do you notice about the graph?"

- Encourage your child to offer as many observations as possible.

"What time do most of the people wake up in the morning?"

"What is the earliest time that someone wakes up?"

"How many people wake up at seven or half past seven?"

"Who wakes up an hour before (name)?"

Identifying Missing Addends

"Now you will learn how to identify missing addends."

"Let's act out a story."

"I have two cents."

- Hold up 2 pennies.

"I want to buy a pencil for five cents."

"How many more pennies will I need to buy the pencil?"

"How did you find the answer?"

"Let's write a number sentence for this story."

"I have two cents and I want to buy a pencil for five cents."

- Write the following on the chalkboard:

$$2¢ + \square = 5¢$$

"This is a some, some more story where we have to find how much more money is needed."

"How much more money do I need?"

- Fill in the 3 cents.

"Let's check our answer."

"Two cents and three more cents is five cents."

"Let's act out a different story."

- Put 5 pennies in your left hand. Do not show your child the number of pennies in your hand.

"I have some pennies in my left hand."

"I will put one cent in my right hand."

- Pick up and hold one penny in your right hand.

"I will give you a clue."

"If I put the pennies in my hands together, I will have six cents."

"How many pennies do you think I have in my left hand?"

"Let's write a number sentence for this story."

- Write the following on the chalkboard:

$$\square + 1¢ = 6¢$$

"This is a some, some more story."

"One cent and how much more is six cents?"   *5¢*

- Fill in the 5 cents.

"Let's check our answer."

"Five cents and one cent is six cents."

"Let's count the pennies in my left hand to check."

- Show your child the pennies.
- Repeat with two more examples, if desired.

"Mathematicians have a special name for the numbers we add together."

"They call them addends."

- Point to the example "2¢ + 3¢ = 5¢" on the chalkboard.

"What numbers are we adding in this problem?"   *2, 3*

"We call these numbers addends."

- Write the following on the chalkboard:

$$\underset{2¢}{\text{addend}} \quad + \quad \underset{3¢}{\text{addend}} \quad = \quad 5¢$$

"Mathematicians call the answer to an addition problem the sum."

- Write the following on the chalkboard:

$$\underset{2¢}{\text{addend}} \quad + \quad \underset{3¢}{\text{addend}} \quad = \quad \underset{5¢}{\text{sum}}$$

- Point to the number sentence "1¢ + 5¢ = 6¢."

"What do we call the one and the five?"   *addends*

"What do we call the six?"   *sum*

"Sometimes addition problems are written like this."

- Write the following on the chalkboard:

$$\begin{array}{r} 2 \\ + \ 5 \\ \hline 7 \end{array}$$

"What is the sum in this problem?"   *7*

"What are the addends?"   *2, 5*

- Write the following on the chalkboard:

$$\begin{array}{rl} 2 & \text{addend} \\ + \ 5 & \text{addend} \\ \hline 7 & \text{sum} \end{array}$$

"The number sentences we wrote before had a missing addend."

"Let's practice finding a missing addend."

"I will tell you one of the addends and the sum and you will try to guess the missing addend."

- Write the following on the chalkboard:

$$1¢ \ + \ \square \ = \ 8¢$$

"We will use a box to show the missing addend."

"One cent and how much more is eight cents?"

"What do you think is the missing addend?"

• Write the suggested addend in the box.

"Let's check your answer."

"Were you right?"

• Write the following on the chalkboard:

$$\square + 2¢ = 9¢$$

"Two cents and how much more is nine cents?"

"What do you think is the missing addend?"

• Write the suggested addend in the box.

"Let's check your answer."

"Were you right?"

• Write the following on the chalkboard:

$$\begin{array}{r} 4¢ \\ + \;\square \\ \hline 5¢ \end{array}$$

"Four cents and how much more is five cents?"

"What do you think is the missing addend?"

• Write the suggested addend in the box.

"Let's check your answer."

"Were you right?"

• Write the following on the chalkboard:

$$\begin{array}{r} \square \\ + \;3¢ \\ \hline 4¢ \end{array}$$

"Three cents and how much more is four cents?"

"What do you think is the missing addend?"

• Write the suggested addend in the box.

"Let's check your answer."

"Were you right?"

• Give your child **Fact Sheet MA 4.2**.

"Now you will have a chance to find the missing addends on a fact sheet."

"You will have as much time as you need to find the missing addends on this fact sheet."

"Write the missing addend for each problem in the box."

"How can you check the answer?"   add

• Allow your child as much time as necessary to complete the sheet.

"We will correct this paper just like we correct a fact sheet."

• Correct the fact sheet with your child.

CLASS PRACTICE

number fact practice

• Practice the addition facts with your child.

• Give your child **Fact Sheet A 4.2.**

• Time your child for one minute.

• Correct the fact sheet with your child.

• Record the score.

• Allow your child to complete the unfinished facts.

WRITTEN PRACTICE

• Complete **Worksheet 29A** with your child.

• Complete **Worksheet 29B** with your child later in the day.

Name _____ *LESSON 29A*
                                Math 2
Date _____

Draw a picture and write a number sentence for this story. Write the answer with a label.

1. Tim had ten dimes. He gave three to his brother. How many dimes does Tim have now?

Number sentence _____ 10 dimes − 3 dimes = 7 dimes

Answer _____ 7 dimes

2. Fill in the missing addends.

$5 + \boxed{5} = \boxed{10}$

$\boxed{1} + 7 = \boxed{8}$

$\begin{array}{r} 6 \\ + \boxed{0} \\ \hline 6 \end{array}$ $\begin{array}{r} \boxed{2} \\ + 5 \\ \hline 7 \end{array}$ $\begin{array}{r} 0 \\ + \boxed{9} \\ \hline 9 \end{array}$ $\begin{array}{r} 3 \\ + 3 \\ \hline \boxed{6} \end{array}$

3. Use a crayon to circle all the even numbers in Problem 2.

4. Number the clock face.

   Show half past eleven.

5. How much money is this? _____ 53¢

2-29Wa   Copyright © 1991 by Saxon Publishers, Inc. and Nancy Larson. Reproduction prohibited.

Name _____ *LESSON 29B*
                                Math 2
Date _____

Draw a picture and write a number sentence for this story. Write the answer with a label.

1. Angelo borrowed six books from the library. He read four books and took them back to the library. How many more books does he have left to read?

Number sentence _____ 6 books − 4 books = 2 books

Answer _____ 2 books

2. Fill in the missing addends.

$\boxed{4} + \boxed{4} = \boxed{8}$

$\boxed{1} + \boxed{4} = 5$

$\begin{array}{r} 1 \\ + 7 \\ \hline \boxed{8} \end{array}$ $\begin{array}{r} 3 \\ + \boxed{4} \\ \hline 7 \end{array}$ $\begin{array}{r} 7 \\ + \boxed{2} \\ \hline 9 \end{array}$ $\begin{array}{r} 0 \\ + \boxed{6} \\ \hline 6 \end{array}$

3. Use a crayon to circle all the even numbers in Problem 2.

4. Number the clock face.

   Show half past eight.

5. How much money is this? _____ 44¢

2-29Wb   Copyright © 1991 by Saxon Publishers, Inc. and Nancy Larson. Reproduction prohibited.

Lesson 30

identifying geometric shape pieces that differ in one way

---

lesson preparation

materials

Oral Assessment #3

Written Assessment #5

shape pieces from Lesson 22

Fact Sheet A 5.1

in the morning

• Write the following in the pattern box on the meeting strip:

$$10,\ \square,\ 20,\ \square,\ 30,\ \square,\ \_\_,\ \_\_,\ \_\_,\ \_\_$$

Answer: 10, □, 20, □, 30, □, 40, □, 50, □

• Write 42¢ on the meeting strip. Provide a cup of 10 dimes and a cup of 10 pennies.

THE MEETING

calendar

• Ask your child to write the date on the calendar and meeting strip.

• Ask your child the following:

days of the week, weekdays, days of the weekend

day of the week _____ days ago, day of the week _____ days from now

months of the year, _____th month, month before, month after

"What is the sixth day of the week?"

• Record on the meeting strip a special event and the number of days until it occurs.

weather graph

• Ask your child to color the graph and write the temperature to the nearest ten degrees in the box he/she colored.

• Ask questions about the graph.

counting

- Count forward and backward by 1's from 91 to 118 as your child writes these numbers.
- Count by 10's to 200 and backward from 200 by 10's.
- Say the even numbers to 20 and backward from 20.
- Say the odd numbers to 19 and backward from 19.

graph questions

- You and your child each ask a question about any of the graphs.

patterning

- Ask your child to do the following:

  identify the pattern (repeating, continuing, or both)

  identify the numbers and shapes to complete the pattern

  read the pattern

money

"Today you will put coins in the coin cup to show 42¢."

"Show 42¢ using the fewest number of pennies."

- Allow time for your child to do this.

  "How many dimes and pennies did you use?"

  "Let's count to check to see if this is 42¢."

- Count the money with your child.

clock

- Ask your child to set the clock on the half hour or hour.
- Ask the following:

  time shown on the clock

  time one hour ago and time one hour from now

- Ask your child to write the digital time on the meeting strip.
- Record on the meeting strip the time an activity will occur.

number of the day

- Write three number sentences for the number of the day on the meeting strip.

ASSESSMENT

Written Assessment

- All of the questions on the assessment are based on concepts and skills presented at least five lessons ago. It is expected that your child will master at least 80% of the concepts on the assessment. If your child is having difficulty with a specific concept, reteach the concept the following day.

  "Today I would like to see what you remember from what we have been practicing."

- Give your child **Written Assessment #5.**

- Read the directions for each problem. Allow time for your child to complete that problem before continuing.

- Correct the paper, noting your child's mistakes on the **Individual Recording Form** Review the errors with your child.

Oral Assessment

- Record your child's response(s) to the oral interview questions on the interview sheet.

THE LESSON

Identifying Geometric Shape Pieces That Differ in One Way

"Today you will learn how to find a shape piece that is different in only one way from another piece."

- Spread the shape pieces from Lesson 22 on the floor.

- Hold up the large red circle.

  "What piece is this?"

- Ask your child to name the piece using all three attributes.

  "Are there any other pieces in this set just like this piece?"   no

  "What piece is almost like this piece?"

- Hold up the piece your child selects.

  "How are these pieces the same?"

- Name the two similar attributes.

  "How are these pieces different?"

- Name the one difference.

  "Is there another piece that is almost like the large red circle?"

- Hold up the piece your child selects.

  "How are these pieces the same?"

"How are these pieces different?"

- Repeat until your child has named all the pieces that are the same in two ways and different in only one way.

- Put all the pieces in a group.

- Hold up the small yellow square.

"What piece is this?"

"What piece is almost like this piece?"

"How are these pieces the same?"

"How are these pieces different?"

- Repeat until your child has named all the pieces that are the same in two ways and different in only one way.

"We are going to play a game."

"This game is called the 'Attribute Train Game.' "

- Place the large green circle in the center of the table.

"We are going to pretend that our pieces are cars of a train."

"Are any of the shape pieces exactly like this piece?" no

"We will add cars to the train using a special rule."

"The special rule for putting new cars on the train is that the next car must be like the car before it in two ways."

"What piece did I put first?" large green circle

"What piece could we put next?" there are many possibilities, including the small green circle, the large green rectangle, the large red circle, etc.

"How are these pieces the same?"

"How are they different?"

"Now we will add another car to the train."

"It must be like the car before it in two ways."

"What piece could we put next?"

"How are these pieces the same?"

"How are they different?"

- Continue, adding at least five more pieces.

"Now I will divide the pieces that are left between us."

- Divide the pieces.

"Let's take turns adding cars to the train using the rule that the new piece must be different from the piece before it in only one way."

"We will try to make our train as long as possible."

- Take turns adding pieces to the train.

Class Practice

number fact practice

- Practice the doubles plus one facts using the pink fact cards with your child.
- Give your child **Fact Sheet A 5.1.**
- Time your child for one minute.
- Correct the fact sheet with your child.
- Record the score.
- Allow time for your child to complete the unfinished facts.

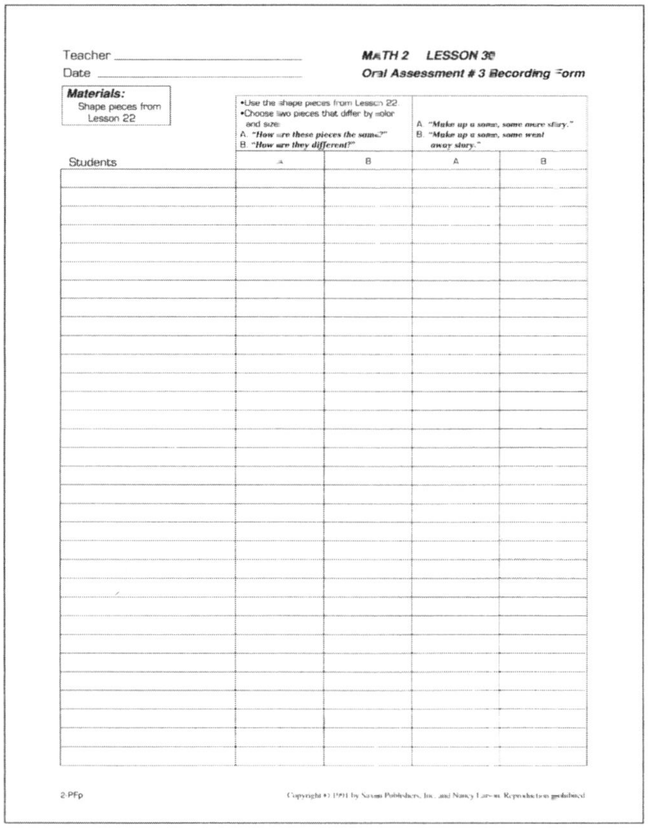

Teacher ______________________

Date ______________________

MATH 2 LESSON 30

Oral Assessment # 3 Recording Form

Materials:
Shape pieces from Lesson 22

Students	•Use the shape pieces from Lesson 22. •Choose two pieces that differ by color and size. A. *"How are these pieces the same?"* B. *"How are they different?"*		A. *"Make up a some, some more story."* B. *"Make up a some, some went away story."*	
	A	B	A	B

2-PFp

Copyright © 1991 by Saxon Publishers, Inc. and Nancy Larson. Reproduction prohibited.

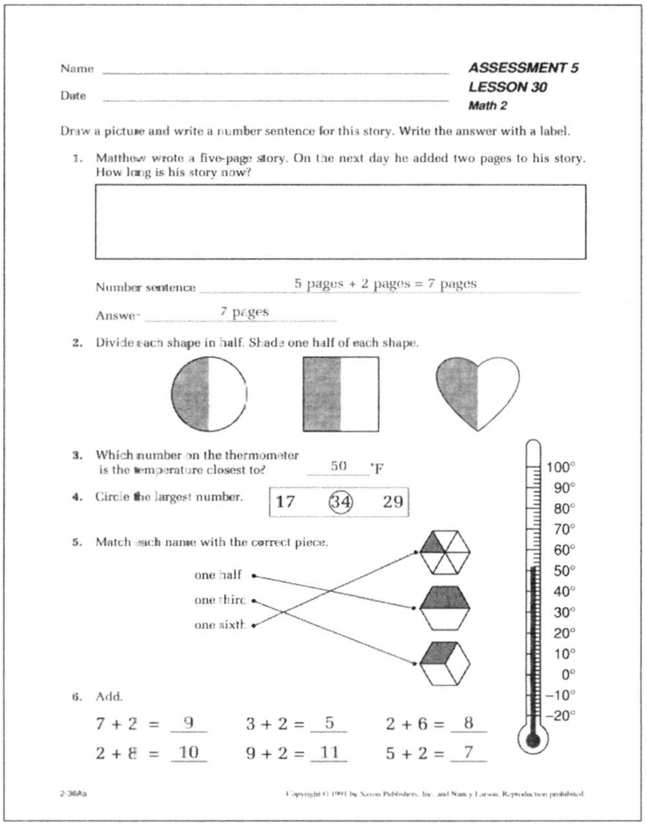

Name ______________________

Date ______________________

ASSESSMENT 5
LESSON 30
Math 2

Draw a picture and write a number sentence for this story. Write the answer with a label.

1. Matthew wrote a five-page story. On the next day he added two pages to his story. How long is his story now?

   Number sentence 5 pages + 2 pages = 7 pages

   Answer 7 pages

2. Divide each shape in half. Shade one half of each shape.

3. Which number on the thermometer is the temperature closest to? 50 °F

4. Circle the largest number. 17 (34) 29

5. Match each name with the correct piece.

   one half
   one third
   one sixth

6. Add.

   7 + 2 = 9 3 + 2 = 5 2 + 6 = 8

   2 + 8 = 10 9 + 2 = 11 5 + 2 = 7

2-30Aa

Copyright © 1991 by Saxon Publishers, Inc. and Nancy Larson. Reproduction prohibited.

Lesson 31

tallying
counting by fives

---

lesson preparation

materials

5 pennies

construction paper mat

Fact Sheet A 5.1

in the morning

• Write the following in the pattern box on the meeting strip:

⊙, △, ◯, ⊙, △, ◯, ⊙, ____, ____, ____, ____, ____

Answer: ⊙, △, ◯, ⊙, △, ◯, ⊙, △, ◯, ⊙, △, ◯

• Write | 17¢ | on the meeting strip. Provide a cup of 10 dimes and a cup of 10 pennies.

THE MEETING

calendar

• Ask your child to write the date on the calendar and meeting strip.

• Ask your child the following:

> day of the week _____ days ago, day of the week _____ days from now
>
> months of the year, _____th month, month before, month after

"What was the date two days ago?"

"What will be the date tomorrow?"

"What is the second day of the week?"

• Record on the meeting strip a special event and the number of days until it occurs.

weather graph

• Ask your child to color the graph and write the temperature to the nearest ten degrees in the box he/she colored.

• Ask questions about the graph.

counting

- Count forward and backward by 1's from 142 to 176 as your child writes these numbers.
- Count by 10's to 200 and backward from 200 by 10's.
- Say the even numbers to 20 and backward from 20.
- Say the odd numbers to 19 and backward from 19.

graph questions

- You and your child each ask a question about any of the graphs.

patterning

- Ask your child to do the following:

    identify the pattern (repeating, continuing, or both)

    identify shapes to complete the pattern

    read the pattern

money

"Today you will put coins in the coin cup to show 17¢."

"Show 17¢ using the fewest number of pennies."

- Allow time for your child to do this.

"How many dimes and pennies did you use?"

"Let's count to check to see if this is 17¢."

- Count the money with your child.

clock

- Ask your child to set the clock on the half hour or hour.
- Ask the following:

    time shown on the clock

    time one hour ago and time one hour from now

- Ask your child to write the digital time on the meeting strip.
- Record on the meeting strip the time an activity will occur.

number of the day

- Write three number sentences for the number of the day on the meeting strip.

THE LESSON

Tallying
Counting by Fives

> *"Today you will learn how to tally and how to count by fives."*

- Give your child a penny.

> *"Look at your penny very carefully."*

> *"What do you notice about the penny?"*

- Allow time for your child to offer as many observations as possible.

> *"Each penny has two sides."*

> *"Do you know what they are called?"*   heads, tails

> *"Why do you think they are called that?"*

> *"Today we are going to try an experiment."*

> *"You are going to shake five pennies in your hands."*

> *"You will open your hands and gently put the pennies on the table."*

> *"We will count how many pennies land heads up and how many pennies land tails up."*

> *"This is how you will shake the pennies."*

- Demonstrate how to carefully shake the pennies and put them gently on the table. Use a piece of construction paper as a mat.

> *"I will keep track of the number of heads and the number of tails on the chalkboard."*

- Write the following on the chalkboard:

heads	tails

- Ask your child to shake the pennies.

> *"How many heads do you see?"*

> *"I will write one tally mark for each head."*

> *"How many tails do you see?"*

> *"I will write one tally mark for each tail."*

- When you reach five, write the tally marks in the following way: ||||
- Write each group of five tally marks below the others in the following way:

heads	tails

- Continue tallying as your child shakes the pennies. Ask your child to shake the pennies 15–20 times.

  "Do you think the pennies landed heads up more often or tails up more often?"

  "How do you know?"

  "Let's count to check how many tally marks are in each group."

  "Let's count the tally marks in the heads column first."

- Write 5, 10, 15, etc., next to each group of five tally marks as you count by fives.

- Repeat for the tails column.

- The chart should look like the following:

heads	tails
JHT 5	JHT 5
JHT 10	JHT 10
JHT 15	JHT 15
JHT 20	JHT 20
JHT 25	JHT 25
JHT 30	JHT 30
JHT 35	JHT 35
JHT 40	

  "What do you notice about our chart?"

- Allow time for your child to offer observations.

  "When we count tally marks, we are counting by fives."

  "Let's count by fives together as I write the numbers we say on a counting strip in your Meeting Book."

- Open the Meeting Book to page 38. Begin at the bottom. Write 0 below the 5. The strip should look like the following:

```
50
45
40
35
30
25
20
15
10
 5
 0
```

  "We will use this counting strip to help us count forward and backward by fives each morning."

  "How will we write '15' using tally marks?"

- Ask your child to use tally marks to show 15 on the chalkboard.

  "Let's count by fives to check."

  "How many groups of five tally marks do we have?"   three

  "How many extra tally marks do we have?"   none

- Repeat with 7, 25, 16, 38, and 41.

CLASS PRACTICE

number fact practice

- Practice the doubles plus one facts using the pink fact cards with your child.

- Give your child **Fact Sheet A 5.1.**

- Time your child for one minute.

- Correct the fact sheet with your child.

- Record the score.

- Allow time for your child to complete the unfinished facts.

WRITTEN PRACTICE

- Complete **Worksheet 31A** with your child.

- Complete **Worksheet 31B** with your child later in the day.

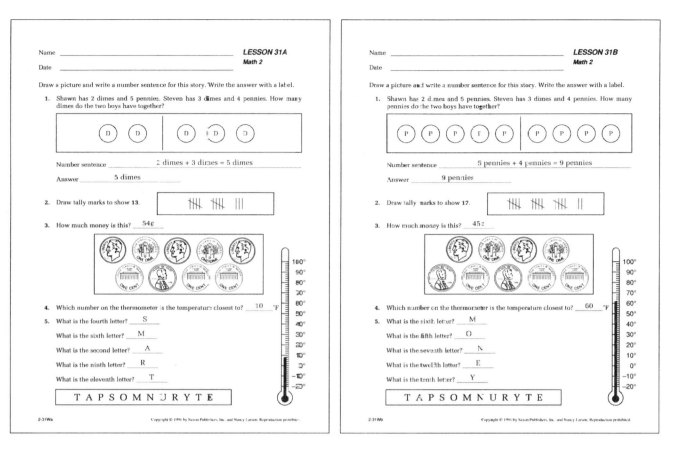

Lesson 31A

Name _____   **LESSON 31A**
Date _____   **Math 2**

Draw a picture and write a number sentence for this story. Write the answer with a label.

1. Shawn has 2 dimes and 5 pennies. Steven has 3 dimes and 4 pennies. How many dimes do the two boys have together?

Number sentence _____ 2 dimes + 3 dimes = 5 dimes _____

Answer _____ 5 dimes

2. Draw tally marks to show **13**.

3. How much money is this? _____ 54¢

4. Which number on the thermometer is the temperature closest to? _____ 10 _____ °F

5. What is the fourth letter? _____ S

What is the sixth letter? _____ M

What is the second letter? _____ A

What is the ninth letter? _____ R

What is the eleventh letter? _____ T

T A P S O M N U R Y T E

2-31Wa

Lesson 31B

Name _____   **LESSON 31B**
Date _____   **Math 2**

Draw a picture and write a number sentence for this story. Write the answer with a label.

1. Shawn has 2 dimes and 5 pennies. Steven has 3 dimes and 4 pennies. How many pennies do the two boys have together?

Number sentence _____ 5 pennies + 4 pennies = 9 pennies _____

Answer _____ 9 pennies

2. Draw tally marks to show **17**.

3. How much money is this? _____ 45¢

4. Which number on the thermometer is the temperature closest to? _____ 60 _____ °F

5. What is the sixth letter? _____ M

What is the fifth letter? _____ O

What is the seventh letter? _____ N

What is the twelfth letter? _____ E

What is the tenth letter? _____ Y

T A P S O M N U R Y T E

2-31Wb

Lesson 32

identifying horizontal, vertical, and oblique lines

---

lesson preparation

materials

hundred number chart

10 color tiles

Master 2-32

scissors

small envelope

Fact Sheet A 4.2

in the morning

• Write the following in the pattern box on the meeting strip:

→, →, ↑, →, →, ↑, →, ___, ___, ___, ___, ___

Answer: →, →, ↑, →, →, ↑, →, →, ↑, →, →, ↑

• Write [26¢] on the meeting strip. Provide a cup of 10 dimes and a cup of 10 pennies.

THE MEETING

calendar

• Ask your child to write the date on the calendar and meeting strip.

• Ask your child the following:

day of the week _____ days ago, day of the week _____ days from now

months of the year, _____th month, month before, month after

"What was the date three days ago?"

"What will be the date two days from now?"

"What is the fourth day of the week?"

• Record on the meeting strip a special event and the number of days until it occurs.

---

weather graph

- Ask your child to color the graph and write the temperature to the nearest ten degrees in the box he/she colored.

- Ask questions about the graph.

counting

- Count forward and backward by 1's from 192 to 209 as your child writes these numbers.

- Count by 10's to 200 and backward from 200 by 10's.

- Say the even numbers to 20 and backward from 20.

- Say the odd numbers to 19 and backward from 19.

graph questions

- You and your child each ask a question about any of the graphs.

patterning

- Ask your child to do the following:

    identify the pattern (repeating, continuing, or both)

    identify the shapes to complete the pattern

    read the pattern

money

"Today you will put coins in the coin cup to show 26¢."

"Show 26¢ using the fewest number of pennies."

- Allow time for your child to do this.

    "How many dimes and pennies did you use?"

    "Let's count to check to see if this is 26¢."

- Count the money with your child.

clock

- Ask your child to set the clock on the half hour or hour.

- Ask the following:

    time shown on the clock

    time one hour ago and time one hour from now

- Ask your child to write the digital time on the meeting strip.

- Record on the meeting strip the time an activity will occur.

number of the day

- Write three number sentences for the number of the day on the meeting strip.

THE LESSON

Identifying Horizontal, Vertical, and Oblique Lines

"Today you will learn how to identify horizontal, vertical, and oblique lines."

- Draw a long horizontal line on the chalkboard.

"There is a special name that mathematicians use to describe a line that goes across like this."

"They say that this is a horizontal line."

- Write the word "horizontal" on the line.

"If I asked you to put your body in a horizontal position, what would you do?"   lie on the floor

"Where do you see horizontal lines in this room?"

- Allow time for your child to identify horizontal lines.

- Draw a long vertical line on the chalkboard.

"There is a special name that mathematicians use to describe a line that goes up and down like this."

"They say that this is a vertical line."

- Write the word "vertical" next to the line.

"If I asked you to put your body in a vertical position, what would you do?"   stand up

"Where do you see vertical lines in this room?"

- Allow time for your child to identify vertical lines.

"If lines are not horizontal or vertical, we say they are oblique."

"These are some examples of oblique lines."

- Draw the following on the chalkboard:

"Where do you see oblique lines in this room?"

- Allow time for your child to identify oblique lines.

- Give your child a hundred number chart.

"We have practiced finding numbers on the hundred number chart before."

"I will say some numbers that will form a line on the hundred number chart."

"See if you can tell what type of line the numbers will make."

"10, 20, 30, 40, 50, 60, 70, 80, 90, 100."

"What type of line do you think these numbers will make?"   vertical

"Let's check this by putting the tiles on the numbers."

• Ask your child to put tiles on the chart as you reread the numbers.

"Were you correct?"

• Remove the tiles from the chart.

"I will say some different numbers that will form a line on the hundred number chart."

"See if you can tell what type of line these numbers will make."

"61, 62, 63, 64, 65, 66, 67, 68, 69, 70."

"What type of line do you think these numbers will make?"   horizontal

"Let's check this by putting tiles on the numbers."

• Ask your child to put the tiles on the chart as you reread the numbers.

"Were you correct?"

• Remove the tiles from the chart.

"See if you can tell what type of line these numbers will make."

"5, 15, 25, 35, 45, 55, 65, 75, 85, 95."

"What type of line do you think these numbers will make?"   vertical

"Let's check this by putting tiles on the numbers."

"Were you correct?"

• Remove the tiles from the chart.

"See if you can tell what type of line these numbers will make."

"9, 18, 27, 36, 45, 54, 63, 72, 81."

"What type of line do you think these numbers will make?"   oblique

"Let's check this by putting tiles on the numbers."

"Were you correct?"

• Remove the tiles from the chart.

"Let's see if you can tell me some numbers that make a horizontal line."

• List the numbers on the chalkboard.

"Use the tiles to cover these numbers."

- Allow time for your child to do this.

  "Were you correct?"

  "Let's see if you can tell me some numbers that make a vertical line."

- List the numbers on the chalkboard.

  "Use the tiles to cover these numbers."

- Allow time for your child to do this.

  "Were you correct?"

  "Let's see if you can tell me some numbers that make an oblique line."

- List the numbers on the chalkboard.

  "Use the tiles to cover these numbers."

- Allow time for your child to do this.

  "Were you correct?"

- Give your child **Master 2-32** and scissors.

  "A hundred number chart has been divided into nine pieces."

  "Today you are going to put the hundred number chart back together."

  "Use the scissors to carefully cut apart the nine large pieces."

  "Cut along the dotted lines."

- Check to make sure that your child does not cut apart the individual number squares.

  "Put the nine pieces together to make a hundred number chart."

- Give your child a hundred number chart if your child is having difficulty.

- Give your child an envelope for the pieces. Save the envelope of pieces for Lesson 35.

CLASS PRACTICE

number fact practice

- Practice the adding zero, adding one, and adding two facts with your child.
- Give your child **Fact Sheet A 4.2**.
- Time your child for one minute.
- Correct the fact sheet with your child.
- Record the score.
- Allow time for your child to complete the unfinished facts.

WRITTEN PRACTICE

- Complete **Worksheet 32A** with your child.

- Complete **Worksheet 32B** with your child later in the day.

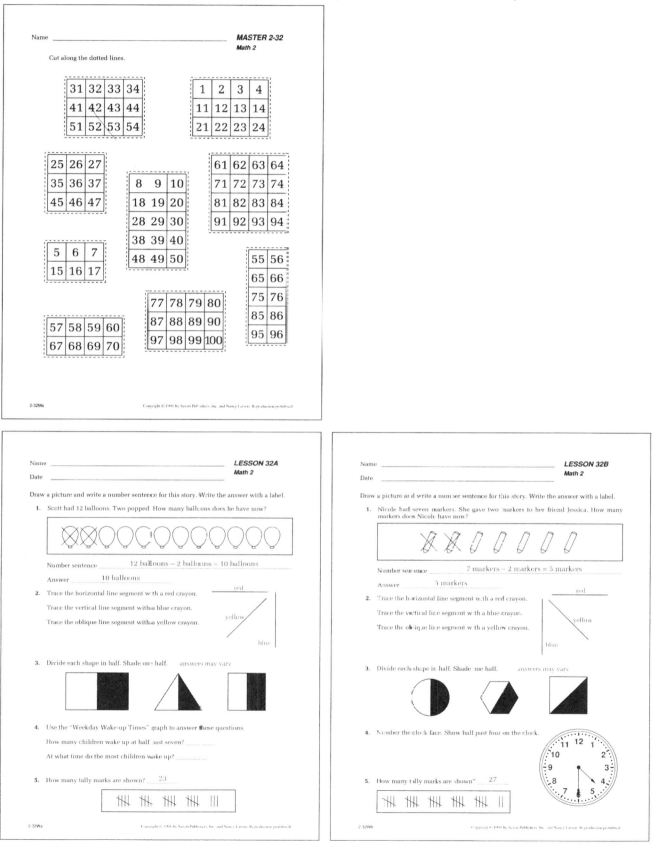

Lesson 33

addition facts—sums of ten

lesson preparation

materials

Meeting Book

1 deck of playing cards

addition fact cards — blue

Fact Sheet MA 6.0

the night before

• Remove the tens, jacks, queens, and kings from the deck of cards. The ace will be used for the number one.

• Separate the blue addition fact cards.

in the morning

• Write the following in the pattern box on the meeting strip:

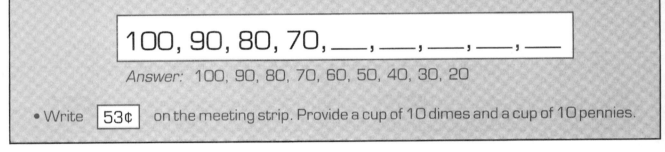

100, 90, 80, 70, ____, ____, ____, ____, ____

Answer: 100, 90, 80, 70, 60, 50, 40, 30, 20

• Write [53¢] on the meeting strip. Provide a cup of 10 dimes and a cup of 10 pennies.

THE MEETING

calendar

• Ask your child to write the date on the calendar and meeting strip.

• Ask your child the following:

day of the week _____ days ago, day of the week _____ days from now

months of the year, _____th month, month before, month after

"What was the date two days ago?"

"What will be the date two days from now?"

"What is the third day of the week?"

• Record on the meeting strip a special event and the number of days until it occurs.

weather graph

- Ask your child to color the graph and write the temperature to the nearest ten degrees in the box he/she colored.
- Ask questions about the graph.

counting

- Count forward and backward by 1's from 354 to 379 as your child writes these numbers.
- Count by 10's to 200 and backward from 200 by 10's.
- Count by 5's to 100 and backward from 50 by 5's.
- Say the even numbers to 20 and backward from 20.
- Say the odd numbers to 19 and backward from 19.

graph questions

- You and your child each ask a question about any of the graphs.

patterning

- Ask your child to do the following:

    identify the pattern (repeating, continuing, or both)

    identify the numbers to complete the pattern

    read the pattern

money

"Today you will put coins in the coin cup to show 53¢."

"Show 53¢ using the fewest number of pennies."

- Allow time for your child to do this.

    "How many dimes and pennies did you use?"

    "Let's count to check to see if this is 53¢."

- Count the money with your child.

clock

- Ask your child to set the clock on the half hour or hour.
- Ask the following:

    time shown on the clock

    time one hour ago and time one hour from now

- Ask your child to write the digital time on the meeting strip.
- Record on the meeting strip the time an activity will occur.

number of the day

- Write three number sentences for the number of the day on the meeting strip.

THE LESSON

Addition Facts — Sums of Ten

"Today you will learn the addition facts that are called the sums of ten facts."

"Each morning when we graph the weather, we talk about how many hot, warm, cool, or cold days we have had."

- Open the Meeting Book to this month's weather graph.

"How many spaces do we have on our graph for hot days?" **10**

"According to our graph, how many hot days have we had so far?"

"How many more spaces do we have left for hot days?"

- Write a number sentence such as 2 + 8 = 10 to represent the used and unused spaces on the graph.

"How many spaces do we have on our graph for warm days?" **10**

"According to our graph, how many warm days have we had so far?"

"How many more spaces do we have left for warm days?"

- Write a number sentence such as 2 + 8 = 10 to represent the used and unused spaces on the graph.

- Repeat for cool and cold days.

"These are some of the number combinations for ten."

"Let's try to list all of the combinations we could have found for used and unused spaces on our graph."

- List your child's responses in an organized list similar to the following. Fill in the combinations as they are given.

0 + 10 = 10	6 + 4 = 10
1 + 9 = 10	7 + 3 = 10
2 + 8 = 10	8 + 2 = 10
3 + 7 = 10	9 + 1 = 10
4 + 6 = 10	10 + 0 = 10
5 + 5 = 10	

"We could write these number facts in another way."

- Write the following on the chalkboard:

0	1	2	3	4	5	6	7	8	9	10
+ 10	+ 9	+ 8	+ 7	+ 6	+ 5	+ 4	+ 3	+ 2	+ 1	+ 0

"Do you see a pattern in these numbers?"

"What is it?"

"We call these the sums of ten because the answer is always ten."

"Do you remember what we call the three and the seven, or the four and the six, in the problems?"   addends

"Today we will learn a card game to help us practice the sums of ten."

"This game is called 'Making 10.' "

"First I will put an ace, 2, 3, 4, and 5 face up in the center of the circle."

"We will count the ace as one."

"Now I will mix and deal seven cards to each of us."

- Deal the cards. Put the rest of the cards in a pile at the side of the table.

"Hold your cards in your hand."

"Now we will take turns playing one card at a time."

"We can put one of our cards on top of another card on the table if the sum of the two numbers is ten."

- For example, 7 can be put on top of 3, 6 on top of 4, 9 on top of the ace, etc.

"If you cannot play, you skip your turn, draw two cards from the pile, and wait for your next turn."

"The object of the game is to play all of your cards."

- Play the game with your child.

- Optional: Another card game for practicing sums of ten is "Sums of Ten Concentration."

- Mix the cards. Place them face down in six rows of six.

- Players take turns turning over two cards.

- If a sum of ten is found, the player keeps the cards and has another turn.

- Play continues until all the cards are paired.

CLASS PRACTICE

number fact practice

- Give your child the blue addition fact cards.

- Give your child **Fact Sheet MA 6.0.**

- Time your child for one minute.
- Correct the fact sheet with your child.
- Record the score.
- Allow time for your child to complete the unfinished facts.

  "On the back of this fact sheet we will practice tallying."

  "Show six using tally marks."

  "How many groups of five tally marks did you draw?"

  "How many extra tally marks did you make?"

- Repeat with 12, 20, 18, and 36.

WRITTEN PRACTICE

- Complete **Worksheet 33A** with your child.
- Complete **Worksheet 33B** with your child later in the day.

Name _____ **LESSON 33A**
Date _____ Math 2

Draw a picture and write a number sentence for this story. Write the answer with a label.

1. Six children got on Bus A at the first stop. Five more children got on Bus A at the second stop. How many children are on Bus A now?

Number sentence _____ 6 children + 5 children = 11 children _____

Answer _____ 11 children _____

2. Use a red crayon to trace the horizontal line segments in these letters.
Use a blue crayon to trace the vertical line segments in these letters.
Use a yellow crayon to trace the oblique line segments in these letters.

M A T H

3. Draw a tally mark for each letter of the alphabet.

How many tally marks did you draw? __26__

4. Fill in the missing addends for the sums of 10.

$\boxed{8} + 2 = 10$ $1 + \boxed{9} = 10$ $\boxed{6} + 4 = 10$

$\boxed{3} + 7 = 10$ $5 + \boxed{5} = 10$ $0 + \boxed{10} = 10$

5. How much money is 6 dimes? __(60¢)__

How much money is 54 pennies? __54¢__

Circle the one that is worth the most.

2-33Wa   Copyright © 1991 by Saxon Publishers, Inc. and Nancy Larson. Reproduction prohibited.

Name _____ **LESSON 33B**
Date _____ Math 2

Draw a picture and write a number sentence for this story. Write the answer with a label.

1. Seven children got on Bus B at the first stop. Six more children got on Bus B at the second stop. How many children are on Bus B now?

Number sentence _____ 7 children + 6 children = 13 children _____

Answer _____ 13 children _____

2. Use a red crayon to trace the horizontal line segments in these letters.
Use a blue crayon to trace the vertical line segments in these letters.
Use a yellow crayon to trace the oblique line segments in these letters.

N A M E

3. Ask someone in your family to let you tally the number of coins in their pocket or wallet. Draw one tally mark for each coin.

How many coins is this? _____

4. Fill in the missing addends for the sums of 10.

$\boxed{8} + 2 = 10$ $1 + \boxed{9} = 10$ $\boxed{6} + 4 = 10$

$\boxed{3} + 7 = 10$ $5 + \boxed{5} = 10$ $0 + \boxed{10} = 10$

5. How much money is 7 dimes? __(70¢)__

How much money is 39 pennies? __39¢__

Circle the one that is worth the most.

2-33Wb   Copyright © 1991 by Saxon Publishers, Inc. and Nancy Larson. Reproduction prohibited.

Lesson 34

dividing a whole into halves, fourths, and eighths

lesson preparation

materials

four 8" construction paper circles (each of yellow, blue, red, and green)

plastic storage bag (at least 8" × 8")

scissors

a deck of playing cards (tens, jacks, queens, and kings removed)

Fact Sheet MA 6.0

the night before

• Cut out construction paper circles

in the morning

• Write the following in the pattern box on the meeting strip:

$$1, 2, \_\_, 4, \_\_, 6, 7, 8, \_\_, 10$$

Answer: 1, 2, 3, 4, 5, 6, 7, 8, 9, 10

• Write   74¢   on the meeting strip. Provide a cup of 10 dimes and a cup of 10 pennies.

THE MEETING

calendar

• Ask your child to write the date on the calendar and meeting strip.

• Ask your child the following:

day of the week _____ days ago, day of the week _____ days from now

months of the year, _____th month, month before, month after

"What was the date three days ago?"

"What will be the date three days from now?"

• Record on the meeting strip a special event and the number of days until it occurs.

weather graph

- Ask your child to color the graph and write the temperature to the nearest ten degrees in the box he/she colored.
- Ask questions about the graph.

counting

- Count forward and backward by 1's from 546 to 573 as your child writes these numbers.
- Count by 10's to 200 and backward from 200 by 10's.
- Count by 5's to 100 and backward from 50 by 5's.
- Say the even numbers to 20 and backward from 20.
- Say the odd numbers to 19 and backward from 19.

graph questions

- You and your child each ask a question about any of the graphs.

patterning

- Ask your child to do the following:

  identify the pattern (repeating, continuing, or both)

  identify the numbers to complete the pattern

  read the pattern

money

"Today you will put coins in the coin cup to show 74¢."

"Show 74¢ using the fewest number of pennies."

- Allow time for your child to do this.

"How many dimes and pennies did you use?"

"Let's count to check to see if this is 74¢."

- Count the money with your child.

clock

- Ask your child to set the clock on the half hour or hour.
- Ask the following:

  time shown on the clock

  time one hour ago and time one hour from now

- Ask your child to write the digital time on the meeting strip.
- Record on the meeting strip the time an activity will occur.

number of the day

- Write three number sentences for the number of the day on the meeting strip.

THE LESSON

Dividing a Whole into Halves, Fourths, and Eighths

"We practiced dividing squares and other shapes in half."

"Today you will learn how to divide circles into halves, fourths, and eighths."

- Give your child four construction paper circles (one each of yellow, blue, red, and green) and a pair of scissors.

"The yellow circle is the whole."

"Are the other circles the same size?"

"Fold the blue circle in half."

"How many pieces do you have?" 2

"Are they the same size?"

"How do you know?"

"Cut along the fold."

"Now fold the red circle in half."

"Fold it in half again."

"How many pieces do you think you will have?"

"Open the circle and count the pieces."

"How many equal pieces do you have?" 4

"What is each piece called?"   one fourth

"How many fourths are in one whole?" 4

"Cut along the folds."

"Now fold the green circle in half."

"Fold it in half again."

"How many pieces do you have now?" 4

"Fold it in half again."

"How many pieces do you have now?" 8

"When we had four equal pieces, we called each piece one fourth."

"What do you think we will call each piece when we have eight equal pieces?"   one eighth

"How many eighths are in the circle?"   8

"Cut along the folds."

"Cover the yellow circle with the blue pieces."

"How many pieces did you use?"   2

"What is each piece called?"   one half

"Cover the yellow circle with the red pieces."

"How many pieces did you use?"   4

"What is each piece called?"   one fourth

"Cover the yellow circle with the green pieces."

"How many pieces did you use?"   8

"What is each piece called?"   one eighth

"Hold up the piece that is one fourth of the circle."

"Hold up the piece that is one half of the circle."

"Hold up the piece that is one eighth of the circle."

"We will use these pieces again."

- Give your child a plastic bag.
- Save the bag of pieces for use in Lessons 41, 50, and 65.

CLASS PRACTICE

number fact practice

- Play the card game "Making 10" or "Sums of 10 Concentration."
- Give your child **Fact Sheet MA 6.0.**
- Time your child for one minute.
- Correct the fact sheet with your child.
- Record the score.
- Allow time for your child to complete the unfinished facts.

WRITTEN PRACTICE

- Complete **Worksheet 34A** with your child.
- Complete **Worksheet 34B** with your child later in the day.

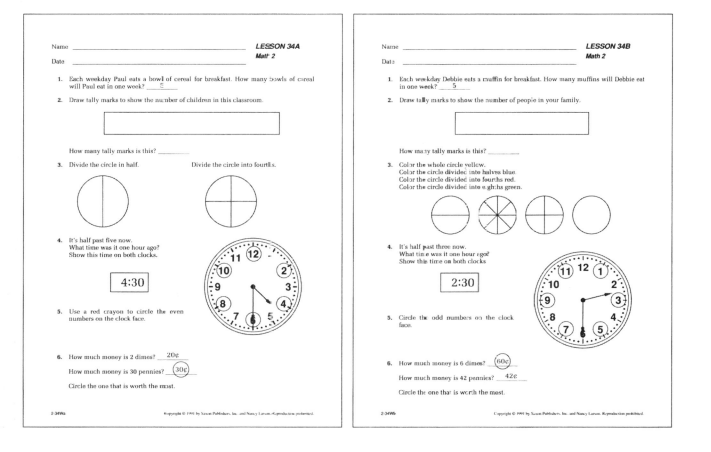

Name _____ **LESSON 34A**
Math 2
Date _____

1. Each weekday Paul eats a bowl of cereal for breakfast. How many bowls of cereal will Paul eat in one week? ____5____

2. Draw tally marks to show the number of children in this classroom.

How many tally marks is this? _____

3. Divide the circle in half.   Divide the circle into fourths.

4. It's half past five now.
   What time was it one hour ago?
   Show this time on both clocks.

   4:30

5. Use a red crayon to circle the even numbers on the clock face.

6. How much money is 2 dimes? ___20¢___

   How much money is 30 pennies? ___30¢___

   Circle the one that is worth the most.

2-34Wa

Name _____ **LESSON 34B**
Math 2
Date _____

1. Each weekday Debbie eats a muffin for breakfast. How many muffins will Debbie eat in one week? ____5____

2. Draw tally marks to show the number of people in your family.

How many tally marks is this? _____

3. Color the whole circle yellow.
   Color the circle divided into halves blue.
   Color the circle divided into fourths red.
   Color the circle divided into eighths green.

4. It's half past three now.
   What time was it one hour ago?
   Show this time on both clocks

   2:30

5. Circle the odd numbers on the clock face.

6. How much money is 6 dimes? ___60¢___

   How much money is 42 pennies? ___42¢___

   Circle the one that is worth the most.

2-34Wb

esson 35

adding ten to a multiple of ten
finding missing numbers on a piece of the
hundred number chart

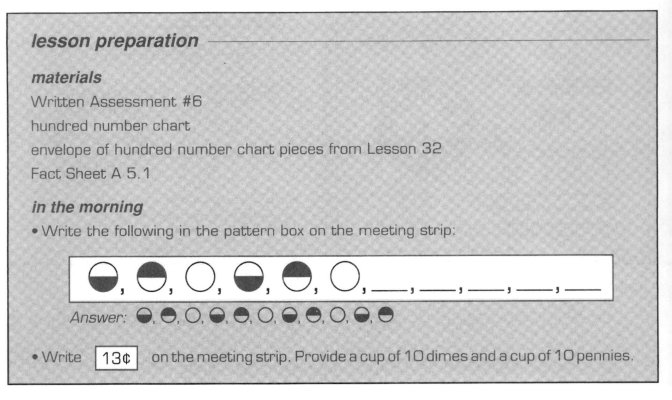

THE MEETING

calendar

- Ask your child to write the date on the calendar and meeting strip.

- Ask your child the following:

    day of the week _____ days ago, day of the week _____ days from now

    months of the year, _____th month, month before, month after

    "What was the date yesterday?"

    "What will be the date tomorrow?"

- Record on the meeting strip a special event and the number of days until it occurs.

weather graph

- Ask your child to color the graph and write the temperature to the nearest ten degrees in the box he/she colored.

• Ask questions about the graph.

counting

- Count forward and backward by 1's twice a week. Select starting and ending numbers, (e.g., count from 679 to 723). Ask your child to write the numbers on the chalkboard.
- Count by 10's to 200 and backward from 200 by 10's.
- Count by 5's to 100 and backward from 50 by 5's.
- Say the even numbers to 30 and backward from 30.
- Say the odd numbers to 29 and backward from 29.

graph questions

- You and your child each ask a question about any of the graphs.

patterning

- Ask your child to do the following:

  identify the pattern (repeating, continuing, or both)

  identify the shapes to complete the pattern

  read the pattern

money

"Put coins in the coin cup to show 13¢."

"Show 13¢ using the fewest number of pennies."

- Allow time for your child to do this.

"How many dimes and pennies did you use?"

"Let's count to check to see if this is 13¢."

- Count the money with your child.

clock

- Ask your child to set the clock on the half hour or hour.
- Ask the following:

  time shown on the clock

  time one hour ago and time one hour from now

- Ask your child to write the digital time on the meeting strip.
- Record on the meeting strip the time an activity will occur.

number of the day

- Write three number sentences for the number of the day on the meeting strip.

ASSESSMENT

Written Assessment

"Today I would like to see what you remember from what we have been practicing."

- Give your child **Written Assessment #6**.

- Read the directions for each problem. Allow time for your child to complete that problem before continuing.

- Correct the paper, noting your child's mistakes on the **Individual Recording Form**. Review the errors with your child.

THE LESSON

Adding Ten to a Multiple of Ten

"Today you will learn how to add ten to a multiple of ten."

- Give your child a hundred number chart.

    "Point to 20 on the hundred number chart."

    "Point to the number that is ten more."

    "What number is ten more than 20?"   30

    "Point to the number that is ten more than 30."

    "What number is it?"   40

    "Point to the number that is ten more than 40."

    "What number is it?"   50

    "Point to the number that is ten more than 70."

    "What number is it?"   80

    "We can write twenty and ten more like this."

- Write "20 + 10 =" on the chalkboard.

    "We can also write it like this."

- Write the following on the chalkboard: 20
                                            + 10

    "Twenty and ten more is equal to what?"

- Write "30" next to and below the examples.

    "Find 50 on the hundred number chart."

    "What is ten more?"

    "How will we write that?"

- Ask your child to write the problem on the chalkboard.

  "Is there another way to write that?"

- Ask your child to write the problem on the chalkboard.

- Repeat with 80.

- Write the following on the chalkboard:

$$40 + 10 = \qquad \begin{array}{r} 80 \\ + 10 \\ \hline \end{array} \qquad 30 + 10 = \qquad \begin{array}{r} 10 \\ + 50 \\ \hline \end{array} \qquad 10 + 60 =$$

  "What are the answers for these problems?"

  "How do you know?"

Finding Missing Numbers on a Piece of the Hundred Number Chart

  "Let's look at the hundred number chart."

  "What number is just above 35?"

  "What number is just below 35?"

  "What do you notice?"

- Draw the following on the chalkboard:

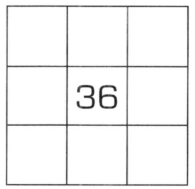

  "I have drawn a piece of the hundred number chart on the chalkboard."

  "It matches one of the pieces of your hundred number chart puzzle."

  "All of the numbers are missing except one."

  "How can you find the missing numbers?"

  "What is one of the missing numbers?"

- Fill in the numbers as your child suggests them. Repeat until all the numbers are listed.

- Give your child the envelope of hundred number chart puzzle pieces.

  "Find the piece that matches my chalkboard piece."

  "Did you guess the correct numbers?"

  "What is the same about the numbers in the horizontal rows?"

"What is the same about the numbers in the vertical columns?"

"Put your pieces in the envelope."

- Draw the following on the chalkboard:

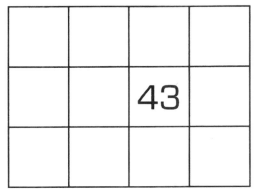

"I have drawn another piece of the hundred number chart on the chalkboard."

"It matches one of the pieces of your hundred number chart puzzle."

"All of the numbers are missing except one."

"How can you find the missing numbers?"

"What is one of the missing numbers?"

- Fill in the number your child suggests. Repeat until all the numbers are listed.

"Look in your envelope and find the piece that matches my chalkboard piece."

"Did you guess the correct numbers?"

"Put your pieces in the envelope."

- Repeat with several more pieces, if desired.

CLASS PRACTICE

number fact practice

- Use the pink fact cards to practice the doubles plus one facts together.
- Give your child **Fact Sheet A 5.1**.
- Time your child for one minute.
- Correct the fact sheet with your child.
- Record the score.
- Allow time for your child to complete the unfinished facts.

WRITTEN PRACTICE

- Complete **Worksheet 35A** with your child.

- Complete **Worksheet 35B** with your child later in the day.

Name _____ **ASSESSMENT 6**

Date _____ **LESSON 35**

Math 2

Draw a picture and write a number sentence for this story. Write the answer with a label.

1. Kathy had eight pencils. Her sister gave her two more. How many pencils does Kathy have now?

   Number sentence _____ 8 pencils + 2 pencils = 10 pencils

   Answer _____ 10 pencils

2. Show two different ways to divide a square in half. Color one half of each square.

   answers may vary

3. How much money is this? _____ 70¢

4. What is the second day of the week? _____ Monday

   What is the fifth month of the year? _____ May

   What is the last month of the year? _____ December

   What is the sixth day of the week? _____ Friday

5. Circle the odd numbers. 4 ⑨ 6 ③ ⑪ 8 12 ⑦

6. Find the sums.

   0 + 4 = __4__   4 + 2 = __6__   2 + 9 = __11__   8 + 8 = __16__

   6 + 6 = __12__   8 + 1 = __9__   1 + 5 = __6__   6 + 2 = __8__

2-35Aa

Name _____ **LESSON 35A**

Date _____ **Math 2**

1. What is the fifth day of the week? _____ Thursday

   What day of the week was yesterday? _____

   What is the eleventh month of the year? _____ November

2. Use the classroom graphs to answer these questions.

   How many children wake up at 6:30? _____

   How many children have a birthday in July or August? _____

3. Fill in the missing numbers on this piece of a hundred number chart.

61	62	63	64
71	72	73	74
81	82	83	84
91	92	93	94

4. Draw a horizontal line.   Draw a vertical line.   Draw an oblique line.

5. Find each sum.

   60 + 10 = __70__   40 + 10 = __50__   10 + 80 = __90__

6. Write three addition facts that have sums that are even numbers.

2-35Wa

Name _____ **LESSON 35B**

Date _____ **Math 2**

1. What is the second day of the week? _____ Monday

   What day of the week will it be tomorrow? _____

   What is the tenth month of the year? _____ October

2. Draw a horizontal line.   Draw a vertical line.   Draw an oblique line.

3. Fill in the missing numbers on this piece of a hundred number chart.

77	78	79	80
87	88	89	90
97	98	99	100

4. Which number on the thermometer is the temperature closest to? _____ 90 __°F

5. Find each sum.

   30 + 10 = __40__   10 + 70 = __80__   20 + 10 = __30__

6. Write three addition facts that have sums that are odd numbers.

2-35Wb

Lesson 36

identifying pairs

---

lesson preparation

materials

7 socks for use as props (3 matching pairs)

100 color tiles

7 small bags

Fact Sheet A 5.1

the night before

• Put the following number of tiles in each bag: 10, 11, 12, 13, 14, 15, and 16.

in the morning

• Write the following in the pattern box on the meeting strip:

> ## A, ___, C, ___, E, ___, G, ___, I, ___

Answer: A, B, C, D, E, F, G, H, I, J

• Write | 38¢ | on the meeting strip. Provide a cup of 10 dimes and a cup of 10 pennies.

THE MEETING

calendar

• Ask your child to write the date on the calendar and meeting strip.

• Ask your child the following:

> day of the week ____ days ago, day of the week ____ days from now
>
> months of the year, ____th month, month before, month after

"What was the date two days ago?"

"What will be the date two days from now?"

• Record on the meeting strip a special event and the number of days until it occurs.

weather graph

- Ask your child to color the graph and write the temperature to the nearest ten degrees in the box he/she colored.

- Ask questions about the graph.

counting

- Count forward and backward by 1's twice a week. Select starting and ending numbers (e.g., count from 679 to 723). Ask your child to write the numbers on the chalkboard.

- Count by 10's to 200 and backward from 200 by 10's.

- Count by 5's to 100 and backward from 50 by 5's.

- Say the even numbers to 30 and backward from 30.

- Say the odd numbers to 29 and backward from 29.

graph questions

- You and your child each ask a question about any of the graphs.

patterning

- Ask your child to do the following:

  identify the pattern (repeating, continuing, or both)

  identify the letters to complete the pattern

  read the pattern

money

- Ask your child to put the dimes and pennies in the coin cup.

- Count the money in the coin cup together.

clock

- Ask your child to set the clock on the half hour or hour.

- Ask the following:

  time shown on the clock

  time one hour ago and time one hour from now

- Ask your child to write the digital time on the meeting strip.

- Record on the meeting strip the time an activity will occur.

number of the day

- Write three number sentences for the number of the day on the meeting strip.

THE LESSON

Identifying Pairs

"Today you will learn how to put things in pairs."

"What things come in pairs?"   socks, shoes, mittens, earrings

"What does the word 'pair' mean?"   two matched objects; each item has a partner

"When we have a pair, we have two matched objects."

"Each object has a partner."

- Use seven socks as props.

  "If I have one sock, do I have a pair?"   no

  "Why not?"   because it doesn't have a partner

  "If I have two socks, do I have a pair?"   yes

  "Why?"   because it does have a partner

- Roll them together.

  "If I have three socks, can I make a pair?"   yes

  "Will I have any socks left over?"

- Roll two socks together and show the extra sock.

- Repeat with 4, 5, 6, and 7 socks.

  "If we can make partners, we have a pair."

- Hold up a color tile.

  "We will pretend that these are socks."

  "I will give you some pretend socks."

  "Put your pretend socks in pairs."

- Give your child the bag of 10 color tiles.

- Ask your child the following questions as you make a chart on the chalkboard:

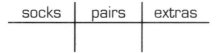

socks	pairs	extras

"How many socks did I give you?"

"How many pairs did you make?"

"Did you have any extras?"

- Repeat with the bags of 13, 16, 12, 15, 11, and 14 tiles.

"Let's check to see which numbers of socks will give us pairs without any socks left over."

- Circle the even numbers of socks in the list.

  "What kind of numbers are these?" even numbers

  "I will draw a picture of some socks on the chalkboard."

- Draw the following on the chalkboard:

  "Use your color tiles to show this number of socks."

  "Come to the chalkboard and circle pairs of socks."

  "We can show pairs by drawing a circle around a group of two."

  "How many pairs of socks do we have?" 4

  "How many extra socks do we have?" 1

- Repeat at least once, using eight socks.

CLASS PRACTICE

number fact practice

- Use the pink fact cards to practice the doubles plus one facts together.
- Give your child **Fact Sheet A 5.1**.
- Time your child for one minute.
- Correct the fact sheet with your child.
- Record the score.
- Allow time for your child to complete the unfinished facts.

WRITTEN PRACTICE

- Complete **Worksheet 36A** with your child.
- Complete **Worksheet 36B** with your child later in the day.

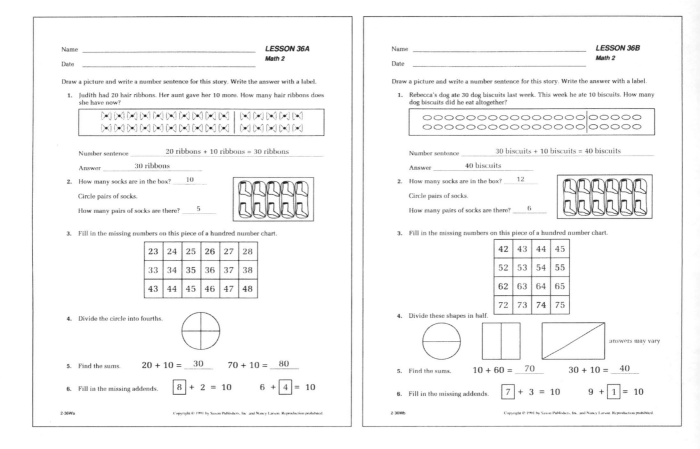

LESSON 36A
Math 2

Name _____

Date _____

Draw a picture and write a number sentence for this story. Write the answer with a label.

1. Judith had 20 hair ribbons. Her aunt gave her 10 more. How many hair ribbons does she have now?

Number sentence _____ 20 ribbons + 10 ribbons = 30 ribbons _____

Answer _____ 30 ribbons _____

2. How many socks are in the box? ____ 10 ____

Circle pairs of socks.

How many pairs of socks are there? ____ 5 ____

3. Fill in the missing numbers on this piece of a hundred number chart.

23	24	25	26	27	28
33	34	35	36	37	38
43	44	45	46	47	48

4. Divide the circle into fourths.

5. Find the sums. $20 + 10 =$ ____30____ $70 + 10 =$ ____80____

6. Fill in the missing addends. $\boxed{8} + 2 = 10$ $6 + \boxed{4} = 10$

2-36Wa

LESSON 36B
Math 2

Name _____

Date _____

Draw a picture and write a number sentence for this story. Write the answer with a label.

1. Rebecca's dog ate 30 dog biscuits last week. This week he ate 10 biscuits. How many dog biscuits did he eat altogether?

Number sentence _____ 30 biscuits + 10 biscuits = 40 biscuits _____

Answer _____ 40 biscuits _____

2. How many socks are in the box? ____ 12 ____

Circle pairs of socks.

How many pairs of socks are there? ____ 6 ____

3. Fill in the missing numbers on this piece of a hundred number chart.

42	43	44	45
52	53	54	55
62	63	64	65
72	73	74	75

4. Divide these shapes in half.

answers may vary

5. Find the sums. $10 + 60 =$ ____70____ $30 + 10 =$ ____40____

6. Fill in the missing addends. $\boxed{7} + 3 = 10$ $9 + \boxed{1} = 10$

2-36Wb

L esson 37

measuring with one-inch tiles

lesson preparation

materials

pencil (unsharpened)

color tiles

book

ruler

Master 2-37

Fact Sheet A 5.2

in the morning

• Write the following in the pattern box on the meeting strip:

> A, b, C, d, ___, ___, ___, ___, ___

Answer: A, b, C, d, E, f, G, h, I

• Write ⟨56¢⟩ on the meeting strip. Provide a cup of 10 dimes and a cup of 10 pennies.

THE MEETING

calendar

• Ask your child to write the date on the calendar and meeting strip.

• Ask your child the following:

day of the week ____ days ago, day of the week ____ days from now

months of the year, ____th month, month before, month after

"What was the date four days ago?"

"What will be the date four days from now?"

• Record on the meeting strip a special event and the number of days until it occurs.

weather graph

• Ask your child to color the graph and write the temperature to the nearest ten degrees in the box he/she colored.

• Ask questions about the graph.

counting

- Count forward and backward by 1's twice a week. Select starting and ending numbers (e.g., count from 679 to 723). Ask your child to write the numbers on the chalkboard.

- Count by 10's to 200 and backward from 200 by 10's.

- Count by 5's to 100 and backward from 50 by 5's.

- Say the even numbers to 30 and backward from 30.

- Say the odd numbers to 29 and backward from 29.

graph questions

- You and your child each ask a question about any of the graphs.

patterning

- Ask your child to do the following:

  identify the pattern (repeating, continuing, or both)

  identify the letters to complete the pattern

  read the pattern

money

- Ask your child to put the dimes and pennies in the coin cup.

- Count the money in the coin cup together.

clock

- Ask your child to set the clock on the half hour or hour.

- Ask the following:

  time shown on the clock

  time one hour ago and time one hour from now

- Ask your child to write the digital time on the meeting strip.

- Record on the meeting strip the time an activity will occur.

number of the day

- Write three number sentences for the number of the day on the meeting strip.

THE LESSON

Measuring With One-Inch Tiles

"We can use color tiles for many things."

"We can use them to cover numbers on the hundred number chart, we can use them to make pairs, and we can use them to help us add."

"Today we will use them in a different way."

"Today you will learn how to measure objects using color tiles."

- Hold up a pencil (preferably unsharpened).

"How can we use the color tiles to measure this pencil?"

- Put the pencil on the table in front of your child.

"Put the beginning of the first tile even with the end of the pencil."

"Now put a second tile next to the first tile."

"It should completely touch the tile next to it."

"Keep adding tiles until you reach the end of the pencil."

"The tiles should be in a straight line."

"About how many tiles long is the pencil?"

- If the pencil (or any other object used in this lesson) is not an exact number of tiles long, accept as an answer either the smaller or the larger number of tiles. For example, if the pencil is a little more than 7 tiles long, accept either 7 or 8 tiles as the answer. Encourage your child to estimate the length to the nearest tile.

"Each of the sides of a color tile is one inch long, so we can say that the pencil is about _____ inches long."

- Choose another object in the room.

"Let's measure _____ with the color tiles."

"How will we do that?"

- Ask your child to demonstrate.

"What do we have to remember to do when we are measuring something with tiles?"   put the tiles next to each other; put the first tile even with the end of the object; keep them straight along the edge of the object

"About how many tiles long is _____?"

"Each of the sides of a color tile is one inch long, so we can say that the _____ is about _____ inches long."

"Now you will have a chance to measure some more objects using color tiles."

"You will use a book, a piece of paper, and a ruler."

- Give your child **Master 2-37**, color tiles, a book, and a ruler.

"Look at example 1 on Master 2-37."

"The small picture on Master 2-37 shows how you will measure."

"This picture tells you to use the color tiles to find how many color tiles wide the book is."

"The width is the way the book opens."

"Measure the width of the book using color tiles."

"Write the answer on Master 2-37."

- Allow time for your child to measure the book with the tiles.

"How many color tiles wide is the book?"

"Each side of a color tile is one inch, so we can say that the book is _____ inches wide."

"Look at example 2 on Master 2-37."

"This picture tells you to use the color tiles to measure Master 2-37 vertically."

"Use your color tiles to measure this paper vertically."

"Write your answer on the paper."

- Allow time for your child to measure the paper with the tiles.

"How many color tiles long is this paper?"

"Each side of a color tile is one inch, so we can say that the paper is _____ inches long."

"Look at example 3 on Master 2-37."

"This picture tells you to use the color tiles to measure the ruler."

"Use your color tiles to measure the ruler."

"Write your answer on the paper."

- Allow time for your child to measure the ruler with the tiles.

"How many color tiles long is the ruler?"   12

"Each side of a color tile is one inch, so we can say that the ruler is 12 inches long."

"Point to the line segment that has an A on one end and a B on the other end."

"Use your color tiles to measure line segment AB on this paper."

- Allow time for your child to measure the line segment.

"How many inches long is the line segment?"   5 inches

"Point to the line segment that has a C on one end and a D on the other end."

"Use your color tiles to measure line segment CD on this paper."

• Allow time for your child to measure the line segment.

"How many inches long is the line segment?"   *7 inches*

CLASS PRACTICE

number fact practice

• Use the fact cards to practice the addition facts with your child.

• Give your child **Fact Sheet A 5.2.**

• Time your child for one minute.

• Correct the fact sheet with your child.

• Record the score.

• Allow time for your child to complete the unfinished facts.

WRITTEN PRACTICE

• Complete **Worksheet 37A** with your child.

• Complete **Worksheet 37B** with your child later in the day.

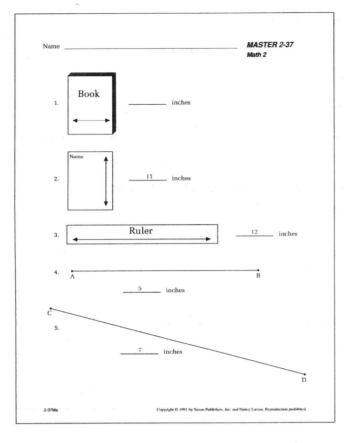

Name _____ **MASTER 2-37**
Math 2

1. Book _____ inches

2. Name _____11_____ inches

3. Ruler _____12_____ inches

4. A————————B _____5_____ inches

5. C————————D _____7_____ inches

Name _____ **LESSON 37A**
Math 2
Date _____

Draw a picture and write a number sentence for this story. Write the answer with a label.

1. Leroy ate a peanut butter sandwich every day last week. This week he ate two peanut butter sandwiches. How many peanut butter sandwiches did he eat?

Number sentence ___ 7 sandwiches + 2 sandwiches = 9 sandwiches ___

Answer ___ 9 sandwiches ___

2. How many shoes are in the box? ___12___

Circle pairs of shoes.

How many pairs of shoes are there? ___6___

3. How many tally marks are shown? ___21___

‖‖‖ ‖‖‖ ‖‖‖ ‖‖‖ |

4. Write the names of the first six months.

1. ___January___ 2. ___February___ 3. ___March___

4. ___April___ 5. ___May___ 6. ___June___

5. Color the even numbers red. Color the odd numbers yellow.

1	2	3	4	5	6	7	8	9	10
11	12	13	14	15	16	17	18	19	20

Name _____ **LESSON 37B**
Math 2
Date _____

Draw a picture and write a number sentence for this story. Write the answer with a label.

1. There were 9 pickles in the jar. Lynn ate three pickles. How many pickles are left?

Number sentence ___ 9 pickles – 3 pickles = 6 pickles ___

Answer ___ 6 pickles ___

2. How many shoes are in the box? ___8___

Circle pairs of shoes.

How many pairs of shoes are there? ___4___

3. How many tally marks are shown? ___19___

‖‖‖ ‖‖‖ ‖‖‖ ||||

4. Fill in the missing months.

___February___ , March, ___April___

5. Color the answers to these examples.

$1 + 1 =$ $2 + 2 =$ $3 + 3 =$ $4 + 4 =$ $5 + 5 =$
$6 + 6 =$ $7 + 7 =$ $8 + 8 =$ $9 + 9 =$ $10 + 10 =$

1	3	5	7	9
11	13	15	17	19

L esson 38

identifying tens and ones

lesson preparation

materials

a deck of playing cards (tens, jacks, queens, and kings removed)

Fact Sheet MA 6.0

in the morning

• Write the following in the pattern box on the meeting strip:

○, △, △, □, ○, △, △, □, ___, ___, ___, ___, ___

Answer: ○, △, △, □, ○, △, △, □, ○, △, △, □, ○

• Write 61¢ on the meeting strip. Provide a cup of 10 dimes and a cup of 10 pennies.

THE MEETING

calendar

• Ask your child to write the date on the calendar and meeting strip.

• Ask your child the following:

day of the week ____ days ago, day of the week ____ days from now

months of the year, ____th month, month before, month after

"What was the date three days ago?"

"What will be the date three days from now?"

• Record on the meeting strip a special event and the number of days until it occurs.

weather graph

• Ask your child to color the graph and write the temperature to the nearest ten degrees in the box he/she colored.

• Ask questions about the graph.

counting

- Count forward and backward by 1's twice a week. Select starting and ending numbers (e.g., count from 679 to 723). Ask your child to write the numbers on the chalkboard.
- Count by 10's to 200 and backward from 200 by 10's.
- Count by 5's to 100 and backward from 50 by 5's.
- Say the even numbers to 30 and backward from 30.
- Say the odd numbers to 29 and backward from 29.

graph questions

- You and your child each ask a question about any of the graphs.

patterning

- Ask your child to do the following:

  identify the pattern (repeating, continuing, or both)

  identify the shapes to complete the pattern

  read the pattern

money

- Ask your child to put the dimes and pennies in the coin cup.
- Count the money in the coin cup together.

clock

- Ask your child to set the clock on the half hour or hour.
- Ask the following:

  time shown on the clock

  time one hour ago and time one hour from now
- Ask your child to write the digital time on the meeting strip.
- Record on the meeting strip the time an activity will occur.

number of the day

- Write three number sentences for the number of the day on the meeting strip.

THE LESSON

Identifying Tens and Ones

"Today you will learn how to find the number of tens and ones in a number."

"Each day I write an amount of money on the meeting strip."

"Then you use dimes and pennies to make that amount of money."

"Today's amount of money was 61¢."

"How many dimes did you use?"   6

"How many pennies did you use?"   1

- Record "6 dimes, 1 penny" on the chalkboard.

"Which digit in the number 61 tells us the number of dimes to use?"

"Which digit in the number 61 tells us the number of pennies to use?"

- Write "47¢" on the chalkboard.

"Let's make 47¢ using only dimes and pennies."

"How could we do that?"

- Record "4 dimes, 7 pennies" on the chalkboard.

"Which digit in the number 47 tells us the number of dimes to use?"

"Which digit in the number 47 tells us the number of pennies to use?"

"Pennies are each worth one cent."

"The digit on the right tells us the number of pennies or ones we have."

"Dimes are each worth ten cents."

"What do you think the digit on the left will tell us?"

"The digit on the left tells us the number of dimes or tens we have."

- Write the following on the chalkboard:

82¢   70¢

"How many dimes and pennies will we need to make 82¢?"

"Which digit tells us the number of dimes or tens?"

"Which digit tells us the number of pennies or ones?"

"How many dimes and pennies will we need to make 70¢?"

"Which digit tells us the number of dimes or tens?"

"Which digit tells us the number of pennies or ones?"

- Write the following on the chalkboard:

93   69

"How many tens and ones will we need to make 93?"

- Write "9 tens, 3 ones" on the chalkboard.

"How many tens and ones will we need to make 69?"

- Write "6 tens, 9 ones" on the chalkboard.

- Repeat with 17, 84, 12, and 58.

CLASS PRACTICE

number fact practice

- Play the "Sums of 10 Concentration" card game.
- Give your child **Fact Sheet MA 6.0.**
- Time your child for one minute.
- Correct the fact sheet with your child.
- Record the score.
- Allow time for your child to complete the unfinished facts.

WRITTEN PRACTICE

- Complete **Worksheet 38A** with your child.
- Complete **Worksheet 38B** with your child later in the day.

Name _____   **LESSON 38A**
Date _____   Math 2

Draw a picture and write a number sentence for this story. Write the answer with a label.

1. Tom has 3 red markers, 4 green markers, and 2 pencils. How many markers does Tom have?

   Number sentence _____3 markers + 4 markers = 7 markers_____

   Answer _____7 markers_____

2. How many tens and ones are in 49? __4__ tens __9__ ones

3. Write the names of the days of the week.

   1. _Sunday_ 2. _Monday_ 3. _Tuesday_ 4. _Wednesday_

   5. _Thursday_ 6. _Friday_ 7. _Saturday_

4. Circle all the squares that are divided into fourths (four equal parts).

5. Number the clock face. Show half past twelve on the clock face.

6. Circle the odd numbers on the clock face.

2-38Wa   Copyright © 1991 by Saxon Publishers, Inc. and Nancy Larson. Reproduction prohibited.

Name _____   **LESSON 38B**
Date _____   Math 2

Draw a picture and write a number sentence for this story. Write the answer with a label.

1. Ken's mom gave him 4 chocolate chip cookies, 1 apple, 2 peanut butter cookies, and a drink box for the field trip. How many cookies did she give him?

   Number sentence _____4 cookies + 2 cookies = 6 cookies_____

   Answer _____6 cookies_____

2. How many tens and ones are in 68? __6__ tens __8__ ones

3. What is the second day of the week? _Monday_

   What is the fourth day of the week? _Wednesday_

   What is the first day of the week? _Sunday_

   What is the sixth day of the week? _Friday_

4. Underline all the circles that are divided into fourths (four equal parts).

5. Number the clock face. Show half past ten on the clock face.

6. Circle the even numbers on the clock face.

2-38Wb   Copyright © 1991 by Saxon Publishers, Inc. and Nancy Larson. Reproduction prohibited.

Lesson 39

identifying halves, fourths, and eighths of a whole
creating and reading a bar graph

---

lesson preparation

materials

3 apples (different types)
3 plates
knife and cutting board
Meeting Book
three 2" construction paper tags (See *the night before.*)
Fact Sheet A 5.2

the night before

- Obtain three different types of apples. Try to use apples that are visibly different in color.
- Cut 2" square construction paper tags in colors that correspond, if possible, to the colors of the apples (e.g., red, green, and yellow). Place each tag on a plate with the apple it corresponds to.

in the morning

- Write the following in the pattern box on the meeting strip:

| 5, 10, 15, 20, ____, ____, ____, ____, ____ |

Answer: 5, 10, 15, 20 25, 30, 35, 40, 45

- Write | 28¢ | on the meeting strip. Provide a cup of 10 dimes and a cup of 10 pennies.

THE MEETING

calendar

- Ask your child to write the date on the calendar and meeting strip.
- Ask your child the following:

    day of the week _____ days ago, day of the week _____ days from now

    months of the year, _____th month, month before, month after

---

"What was the date two days ago?"

"What will be the date two days from now?"

- Record on the meeting strip a special event and the number of days until it occurs.

weather graph

- Ask your child to color the graph and write the temperature to the nearest ten degrees in the box he/she colored.
- Ask questions about the graph.

counting

- Count forward and backward by 1's twice a week. Select starting and ending numbers (e.g., count from 679 to 723). Ask your child to write the numbers on the chalkboard.
- Count by 10's to 200 and backward from 200 by 10's.
- Count by 5's to 100 and backward from 50 by 5's.
- Say the even numbers to 30 and backward from 30.
- Say the odd numbers to 29 and backward from 29.

graph questions

- You and your child each ask a question about any of the graphs.

patterning

- Ask your child to do the following:

    identify the pattern (repeating, continuing, or both)

    identify the numbers to complete the pattern

    read the pattern

money

- Ask your child to put the dimes and pennies in the coin cup.
- Count the money in the coin cup together.

clock

- Ask your child to set the clock on the half hour or hour.
- Ask the following:

    time shown on the clock

    time one hour ago and time one hour from now

- Ask your child to write the digital time on the meeting strip.
- Record on the meeting strip the time an activity will occur.

number of the day

- Write three number sentences for the number of the day on the meeting strip.

THE LESSON

Identifying Halves, Fourths, and Eighths of a Whole
Creating and Reading a Bar Graph

"Today you will learn how to divide a solid into halves, fourths, and eighths."

"You will also learn how to make and read a bar graph."

"Today we will use apples for our math lesson."

- Cut apple A n half vertically.

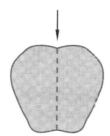

"When I cut the apple like this, I am cutting it in half."

"How many pieces do we have?"

"What do we call each piece?" one half

"How many halves are in one whole?" 2

"I will cut each half of the apple in half again."

"How many pieces do you think we will have?"

- Cut each half of the apple like this

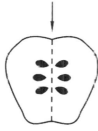

"How many pieces do we have now?" 4

"What is each piece called?" one fourth

"I will cut each piece of the apple in half again."

"How many pieces do you think we will have?"

• Cut the apple fourths in half like this:

"How many pieces do we have now?"   8

"What is each piece called?"   one eighth

• Put the pieces of apple A on a plate with the corresponding construction paper tag.

• Repeat with apple B and apple C.

"Are all apples the same?"

"How are they different?"

"Do all apples taste the same?"

"Today we will taste a piece of each apple to see which one we like best."

"We will also ask some friends, family members, and neighbors to taste a piece of each apple to see which one they like best."

"I will cut the apples into small pieces so that everyone can have a taste of each apple."

• Cut each eighth in half.

"Try each apple."

• Give your child a small piece of apple A, apple B, and apple C.

"Which apple do you like best?"

• Open the Meeting Book to page 33.

"Color a box on the graph to show the apple you liked best."

"Look at that apple's tag to see which color to color the box."

"What box will you color?"

• Allow time for your child to do this.

"I liked the _____ apple best."

"What box will you color to show the apple I liked best?"

• Allow time for your child to do this.

"Now you will ask some other people to taste the apples and choose their favorite."

"Who would you like to ask?"

- List 8–10 names on the chalkboard.
- Give each of these people a small piece of apple A, apple B, and apple C.
- Ask your child to show the choices on the graph.
- When the graph is finished, continue.

    "What do you notice about our graph?"

- Encourage your child to offer as many observations as possible.

    "How many people liked apple A best?"

    "How many people liked apple B best?"

    "How many people liked apple C best?"

    "Which apple is the favorite?"

    "How many more people chose _____ than chose _____?"

- This graph will be used during The Meeting.

CLASS PRACTICE

number fact practice

- Use the fact cards to practice all the addition facts with your child.
- Give your child **Fact Sheet A 5.2.**
- Time your child for one minute.
- Correct the fact sheet with your child.
- Record the score.
- Allow time for your child to complete the unfinished facts.

WRITTEN PRACTICE

- Complete **Worksheet 39A** with your child.
- Complete **Worksheet 39B** with your child later in the day.

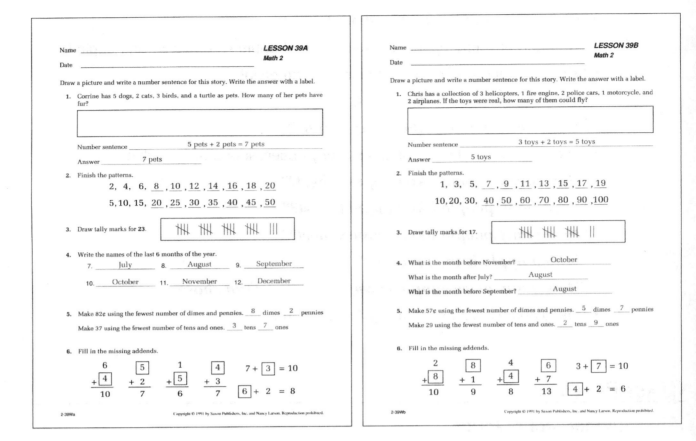

Name _____ **LESSON 39A**
Math 2
Date _____

Draw a picture and write a number sentence for this story. Write the answer with a label.

1. Corrine has 5 dogs, 2 cats, 3 birds, and a turtle as pets. How many of her pets have fur?

Number sentence _____5 pets + 2 pets = 7 pets_____

Answer _____7 pets_____

2. Finish the patterns.

2, 4, 6, _8_ , _10_ , _12_ , _14_ , _16_ , _18_ , _20_

5, 10, 15, _20_ , _25_ , _30_ , _35_ , _40_ , _45_ , _50_

3. Draw tally marks for **23**.

4. Write the names of the last 6 months of the year.

7. ___July___   8. ___August___   9. ___September___

10. ___October___   11. ___November___   12. ___December___

5. Make 82¢ using the fewest number of dimes and pennies. _8_ dimes _2_ pennies

Make 37 using the fewest number of tens and ones. _3_ tens _7_ ones

6. Fill in the missing addends.

```
  6        5        1        4      7 + 3 = 10
+ 4      + 2      + 5      + 3
----     ----     ----     ----     6 + 2 = 8
 10        7        6        7
```

Name _____ **LESSON 39B**
Math 2
Date _____

Draw a picture and write a number sentence for this story. Write the answer with a label.

1. Chris has a collection of 3 helicopters, 1 fire engine, 2 police cars, 1 motorcycle, and 2 airplanes. If the toys were real, how many of them could fly?

Number sentence _____3 toys + 2 toys = 5 toys_____

Answer _____5 toys_____

2. Finish the patterns.

1, 3, 5, _7_ , _9_ , _11_ , _13_ , _15_ , _17_ , _19_

10, 20, 30, _40_ , _50_ , _60_ , _70_ , _80_ , _90_ , _100_

3. Draw tally marks for **17**.

4. What is the month before November? ___October___

What is the month after July? ___August___

What is the month before September? ___August___

5. Make 57¢ using the fewest number of dimes and pennies. _5_ dimes _7_ pennies

Make 29 using the fewest number of tens and ones. _2_ tens _9_ ones

6. Fill in the missing addends.

```
  2        8        4        6      3 + 7 = 10
+ 8      + 1      + 4      + 7
----     ----     ----     ----     4 + 2 = 6
 10        9        8       13
```

L esson 40

identifying geometric shape pieces that are alike in only one way

lesson preparation

materials

Written Assessment #7

Oral Assessment #4

shape pieces from Lesson 22

10 dimes, 10 pennies

Fact Sheet A 5.2

in the morning

• Write the following in the pattern box on the meeting strip:

$$2, \ 4, \ 6, \ 8,\underline{\quad},\underline{\quad},\underline{\quad},\underline{\quad},\underline{\quad}$$

Answer: 2, 4, 6, 8, 10, 12, 14, 16, 18

• Write [34¢] on the meeting strip. Provide a cup of 10 dimes and a cup of 10 pennies.

THE MEETING

calendar

• Ask your child to write the date on the calendar and meeting strip.

• Ask your child the following:

date _____ days ago, date _____ days from now

day of the week _____ days ago, day of the week _____ days from now

months of the year, _____th month, month before, month after

• Record on the meeting strip a special event and the number of days until it occurs.

weather graph

• Ask your child to color the graph and write the temperature to the nearest ten degrees in the box he/she colored.

• Ask questions about the graph.

counting

- Count by 10's to 200 and backward from 200 by 10's.
- Count by 5's to 100 and backward from 50 by 5's.
- Say the even numbers to 30 and backward from 30.
- Say the odd numbers to 29 and backward from 29.

graph questions

- You and your child each ask a question about any of the graphs.

patterning

- Ask your child to do the following:

  identify the pattern (repeating, continuing, or both)

  identify the numbers to complete the pattern

  read the pattern

money

- Ask your child to put the dimes and pennies in the coin cup.
- Count the money in the coin cup together.

clock

- Ask your child to set the clock on the half hour or hour.
- Ask the following:

  time shown on the clock

  time one hour ago and time one hour from now

- Ask your child to write the digital time on the meeting strip.
- Record on the meeting strip the time an activity will occur.

number of the day

- Write three number sentences for the number of the day on the meeting strip.

ASSESSMENT

Written Assessment

- All of the questions on the assessment are based on concepts and skills presented at least five lessons ago. It is expected that your child will master at least 80% of the concepts on the assessment. If your child is having difficulty with a specific concept, reteach the concept the following day.

"Today I would like to see what you remember from what we have been practicing."

- Give your child **Written Assessment #7**.

- Read the directions for each problem. Allow time for your child to complete that problem before continuing.

- Correct the paper, noting your child's mistakes on the **Individual Recording Form**. Review the errors with your child.

Oral Assessment

- Record your child's response(s) to the oral interview questions on the interview sheet.

THE LESSON

Identifying Geometric Shape Pieces that are Alike in Only One Way

"Today we will use our shape pieces to play the 'Attribute Train Game.'"

"How did we play this game when we played it before?"

"Let's try it."

- Place the large red triangle in the center of the table.

"We are going to pretend that our pieces are cars of a train."

"What is the special rule for putting new cars on the train?"

"The special rule for putting new cars on the train is that the next car must be like the one before in two ways."

"What piece did I put first?" large red triangle

"What piece could we put next?" there are many possibilities, including the small red triangle, the large green triangle, the large red circle, etc.

"How are these pieces the same?"

"How are they different?"

"Now we will add another car to the train."

"It must be like the car before it in two ways."

"What piece could we put next?"

"How are these pieces the same?"

"How are they different?"

- Continue, adding at least ten more pieces.

"Today we are going to play the game using a different rule for putting new cars on the train."

"The new rule for putting cars on the train is that the next car must be the same in only one way."

- Place the large yellow square in the center of the table.

"What piece did I put first?"   large yellow square

"What piece could we put next if it must be the same in only one way?"   there are many possibilities, including the small red square, the large green triangle, the small yellow circle, etc.

"In what way are the pieces the same?"

"How are the pieces different?"

"Now we will add another car to the train."

"It must be the same as the car before it in only one way."

"What piece could we put next?"

"In what way are the pieces the same?"

"How are the pieces different?"

- Continue, adding at least five more pieces.

"Now we'll divide the pieces that are left between us."

- Divide the pieces.

"Let's take turns adding cars to the train, using the rule that the new piece must be the same as the piece before it in only one way."

"We will try to make our train as long as possible."

- Take turns adding pieces to the train.

CLASS PRACTICE

number fact practice

- Use the fact cards to practice all the addition facts with your child.
- Give your child **Fact Sheet A 5.2**.
- Time your child for one minute.
- Correct the fact sheet with your child.
- Record the score.
- Allow time for your child to complete the unfinished facts.

Left Form

Teacher _____

Date _____

MATH 2   LESSON 40
Oral Assessment # 4 Recording Form

Materials:
10 dimes
10 pennies

Students	A. *"Count backward from 100 by 10's."*	B. *"Count by 5's to 50."*	A. *"Show 13¢."*	B. *"Show 40¢."* C. *"Show 67¢."*		
	A	B	A	B	C	

2-PFq

Right Form

Name _____

Date _____

ASSESSMENT 7
LESSON 40
Math 2

Draw a picture and write a number sentence for this story. Write the answer with a label.

1. Simone helped her brother bake cupcakes. There were 12 cupcakes. They gave five cupcakes to their neighbors. How many cupcakes do they have left?

   Number sentence _____ 12 cupcakes – 5 cupcakes = 7 cupcakes

   Answer _____ 7 cupcakes

2. Show 32 using tally marks.

3. How much money is this? __46¢__

4. Number the clock face. Show half past eight on both clocks.

   8:30

5. Use "The Weekday Wake-up Time" class graph to answer these questions.

   How many children wake up at 7:00? _____

   At what time do most children wake up? _____

   What is the latest time that someone in this class wakes up? _____

6. Find the sums.

   $4 + 5 =$ __9__ $7 + 6 =$ __13__ $8 + 9 =$ __17__ $3 + 4 =$ __7__

2-40Aa

esson 41

naming fractional parts of a whole

---

lesson preparation

materials
fraction pieces from Lesson 34

Fact Sheet A 6.2

in the morning
• Write the following in the pattern box on the meeting strip:

◸, ◺, ◸, ◺, ____, ____, ____, ____, ____

Answer: ◸, ◺, ◸, ◺, ◸, ◺, ◸, ◺, ◸

• Write ┃ 47¢ ┃ on the meeting strip. Provide a cup of 10 dimes and a cup of 10 pennies.

THE MEETING

calendar

> • Ask your child to write the date on the calendar and meeting strip.
>
> • Ask your child the following:
>
>> date _____ days ago, date _____ days from now
>>
>> day of the week _____ days ago, day of the week _____ days from now
>>
>> months of the year, _____th month, month before, month after
>
> • Record on the meeting strip a special event and the number of days until it occurs.

weather graph

> • Ask your child to color the graph and write the temperature to the nearest ten degrees in the box he/she colored.
>
> • Ask questions about the graph.

counting

> • Count by 10's to 200 and backward from 200 by 10's.
>
> • Count by 5's to 100 and backward from 50 by 5's.

- Say the even numbers to 30 and backward from 30.
- Say the odd numbers to 29 and backward from 29.

graph questions

- You and your child each ask a question about any of the graphs.

patterning

- Ask your child to do the following:

  identify the pattern (repeating, continuing, or both)

  identify the shapes to complete the pattern

  read the pattern

money

- Ask your child to put the dimes and pennies in the coin cup.
- Count the money in the coin cup together.

clock

- Ask your child to set the clock on the half hour or hour.
- Ask the following:

  time shown on the clock

  time one hour ago and time one hour from now

- Ask your child to write the digital time on the meeting strip.
- Record on the meeting strip the time an activity will occur.

number of the day

- Write three number sentences for the number of the day on the meeting strip.

THE LESSON

Naming Fractional Parts of a Whole

"Today you will learn how to name fractional parts of a whole."

"A few days ago we cut and tasted apples."

"When we cut our apples, how did we cut them?"

"Today we will pretend that our circle fraction pieces are apples."

- Give your child the circle fraction pieces from Lesson 34.

"Let's pretend that the yellow circle is the whole apple."

"Which color piece can we use to show one half of the apple?"   blue

"We call each blue piece one half."

"Which color piece can we use to show one fourth of the apple?"   red

"We call each red piece one fourth."

"What are two red pieces called?"   two fourths

"What are three red pieces called?"   three fourths

"What are four red pieces called?"   four fourths

"How many green pieces will we need to cover the apple?"   8

"Use the green pieces to cover the apple."

"What is one green piece called?"   one eighth

"What are two green pieces called?"   two eighths

"What are three green pieces called?"   three eighths

• Continue to eight pieces.

"Cover the yellow apple with two red pieces."

"How much of the whole apple is covered?"   two fourths

"What other piece is the same size?"   one half (blue)

"Cover the two fourths with the one half."

"One half of an apple is the same amount as two fourths of an apple."

"Cover the yellow apple with two green pieces."

"How much of the whole apple is covered?"   two eighths

"What other piece is the same size?"   one red piece

"Cover the two eighths with the one fourth."

"One fourth of an apple is the same amount as two eighths of an apple."

"Cover the yellow apple with four green pieces."

"How much of the whole apple is covered?"   four eighths

"What other piece is the same size?"   one blue piece

"Cover the four eighths with the one half."

"One half of an apple is the same amount as four eighths of an apple."

"We will use the fraction pieces again."

"Put them back in the bag."

• Draw the following on the chalkboard:

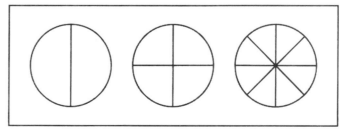

"How many pieces are in my first circle?"   2

"What will we call each piece?"   *one half*

- Shade one half of the circle.

"How much of my circle is shaded?"   *one half*

- Write "one half" below the first circle.

"How many pieces are in my second circle?"   *4*

"What will we call each piece?"   *one fourth*

- Shade two fourths of the circle.

"How much of my circle is shaded?"   *two fourths*

- Write "two fourths" below the second circle.

"How many pieces are in my third circle?"   *8*

- Shade four eighths of the circle.

"What will we call each piece?"   *one eighth*

"How much of my circle is shaded?"   *four eighths*

- Write "four eighths" below the third circle.

- Repeat with different amounts shaded, if desired.

- Save the circle fraction pieces for use in Lesson 65.

CLASS PRACTICE

number fact practice

- Use the fact cards to practice the addition facts with your child.
- Give your child **Fact Sheet A 6.2.**
- Time your child for one minute.
- Correct the fact sheet with your child.
- Record the score.
- Allow time for your child to complete the unfinished facts.

"On the back of the fact sheet, write the numbers from 95 to 120."

WRITTEN PRACTICE

- Complete **Worksheet 41A** with your child.
- Complete **Worksheet 41B** with your child later in the day.

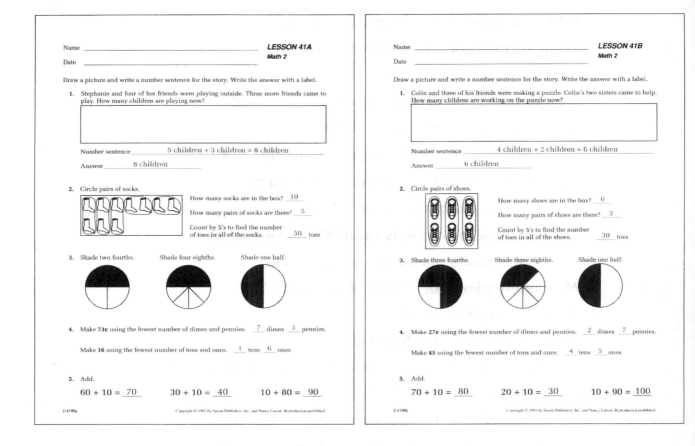

Draw a picture and write a number sentence for the story. Write the answer with a label.

1. Stephanie and four of her friends were playing outside. Three more friends came to play. How many children are playing now?

   Number sentence _____ 5 children + 3 children = 8 children _____

   Answer _____ 8 children _____

2. Circle pairs of socks.

   How many socks are in the box? __10__

   How many pairs of socks are there? __5__

   Count by 5's to find the number of toes in all of the socks. __50__ toes

3. Shade two fourths.   Shade four eighths.   Shade one half.

4. Make 73¢ using the fewest number of dimes and pennies. __7__ dimes __3__ pennies.

   Make 16 using the fewest number of tens and ones. __1__ tens __6__ ones

5. Add.

   60 + 10 = __70__   30 + 10 = __40__   10 + 80 = __90__

Draw a picture and write a number sentence for the story. Write the answer with a label.

1. Colin and three of his friends were making a puzzle. Colin's two sisters came to help. How many children are working on the puzzle now?

   Number sentence _____ 4 children + 2 children = 6 children _____

   Answer _____ 6 children _____

2. Circle pairs of shoes.

   How many shoes are in the box? __6__

   How many pairs of shoes are there? __3__

   Count by 5's to find the number of toes in all of the shoes. __30__ toes

3. Shade three fourths.   Shade three eighths.   Shade one half.

4. Make 27¢ using the fewest number of dimes and pennies. __2__ dimes __7__ pennies.

   Make 45 using the fewest number of tens and ones. __4__ tens __5__ ones

5. Add.

   70 + 10 = __80__   20 + 10 = __30__   10 + 90 = __100__

esson 42

addition facts—adding nine

---

lesson preparation

materials
addition fact cards — yellow

Fact Sheet A 7.1

the night before
• Separate the yellow addition fact cards.

in the morning
• Write the following in the pattern box on the meeting strip:

> # 50, 45, 40, 35, ___, ___, ___, ___, ___, ___

Answer: 50, 45, 40, 35, 30, 25, 20, 15, 10, 5

• Write | 19¢ | on the meeting strip. Provide a cup of 10 dimes and a cup of 10 pennies.

THE MEETING

calendar

• Ask your child to write the date on the calendar and meeting strip.

• Ask your child the following:

> date _____ days ago, date _____ days from now
>
> day of the week _____ days ago, day of the week _____ days from now
>
> months of the year, _____th month, month before, month after

• Record on the meeting strip a special event and the number of days until it occurs.

weather graph

• Ask your child to color the graph and write the temperature to the nearest ten degrees in the box he/she colored.

• Ask questions about the graph.

---

counting

- Count by 10's to 200 and backward from 200 by 10's.
- Count by 5's to 100 and backward from 50 by 5's.
- Say the even numbers to 30 and backward from 30.
- Say the odd numbers to 29 and backward from 29.

graph questions

- You and your child each ask a question about any of the graphs.

patterning

- Ask your child to do the following:

    identify the pattern (repeating, continuing, or both)

    identify the numbers to complete the pattern

    read the pattern

money

- Ask your child to put the dimes and pennies in the coin cup.
- Count the money in the coin cup together.

clock

- Ask your child to set the clock on the half hour or hour.
- Ask the following:

    time shown on the clock

    time one hour ago and time one hour from now

- Ask your child to write the digital time on the meeting strip.
- Record on the meeting strip the time an activity will occur.

number of the day

- Write three number sentences for the number of the day on the meeting strip.

THE LESSON

Addition Facts—Adding Nine

"We have been adding ten to a number."

"Today you will learn how to add nine to a number."

- Write the following on the chalkboard:

8 + 10	3 + 10
6 + 10	4 + 10
5 + 10	9 + 10
7 + 10	1 + 10
2 + 10	

"What is the same about all of these problems?"   adding ten

"What is the answer when we add ten to each of these numbers?"

- Record each answer next to the problem.

"I changed my mind. I don't want to add ten, I only want to add nine. How will that change each answer?"

- Write the following next to each of the previous problems:

8 + 9	3 + 9
6 + 9	4 + 9
5 + 9	9 + 9
7 + 9	1 + 9
2 + 9	

"These answers are almost the same as adding ten except the answers are one less."

"To find the answers, we will add ten and then count back or subtract one."

- Record each answer next to the problem

- Write the following on the chalkboard:

$$
\begin{array}{ccccccccc}
1 & 2 & 3 & 4 & 5 & 6 & 7 & 8 & 9 \\
+10 & +10 & +10 & +10 & -10 & +10 & +10 & +10 & +10
\end{array}
$$

$$
\begin{array}{ccccccccc}
1 & 2 & 3 & 4 & 5 & 6 & 7 & 8 & 9 \\
+9 & +9 & +9 & +9 & +9 & +9 & +9 & +9 & +9
\end{array}
$$

"We can use adding ten to help us find the answers for our adding nine facts."

"When we are adding nine to a number, we can add ten and then count back or subtract one."

"Tell me the answers as I write them on the chalkboard."

- Write the answers on the chalkboard.

"What do you notice about all of the answers when we add nine?"

"Now I'll say the problems in random order."

"Let's see how fast you can say the answers."

- Allow your child to refer to the chalkboard problems, if necessary.
- Read problems in random order.
- Erase the answers for the adding nine facts only. (Do not erase the adding ten problems or answers.)
- Continue giving problems in random order.

  "Now I will say the answer to an adding nine problem."

  "See if you can tell me the problem."

  "Eleven."   two plus nine

- Repeat with all problems.
- If your child is having difficulty, rewrite the answers on the chalkboard.
- Give your child the yellow fact cards.

  "Use the fact cards to practice the adding nine facts."

- Allow time for your child to do this.

CLASS PRACTICE

number fact practice

- Give your child **Fact Sheet A 7.1.**
- Time your child for one minute.
- Correct the fact sheet with your child and record the score.
- Allow time for your child to complete the unfinished facts.

  "On the back of the fact sheet, write the numbers from 95 to 120."

WRITTEN PRACTICE

- Complete **Worksheet 42A** with your child.
- Complete **Worksheet 42B** with your child later in the day.

LESSON 42A

Name _____
Date _____

Math 2

Draw a picture and write a number sentence for the story. Write the answer with a label.

1. Marcus had 40¢. His mother gave him 10¢. How much money does Marcus have now?

   [ ]

   Number sentence _____ 40¢ + 10¢ = 50¢ _____

   Answer _____ 50¢ _____

2. Shade one fourth.   Shade three sixths.   Shade two eighths.

3. Use the favorite apple graph to answer these questions.

   How many children voted? _____

   What was the favorite apple? _____

   How many children chose that apple? _____

4. I have 3 dimes and 4 pennies. How much money is that? __34¢__

   I have 5 tens and 6 ones. How much is that? __56__

5. Find each sum.

10	9		10	9		5	5
+ 4	+ 4		+ 7	+ 7		+ 10	+ 9
14	13		17	16		15	14

LESSON 42B

Name _____
Date _____

Math 2

Draw a picture and write a number sentence for the story. Write the answer with a label.

1. Althea had 60¢. She spent 10¢. How much money does she have now?

   [ ]

   Number sentence _____ 60¢ − 10¢ = 50¢ _____

   Answer _____ 50¢ _____

2. Shade three fourths.   Shade one half.   Shade six eighths.

3. Fill in the missing numbers.

   5, 10, 15, __20__ , __25__ , __30__ , __35__ , __40__ , __45__ , __50__

   1, 3, 5, __7__ , __9__ , __11__ , __13__ , __15__ , __17__ , __19__

4. I have 2 dimes and 5 pennies. How much money is that? __25¢__

   I have 9 tens and 7 ones. How much is that? __97__

5. Find each sum.

10	9		6	6		10	9
+ 3	+ 3		+ 10	+ 9		+ 4	+4
13	12		16	15		14	13

Lesson 43

trading pennies for dimes

lesson preparation

materials

cup of 20 pennies

cup of 10 dimes

construction paper work mat

Fact Sheet A 6.2

in the morning

Note: This lesson may take two days.

• Write the following in the pattern box on the meeting strip:

| 11, ___, 13, ___, 15, ___, 17, ___, 19, ___, 21 |

Answer: 11, 12, 13, 14, 15, 16, 17, 18, 19, 20, 21

• Write ⟨27¢⟩ on the meeting strip. Provide a cup of 10 dimes and a cup of 10 pennies.

THE MEETING

calendar

- • Ask your child to write the date on the calendar and meeting strip.
- • Ask your child the following:

  date _____ days ago, date _____ days from now

  day of the week _____ days ago, day of the week _____ days from now

  months of the year, _____th month, month before, month after

- • Record on the meeting strip a special event and the number of days until it occurs.

weather graph

- • Ask your child to color the graph and write the temperature to the nearest ten degrees in the box he/she colored.
- • Ask questions about the graph.

counting

- Count by 10's to 200 and backward from 200 by 10's.
- Count by 5's to 100 and backward from 50 by 5's.
- Say the even numbers to 30 and backward from 30.
- Say the odd numbers to 29 and backward from 29.

graph questions

- You and your child each ask a question about any of the graphs.

patterning

- Ask your child to do the following:

  identify the pattern (repeating, continuing, or both)

  identify the numbers to complete the pattern

  read the pattern

money

- Ask your child to put the dimes and pennies in the coin cup.
- Count the money in the coin cup together.

clock

- Ask your child to set the clock on the half hour or hour.
- Ask the following:

  time shown on the clock

  time one hour ago and time one hour from now
- Ask your child to write the digital time on the meeting strip.
- Record on the meeting strip the time an activity will occur.

number of the day

- Write three number sentences for the number of the day on the meeting strip.

THE LESSON

Trading Pennies for Dimes

"Today you will learn how to trade pennies for dimes."

- Give your child a cup of 20 pennies, a cup of 10 dimes, and the work mat.

"Show 14¢ using just pennies."

"How many pennies did you use?"   14

"How many pennies can we trade for a dime?"   10

"Trade ten pennies for a dime."

"How many dimes and pennies do you have now?"   1 dime, 4 pennies

"Fourteen pennies is the same as one dime and four pennies."

- Write "14 pennies = 1 dime + 4 pennies" on the chalkboard.

- Repeat with 18 pennies and 12 pennies.

"Show 20¢ using just pennies."

"How many pennies did you use?"   20

"Trade the pennies for dimes."

"How many dimes and pennies do you have now?"   2 dimes, 0 pennies

"Twenty pennies is the same as two dimes and zero pennies."

- Write "20 pennies = 2 dimes + 0 pennies" on the chalkboard.

"If we had 36 pennies and we traded the pennies for dimes, how many dimes and pennies would we have?"

- Write "36 pennies = 3 dimes + 6 pennies" on the chalkboard.

"What digit in the number 36 tells us the number of dimes?"   3

"What digit in the number 36 tells us the number of pennies?"   6

"If we had 52 pennies and we traded the pennies for dimes, how many dimes and pennies would we have?"

- Write "52 pennies = 5 dimes + 2 pennies" on the chalkboard.

"What digit in the number 52 tells us the number of dimes?"   5

"What digit in the number 52 tells us the number of pennies?"   2

- Write "64¢" on the chalkboard.

"How many dimes and pennies will we need to make 64¢?"   6 dimes, 4 pennies

"Which digit tells us the number of dimes?"

"Which digit tells us the number of pennies?"

"Put 3 dimes and 17 pennies on your mat."

"Count the money."

"How much money is this?"   47¢

"Trade ten pennies for a dime."

"How many dimes and pennies do you have now?"   4 dimes, 7 pennies

"How much money is this?"   47¢

- Write "3 dimes + 17 pennies = 4 dimes + 7 pennies = 47¢" on the chalkboard.

"Put seven dimes and twelve pennies on your mat."

"Count the money."

"How much money is this?" 82¢

"Trade ten pennies for a dime."

"How many dimes and pennies do you have now?" 8 dimes, 2 pennies

"How much money is this?" 82¢

- Write "7 dimes + 12 pennies = 8 dimes + 2 pennies = 82¢" on the chalkboard.

- Repeat with 4 dimes and 16 pennies if your child needs additional practice.

"Put the dimes in one cup and the pennies in the other cup."

"Now you will have a chance to practice some problems."

"I will write a problem on the chalkboard."

"You will write the answer using the fewest possible number of pennies."

- Write each of the following on the chalkboard one at a time. Allow your child to check his/her answer using the dimes and pennies, if necessary.

  24 pennies = _____ dimes + _____ pennies

  30 pennies = _____ dimes + _____ pennies

  79 pennies = _____ dimes + _____ pennies

  2 dimes + 18 pennies = _____ dimes + _____ pennies = _____ ¢

  7 dimes – 14 pennies = _____ dimes + _____ pennies = _____ ¢

  5 dimes – 19 pennies = _____ dimes + _____ pennies = _____ ¢

CLASS PRACTICE

number fact practice

- Use the fact cards to practice the addition facts with your child.

- Give your child **Fact Sheet A 6.2.**

- Time your child for one minute.

- Correct the fact sheet with your child and record the score.

- Allow time for your child to complete the unfinished facts.

"On the back of the fact sheet, write the numbers from 142 to 167."

WRITTEN PRACTICE

- Complete **Worksheet 43A** with your child.

- Complete **Worksheet 43B** with your child later in the day.

LESSON 43A
Math 2

Name _____

Date _____

Draw a picture and write a number sentence for this story. Write the answer with a label.

1. Anna has 1 ruler, 5 pencils, 2 notebooks, and 4 markers. How many of these things can she use to write with?

   [blank box]

   Number sentence _____ 5 things + 4 things = 9 things _____

   Answer _____ 9 things _____

2. Draw a picture of the favorite apple graph your class made.   answers may vary

   _____
   _____
   _____

   [grid]

   What apple was chosen by the fewest number of children? _____

   How many children chose that apple? _____

3. How much money is 2 pennies and 6 dimes? _____ 62¢

   How much is 5 ones and 3 tens? _____ 35

4. Use the fewest number of dimes and pennies.

   3 dimes and 14 pennies = __4__ dimes + __4__ pennies = __44__ ¢

5. Fill in the missing numbers on this piece of a hundred number chart.

43	44	45	46	47
53	54	55	56	57
63	64	65	66	67

LESSON 43B
Math 2

Name _____

Date _____

Draw a picture and write a number sentence for this story. Write the answer with a label.

1. Mike counted three cans of corn, four jars of beans, a package of napkins, and two jars of plant food on the shelf. How many containers of food are safe to eat?

   [blank box]

   Number sentence _____ 3 containers + 4 containers = 7 containers _____

   Answer _____ 7 containers _____

2. Fill in the missing addends.

   $\boxed{2} + 8 = 10$ $\boxed{5} + 4 = 9$ $\boxed{6} + 6 = 12$

   $5 + \boxed{2} = 7$ $3 + \boxed{0} = 3$ $7 + \boxed{1} = 8$

3. How much money is 3 pennies and 9 dimes? _____ 93¢

   How much is 6 ones and 2 tens? _____ 26

4. Use the fewest number of dimes and pennies.

   5 dimes and 12 pennies = __6__ dimes + __2__ pennies = __62__ ¢

5. Fill in the missing numbers on this piece of a hundred number chart.

72	73	74	75	76	77	78	79
82	83	84	85	86	87	88	89
92	93	94	95	96	97	98	99

Lesson 44

weighing objects using nonstandard units

---

lesson preparation

materials

balance (scale)

pencil, scissors, marker

color tiles

Master 2-44

Fact Sheet A 7.1

in the morning

• Write the following in the pattern box on the meeting strip:

◺, ◹, ◺, ◹, ◺, _____, _____, _____, _____, _____

Answer: ◺, ◹, ◺, ◹, ◺, ◹, ◺, ◹, ◺, ◹

• Write 65¢ on the meeting strip. Provide a cup of 10 dimes and a cup of 20 pennies.

THE MEETING

calendar

• Ask your child to write the date on the calendar and meeting strip.

• Ask your child the following:

   date _____ days ago, date _____ days from now

   day of the week _____ days ago, day of the week _____ days from now

   months of the year, _____th month, month before, month after

• Record on the meeting strip a special event and the number of days until it occurs.

weather graph

• Ask your child to color the graph and write the temperature to the nearest ten degrees in the box he/she colored.

• Ask questions about the graph.

counting

- Count by 10's to 200 and backward from 200 by 10's.
- Count by 5's to 100 and backward from 50 by 5's.
- Say the even numbers to 30 and backward from 30.
- Say the odd numbers to 29 and backward from 29.

graph questions

- You and your child each ask a question about any of the graphs.

patterning

- Ask your child to do the following:

     identify the pattern (repeating, continuing, or both)

     identify the shapes to complete the pattern

     read the pattern

money

- Ask your child to put the dimes and pennies in the coin cup.
- Count the money in the coin cup together.

clock

- Ask your child to set the clock on the half hour or hour.
- Ask the following:

     time shown on the clock

     time one hour ago and time one hour from now

- Ask your child to write the digital time on the meeting strip.
- Record on the meeting strip the time an activity will occur.

number of the day

- Write three number sentences for the number of the day on the meeting strip.

THE LESSON

Weighing Objects Using Nonstandard Units

"Today you will learn how to compare and weigh objects using a balance."

"What do you notice about the balance?"

- Allow time for your child to offer observations about the balance.

"If we put a color tile on each side of the balance, what do you think is going to happen?" the balance will be level

- Put a color tile on each side of the balance.

  "What happened?"

  "If we put an extra color tile on the right side, what do you think is going to happen?"

- Put one more color tile on the right side.

  "What happened?" the right side went down

- Show your child a pencil and a pair of scissors.

  "Hold these objects."

  "Which feels heavier?"

  "How can we use the balance to check?"

- Put the pencil in one pan of the balance and the scissors in the other pan.

  "What happened?"

  "Which object is heavier?"

- Remove the pencil and scissors from the balance.

  "Let's use color tiles to weigh these objects."

  "How many color tiles do you think we will need to balance the pencil?"

- Write the following on the chalkboard:

	estimate	actual
pencil		
scissors		

  "Let's try it to see."

  "Put the pencil on the left side of the balance."

  "Now add color tiles to the right side one at a time until the scale is level."

  "Count the number of tiles you use."

  "The pencil weighs the same as _____ color tiles."

- Record the amount on the chalkboard chart.

- Remove the pencil and the tiles from the balance.

  "Let's try weighing the scissors using the color tiles."

  "Do you think we will use more tiles or fewer tiles than we used to balance the pencil?"

  "Why?"

  "How many tiles do you think we will use?"

- Record the estimate on the chalkboard chart.

"Put the scissors on the left side of the balance."

"Now add color tiles to the right side one at a time until the scale is level."

"Count the number of tiles you use."

"The scissors weigh the same as _____ color tiles."

- Record the amount on the chalkboard chart.

"Let's compare a marker to the scissors."

"How will we do that?"

- Put the marker on one side of the balance and the scissors on the other side.

"Which is heavier?"

"If the scissors weighed _____ color tiles, how much do you think the marker will weigh?"

- Record the estimate on the chalkboard chart.

"Put the marker on the left side of the balance."

"Now add color tiles to the right side one at a time until the scale is level."

"Count the number of tiles you use."

"The marker weighs the same as _____ color tiles."

"Now you will have a chance to weigh some more objects using color tiles."

- Give your child **Master 2-44**.

"Choose an object in our house to weigh."

"Estimate how many color tiles you will need to make the scale level."

"Write your estimate next to the name of the object on Master 2-44."

"Put the object in one pan of the balance."

"Put the color tiles on the other side, one at a time, until the scale is level."

"Record the number of color tiles you used in the last column."

"When you finish, choose four more objects to weigh."

CLASS PRACTICE

number fact practice

- Practice the adding nine facts together.
- Give your child **Fact Sheet A 7.1**.

- Time your child for one minute.

- Correct the fact sheet with your child and record the score.

- Allow time for your child to complete the unfinished facts.

   "On the back of the fact sheet, write the numbers from 95 to 120."

WRITTEN PRACTICE

- Complete **Worksheet 44A** with your child.

- Complete **Worksheet 44B** with your child later in the day.

Name _____ **MASTER 2-44**
                                    Math 2

Object Weighed	Estimate	Actual

answers may vary

2-44Ma   Copyright © 1991 by Saxon Publishers, Inc. and Nancy Larson. Reproduction prohibited.

Name _____ **LESSON 44A**
                                    Math 2
Date _____

Write a number sentence for the story. Write the answer with a label.

1. On Fridays, the children can buy pizza or hot dogs for lunch. Last Friday, eighty children bought pizza and ten children bought hot dogs. How many children bought lunch last Friday?

   Number sentence ____ 80 children + 10 children = 90 children

   Answer ____ 90 children

2. Divide the square in half using a vertical line segment.   Divide the square in half using a horizontal line segment.   Divide the square in half using an oblique line segment.

   Shade one half of each square.
                      answers may vary

3. How much money is 2 dimes and 6 pennies? 26¢

   How much money is 2 pennies and 6 dimes? (62¢)

   Circle the one that is more.

4. Find each sum.

   7 + 8 = 15   5 + 6 = 11   9 + 8 17

   4 + 9 = 13   5 + 9 = 14   9 + 6 15

5. Count by 5's. Write the numbers.

   5, 10 , 15 , 20 , 25 , 30 , 35 , 40 , 45 , 50

2-44Wa   Copyright © 1991 by Saxon Publishers, Inc. and Nancy Larson. Reproduction prohibited.

Name _____ **LESSON 44B**
                                    Math 2
Date _____

Write a number sentence for the story. Write the answer with a label.

1. Crystal read the first 40 pages of the book on Saturday. On Sunday, she read 10 more pages. How many pages did she read altogether?

   Number sentence ____ 40 pages + 10 pages = 50 pages

   Answer ____ 50 pages

2. Divide the circle in half using a vertical line segment.   Divide the circle in half using a horizontal line segment.   Divide the circle in half using an oblique line segment.

   Shade one half of each circle.
                      answers may vary

3. How much money is 3 dimes and 5 pennies? 35¢

   How much money is 3 pennies and 5 dimes? (53¢)

   Circle the one that is more.

4. Find each sum.

   3 + 4 = 7   6 + 7 = 13   5 + 6 11

   3 + 9 = 12   5 + 9 = 14   9 + 7 16

5. Count backwards by 5's. Write the numbers.

   50, 45 , 40 , 35 , 30 , 25 , 20 , 15 , 10 , 5

2-44Wb   Copyright © 1991 by Saxon Publishers, Inc. and Nancy Larson. Reproduction prohibited.

Lesson 45

subtracting half of a double

---

lesson preparation

materials

Written Assessment #8

balance

20 color tiles

Fact Sheet S 1.2

in the morning

• Write the following in the pattern box on the meeting strip:

↓ , ↑ , → , ← , ↓ , ↑ , → , ___ , ___ , ___ , ___ , ___

Answer: ↓, ↑, →, ←, ↓, ↑, →, ←, ↓, ↑, →, ←

• Write [81¢] on the meeting strip. Provide a cup of 10 dimes and a cup of 20 pennies.

THE MEETING

calendar

• Ask your child to write the date on the calendar and meeting strip.

• Ask your child the following:

> date _____ days ago, date _____ days from now
>
> day of the week _____ days ago, day of the week _____ days from now
>
> months of the year, _____th month, month before, month after

• Record on the meeting strip a special event and the number of days until it occurs.

weather graph

• Ask your child to color the graph and write the temperature to the nearest ten degrees in the box he/she colored.

• Ask questions about the graph.

counting

• Count by 10's to 200 and backward from 200 by 10's.

- Count by 5's to 100 and backward from 50 by 5's.
- Say the even numbers to 30 and backward from 30.
- Say the odd numbers to 29 and backward from 29.

graph questions

- You and your child each ask a question about any of the graphs.

patterning

- Ask your child to do the following:

    identify the pattern (repeating, continuing, or both)

    identify the shapes to complete the pattern

    read the pattern

money

- Ask your child to put the dimes and pennies in the coin cup.
- Count the money in the coin cup together.

clock

- Ask your child to set the clock on the half hour or hour.
- Ask the following:

    time shown on the clock

    time one hour ago and time one hour from now

- Ask your child to write the digital time on the meeting strip.
- Record on the meeting strip the time an activity will occur.

number of the day

- Write three number sentences for the number of the day on the meeting strip.

Assessment

Written Assessment

"Today I would like to see what you remember from what we have been practicing."

- Give your child **Written Assessment #8.**
- Read the directions for each problem. Allow time for your child to complete that problem before continuing.
- Correct the paper, noting your child's mistakes on the **Individual Recording Form.** Review the errors with your child.

THE LESSON

Subtracting Half of a Double

"Today you will learn how to subtract half of a double."

- Point to the balance.

  "What do we call this?"   balance

  "What do we use it for?"   to compare the weights of objects; to weigh things

  "How can we use it to tell if objects have the same weight?"   if the balance is level, the objects have the same weight

  "I have two color tiles."

  "How many should I put in each side to make the balance level?"

- Put the tiles on the balance.

- Record on the chalkboard

  "I have four color tiles."

  "How many should I put in each side to make the balance level?"

- Put the tiles on the balance.

- Record on the chalkboard:

  "I have six color tiles."

  "How many should I put in each side to make the balance level?"

- Put the tiles on the balance.

- Record on the chalkboard:

- Continue with 8, 10, 12, 14, 16, 18, and 20 color tiles, if necessary.

  "What do you notice?"   these are the doubles

  "If we had 40 color tiles, how many would we put on each side to make the balance level?"

  "How do you know?"

  "If we had 100 color tiles, how many would we put on each side to make the balance level?"

"How do you know?"

"When we want to make the balance level, we put the same number of color tiles on each side."

"We put half of the color tiles on one side of the balance and half on the other side of the balance."

"If we had 24 color tiles, how many color tiles would we put on each side?"

"Half of 24 is 12."

"Will the balance be level if we have seven color tiles to use?"   no

- Show this using the balance.

- Repeat with 3, 9, 13, and 5 color tiles.

"Why won't these tiles make the balance level?"   there is an odd *number of color tiles; there is one left over*

"When we divide a set of objects in half, both halves must have the same number of objects."

"I have six pennies."

"If I give half of the pennies to you, how many will I give away?"

"How many will I have left?"

"I have ten pennies."

"If I give half of the pennies to you, how many will I give away?"

"How many will I have left?"

"Let's make up some problems about giving away half."

- Write the following on the chalkboard:

2   4   6   8   10   12   14   16   18   20

"If I have two pennies and give half away, how many do I have left?"

- Record: 2
             -1
             1

"If I have four pennies and give half away, how many do I have left?"

- Record: 4
             -2
             2

"If I have six pennies and give half away, how many do I have left?"

- Record: 6
             -3
             3

- Repeat with 8, 10, 12, 14, 16, 18, and 20 pennies.

"What do you notice about each problem?"

• Allow time for your child to offer observations about the problems.

"We will call these problems 'subtracting half of a double facts.' "

"Why do you think we will call them that?"

"There is always a way to check subtraction."

"We can check subtraction by adding up."

• Demonstrate.

"Today you will practice the subtracting half of a double facts."

"Take out your tan addition fact cards."

"What do we call these facts?"   doubles

"What do you notice about the facts on the other side?"   they are subtracting half of a double facts

"Practice these subtraction number facts using your fact cards."

CLASS PRACTICE

number fact practice

• Give your child **Fact Sheet S 1.2.**

• Time your child for one minute.

• Correct the fact sheet with your child and record the score.

• Allow time for your child to complete the unfinished facts.

WRITTEN PRACTICE

• Complete **Worksheet 45A** with your child.

• Complete **Worksheet 45B** with your child later in the day.

Name _____

Date _____

Draw a picture and write a number sentence for this story. Write the answer with a label.

1. Steven has four dimes. Mark has two dimes. How many dimes do they have together?

   [ ]

   Number sentence _____ 4 dimes + 2 dimes = 6 dimes _____

   Answer _____ 6 dimes _____

   How much money is that? __60¢__

2. I have 7 dimes and 2 pennies. How much money is that? __72¢__

   I have 3 tens and 8 ones. How much is that? __38__

3. Count by 10's.

   _10_, _20_, _30_, _40_, _50_, _60_, _70_, _80_, _90_, _100_

   Count by 5's.

   _5_, _10_, _15_, _20_, _25_, _30_, _35_, _40_, _45_, _50_

4. Color the circle divided into fourths red.

   Color the circle divided into halves blue.

   Color the circle divided into eighths green.

   green   red   blue

5. Find the sums.

   40 + 10 = __50__   10 + 80 = __90__   10 + 70 = __80__

6. Fill in the missing addends.

   [3] + 7 = 10   4 + [6] = 10   [8] + 2 = 10

Name _____

Date _____

Draw a picture and write a number sentence for this story. Write the answer with a label.

1. Jay counted nine sharpened pencils and six unsharpened pencils in the pencil can. How many pencils are in the can?

   [ ]

   Number sentence _____ 9 pencils + 6 pencils = 15 pencils _____

   Answer _____ 15 pencils _____

2. How many mittens are in the box? __8__
   Circle pairs of mittens.

   How many pairs of mittens are there? __4__

   Count by 5's to find the number of fingers in all the mittens.

   __40__ fingers

3. What is an even number greater than 6? _____

   What is an odd number less than 6? _____   answers may vary

4. Which number on the thermometer is the temperature closest to? __20__ °F

5. How many dimes are there? __6__   How many pennies are there? __4__

   How much money is this? __64¢__

Name _____

Date _____

Draw a picture and write a number sentence for this story. Write the answer with a label.

1. Ferma has a set of 10 markers. She threw away 3 markers because they were dry. How many markers does she have now?

   [ ]

   Number sentence _____ 10 markers – 3 markers = 7 markers _____

   Answer _____ 7 markers _____

2. How many gloves are in the box? __6__
   Circle pairs of gloves.

   How many pairs of gloves are there? __3__

   Count by 5's to find the number of fingers in all the gloves.

   __30__ fingers

3. What is an even number less than 6? _____

   What is an odd number greater than 6? _____   answers may vary

4. Which number on the thermometer is the temperature closest to? __40__ °F

5. How many dimes are there? __5__   How many pennies are there? __6__

   How much money is this? __56¢__

Lesson 46

measuring to the nearest inch using a ruler

---

lesson preparation

materials

ruler

yardstick

tape measure (optional)

one color tile

2" × 8" construction paper rectangle

3" × 9" construction paper rectangle

Master 2-46

Fact Sheet A 7.1

in the morning

• Write the following in the pattern box on the meeting strip:

> 5, __, 15, __, 25, __, 35, __, 45

Answer: 5, 10, 15, 20, 25, 30, 35, 40, 45

• Write 93¢ on the meeting strip. Provide a cup of 10 dimes and a cup of 20 pennies.

THE MEETING

calendar

• Ask your child to write the date on the calendar and meeting strip.

• Ask your child the following:

   date _____ days ago, date _____ days from now

   day of the week _____ days ago, day of the week _____ days from now

   months of the year, _____th month, month before, month after

• Record on the meeting strip a special event and the number of days until it occurs.

weather graph

- Ask your child to color the graph and write the temperature to the nearest ten degrees in the box he/she colored.
- Ask questions about the graph.

counting

- Count by 10's to 200 and backward from 200 by 10's.
- Count by 5's to 100 and backward from 50 by 5's.
- Say the even numbers to 30 and backward from 30.
- Say the odd numbers to 29 and backward from 29.

graph questions

- You and your child each ask a question about any of the graphs.

patterning

- Ask your child to do the following:

    identify the pattern (repeating, continuing, or both)

    identify the numbers to complete the pattern

    read the pattern

money

- Ask your child to put the dimes and pennies in the coin cup.
- Count the money in the coin cup together.

clock

- Ask your child to set the clock on the half hour or hour.
- Ask the following:

    time shown on the clock

    time one hour ago and time one hour from now

- Ask your child to write the digital time on the meeting strip.
- Record on the meeting strip the time an activity will occur.

number of the day

- Write three number sentences for the number of the day on the meeting strip.

THE LESSON

Measuring to the Nearest Inch Using a Ruler

"Last week we measured using inch tiles."

"What did we measure?"   book, paper, ruler, line segments

"It would be difficult for us to carry inch tiles to measure things."

"Instead, we will use a ruler, a yardstick (meterstick), or a tape measure."

• If possible, show your child each of these.

"These are tools we use to tell us how long something is."

"Today you will learn how to measure to the nearest inch using a ruler."

• Hold up an inch tile and a ruler.

"How many inch tiles long is this ruler?"

• Give your child a 12 inch/centimeter ruler and an inch tile.

"Which side of the ruler do you think is the inch side?"

"How do you know?"

"Let's check it by using an inch tile."

"Does the inch tile fit between two numbers on one side of the ruler?"

"This is the inch side of the ruler."

"How can we tell, without using tiles, that the ruler is 12 inches long?"

• Show your child a yardstick.

"How many inches long do you think a yardstick is?"

"How do you know?"   numbers on the yardstick

"We can use the ruler, yardstick, or tape measure to measure things to the nearest inch."

"Why do you think it is easier to measure using a ruler than using color tiles?"

• Show your child the 2" × 8" construction paper rectangle.

"Let's measure how many inches long this piece of paper is, using a ruler."

• Demonstrate as you say the following:

"First I will put the inch side of the ruler next to the bottom edge of the rectangle."

"I will put the beginning of the ruler (the 0 on the ruler) at the left edge of the rectangle."

"Now I will read the number on the ruler beneath the right edge of the rectangle."

"The rectangle is eight inches long."

"We can write eight inches like this."

• Write "8 inches = 8" " on the chalkboard.

"We can use this symbol instead of the word inch."

• Repeat with the 2" side.

• Give your child the 9" × 3" construction paper rectangle.

"Use your ruler to measure the length of this rectangle."

"What will you have to remember when you measure?" put the inch side of the ruler next to the bottom edge of the paper; put the beginning of the ruler (the 0 on the ruler) at the left edge of the paper; read the number on the ruler beneath the right edge of the paper

• Repeat with the 3" side.

"Now you will have a chance to measure some line segments to the nearest inch."

• Give your child **Master 2-46**.

"Use your ruler to measure each line segment to the nearest inch."

"Write the length next to each segment."

CLASS PRACTICE

number fact practice

• Practice the adding nine facts together.

• Give your child **Fact Sheet A 7.1**.

• Time your child for one minute.

• Correct the fact sheet with your child and record the score.

• Allow time for your child to complete the unfinished facts.

"On the back of the fact sheet, write the numbers from 180 to 210."

WRITTEN PRACTICE

• Complete **Worksheet 46A** with your child.

• Complete **Worksheet 46B** with your child later in the day.

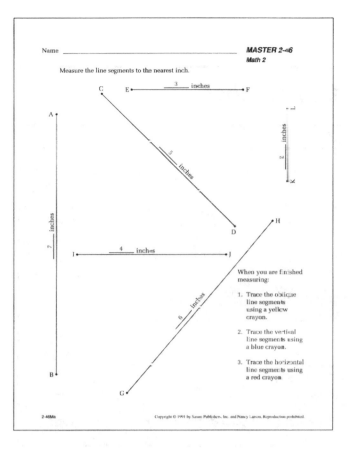

Measure the line segments to the nearest inch.

C • E •———— 3 ————• F inches

A •

inches

5 inches

inches

L

K

H

D •

I •———— 4 ————• J inches

6 inches

When you are finished measuring:

1. Trace the oblique line segments using a yellow crayon.

2. Trace the vertical line segments using a blue crayon.

3. Trace the horizontal line segments using a red crayon.

B •

G •

Name _____   **LESSON 46A**
Date _____   **Math 2**

Draw a picture and write a number sentence for this story. Write the answer with a label.

1. Amy's dog had 8 puppies. She gave five puppies to her friends and one to her grandmother. How many puppies does Amy have now?

   [ ]

   Number sentence _____ 8 puppies – 6 puppies = 2 puppies
   Answer _____ 2 puppies

2. Measure this line segment using inches.

   •———————————————•   5

3. Fill in the missing numbers. Use the fewest possible number of pennies.

   3 dimes and 16 pennies = __4__ dimes + __6__ pennies = __46__ ¢

4. I have 4 pennies and 9 dimes. How much money is that? __94¢__

   I have 6 ones and 7 tens. How much is that? __76__

5. Write the digital time.   What time will it be one hour later?
                                      Show the time on the clocks.

   [3:30]   [4:30]

Name _____   **LESSON 46B**
Date _____   **Math 2**

Draw a picture and write a number sentence for this story. Write the answer with a label.

1. Beth's cat had 5 kittens. She gave 2 kittens to her teacher and one to her aunt. How many kittens does Beth have now?

   [ ]

   Number sentence _____ 5 kittens – 3 kittens = 2 kittens
   Answer _____ 2 kittens

2. This is a six-inch line segment. Find something at home that is 6" long.

   •———————————————————•

   What did you find? _____ answers may vary

3. Fill in the missing numbers. Use the fewest possible number of pennies.

   5 dimes and 12 pennies = __6__ dimes + __2__ pennies = __62__ ¢

4. I have 9 pennies and 4 dimes. How much money is that? __49¢__

   I have 7 ones and 6 tens. How much is that? __67__

5. Write the digital time.   What time will it be one hour later?
                                      Show the time on the clocks.

   [7:30]   [8:30]

 esson 47

adding ten to a two-digit number

---

lesson preparation

materials

hundred number chart

Fact Sheet A 6.2

in the morning

- Write the following in the pattern box on the meeting strip:

> # A, D, G, J, __, __, __, __, __

Answer: A, D, G, J, M, P, S, V, Y

- Write 78¢ on the meeting strip. Provide a cup of 10 dimes and a cup of 20 pennies.

THE MEETING

calendar

- Ask your child to write the date on the calendar and meeting strip.
- Ask your child the following:

  date _____ days ago, date _____ days from now

  day of the week _____ days ago, day of the week _____ days from now

  months of the year, _____th month, month before, month after

- Record on the meeting strip a special event and the number of days until it occurs.

weather graph

- Ask your child to color the graph and write the temperature to the nearest ten degrees in the box he/she colored.
- Ask questions about the graph.

counting

- Count by 10's to 200 and backward from 200 by 10's.
- Count by 5's to 100 and backward from 50 by 5's.

- Say the even numbers to 30 and backward from 30.
- Say the odd numbers to 29 and backward from 29.

graph questions

- You and your child each ask a question about any of the graphs.

patterning

- Ask your child to do the following:

    identify the pattern (repeating, continuing, or both)

    identify the letters to complete the pattern

    read the pattern

money

- Ask your child to put the dimes and pennies in the coin cup.
- Count the money in the coin cup together.

clock

- Ask your child to set the clock on the half hour or hour.
- Ask the following:

    time shown on the clock

    time one hour ago and time one hour from now

- Ask your child to write the digital time on the meeting strip.
- Record on the meeting strip the time an activity will occur.

number of the day

- Write three number sentences for the number of the day on the meeting strip.

THE LESSON

Adding Ten to a Two-Digit Number

"We have been adding ten to numbers like 20, 40, 70, 10, and 50."

- Write the following problems on the chalkboard:

    10 + 40 =   20 + 10 =   80 + 10 =

"What are these answers?"

"How do you know?"

"Today you will learn how to add ten to any two-digit number."

- Give your child a hundred number chart.

- Write "27 + 10" on the chalkboard.

  "We will find answers to problems such as 27 plus 10."

  "What do you think this answer will be?"

  "Let's check the answer using the hundred number chart."

  "How will we do this?"   *find 27 on the hundred number chart and count ten more*

- Record: 27 + 10 = 37

- Write "64 + 10" on the chalkboard.

  "What do you think this answer will be?"

  "Let's check the answer using the hundred number chart."

  "How will we do this?"

- Record: 64 + 10 = 74

- Write "42 + 10" on the chalkboard.

  "What do you think this answer will be?"

  "Let's check the answer using the hundred number chart."

- Record: 42 + 10 = 52

  "What happens when we add ten to a two-digit number?"   *the tens' digit becomes one more, the ones' digit stays the same*

- Write the following problems on the chalkboard:

  75 + 10 =   59 + 10 =   83 + 10 =   31 + 10 =

  "What is the answer to each of these problems?"

  "How do you know?"

  "Which digits tell us the number of tens?"

  "Which digits tell us the number of ones?"

CLASS PRACTICE

number fact practice

- Use the fact cards to practice the addition facts with your child.
- Give your child **Fact Sheet A 6.2.**
- Time your child for one minute.
- Correct the fact sheet with your child and record the score.
- Allow time for your child to complete the unfinished facts.

  "On the back of the fact sheet, write the numbers from 95 to 120."

WRITTEN PRACTICE

- Complete **Worksheet 47A** with your child.
- Complete **Worksheet 47B** with your child later in the day.

Name _____   ***LESSON 47A***
                                                   Math 2
Date _____

Write a number sentence for this story. Write the answer with a label.

1. Stuart has 10 large marbles and 24 small marbles. How many marbles does he have?

    Number sentence _____10 marbles + 24 marbles = 34 marbles_____

    Answer _____34 marbles_____

2. Measure this line segment to the nearest inch.

    •————————————————————•   ____4___ "

3. Fill in the missing numbers. Use the fewest possible number of pennies.

    4 dimes and 17 pennies = __5__ dimes + __7__ pennies = __57__ ¢

4. Color one half using a blue crayon.

    Color one fourth using a red crayon.

    Color one eighth using a green crayon.

    Color one sixth using a purple crayon.

5. Find each sum.

    36 + 10 = __46__   28 + 10 = __38__

    10 + 42 = __52__   73 + 10 = __83__

6. Fill in the missing numbers on this piece of a hundred number chart.

62	63	64	65	66
72	73	74	75	76
82	83	84	85	86

Name _____   ***LESSON 47B***
                                                   Math 2
Date _____

Write a number sentence for this story. Write the answer with a label.

1. Clara has 35 pennies. Her brother gave her 10 more pennies. How many pennies does she have now?

    Number sentence _____35 pennies + 10 pennies = 45 pennies_____

    Answer _____45 pennies_____

2. Which number on the thermometer is the temperature closest to? __60__ °F

3. Fill in the missing numbers. Use the fewest possible number of pennies.

    6 dimes and 18 pennies = __7__ dimes + __8__ pennies = __78__ ¢

4. Color one half using a blue crayon.

    Color one fourth using an orange crayon.

    Color one eighth using a red crayon.

    Color one third using a green crayon.

5. Find each sum.

    85 + 10 = __95__   10 + 37 = __47__

    29 + 10 = __39__   10 + 21 = __31__

6. Fill in the missing numbers on this piece of a hundred number chart.

23	24	25	26	27
33	34	35	36	37
43	44	45	46	47

esson 48

counting nickels
identifying similarities and differences of coins

---

lesson preparation

materials

penny, nickel, dime
cup of 4 dimes and 10 nickels
Fact Sheet A 7.1

in the morning

• Write the following in the pattern box on the meeting strip:

> 100, __, 80, __, 60, __, 40, __, 20, __, 0

Answer: 100, 90, 80, 70, 60, 50, 40, 30, 20, 10, 0

• Write | 45¢ | on the meeting strip. Provide a cup of 10 dimes and a cup of 20 pennies.

THE MEETING

calendar

• Ask your child to write the date on the calendar and meeting strip.

• Ask your child the following:

    date _____ days ago, date _____ days from now

    day of the week _____ days ago, day of the week _____ days from now

    months of the year, _____th month, month before, month after

• Record on the meeting strip a special event and the number of days until it occurs.

weather graph

• Ask your child to color the graph and write the temperature to the nearest ten degrees in the box he/she colored.

• Ask questions about the graph.

counting

- Count by 10's to 200 and backward from 200 by 10's.
- Count by 5's to 100 and backward from 50 by 5's.
- Say the even numbers to 30 and backward from 30.
- Say the odd numbers to 29 and backward from 29.

graph questions

- You and your child each ask a question about any of the graphs.

patterning

- Ask your child to do the following:

  identify the pattern (repeating, continuing, or both)

  identify the numbers to complete the pattern

  read the pattern

money

- Ask your child to put the dimes and pennies in the coin cup.
- Count the money in the coin cup together.

clock

- Ask your child to set the clock on the half hour or hour.
- Ask the following:

  time shown on the clock

  time one hour ago and time one hour from now
- Ask your child to write the digital time on the meeting strip.
- Record on the meeting strip the time an activity will occur.

number of the day

- Write three number sentences for the number of the day on the meeting strip.

THE LESSON

Counting Nickels
Identifying Similarities and Differences of Coins

- Give your child a nickel, a penny, and a dime.

  "Point to the dime."

  "Point to the penny."

"What do we call the other coin?"   nickel

"Look carefully at all the coins."

"What is the same about all of the coins?"

- Encourage your child to find as many similarities as possible.

"What is different about the coins?"

- Encourage your child to find as many differences as possible.

- Optional: Encourage your child to research various aspects of the coins. For example: What does the date on the coin mean? Who are the people and the buildings on the coins? Where and how are coins made?

"We have been counting dimes and pennies each day."

"Today you will learn how to count nickels."

"On the back of the nickel are the words 'five cents.' "

"How many pennies are the same as one nickel?"

"Five pennies are the same as one nickel."

"When we count nickels, what do you think we will count by?"   fives

- Hold up ten nickels.

"I have ten nickels."

"Let's count by fives to find out how much money this is."

- Hold up the nickels one at a time as your child counts by fives.

"Ten nickels are the same as 50 cents."

- Hold up three nickels.

"I have three nickels."

"Let's count by fives as I hold up each nickel."

- Hold up the nickels one at a time as your child counts by fives.

"Three nickels are the same as 15 cents."

- Repeat with 6 nickels, 8 nickels, and 5 nickels.

- Give your child a cup of 4 dimes and 8 nickels.

"Sort your coins."

- Allow time for your child to do this.

"Show 20¢ using dimes."

"Show 20¢ using nickels."

"How many dimes did you use?"   2

"How many nickels did you use?"   4

- Repeat with 40¢, 10¢, and 30¢.

"Show 25¢ using nickels."

"How many nickels did you use?" 5

• Repeat with 15¢, 35¢, and 50¢.

CLASS PRACTICE

number fact practice

• Practice the adding nine facts together.

• Give your child **Fact Sheet A 7.1.**

• Time your child for one minute.

• Correct the fact sheet with your child and record the score.

• Allow time for your child to complete the unfinished facts.

WRITTEN PRACTICE

• Complete **Worksheet 48A** with your child.

• Complete **Worksheet 48B** with your child later in the day.

Name _____   **LESSON 48A**
                                     Math 2
Date _____

Write a number sentence for the story. Write the answer with a label.

1. The children in Room 2 collected 83 cans of food for the food drive. Ms. Roman brought in 10 more cans of food. How many cans of food did they collect?

   Number sentence   83 cans + 10 cans = 93 cans

   Answer   93 cans

2. Measure each line segment using inches.

   How long is the horizontal line segment?   2 "

   How long is the vertical line segment?   3 "

3. How much money is this?   35¢

4. Chris has nine white socks.
   Draw the socks and circle the pairs.

   How many pairs are there?   4

   How many extras are there?   1

5. Find each sum.

   10 + 36 = __46__   8 + 9 = __17__   3 + 9 = __12__

   13 + 10 = __23__   5 + 6 = __11__   9 + 7 = __16__

Name _____   **LESSON 48B**
                                     Math 2
Date _____

Write a number sentence for the story. Write the answer with a label.

1. Ten children voted yes and 31 children voted no. How many children voted?

   Number sentence   10 children + 31 children = 41 children

   Answer   41 children

2. Color the fifth triangle blue.
   Color the sixth triangle red.
   Color the middle triangle yellow.

3. How much money is this?   25¢

4. Missy has seven white socks.
   Draw the socks and circle the pairs.

   How many pairs are there?   3

   How many extras are there?   1

5. Find each sum.

   10 + 43 = __53__   6 + 7 = __13__   4 + 9 = __13__

   16 + 10 = __26__   5 + 4 = __9__   9 + 6 = __15__

Lesson 49

subtracting a number from itself
subtracting one
subtracting zero

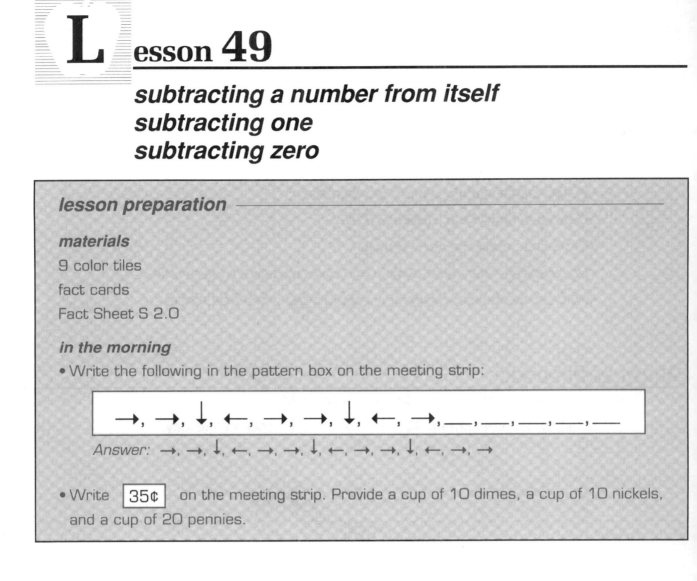

lesson preparation

materials

9 color tiles

fact cards

Fact Sheet S 2.0

in the morning

• Write the following in the pattern box on the meeting strip:

→, →, ↓, ←, →, →, ↓, ←, →, ___, ___, ___, ___, ___

Answer: →, →, ↓, ←, →, →, ↓, ←, →, →, ↓, ←, →, →

• Write ⟨35¢⟩ on the meeting strip. Provide a cup of 10 dimes, a cup of 10 nickels, and a cup of 20 pennies.

THE MEETING

calendar

• Ask your child to write the date on the calendar and meeting strip.

• Ask your child the following:

date _____ days ago, date _____ days from now

day of the week _____ days ago, day of the week _____ days from now

months of the year, _____th month, month before, month after

• Record on the meeting strip a special event and the number of days until it occurs.

weather graph

• Ask your child to color the graph and write the temperature to the nearest ten degrees in the box he/she colored.

• Ask questions about the graph.

counting

- Count by 10's to 200 and backward from 200 by 10's.
- Count by 5's to 100 and backward from 50 by 5's.
- Say the even numbers to 30 and backward from 30.
- Say the odd numbers to 29 and backward from 29.

graph questions

- You and your child each ask a question about any of the graphs.

patterning

- Ask your child to do the following:

  identify the pattern (repeating continuing, or both)

  identify the shapes to complete the pattern

  read the pattern

money

"Today you will put coins in the coin cup to show 35¢."

"You may use dimes, nickels, and pennies to do this."

"Try to use the fewest number of pennies and nickels."

- Allow time for your child to do this

"What coins did you use?"

"Let's count to check to see if this is 35¢."

"When we count the coins, we will count the dimes first, then the nickels, and then the pennies."

- Count the money in the coin cup with your child.

clock

- Ask your child to set the clock on the half hour or hour.
- Ask the following:

  time shown on the clock

  time one hour ago and time one hour from now

- Ask your child to write the digital time on the meeting strip.
- Record on the meeting strip the time an activity will occur.

number of the day

- Write three number sentences for the number of the day on the meeting strip.

THE LESSON

Subtracting a Number from Itself
Subtracting One
Subtracting Zero

"Today you will learn some more subtraction facts."

"In today's lesson we will pretend that color tiles are cookies."

"I have five cookies."

• Use the color tiles to act out the story.

"If I eat the five cookies, how many cookies will I have left?"   zero

"What type of story is this?"   some, some went away

"How would we write a number sentence for this story?"

• Write "5 cookies – 5 cookies = 0 cookies" on the chalkboard.

"What will happen if I have six cookies and eat six cookies? . . . if I have nine cookies and eat nine cookies?"   there are none left

• Write the following on the chalkboard:

$$
\begin{array}{cccccccccc}
1 & 2 & 3 & 4 & 5 & 6 & 7 & 8 & 9 & 10 \\
\underline{-1} & \underline{-2} & \underline{-3} & \underline{-4} & \underline{-5} & \underline{-6} & \underline{-7} & \underline{-8} & \underline{-9} & \underline{-10}
\end{array}
$$

"Make up a story about one of these problems."

"What is the answer?"   zero

• Record the answer for each problem.

"What's the same about all of these problems?"   the numbers in the problem are the same; the answers are all zero

• Record the answers below the problems.

"Now let's try a different story."

• Use the color tiles to act out the story.

"I had seven cookies."

"I ate one of them for lunch."

"How many cookies do I have left?"   6

"What happened in this story?"

"What type of story is this?"   some, some went away

"How many cookies did I have?"   7

"How many went away?"   1

"How many are left?"   6

"How would we write a number sentence for this story?"

• Write "7 cookies – 1 cookie = 6 cookies" on the chalkboard.

"What will happen if I have four cookies and eat one cookie? . . . if I have nine cookies and eat one cookie?"

• Write the following on the chalkboard. Do not erase the previous facts.

$$\begin{array}{ccccccccc} 2 & 3 & 4 & 5 & 6 & 7 & 8 & 9 & 10 \\ -1 & -1 & -1 & -1 & -1 & -1 & -1 & -1 & -1 \end{array}$$

"Make up a story about one of these problems."

"What is the answer?"

• Record the answer on the chalkboard.

"What's the same about all of these problems?"

"What are the answers?"

• Record the answers below the problems.

"What do you notice about the answers?"

"Now let's try a different problem."

• Use the color tiles to act out the story.

"I had five cookies."

"I didn't eat any of them for lunch."

"How many cookies do I have left?"   5

"What happened in this story?"

"This is a special some went away story."

"How many cookies did I have?"   5

"How many went away?"   none

"How many are left?"   5

• Write the following on the chalkboard. Do not erase the previous facts.

$$\begin{array}{cccccccccc} 1 & 2 & 3 & 4 & 5 & 6 & 7 & 8 & 9 & 10 \\ -0 & -0 & -0 & -0 & -0 & -0 & -0 & -0 & -0 & -0 \end{array}$$

"Make up a story about one of these problems."

"What is the answer?"

• Record the answer on the chalkboard.

"What's the same about all of these problems?"

"What are the answers?"

• Record the answers below the problems.

"What do you notice about the answers?"

"Now you will use your fact cards."

"Find all the fact cards that match the number facts we learned today."

"Put them in rows on the table like I've written the facts on the chalkboard."

"Try to say each answer to yourself."

"The answer is the top number on the other side."

"Turn over each card to check your answer."

Class Practice

number fact practice

- Give your child **Fact Sheet S 2.0.**
- Time your child for one minute.
- Correct the fact sheet with your child and record the score.
- Allow time for your child to complete the unfinished facts.

Written Practice

- Complete **Worksheet 49A** with your child.
- Complete **Worksheet 49B** with your child later in the day.

Name _____   **LESSON 49A**
                                               Math 2
Date _____

1. Steven had 6 nickels. Michelle gave him 2 more nickels. Write a number sentence to show how many nickels he has now.

   Number sentence _____6 nickels + 2 nickels = 8 nickels_____

   Answer ___8 nickels___   How much money is that? __40¢__

2. Measure each line segment using inches.

   horizontal line segment ___3 inches___

   oblique line segment ___2 inches___

   vertical line segment ___5 inches___

3. What is the fifth day of the week?   Thursday

   What is the second day of the week?   Monday

   What is the seventh day of the week?   Saturday

4. Number the clock face.
   Show half past five on the clock face.

   Write the digital time one hour from now.

   6:30

5. Find each answer.

   26 + 10 = __36__   10 + 72 = __82__   7 + 8 = __15__

   6 − 6 = __0__   9 + 5 = __14__   50 + 10 = __60__

2-49Wa   Copyright © 1991 by Saxon Publishers, Inc. and Nancy Larson. Reproduction prohibited.

Name _____   **LESSON 49B**
                                               Math 2
Date _____

1. Marsha and Sandy are saving nickels. Marsha has 3 nickels and Sandy has 7 nickels. Write a number sentence to show how many nickels they have.

   Number sentence _____3 nickels + 7 nickels = 10 nickels_____

   Answer ___10 nickels___   How much money is that? __50¢__

2. Finish the number patterns.

   85, 84, 83, __82__, __81__, __80__, __79__, __78__, __77__, __76__

   1, 3, 5, __7__, __9__, __11__, __13__, __15__, __17__, __19__

   2, 4, 6, __8__, __10__, __12__, __14__, __16__, __18__, __20__

   100, 90, 80, __70__, __60__, __50__, __40__, __30__, __20__, __10__

3. Write the letter E in the fourth square.
   Write the letter F in the first square.
   Write the letter D in the sixth square.
   Write the letter I in the third square.
   Write the letter N in the fifth square.
   Write the letter R in the second square.

   F | R | I | E | N | D

4. Number the clock face.
   Show half past nine on the clock face.

   Write the digital time one hour from now.

   10:30

5. Find each answer.

   39 + 10 = __49__   6 + 9 = __15__   30 + 10 = __40__

   5 − 5 = 0   9 + 8 = __17__   10 + 19 = __29__

2-49Wb   Copyright © 1991 by Saxon Publishers, Inc. and Nancy Larson. Reproduction prohibited.

Lesson 50

finding the area of shapes using pattern blocks

lesson preparation

materials

Written Assessment #9

Oral Assessment #5

fraction piece set from Lesson 34

pattern blocks

Master 2-50

Fact Sheet S 3.2

in the morning

• Write the following in the pattern box on the meeting strip:

> ___, ___, ___, 72, 73, 74, ___, ___, ___

Answer: 69, 70, 71, 72, 73, 74, 75, 76, 77

• Write ⎡40¢⎤ on the meeting strip. Provide a cup of 10 dimes, a cup of 10 nickels, and a cup of 20 pennies.

THE MEETING

calendar

• Ask your child to write the date on the calendar and meeting strip.

• Ask your child the following:

  date _____ days ago, date _____ days from now

  day of the week _____ days ago, day of the week _____ days from now

  months of the year, _____th month, month before, month after

• Record on the meeting strip a special event and the number of days until it occurs.

weather graph

• Ask your child to color the graph and write the temperature to the nearest ten degrees in the box he/she colored.

• Ask questions about the graph.

counting

- Count by 10's to 200 and backward from 200 by 10's.
- Count by 5's to 100 and backward from 50 by 5's.
- Say the even numbers to 100 and backward from 100.
- Say the odd numbers to 49 and backward from 49.

graph questions

- You and your child each ask a question about any of the graphs.

patterning

- Ask your child to do the following:

    identify the pattern (repeating, continuing, or both)

    identify the numbers to complete the pattern

    read the pattern

money

- Ask your child to put the dimes, nickels, and/or pennies in the coin cup.
- Count the money in the coin cup together.

clock

- Ask your child to set the clock on the half hour or hour.
- Ask the following:

    time shown on the clock

    time one hour ago and time one hour from now
- Ask your child to write the digital time on the meeting strip.
- Record on the meeting strip the time an activity will occur.

number of the day

- Write three number sentences for the number of the day on the meeting strip.

ASSESSMENT

Written Assessment

"Today I would like to see what you remember from what we have been practicing."

- Give your child **Written Assessment #9.**

- Read the directions for each problem. Allow time for your child to complete that problem before continuing.
- Correct the paper, noting your child's mistakes on the **Individual Recording Form.** Review the errors with your child.

Oral Assessment

- Record your child's response(s) to the oral interview questions on the interview sheet.

THE LESSON

Finding the Area of Shapes Using Pattern Blocks

- Give your child **Master 2-50.**

  "Today you will learn how to find the area of each of these shapes."

  "What math material do you think you will use to do this?" pattern **blocks**

  "How do you know?" the shape of the pieces at the top of the paper

- Give your child a container of pattern blocks.

  "Cover the shapes at the top of the paper using one pattern block for each."

  "What colors did you use?" green, blue, red, yellow

  "We will say that the green pattern block has an area of one unit."

  "Cover the blue pattern block using green pattern blocks."

  "How many green pattern blocks did you use to cover the blue pattern block?" 2

  "What is the area of the blue pattern block?" 2 units

  "Write that inside the shape."

- Repeat with the red and the yellow pattern blocks.

  "Now you will use pattern blocks to find the area of each design at the bottom of the paper."

  "Which design do you think has the largest area?"

  "Which design do you think has the smallest area?"

  "You will use the pattern blocks to find the areas."

  "Cover design A using pattern blocks."

  "You may use yellow, red, blue, and green pattern blocks."

- Allow time for your child to cover the design.

"Now I will write the value of the pattern blocks you used to cover design A on the chalkboard."

- Write the values on the chalkboard.

"We can find the area of the design by adding the value of these pattern blocks."

"What is the area of the design?"

- Repeat with designs B and C.

"On the back of the paper, make a design with an area of twelve units."

CLASS PRACTICE

number fact practice

- Give your child **Fact Sheet S 3.2.**
- Time your child for one minute.
- Correct the fact sheet with your child and record the answer.
- Allow time for your child to complete the unfinished facts.

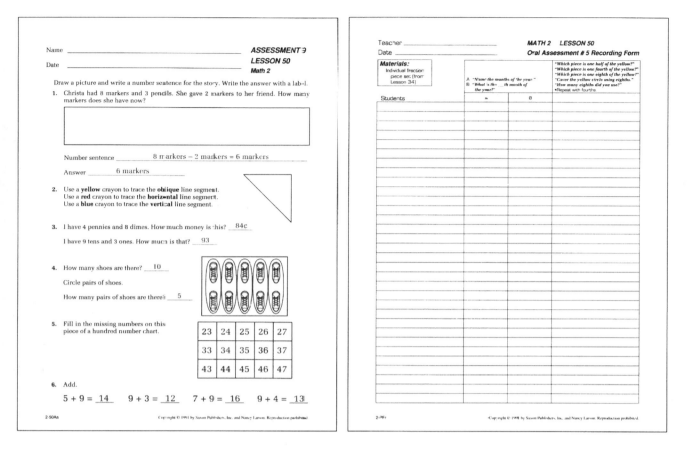

Name _____   **ASSESSMENT 9**
Date _____   **LESSON 50**
                            Math 2

Draw a picture and write a number sentence for the story. Write the answer with a label.

1. Christa had 8 markers and 3 pencils. She gave 2 markers to her friend. How many markers does she have now?

   Number sentence ____ 8 markers − 2 markers = 6 markers
   Answer ____ 6 markers

2. Use a **yellow** crayon to trace the **oblique** line segment.
   Use a **red** crayon to trace the **horizontal** line segment.
   Use a **blue** crayon to trace the **vertical** line segment.

3. I have 4 pennies and 8 dimes. How much money is this? __84¢__
   I have 9 tens and 3 ones. How much is that? __93__

4. How many shoes are there? __10__
   Circle pairs of shoes.
   How many pairs of shoes are there? __5__

5. Fill in the missing numbers on this piece of a hundred number chart.

23	24	25	26	27
33	34	35	36	37
43	44	45	46	47

6. Add.
   5 + 9 = __14__   9 + 3 = __12__   7 + 9 = __16__   9 + 4 = __13__

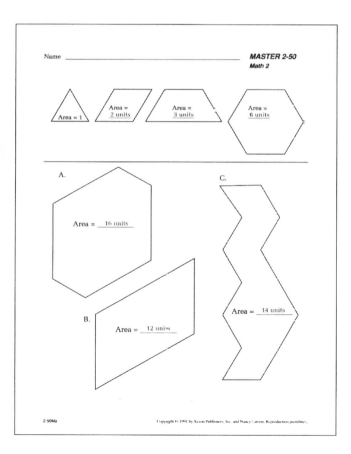

Name _____   **MASTER 2-50**
                            Math 2

Area = 1   Area = 2 units   Area = 3 units   Area = 6 units

A. Area = __16 units__
B. Area = __12 units__
C. Area = __14 units__

Lesson 51

creating and reading a Venn diagram

---

lesson preparation ─────────────────────────────

materials

construction paper

ten 2" yellow squares

red and green crayons

2 yarn or string loops (see below)

Fact Sheet A 7.1

the night before

• Make two yarn or heavy string loops. Use approximately three feet of yarn for each loop. If possible, make one loop green and the other red. Tape a red construction paper tag to one loop and a green construction paper tag to the other.

in the morning

• Write the following in the pattern box on the meeting strip:

$$10, \underline{\quad}, 20, \underline{\quad}, 30, \underline{\quad}, 40, \underline{\quad}, 50, \underline{\quad}, 60$$

Answer: 10, 15, 20, 25, 30, 35, 40, 45, 50, 55, 60

• Write 31¢ on the meeting strip. Provide a cup of 10 dimes, a cup of 10 nickels, and a cup of 20 pennies.

THE MEETING

calendar

• Ask your child to write the date on the calendar and meeting strip.

• Ask your child the following:

date _____ days ago, date _____ days from now

day of the week _____ days ago, day of the week _____ days from now

months of the year, _____th month, month before, month after

• Record on the meeting strip a special event and the number of days until it occurs.

---

weather graph

- Ask your child to color the graph and write the temperature to the nearest ten degrees in the box he/she colored.
- Ask questions about the graph.

counting

- Count by 10's to 200 and backward from 200 by 10's.
- Count by 5's to 100 and backward from 50 by 5's.
- Say the even numbers to 100 and backward from 100.
- Say the odd numbers to 49 and backward from 49.

graph questions

- You and your child each ask a question about any of the graphs.

patterning

- Ask your child to do the following:

    identify the pattern (repeating continuing, or both)

    identify the numbers to complete the pattern

    read the pattern

money

- Ask your child to put the dimes, nickels, and/or pennies in the coin cup.
- Count the money in the coin cup together.

clock

- Ask your child to set the clock on the half hour or hour.
- Ask the following:

    time shown on the clock

    time one hour ago and time one hour from now

- Ask your child to write the digital time on the meeting strip.
- Record on the meeting strip the time an activity will occur.

number of the day

- Write three number sentences for the number of the day on the meeting strip.

THE LESSON

Creating and Reading a Venn Diagram

> *"Today you will learn how to make and read a different type of graph."*

- Give your child a 2" paper square and a red and a green crayon.

> *"Write the names of all the children in our family on this piece of paper."*
>
> *"Will you write your own name?"*
>
> *"Why?"*   because I am a child in the family
>
> *"Circle the boys' names in red and the girls' names in green."*

- Repeat for 5–9 more families you know.

> *"Which tags have red circles?"*
>
> *"What kind of names are these?"*   boys'
>
> *"Which tags have green circles?"*
>
> *"What kind of names are these?"*   girls'

- Draw the following on the chalkboard:

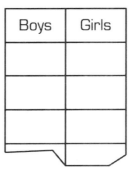

> *"Let's make a graph to show the boys and girls in the families."*
>
> *"Let's try to place the tags on this graph."*
>
> *"Can we use this graph to graph our tags?"*

- Choose a family with only boys.

> *"Where will we graph this tag?"*

- Choose a family with only girls.

> *"Where will we graph this tag?"*

- Choose a family with both boys and girls.

> *"Where will we graph this tag?"*

- Discuss the dilemma with your child.

> *"We will need to use a different type of graph for our tags."*
>
> *"It is called a Venn diagram."*
>
> *"A Venn diagram is made using circles."*

• Place the two yarn circles on the floor. Label the red circle "Boys" and the green circle "Girls." Arrange the circles like this:

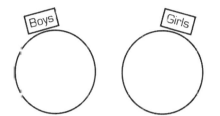

"How do you think we will graph the tags?"

"Where will we put the tags of the families with only boys?"

"Where will we put the tags of the families with only girls?"

"Let's put those tags in the circles now."

"Now watch what I am going to do."

• Overlap the circles so that they look like this:

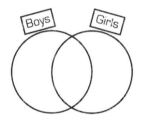

"What did I do?"

"Why did I do this?"

"Which tags will we put in the area where the circles overlap?"

"Which families have both boys and girls?"

"Where will we graph these tags?"

• Place those tags where the circles overlap.

"We say that this is where the circles intersect."

"Now let's answer questions about our new graph."

"How many families have boys?"

• Count all the tags in the red string circle.

"How many families have girls?"

• Count the tags in the green string circle.

"How many families have both boys and girls?"

• Count all the tags in the intersection.

"How many families have only boys?"

• Count the tags in the red string circle but not in the intersection.

"How many families have only girls?"

• Count all the tags in the green string circle but not in the intersection.

"We call this graph a Venn diagram."

CLASS PRACTICE

number fact practice

- Use the yellow fact cards to practice the adding nine facts with your child.
- Give your child **Fact Sheet A 7.1.**
- Time your child for one minute.
- Correct the fact sheet with your child and record the score.
- Allow time for your child to complete the unfinished facts.

WRITTEN PRACTICE

- Complete **Worksheet 51A** with your child.
- Complete **Worksheet 51B** with your child later in the day.

Name _____ **LESSON 51A**
                                        Math 2
Date _____

Write a number sentence for the story. Write the answer with a label.

1. Susan had 12 pencils. Marsha gave Susan two pencils. How many pencils does Susan have now?

   Number sentence _____ 12 pencils + 2 pencils = 14 pencils _____

   Answer _____ 14 pencils _____

2. How much money is in each pocket?

   4 nickels   4 dimes   22 pennies
   20¢   40¢   22¢

   Circle the pocket with the most money.

3. Show **18** using tally marks.

   𝍷𝍷𝍷𝍷 𝍷𝍷𝍷𝍷 𝍷𝍷𝍷𝍷 𝍷𝍷𝍷

4. One of these socks is my favorite sock.
   Use the clues to find my favorite sock.
   It is not the sock with the triangles.
   It is not the sock with the vertical lines.
   It is not first.
   It is not fourth.
   Circle my favorite sock.

5. Six children in Miss Wood's class graphed their tags.

   How many families have boys? __5__

   How many families have only girls? __1__

   How many families have both boys and girls? __3__

6. Find each answer.

   18 + 10 = __28__   77 + 10 = __87__   9 − 1 = __8__   5 − 5 = __0__

2-51Wa

Copyright © 1991 by Saxon Publishers, Inc. and Nancy Larson. Reproduction prohibited.

Name _____ **LESSON 51B**
                                        Math 2
Date _____

Write a number sentence for the story. Write the answer with a label.

1. Miss Allen had 6 games. Miss Allen gave Mrs. Paolino's class 3 games to use. How many games does Miss Allen have now?

   Number sentence _____ 6 games − 3 games = 3 games _____

   Answer _____ 3 games _____

2. How much money is in each pocket?

   7 nickels   3 dimes   18 pennies
   35¢   30¢   18¢

   Circle the pocket with the most money.

3. Show **31** using tally marks.

   𝍷𝍷𝍷𝍷 𝍷𝍷𝍷𝍷 𝍷𝍷𝍷𝍷 𝍷𝍷𝍷𝍷 𝍷𝍷𝍷𝍷 𝍷𝍷𝍷𝍷 𝍷

4. One of these socks is my sister's favorite sock.
   Use the clues to find her favorite sock.
   It is not the sock with the triangles.
   It is not the sock with the vertical lines.
   It is not first.
   It is not second.
   Circle my sister's favorite sock.

5. Eight children in Miss Gen's class graphed their tags.

   How many families have girls? __5__

   How many families have only boys? __3__

   How many families have both boys and girls? __4__

6. Find each answer.

   26 + 10 = __36__   86 + 10 = __96__   7 − 1 = __6__   3 − 0 = __3__

2-51Wb

Copyright © 1991 by Saxon Publishers, Inc. and Nancy Larson. Reproduction prohibited.

Lesson 52

identifying a line of symmetry
creating a symmetrical design

---

lesson preparation

materials

yardstick

poster paint (3 or 4 colors, if possible; see lesson)

white construction paper (4–5 pieces)

Fact Sheet A 7.2

in the morning

• Write the following in the pattern box on the meeting strip:

> A, z, B, y, C, x, ___, ___, ___, ___, ___

Answer: A, z, B, y, C, x, D, w, E, v, F

• Write [15¢] on the meeting strip. Provide a cup of 10 dimes, a cup of 10 nickels, and a cup of 20 pennies.

THE MEETING

calendar

• Ask your child to write the date on the calendar and meeting strip.

• Ask your child the following:

> date _____ days ago, date _____ days from now

> day of the week _____ days ago, day of the week _____ days from now

> months of the year, _____th month, month before, month after

• Record on the meeting strip a special event and the number of days until it occurs.

weather graph

• Ask your child to color the graph and write the temperature to the nearest ten degrees in the box he/she colored.

• Ask questions about the graph.

counting

- Count by 10's to 200 and backward from 200 by 10's.
- Count by 5's to 100 and backward from 50 by 5's.
- Say the even numbers to 100 and backward from 100.
- Say the odd numbers to 49 and backward from 49.

graph questions

- You and your child each ask a question about any of the graphs.

patterning

- Ask your child to do the following:

  identify the pattern (repeating, continuing, or both)

  identify the letters to complete the pattern

  read the pattern

money

- Ask your child to put the dimes, nickels, and/or pennies in the coin cup.
- Count the money in the coin cup together.

clock

- Ask your child to set the clock on the half hour or hour.
- Ask the following:

  time shown on the clock

  time one hour ago and time one hour from now
- Ask your child to write the digital time on the meeting strip.
- Record on the meeting strip the time an activity will occur.

number of the day

- Write three number sentences for the number of the day on the meeting strip.

THE LESSON

Identifying a Line of Symmetry

"Today you will learn about lines of symmetry and how to create a symmetrical design."

- If possible, have your child stand in front of a full-length mirror.

"Our bodies are made in a very special way."

- Hold the yardstick vertically in the center of your child.

   "Let's pretend that we could draw a line down the middle of your body."

   "What do you see that is the same on each side of the line?"

   "You have one eye on each side, one arm on each side, and one ear on each side."

   "We call this line the line of symmetry."

   "This means that what is on one side of the line is the same as what is on the other side of the line."

   "If we could fold you along this line of symmetry, your eyes would match, your hands would match, and your ears would match."

   "When we draw a line of symmetry, we divide a shape in half."

- Draw the following on the chalkboard:

   "Draw the line of symmetry in one of these shapes."

- Ask your child to draw the line.

   "How do we know that this is a line of symmetry?"   both sides are the same

- Repeat with the other chalkboard shapes.

Creating a Symmetrical Design

   "Now we will use paint to make symmetrical designs."

   "I will make one first to show you how we will do this."

   "First I will fold my paper in half vertically."

   "Now I will put a small spoonful of paint near the fold."

- Demonstrate, using 1–2 tbsp. of paint.

   "Now I will fold the paper."

   "Next, I will push the paint away from the fold and towards the outside, like this."

- Demonstrate pushing the paint gently, using the edge of your hand.

   "When I have finished, I will gently open the paper."

- Open the paper.

   "What do you notice about my design?"   both sides match or are the same

   "Where is the line of symmetry?"

"Now you will have a chance to make a symmetrical design using paint."

- Give your child a piece of paper.

"Fold your paper in half vertically."

"When I put a spoonful of paint on your paper, fold your paper and gently push away from the fold toward the outside edge of the paper."

- If possible, allow your child to choose from among three or four colors. Put the spoonful of paint on or near the fold.

- Repeat to make several designs, if desired.

- Optional: Your child can write about his/her symmetrical design.

CLASS PRACTICE

number fact practice

"Use your fact cards to practice all the addition facts."

- Allow time for your child to do this.

- Give your child **Fact Sheet A 7.2.**

- Time your child for one minute.

- Correct the fact sheet with your child and record the score.

- Allow time for your child to complete the unfinished facts.

"On the back of your fact sheet, write the numbers from 195 to 221."

WRITTEN PRACTICE

- Complete **Worksheet 52A** with your child.

- Complete **Worksheet 52B** with your child later in the day.

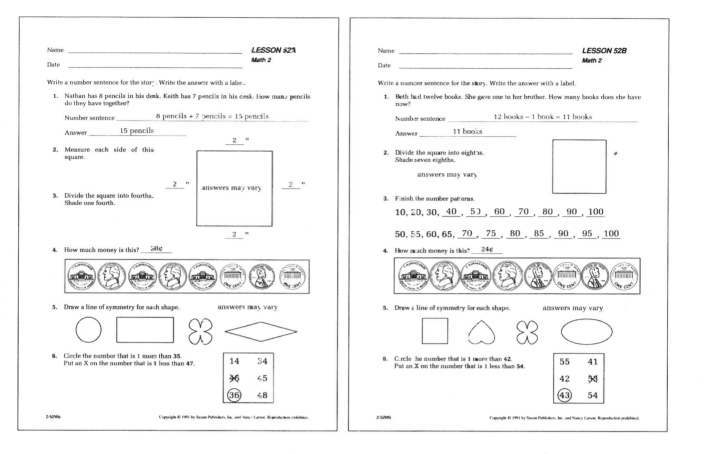

Lesson 53

subtracting ten from a two-digit number

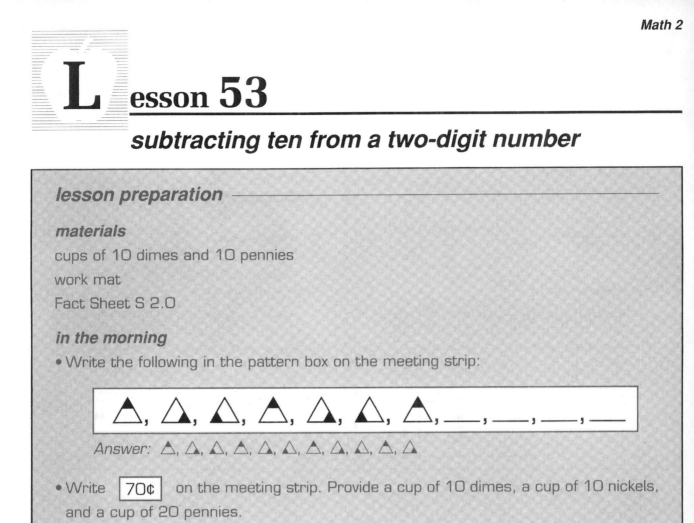

lesson preparation

materials

cups of 10 dimes and 10 pennies

work mat

Fact Sheet S 2.0

in the morning

• Write the following in the pattern box on the meeting strip:

△, △, △, △, △, △, △, ___, ___, ___, ___

Answer: △, △, △, △, △, △, △, △, △, △, △

• Write ☐70¢☐ on the meeting strip. Provide a cup of 10 dimes, a cup of 10 nickels, and a cup of 20 pennies.

THE MEETING

calendar

• Ask your child to write the date on the calendar and meeting strip.

• Ask your child the following:

date _____ days ago, date _____ days from now

day of the week _____ days ago, day of the week _____ days from now

months of the year, _____th month, month before, month after

• Record on the meeting strip a special event and the number of days until it occurs.

weather graph

• Ask your child to color the graph and write the temperature to the nearest ten degrees in the box he/she colored.

• Ask questions about the graph.

counting

• Count by 10's to 200 and backward from 200 by 10's.

- Count by 5's to 100 and backward from 50 by 5's.
- Say the even numbers to 100 and backward from 100.
- Say the odd numbers to 49 and backward from 49.

graph questions

- You and your child each ask a question about any of the graphs.

patterning

- Ask your child to do the following:

    identify the pattern (repeating, continuing, or both)

    identify the shapes to complete the pattern

    read the pattern

money

- Ask your child to put the dimes, nickels, and/or pennies in the coin cup.
- Count the money in the coin cup together.

clock

- Ask your child to set the clock on the half hour or hour.
- Ask the following:

    time shown on the clock

    time one hour ago and time one hour from now

- Ask your child to write the digital time on the meeting strip.
- Record on the meeting strip the time an activity will occur.

number of the day

- Write three number sentences for the number of the day on the meeting strip.

THE LESSON

Subtracting Ten from a Two-Digit Number

"Today you will learn how to subtract ten from a two-digit number."

"We will use dimes and pennies to help us do that."

- Write the following problems on the chalkboard:

    30¢ + 10¢ =   20¢ + 10¢ =   80¢ + 10¢ =   60¢ + 10¢ =

- Give your child a cup of 10 dimes, a cup of 10 pennies, and a work mat.

"Let's find these answers using dimes."

"How will you show 30 cents?"   3 dimes

"Put three dimes on your mat."

"What does the problem tell us to do next?"   take one more dime

"Put one more dime on your mat."

"What type of problem is this?"   some, some more

"How many dimes do you have now?"   4

"How much money is this?"   40 cents

- Record the answer on the chalkboard.

"What does the next problem tell us to do?"   use two dimes and then one more dime

"Show this with your coins."

"How many dimes do you have now?"   3

"How much money is this?"   30 cents

- Record the answer on the chalkboard.
- Repeat with the last two problems.
- Write the following problems on the chalkboard:

  30¢ – 10¢ =   20¢ – 10¢ =   80¢ – 10¢ =   60¢ – 10¢ =

"Let's find these answers using dimes and pennies."

"How many dimes will you use to show 30 cents?"   3

"What does the problem tell us to do next?"   take away one dime

"Take away one dime."

"What type of problem is this?"   some, some went away

"How many dimes do you have now?"   2

"How much money is this?"   20 cents

- Record the answer on the chalkboard.

"What does the next problem tell us to do?"   use two dimes and take away one dime

"Show this using your coins."

"How many dimes do you have now?"   1

"How much money is this?"   10 cents

- Record the answer on the chalkboard.
- Repeat with the remaining problems.
- Write the following problems on the chalkboard:

  75¢ + 10¢ =   59¢ + 10¢ =   83¢ + 10¢ =   31¢ + 10¢ =

"What type of problems are these?"   some, some more

"How could we show the first problem using dimes and pennies?"

- Allow time for your child to put the coins on the mat.

  "How many dimes do you have now?"   8

  "How many pennies?"   5

  "What is the answer?"   85¢

- Record the answer on the chalkboard.
- Repeat with the remaining three problems.
- Write the following problems on the chalkboard:

  75¢ – 10¢ =   59c – 10¢ =   83¢ – 10¢ =   31¢ – 10¢ =

  "What type of problems are these?"   some, some went away

  "How could we show the first problem using dimes and pennies?"

- Allow time for your child to put the coins on the mat.

  "How many dimes do you have now?"   6

  "How many pennies?"   5

  "What is the answer?"   65¢

- Record the answer on the chalkboard.
- Repeat with the remaining three problems.

  "Put your dimes and pennies in the cups."

  "Now you will have a chance to try some problems without using dimes and pennies."

- Write the following problems on the chalkboard:

  42¢ – 10¢ =   36¢ – 10¢ =   54¢ + 10¢ =

  23¢ – 10¢ =   92¢ – 10¢ =   62¢ + 10¢ =

- Ask your child to write each answer on the chalkboard.

  "What changes when we add or subtract ten?"

CLASS PRACTICE

number fact practice

- Use the peach fact cards to practice the subtracting one facts with your child.
- Give your child **Fact Sheet S 2.0**.
- Time your child for one minute.
- Correct the fact sheet with your child and record the score.
- Allow time for your child to complete the unfinished facts.

WRITTEN PRACTICE

- Complete **Worksheet 53A** with your child.
- Complete **Worksheet 53B** with your child later in the day.

Name _____   **LESSON 53A**
                                          Math 2
Date _____

Write a number sentence for the story. Write the answer with a label.

1. Mrs. Hannan's classroom has 32 pairs of right-handed scissors and 10 pairs of left-handed scissors. How many pairs of scissors are there in all?

   Number sentence _____32 pairs + 10 pairs = 42 pairs_____

   Answer _____42 pairs_____

2. Draw a line of symmetry for each letter.

   M   O   D

3. How much money is this? ____51¢____

4. How many earrings are there? ____14____

   Circle the pairs.

   How many pairs of earrings are there? ____7____

5. Find each answer.

   78 + 10 = ____88____   35 − 10 = ____25____   23 + 10 = ____33____

   30 − 10 = ____20____   60 + 10 = ____70____   62 − 10 = ____52____

6. What time is shown on the clock?

   | 7:30 |

   What time will it be one hour later?

   | 8:30 |

2-53Wa   Copyright © 1991 by Saxon Publishers, Inc. and Nancy Larson. Reproduction prohibited.

Name _____   **LESSON 53B**
                                          Math 2
Date _____

Write a number sentence for the story. Write the answer with a label.

1. Ryan had 64 baseball cards. His brother gave him 10 more cards. How many cards does he have now?

   Number sentence _____64 cards + 10 cards = 74 cards_____

   Answer _____74 cards_____

2. Draw a line of symmetry for each letter.

   H   V   X

3. How much money is this? ____47¢____

4. Draw 12 shoes.

   Circle the pairs.

   How many pairs of shoes are there? ____6____

5. Find each answer.

   65 + 10 = ____75____   57 − 10 = ____47____   37 + 10 = ____47____

   40 − 10 = ____30____   20 + 10 = ____30____   84 − 10 = ____74____

6. What time is shown on the clock?

   | 1:30 |

   What time will it be one hour later?

   | 2:30 |

2-53Wb   Copyright © 1991 by Saxon Publishers, Inc. and Nancy Larson. Reproduction prohibited.

L esson 54

ordering two-digit numbers

---

lesson preparation

materials

3 small plastic bags (labeled A, B, and C) filled with coins as follows: Bag A, 4 dimes; Bag B, 23 pennies; Bag C, 10 nickels

fifty 3" × 5" cards

Fact Sheet S 3.2

the night before

- Cut the 3" × 5" cards in half. Write one number from 0–99 on each card. Mix the cards together.

in the morning

- Write the following in the pattern box on the meeting strip:

  94, 95, 96, 97, ___, ___, ___, ___, ___

  Answer: 94, 95, 96, 97, 98, 99, 100, 101, 102

- Write 85¢ on the meeting strip. Provide a cup of 10 dimes, a cup of 10 nickels, and a cup of 20 pennies.

THE MEETING

calendar

- Ask your child to write the date on the calendar and meeting strip.
- Ask your child the following:

  date _____ days ago, date _____ days from now

  day of the week _____ days ago, day of the week _____ days from now

  months of the year, _____th month, month before, month after

- Record on the meeting strip a special event and the number of days until it occurs.

weather graph

- Ask your child to color the graph and write the temperature to the nearest ten degrees in the box he/she colored.
- Ask questions about the graph.

counting

- Count by 10's to 200 and backward from 200 by 10's.
- Count by 5's to 100 and backward from 50 by 5's.
- Say the even numbers to 100 and backward from 100.
- Say the odd numbers to 49 and backward from 49.

graph questions

- You and your child each ask a question about any of the graphs.

patterning

- Ask your child to do the following:

  identify the pattern (repeating, continuing, or both)

  identify the numbers to complete the pattern

  read the pattern

money

- Ask your child to put the dimes, nickels, and/or pennies in the coin cup.
- Count the money in the coin cup together.

clock

- Ask your child to set the clock on the half hour or hour.
- Ask the following:

  time shown on the clock

  time one hour ago and time one hour from now
- Ask your child to write the digital time on the meeting strip.
- Record on the meeting strip the time an activity will occur.

number of the day

- Write three number sentences for the number of the day on the meeting strip.

THE LESSON

Ordering Two-Digit Numbers

"Today you will learn how to order two-digit numbers."

- Show your child Bags A, B, and C.

  "Which bag do you think has the most money?"

  "Let's count to see how much money is in each bag."

  "We have dimes in Bag A."

  "How do we count dimes?"   by tens

- Slide the coins as your child counts. Record on the chalkboard:

  Bag A = 40¢

  "We have pennies in Bag B."

  "How do we count pennies?"   by ones

- Slide the coins as your child counts. Record on the chalkboard:

  Bag B = 23¢

  "We have nickels in bag C."

  "How do we count nickels?"   by fives

- Slide the coins as your child counts. Record on the chalkboard:

  Bag C = 50¢

  "How many pennies are there in 40¢? . . . 23¢? . . . in 50¢?"

  "Which bag has the largest, or greatest, amount of money?"   Bag C

  "Which bag has the smallest, or least, amount of money?"   Bag B

  "Let's put these numbers in order from least to greatest."

- Write the following on the chalkboard:

  —————— , —————— , ——————

  smallest   largest
  least   greatest

  "Which will come first?"

  "Which will come next?"

  "Which will come last?"

  "Now you will have a chance to practice putting numbers in order from least to greatest."

  "I have written the numbers from zero to ninety-nine on these cards."

- Show your child the 3" × $2\frac{1}{2}$" number cards.

  "I will give you three cards."

"Put these number cards in order from least to greatest."

"Put the smallest number on the left."

- Give your child three number cards.

- Repeat several times with different number cards.

  "Now I will give you six number cards."

  "Put these number cards in order from least to greatest."

- When your child finishes, continue.

  "Let's read these number cards from least to greatest."

- Repeat several times with different number cards.

  "Now I will give you 12 number cards."

  "Put these number cards in order from least to greatest."

- When your child finishes, continue.

  "Let's read these number cards from least to greatest."

CLASS PRACTICE

number fact practice

- Give your child **Fact Sheet S 3.2**.

- Time your child for one minute.

- Correct the fact sheet with your child and record the score.

- Allow time for your child to complete the unfinished facts.

WRITTEN PRACTICE

- Complete **Worksheet 54A** with your child.

- Complete **Worksheet 54B** with your child later in the day.

LESSON 54A
Math 2

Name _____

Date _____

Write a number sentence for the story. Write the answer with a label.

1. Weston had 60¢. His sister gave him ten cents. How much money does he have now?

   What kind of problem is this? _____ some, some more

   Number sentence ___ 60¢ + 10¢ = 70¢ ___   Answer ___ 70¢

2. Circle the largest number.
   Put an X on the smallest number.
   Write the numbers in order from smallest to largest.

   (52)   38   2X̶

   ___21___   ___38___   ___52___
   smallest   largest

3.

   How many dimes are there? ___2___   How much money is that? ___20¢___
   How many nickels are there? ___7___   How much money is that? ___35¢___
   How many pennies are there? ___2___   How much money is that? ___2¢___

4. Find each answer.

   55 + 10 = ___65___   27 + 10 = ___37___   40 − 10 = ___30___

   64 − 10 = ___54___   51 − 10 = ___41___   70 + 10 = ___80___

5. Measure this line segment using inches.

   ●———————————————●   ___5___

2-54Wa   Copyright © 1991 by Saxon Publishers, Inc. and Nancy Larson. Reproduction prohibited.

LESSON 54B
Math 2

Name _____

Date _____

Write a number sentence for the story. Write the answer with a label.

1. Stephen had 50¢. He gave his sister ten cents. How much money does he have now?

   What kind of problem is this? _____ some, some went away

   Number sentence ___ 50¢ − 10¢ = 40¢ ___   Answer ___ 40¢

2. Circle the largest number.
   Put an X on the smallest number.
   Write the numbers in order from smallest to largest.

   4X̶9   (73)   61

   ___49___   ___61___   ___73___
   smallest   largest

3. Ask your mom or dad to let you count the change they have in their pocket or purse.
   (Don't count the quarters.)

   How many dimes are there? _____   How much money is that? _____
   How many nickels are there? _____   How much money is that? _____
   How many pennies are there? _____   How much money is that? _____

4. Find each answer.

   24 − 10 = ___14___   46 + 10 = ___56___   61 − 10 = ___51___

   50 + 10 = ___60___   70 − 10 = ___60___   82 + 10 = ___92___

5. Finish the number patterns.

   5, 10, 15, ___20___, ___25___, ___30___, ___35___, ___40___, ___45___, ___50___

   1, 3, 5, ___7___, ___9___, ___11___, ___13___, ___15___, ___17___, ___19___

   56, 57, 58, ___59___, ___60___, ___61___, ___62___, ___63___, ___64___, ___65___

2-54Wb   Copyright © 1991 by Saxon Publishers, Inc. and Nancy Larson. Reproduction prohibited.

Lesson 55

drawing lines using a ruler
drawing a number line

---

lesson preparation

materials

Written Assessment #10

2 pieces of paper

2 rulers

2 color tiles

Fact Sheet A 7.2

in the morning

• Write the following in the pattern box on the meeting strip:

Answer: ▭, ▯, ▯, ▭, ▭, ▯, ▭, ▭, ▯, ▭, ▭

• Write 92¢ on the meeting strip. Provide a cup of 10 dimes, a cup of 10 nickels, and a cup of 20 pennies.

THE MEETING

calendar

• Ask your child to write the date on the calendar and meeting strip.

• Ask your child the following:

date _____ days ago, date _____ days from now

day of the week _____ days ago, day of the week _____ days from now

months of the year, _____th month, month before, month after

• Record on the meeting strip a special event and the number of days until it occurs.

weather graph

• Ask your child to color the graph and write the temperature to the nearest ten degrees in the box he/she colored.

- Ask questions about the graph.

counting

"Today we will practice adding ten, adding one, subtracting ten, and subtracting one from a number."

- Turn to the hundred number chart in the Meeting Book.

"Point to 53."

"Now I will say add ten, subtract ten, add one, or subtract one."

"You will say the new answer and point to the number on the hundred number chart."

"Let's try that."

"Add ten."   63

"Add one."   64

"Subtract ten."   54

"Subtract ten.   44

"Subtract one."   43

"What will we need to do to get back to 53?"   add ten

"Add ten."

"We will practice this every morning."

- Do the following 2–3 times a week:

  count by 10's to 300 and backward from 300 by 10's

  count by 5's to 100 and backward from 50 by 5's

  say the even numbers to 100 and backward from 100

  say the odd numbers to 49 and backward from 49

graph questions

- You and your child each ask a question about any of the graphs.

patterning

- Ask your child to do the following:

  identify the pattern (repeating, continuing, or both)

  identify the shapes to complete the pattern

  read the pattern

money

- Ask your child to put the dimes, nickels, and/or pennies in the coin cup.

- Count the money in the coin cup together.

clock

- Ask your child to set the clock on the half hour or hour.
- Ask the following:

  time shown on the clock

  time one hour ago and time one hour from now
- Ask your child to write the digital time on the meeting strip.
- Record on the meeting strip the time an activity will occur.

number of the day

- Write three number sentences for the number of the day on the meeting strip.

ASSESSMENT

Written Assessment

"Today I would like to see what you remember from what we have been practicing."

- Give your child **Written Assessment #10**.
- Read the directions for each problem. Allow time for your child to complete that problem before continuing.
- Correct the paper, noting your child's mistakes on the **Individual Recording Form**. Review the errors with your child.

THE LESSON

Drawing Lines Using a Ruler
Drawing a Number Line

"Today you will learn how to draw lines using a ruler."

"You will also learn how to make a number line."

"First we will practice drawing lines using a ruler."

"Let me show you how we will do this."

- Sit next to your child to demonstrate.

  "I put my ruler on the paper where I want to draw my line."

  "Now I hold the ruler with the fingers of the hand I don't write with."

  "I spread my fingers like this and press down to keep the ruler steady."

  "I won't move my ruler until I finish drawing my line."

"To show that this is a line, I will put small arrows at each end."

"The small arrows show that my line continues in both directions."

• Give your child a piece of paper, a ruler, and a color tile.

• Position the piece of paper horizontally:

"Draw a horizontal line on the paper."

• Allow time for your child to do this.

"How will we show that it is a line?"   put arrows on the ends

"Now we are going to make a number line."

"Put a small mark at the left end of your line, like this."

• Demonstrate on your number line:

←———|————————————————————→

"We will use the color tile to tell us where to make our next mark."

"Watch as I mark my number line."

←——▢|——————————————————→

"Each time I make a mark, I move the square to the right."

"It must touch the last mark but not cover it."

"Mark your number line so that it looks like mine."

• Allow time for your child to mark his/her number line.

"When we put numbers on our number line, we will start with zero."

• Demonstrate on the chalkboard number line:

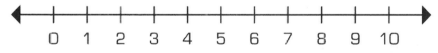

"Number your number line just like mine."

• Allow time for your child to do this.

"Now I will put a point at four on the number line."

"I do this by putting a dot on the number line at four."

"I will call this point A."

"I write the letter A above the point to show where it is."

• Demonstrate on your paper:

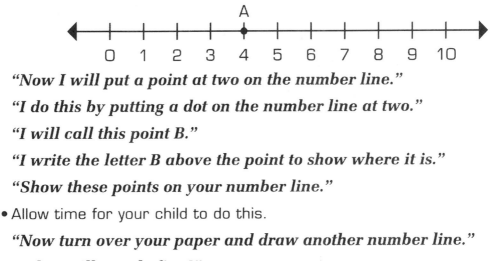

"Now I will put a point at two on the number line."

"I do this by putting a dot on the number line at two."

"I will call this point B."

"I write the letter B above the point to show where it is."

"Show these points on your number line."

- Allow time for your child to do this.

"Now turn over your paper and draw another number line."

"What will you do first?"

"Next?"

- Allow time for your child to draw the number line.

- Put the following points on the number line: C, 6; D, 1; E, 5

Class Practice

number fact practice

"Use your fact cards to practice all the addition facts."

- Allow time for your child to do this.

- Give your child **Fact Sheet A 7.2**.

- Time your child for one minute.

- Correct the fact sheet with your child and record the score.

- Allow time for your child to complete the unfinished facts.

Written Practice

- Complete **Worksheet 55A** with your child.

- Complete **Worksheet 55B** with your child later in the day.

Name _____ **ASSESSMENT 10**
Date _____ **LESSON 55**
                               Math 2

Write a number sentence for the story. Write the answer with a label.

1. Courtney had a box of 64 crayons. She gave Sharon 10 crayons. How many crayons does Courtney have now?

   Number sentence _____64 crayons – 10 crayons = 54 crayons_____

   Answer _____54 crayons_____

2. One of these is my favorite color.
   Use the clues to find my favorite color.

   It is not in the middle.   blue   yellow   red   (green)   purple
   It does not have four letters.
   It is not fifth.
   It is not second.
   Circle my favorite color.

3. What number on the thermometer is the temperature closest to? _80_ °F

4. Measure each line segment using inches.

   ●————————————————● _4_ "

   ●——————————● _2_ "

5. Finish the patterns.

   2, 4, 6, _8_ , _10_ , _12_ , _14_ , _16_ , _18_ , _20_

   1, 3, 5, _7_ , _9_ , _11_ , _13_ , _15_ , _17_ , _19_

6. Find each answer.

   46 + 10 = _56_   10 + 37 = _47_   74 + 10 = _84_

2-55Aa   Copyright © 1991 by Saxon Publishers, Inc. and Nancy Larson. Reproduction prohibited.

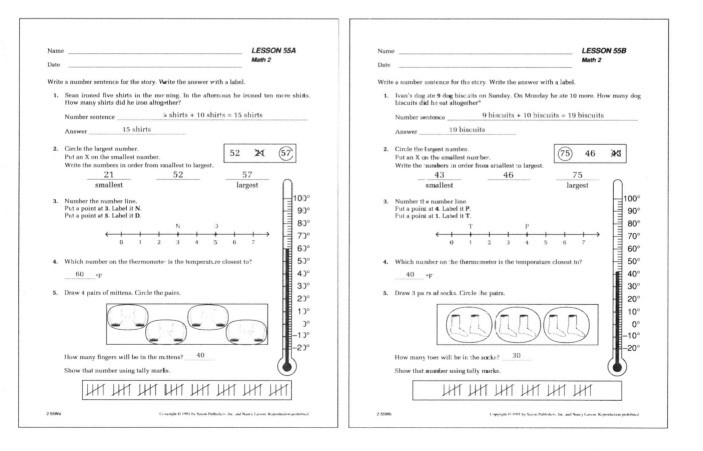

Name _____ **LESSON 55A**
Date _____ **Math 2**

Write a number sentence for the story. Write the answer with a label.

1. Sean ironed five shirts in the morning. In the afternoon he ironed ten more shirts. How many shirts did he iron altogether?

   Number sentence _____5 shirts + 10 shirts = 15 shirts_____

   Answer _____15 shirts_____

2. Circle the largest number.   52   ✗   ⑤⑦
   Put an X on the smallest number.
   Write the numbers in order from smallest to largest.
   _21_   _52_   _57_
   smallest   largest

3. Number the number line.
   Put a point at 3. Label it **N**.
   Put a point at 5. Label it **D**.

   0 1 2 3 4 5 6 7

4. Which number on the thermometer is the temperature closest to?

   _60_ °F

5. Draw 4 pairs of mittens. Circle the pairs.

   How many fingers will be in the mittens? _40_

   Show that number using tally marks.

2-55Wa   Copyright © 1991 by Saxon Publishers, Inc. and Nancy Larson. Reproduction prohibited.

Name _____ **LESSON 55B**
Date _____ **Math 2**

Write a number sentence for the story. Write the answer with a label.

1. Ivan's dog ate 9 dog biscuits on Sunday. On Monday he ate 10 more. How many dog biscuits did he eat altogether?

   Number sentence _____9 biscuits + 10 biscuits = 19 biscuits_____

   Answer _____19 biscuits_____

2. Circle the largest number.   ⑦⑤   46   ✗
   Put an X on the smallest number.
   Write the numbers in order from smallest to largest.
   _43_   _46_   _75_
   smallest   largest

3. Number the number line.
   Put a point at 4. Label it **P**.
   Put a point at 1. Label it **T**.

   0 1 2 3 4 5 6 7

4. Which number on the thermometer is the temperature closest to?

   _40_ °F

5. Draw 3 pairs of socks. Circle the pairs.

   How many toes will be in the socks? _30_

   Show that number using tally marks.

2-55Wb   Copyright © 1991 by Saxon Publishers, Inc. and Nancy Larson. Reproduction prohibited.

esson 56

measuring to the nearest foot

---

lesson preparation

materials

2 rulers

Master 2-56

Fact Sheet S 3.4

in the morning

• Write the following in the pattern box on the meeting strip:

☐, 2, ☐, 4, ☐, 6, ☐, ___, ___, ___, ___

Answer: ☐, 2, ☐, 4, ☐, 6, ☐, 8, ☐, 10, ☐

• Write ⟨37¢⟩ on the meeting strip. Provide a cup of 10 dimes, a cup of 10 nickels, and a cup of 20 pennies.

THE MEETING

calendar

- Ask your child to write the date on the calendar and meeting strip.
- Ask your child the following:

    date _____ days ago, date _____ days from now

    day of the week _____ days ago, day of the week _____ days from now

    months of the year, _____th month, month before, month after

- Record on the meeting strip a special event and the number of days until it occurs.

weather graph

- Ask your child to color the graph and write the temperature to the nearest ten degrees in the box he/she colored.
- Ask questions about the graph.

counting

"We will practice adding and subtracting ten and one again today."

- Ask your child to select a number on the hundred number chart.

"I will say add ten, subtract ten, add one, or subtract one."

"Say the new answer and point to the number on the hundred number chart."

- Do this 6–10 times.

"What will you need to do to get back to (starting number)?"

- Ask your child to give the directions for returning to the starting number.

- Do the following 2–3 times a week:

    count by 10's to 300 and backward from 300 by 10's

    count by 5's to 100 and backward from 50 by 5's

    say the even numbers to 100 and backward from 100

    say the odd numbers to 49 and backward from 49

graph questions

- You and your child each ask a question about any of the graphs.

patterning

- Ask your child to do the following:

    identify the pattern (repeating, continuing, or both)

    identify the numbers or shapes to complete the pattern

    read the pattern

money

- Ask your child to put the dimes, nickels, and/or pennies in the coin cup.

- Count the money in the coin cup together.

clock

- Ask your child to set the clock on the half hour or hour.

- Ask the following:

    time shown on the clock

    time one hour ago and time one hour from now

- Ask your child to write the digital time on the meeting strip.

- Record on the meeting strip the time an activity will occur.

number of the day

- Write three number sentences for the number of the day on the meeting strip.

THE LESSON

Measuring to the Nearest Foot

"We have been measuring line segments using inches."

"Today you will learn how to measure longer distances using feet."

"If we wanted to measure the distance from (choose a place in the room) to the door, we could use our feet to measure the distance."

"We could do this by putting our heel in front of our toes and counting the steps."

"How many steps do you think it will take for me to reach the door?"

- Walk the distance in a heel, toe manner.

- Record the distance on the following chalkboard chart:

	Parent's feet
Distance from _____ to door	

"Now you will measure the distance from the _____ to the door using your feet."

"How many steps do you think it will take for you to reach the door?"

- Ask your child to walk the distance in a heel, toe manner.

- Record the distance on the chalkboard chart:

	Parent's feet	Child's feet
Distance from _____ to door		

"What happened when we used different people's feet to measure the distance?"

"Why are these numbers different?"

"Your 12-inch ruler is the official length of a foot."

"Why do you think we use this instead of our feet?"   people's feet are different lengths

"When we measure in ruler feet, we need to put the rulers end to end, just like we do with our real feet."

- Demonstrate by putting the rulers end to end from the given location to the door.

"Let's count how many ruler feet it is from the _____ to the door."

- Record the distance on the chalkboard chart:

	Parent's feet	Child's feet	Ruler feet
Distance from _____ to door			

"When we tell other people how many feet long something is, we tell them the length in ruler feet."

"We say that the length, or distance, is _____ feet."

- Choose two objects approximately 6 feet apart.

"Now we will measure the distance from _____ to _____ using our feet and ruler feet."

- Give your child **Master 2-56**.

- Measure the distance in a heel, toe manner, first using your feet and then your child's feet. Record the results on **Master 2-56**.

"Now we will measure the distance in feet using a ruler."

"Do you think there will be more or less ruler feet than the number of our feet?"

"Why?"

"Let's try it to see."

"What should we do if we can only use two rulers?" move them like we moved our feet

"This is just like moving our feet."

"I will keep lifting the first ruler and putting it after the last ruler as we count the number of ruler feet we have used."

"How many feet is it from _____ to _____?"

- Repeat two more times. Limit the distance to be measured to 10 feet.

CLASS PRACTICE

number fact practice

- Use the tan, peach, and lavender fact cards to practice the subtraction facts with your child.

- Give your child **Fact Sheet S 3.4**.

- Time your child for one minute.

- Correct the fact sheet with your child and record the score.

- Allow time for your child to complete the unfinished facts.

WRITTEN PRACTICE

• Complete **Worksheet 56A** with your child.

• Complete **Worksheet 56B** with your child later in the day.

Name _____ MASTER 2-56
Math 2

What we measured	_____'s feet	_____'s feet	ruler feet

2-56Ma   Copyright © 1991 by Saxon Publishers, Inc. and Nancy Larson. Reproduction prohibited.

Name _____ LESSON 56A
Math 2
Date _____

Write a number sentence for the story. Write the answer with a label.

1. There were 25 children in Room 12. Ten children left to go to the library. How many children are in Room 12 now?

   Number sentence _____25 children − 10 children = 15 children_____

   Answer _____15 children_____

2. Number the number line.
   Put a point at **4.** Label it **P.**
   Put a point at **1.** Label it **T.**

   Where is point **Z?** ___5___

3. Fill in the missing numbers on this piece of a hundred number chart.

	37	38	39	
45	46	47	48	49
	56	57		

4. Color the even numbers in Problem 3 red.
   Color the odd numbers in Problem 3 yellow.

5. Fill in the missing numbers. Use the fewest number of pennies possible.

   7 dimes + 11 pennies = ___8___ dimes + ___1___ pennies = ___81___ ¢

   3 dimes + 19 pennies = ___4___ dimes + ___9___ pennies = ___49___ ¢

6. Find each answer.

   58 − 10 = ___48___   92 − 10 = ___82___   18 − 10 = ___8___

   40 + 10 = ___50___   35 + 10 = ___45___   50 − 10 = ___40___

2-56Wa   Copyright © 1991 by Saxon Publishers, Inc. and Nancy Larson. Reproduction prohibited.

Name _____ LESSON 56B
Math 2
Date _____

Write a number sentence for the story. Write the answer with a label.

1. Sarah had 65¢. She gave her brother a dime. How much money does she have now?

   Number sentence _____65¢ − 10¢ = 55¢_____

   Answer _____55¢_____

2. Number the number line.
   Put a point at **5.** Label it **A.**
   Put a point at **3.** Label it **B.**

   Where is point **C?** ___4___

3. Fill in the missing numbers on this piece of a hundred number chart.

72	73	74				
82	83	84	85	86	87	88
		94	95	96		

4. Use your feet to measure the length of your bedroom.
   (Walk in a straight line.) _____ feet

   Have someone else in your family measure your bedroom with their feet.
   Name _____ _____ feet
   Who took more steps? _____
   Why? _____

5. Fill in the missing numbers. Use the fewest number of pennies possible.

   2 dimes + 16 pennies = ___3___ dimes + ___6___ pennies = ___36___ ¢

6. Find each answer.

   63 − 10 = ___53___   80 + 10 = ___90___   17 + 10 = ___27___

   24 + 10 = ___34___   26 − 10 = ___16___   32 − 10 = ___22___

2-56Wb   Copyright © 1991 by Saxon Publishers, Inc. and Nancy Larson. Reproduction prohibited.

Lesson 57

making geometric shapes on a geoboard
identifying the angles of a shape

lesson preparation

materials

1 geoboard and geoband

Fact Sheet A 7.2

in the morning

- Write the following in the pattern box on the meeting strip:

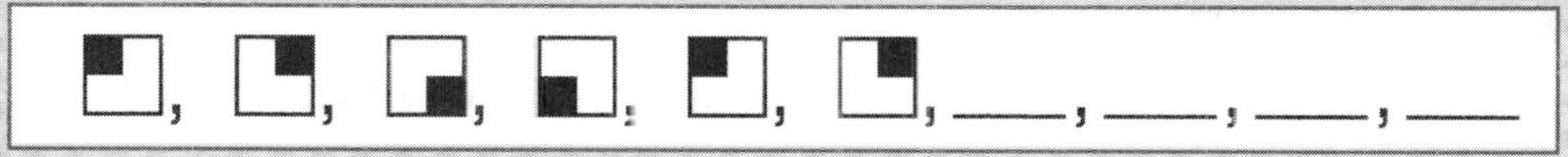

Answer: 

- Write 84¢ on the meeting strip. Provide a cup of 10 dimes, a cup of 10 nickels, and a cup of 20 pennies.

THE MEETING

calendar

- Ask your child to write the date on the calendar and meeting strip.
- Ask your child the following:

  date _____ days ago, date _____ days from now

  day of the week _____ days ago, day of the week _____ days from now

  months of the year, _____th month, month before, month after
- Record on the meeting strip a special event and the number of days until it occurs.

weather graph

- Ask your child to color the graph and write the temperature to the nearest ten degrees in the box he/she colored.
- Ask questions about the graph.

counting

"We will practice adding and subtracting ten and one again today."

- Ask your child to select a number on the hundred number chart.

  "I will say add ten, subtract ten, add one, or subtract one."

  "Say the new answer and point to the number on the hundred number chart."

- Do this 6–10 times.

  "What will you need to do to get back to (starting number)?"

- Ask your child to give the directions for returning to the starting number.

- Do the following 2–3 times a week:

  count by 10's to 300 and backward from 300 by 10's

  count by 5's to 100 and backward from 50 by 5's

  say the even numbers to 100 and backward from 100

  say the odd numbers to 49 and backward from 49

graph questions

- You and your child each ask a question about any of the graphs.

patterning

- Ask your child to do the following:

  identify the pattern (repeating, continuing, or both)

  identify the shapes to complete the pattern

  read the pattern

money

- Ask your child to put the dimes, nickels, and/or pennies in the coin cup.
- Count the money in the coin cup together.

clock

- Ask your child to set the clock on the half hour or hour.
- Ask the following:

  time shown on the clock

  time one hour ago and time one hour from now

- Ask your child to write the digital time on the meeting strip.
- Record on the meeting strip the time an activity will occur.

number of the day

- Write three number sentences for the number of the day on the meeting strip.

THE LESSON

Making Geometric Shapes on a Geoboard
Identifying the Angles of a Shape

"Today you will learn how to identify the angles of a shape."

- Show your child a geoboard.

  "This is called a geoboard."

  "What do you notice about the geoboard?"

- Encourage your child to offer observations.

  "We use geobands on a geoboard."

- Show your child a geoband (rubber band).

  "When we use geoboards and geobands, we will need to use them in a safe way."

  "What do you think is a safe way to use a geoboard and geoband?"

  "I will show you a safe way many people use to put a geoband on the geoboard and to take it off the geoboard."

- Demonstrate as you say the steps.

  "Put your geoband over a peg."

  "Put your finger on that peg."

  "Now, carefully stretch the band to another peg."

  "Test it before you take your finger off the first peg."

- Demonstrate.

  "When you take the geoband off the geoboard, put your finger on one peg and carefully lift the other end of the band."

- Give your child the geoboard and the geoband.

  "Practice putting the geoband on the geoboard and taking it off."

- Allow time for your child to do this.

  "Make something with your geoband."

- Allow time for your child to explore using the geoband on the geoboard.

- Ask your child to describe what he/she did with the geoband.

  "Take the geoband off the geoboard."

  "Now make a shape with three sides."

- Allow time for your child to make the shape.

  "What do we call this shape?"   triangle

  "Put your finger on one of the pegs where the sides meet."

  "Now move your finger off the peg into the inside of the triangle."

"Some people call this a corner of the triangle."

"Mathematicians call this an angle."

"Point to another angle in the triangle."

"Point to another angle in the triangle."

"How many angles does a triangle have?"   3

"Now make a shape with four sides."

- Allow time for your child to make the shape.

"Put your finger on one of the pegs where the sides meet."

"Now move your finger off the peg into the inside of the shape."

"This is one of the angles of the shape."

"How many angles do you see?"   4

"Now make a square."

- Allow time for your child to make the square.

"How do you know that the shape is a square?"

"Now make a rectangle."

- Allow time for your child to make the rectangle.

"How do you know that this shape is a rectangle?"

"How many angles does a four-sided shape have?"   4

"Carefully take your geoband off your geoboard."

CLASS PRACTICE

number fact practice

"Use your fact cards to practice all the addition facts."

- Allow time for your child to do this.
- Give your child **Fact Sheet A 7.2.**
- Time your child for one minute.
- Correct the fact sheet with your child and record the score.
- Allow time for your child to complete the unfinished facts.

WRITTEN PRACTICE

- Complete **Worksheet 57A** with your child.
- Complete **Worksheet 57B** with your child later in the day.

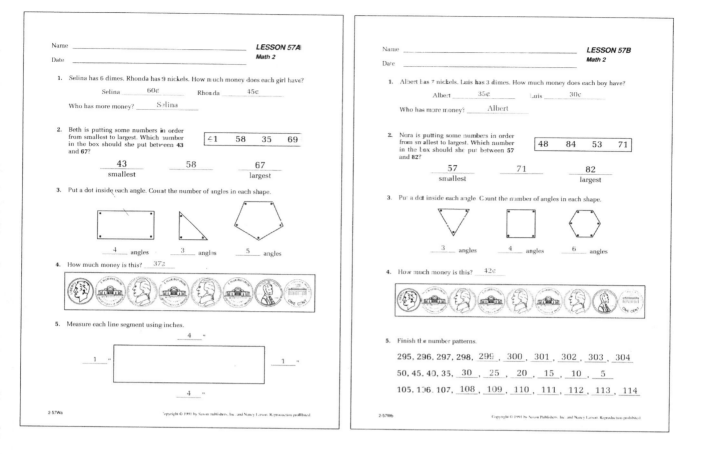

Name _____ **LESSON 57A**
Date _____ **Math 2**

1. Selina has 6 dimes. Rhonda has 9 nickels. How much money does each girl have?

   Selina ____60¢____ Rhonda ____45¢____

   Who has more money? ____Selina____

2. Beth is putting some numbers in order from smallest to largest. Which number in the box should she put between **43** and **67**?

   [41   58   35   69]

   ____43____   ____58____   ____67____
   smallest   largest

3. Put a dot inside each angle. Count the number of angles in each shape.

   ____4____ angles   ____3____ angles   ____5____ angles

4. How much money is this? ___37¢___

5. Measure each line segment using inches.

   4 "
   1 "   1 "
   4 "

2-57Wa

Name _____ **LESSON 57B**
Date _____ **Math 2**

1. Albert has 7 nickels. Luis has 3 dimes. How much money does each boy have?

   Albert ____35¢____ Luis ____30¢____

   Who has more money? ____Albert____

2. Nora is putting some numbers in order from smallest to largest. Which number in the box should she put between **57** and **82**?

   [48   84   53   71]

   ____57____   ____71____   ____82____
   smallest   largest

3. Put a dot inside each angle. Count the number of angles in each shape.

   ____3____ angles   ____4____ angles   ____6____ angles

4. How much money is this? ___42¢___

5. Finish the number patterns.

   295, 296, 297, 298, _299_, _300_, _301_, _302_, _303_, _304_

   50, 45, 40, 35, _30_, _25_, _20_, _15_, _10_, _5_

   105, 106, 107, _108_, _109_, _110_, _111_, _112_, _113_, _114_

2-57Wb

311

Lesson 58

addition facts—last eight facts

lesson preparation

materials

cup of 14 pennies

work mat

addition fact cards — white

Fact Sheet A 8.1

the night before

• Separate the white addition fact cards.

in the morning

• Write the following in the pattern box on the meeting strip:

> 11, 22, 33, 44, ___, ___, ___, ___, ___

Answer: 11, 22, 33, 44, 55, 66, 77, 88, 99

• Write 46¢ on the meeting strip. Provide a cup of 10 dimes, a cup of 10 nickels, and a cup of 20 pennies.

THE MEETING

calendar

• Ask your child to write the date on the calendar and meeting strip.

• Ask your child the following:

    date ____ days ago, date ____ days from now

    day of the week ____ days ago, day of the week ____ days from now

    months of the year, ____th month, month before, month after

• Record on the meeting strip a special event and the number of days until it occurs.

weather graph

• Ask your child to color the graph and write the temperature to the nearest ten degrees in the box he/she colored.

• Ask questions about the graph.

counting

"We will practice adding and subtracting ten and one again today."

• Ask your child to select a number on the hundred number chart.

"I will say add ten, subtract ten, add one, or subtract one."

"Say the new answer and point to the number on the hundred number chart."

• Do this 6–10 times.

"What will you need to do to get back to (starting number)?"

• Ask your child to give the directions for returning to the starting number.

• Do the following 2–3 times a week:

count by 10's to 300 and backward from 300 by 10's

count by 5's to 100 and backward from 50 by 5's

say the even numbers to 100 and backward from 100

say the odd numbers to 49 and backward from 49

graph questions

• You and your child each ask a question about any of the graphs.

patterning

• Ask your child to do the following:

identify the pattern (repeating, continuing, or both)

identify the numbers to complete the pattern

read the pattern

money

• Ask your child to put the dimes, nickels, and/or pennies in the coin cup.

• Count the money in the coin cup together.

clock

• Ask your child to set the clock on the half hour or hour.

• Ask the following:

time shown on the clock

time one hour ago and time one hour from now

• Ask your child to write the digital time on the meeting strip.

• Record on the meeting strip the time an activity will occur.

number of the day

- Write three number sentences for the number of the day on the meeting strip.

THE LESSON

Addition Facts — Last Eight Facts

"We have learned almost all of the addition facts."

"Today you will learn the last eight addition facts."

- Write the following on the chalkboard:

$$\begin{array}{cccccccc} 5 & 6 & 8 & 7 & 7 & 8 & 8 & 8 \\ +\,3 & +\,3 & +\,3 & +\,4 & +\,5 & +\,4 & +\,5 & +\,6 \end{array}$$

"Which of these answers do you think will be more than 10?"

"How do you know?"

- Put a check above each problem your child names.

"Let's use pennies to find these answers."

- Give your child a construction paper work mat and a cup of twenty pennies.

"Let's make up a story for the first problem."

- Ask your child to make up a story to match the problem.

"How will we use the pennies to find the answer to this story?"

"Put five pennies on your mat."

"Now put three more pennies on your mat."

"How many pennies do you have now?" 8

"How did you find the answer?"

- Encourage your child to count on from five.

- Record the answer beneath the chalkboard problem.

"Let's make up a story for the next problem."

- Ask your child to make up a story to match the problem.

"What do you think the answer will be?"

"How do you know?"

"Let's use the pennies to find the answer to this story."

"Put six pennies on the work mat."

"Now put three more pennies on the work mat."

"How many pennies do you have now?" 9

- Record the answer beneath the chalkboard problem.
- Repeat with the other problems.

  "Put the pennies in the cup."

  "Which problems have the same answers?" 8 + 3, 7 + 4; 7 + 5, 8 + 4

  "What do you notice about these problems?"

  "Some people call these eight number facts the oddballs because they don't follow a pattern."

  "Let's practice these number facts together."

  "I'll say a problem."

  "Let's see how fast you can say the answer."

- Allow your child to refer to the chalkboard problems, if necessary.

CLASS PRACTICE

number fact practice

- Give your child the white addition fact cards.
- Use the white fact cards to practice the oddball number facts with your child.
- Give your child **Fact Sheet A 8.1.**
- Time your child for one minute.
- Correct the fact sheet with your child and record the score.
- Allow time for your child to complete the unfinished facts.

WRITTEN PRACTICE

- Complete **Worksheet 58A** with your child.
- Complete **Worksheet 58B** with your child later in the day.

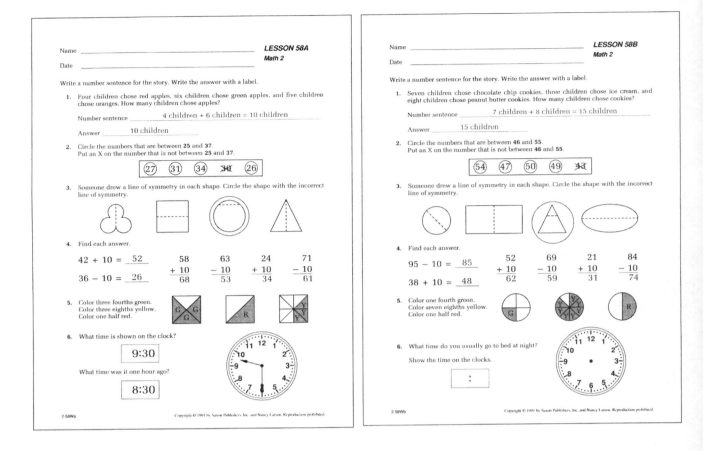

Name _____ **LESSON 58A**
Date _____ Math 2

Write a number sentence for the story. Write the answer with a label.

1. Four children chose red apples, six children chose green apples, and five children chose oranges. How many children chose apples?

   Number sentence _____ 4 children + 6 children = 10 children _____

   Answer _____ 10 children _____

2. Circle the numbers that are between **25** and **37**.
   Put an X on the number that is not between **25** and **37**.

   (27) (31) (34) 38 (26)

3. Someone drew a line of symmetry in each shape. Circle the shape with the incorrect line of symmetry.

4. Find each answer.

   42 + 10 = __52__   58   63   24   71
                     + 10   − 10   + 10   − 10
   36 − 10 = __26__   ‾‾68‾‾ ‾‾53‾‾ ‾‾34‾‾ ‾‾61‾‾

5. Color three fourths green.
   Color three eighths yellow.
   Color one half red.

6. What time is shown on the clock?

   [9:30]

   What time was it one hour ago?

   [8:30]

2-58Wa

Copyright © 1991 by Saxon Publishers, Inc. and Nancy Larson. Reproduction prohibited.

Name _____ **LESSON 58B**
Date _____ Math 2

Write a number sentence for the story. Write the answer with a label.

1. Seven children chose chocolate chip cookies, three children chose ice cream, and eight children chose peanut butter cookies. How many children chose cookies?

   Number sentence _____ 7 children + 8 children = 15 children _____

   Answer _____ 15 children _____

2. Circle the numbers that are between **46** and **55**.
   Put an X on the number that is not between **46** and **55**.

   (54) (47) (50) (49) 43

3. Someone drew a line of symmetry in each shape. Circle the shape with the incorrect line of symmetry.

4. Find each answer.

   95 − 10 = __85__   52   69   21   84
                     + 10   − 10   + 10   − 10
   38 + 10 = __48__   ‾‾62‾‾ ‾‾59‾‾ ‾‾31‾‾ ‾‾74‾‾

5. Color one fourth green.
   Color seven eighths yellow.
   Color one half red.

6. What time do you usually go to bed at night?

   Show the time on the clocks.

   [:]

2-58Wb

Copyright © 1991 by Saxon Publishers, Inc. and Nancy Larson. Reproduction prohibited.

L esson 59

identifying 1-cup and 1/2-cup measuring cups, tablespoons, teaspoons, and 1/2 teaspoons
reading a recipe

---

lesson preparation

materials

recipe written on paper

liquid and dry measuring cups

measuring spoons

Fact Sheet A 8.1

the night before

• Write the following recipe or a recipe of your choice on a piece of paper. In Lesson 62, your child will participate in the mixing and baking of the ingredients.

Apple Jack Cookies

Ingredients:	Directions:
1 cup light brown sugar	Cream together sugar and shortening.
$\frac{1}{2}$ cup shortening	Beat in egg.
1 egg	Sift together dry ingredients and add to mixture.
$1\frac{1}{2}$ cups flour	Beat until well blended.
$\frac{1}{2}$ tsp. baking soda	Stir in apples.
$\frac{1}{2}$ tsp. salt	Drop in the shape of balls on a greased cookie
1 tsp. nutmeg	sheet.
1 cup chopped unpeeled apple	Bake at 375 degrees for 12–15 minutes.

in the morning

• Write the following in the pattern box on the meeting strip:

> ___, ___, ___, 37, 38, 39, ___, ___, ___

Answer: 34, 35, 36, 37, 38, 39, 40, 41, 42

• Write 60¢ on the meeting strip. Provide a cup of 10 dimes, a cup of 10 nickels, and a cup of 20 pennies.

THE MEETING

calendar

- Ask your child to write the date on the calendar and meeting strip.

- Ask your child the following:

   date _____ days ago, date _____ days from now

   day of the week _____ days ago, day of the week _____ days from now

   months of the year, _____th month, month before, month after

- Record on the meeting strip a special event and the number of days until it occurs.

weather graph

- Ask your child to color the graph and write the temperature to the nearest ten degrees in the box he/she colored.

- Ask questions about the graph.

counting

   "We will practice adding and subtracting ten and one again today."

- Ask your child to select a number on the hundred number chart.

   "I will say add ten, subtract ten, add one, or subtract one."

   "Say the new answer and point to the number on the hundred number chart."

- Do this 6–10 times.

   "What will you need to do to get back to (starting number)?"

- Ask your child to give the directions for returning to the starting number.

- Do the following 2–3 times a week:

   count by 10's to 300 and backward from 300 by 10's

   count by 5's to 100 and backward from 50 by 5's

   say the even numbers to 100 and backward from 100

   say the odd numbers to 49 and backward from 49

graph questions

- You and your child each ask a question about any of the graphs.

patterning

- Ask your child to do the following:

   identify the pattern (repeating, continuing, or both)

   identify the numbers to complete the pattern

   read the pattern

money

- Ask your child to put the dimes, nickels, and/or pennies in the coin cup.
- Count the money in the coin cup together.

clock

- Ask your child to set the clock on the half hour or hour.
- Ask the following:

    time shown on the clock

    time one hour ago and time one hour from now

- Ask your child to write the digital time on the meeting strip.
- Record on the meeting strip the time an activity will occur.

number of the day

- Write three number sentences for the number of the day on the meeting strip.

THE LESSON

Identifying 1-Cup and 1/2-Cup Measuring Cups, Tablespoons, Teaspoons, and 1/2 Teaspoons
Reading a Recipe

"Today you will learn how to identify measuring cups and measuring spoons."

"You will also learn how to read a recipe."

"What is a recipe?" list of ingredients and directions for making something

"I have written a recipe on this paper."

"In a few days we will work together to make this recipe."

"Let's read the recipe together."

- Point to the words as your child reads the recipe.

"The things we use to make the recipe are called the ingredients."

"What ingredients will we use for our recipe?"

"How will we know how much of each ingredient to use?"

- Show your child a set of measuring cups (nesting, if possible), a liquid 1-cup measuring cup, and measuring spoons.

"These are called measuring cups and spoons."

- Hold up the liquid 1-cup measuring cup.

"We use this cup when we measure liquids such as water, milk, or oil."

"Will we need to use this cup for our recipe?"   no

"Which measuring cups will we need to use for our recipe?"   1-cup, 1/2-cup

"Which is the 1-cup measuring cup?"

"Which is the 1/2-cup measuring cup?"

• Continue, if other cups are required for your recipe.

"Sometimes recipes use the abbreviations 'tsp.,' 'tbsp.,' 't,' or 'T.' "

• Write the following on the chalkboard:

teaspoon	tablespoon
tsp.	tbsp.
t	T

"These are abbreviations for teaspoon and tablespoon."

• In the original words, underline the letters used in the abbreviations.

"Some people remember that the teaspoon is the small spoon and the tablespoon is the large spoon by remembering that a teabag is smaller than a table."

"Just as there is a 1/2-cup measuring cup, there is also a 1/2-teaspoon measuring spoon."

"Now let's try to find the measuring cup or spoon we will use for each of our ingredients."

• Ask your child to hold up the appropriate cup or spoon for each ingredient.

CLASS PRACTICE

number fact practice

"Today we will play a new fact game."

"This game is called 'Number Fact Jeopardy.' "

"What are the doubles plus one facts?"

• Write them on the chalkboard as your child lists the facts.

• Erase the answers.

"Now I will say the answer to a doubles plus one fact."

"Let's see how quickly you can say the problem."

• Say the answers in random order.

• Repeat several times.

• Repeat with other groups of facts your child needs to practice.

- Use the white fact cards to practice the oddball addition facts with your child.

- Give your child **Fact Sheet A 8.1.**

- Time your child for one minute.

- Correct the fact sheet with your child and record the score.

- Allow time for your child to complete the unfinished facts.

    "On the back of your fact sheet, write the numbers from 285 to 320."

WRITTEN PRACTICE

- Complete **Worksheet 59A** with your child.

- Complete **Worksheet 59B** with your child later in the day.

Name _____   **LESSON 59A**
                                Math 2
Date _____

Write a number sentence for the story. Write the answer with a label.

1. Forty-seven children were on the bus. Ten children got off at the first stop. How many children are on the bus now?

    Number sentence _47 children – 10 children = 37 children_   Answer _37 children_

2. Number the number line using the even numbers.

    Put a point at **12.** Label it **A.**
    Put a point at **8.** Label it **B.**

3. Draw a line to the correct picture.

    one fourth
    five eighths
    one third
    three fourths

4. Find each answer.

    46 + 10 = _56_   63 – 10 = _53_
    27 + 10 = _37_   48 – 10 = _38_
    10 more than 31 = _41_   10 less than 57 = _47_
    10 more than 72 = _82_   10 less than 84 = _74_

5. Using tally marks, show the number of children in your class.

6. Write the numbers in order from smallest to largest.

    | 65 | 54 | 63 | _54_   _63_   _65_
                       smallest   largest

2-59Wa   Copyright © 1994 by Saxon Publishers, Inc. and Nancy Larson. Reproduction prohibited.

Name _____   **LESSON 59B**
                                Math 2
Date _____

Write a number sentence for the story. Write the answer with a label.

1. Thirty-seven children were on the bus. Ten children got off at the first stop. How many children are on the bus now?

    Number sentence _37 children – 10 children = 27 children_   Answer _27 children_

2. Number the number line using the even numbers.

    Put a point at **14.** Label it **C.**
    Put a point at **6.** Label it **D.**

3. Draw a line to the correct picture.

    one fourth
    two thirds
    three eighths
    two fourths

4. Find each answer.

    29 + 10 = _39_   92 – 10 = _82_
    73 + 10 = _83_   71 – 10 = _61_
    10 more than 42 = _52_   10 less than 38 = _28_
    10 more than 55 = _65_   10 less than 56 = _46_

5. Using tally marks, show the number of lights in your house.

6. Write the numbers in order from smallest to largest.

    | 86 | 81 | 74 | _74_   _81_   _86_
                       smallest   largest

2-59Wb   Copyright © 1994 by Saxon Publishers, Inc. and Nancy Larson. Reproduction prohibited.

esson 60

creating congruent shapes

---

lesson preparation ———————————————————

materials

Written Assessment #11

Oral Assessment #6

2 geoboards and geobands

Master 2-60

individual clock

Fact Sheet A 8.1

in the morning

• Write the following in the pattern box on the meeting strip:

$$20, 22, 24, 26, \underline{\quad}, \underline{\quad}, \underline{\quad}, \underline{\quad}, \underline{\quad}$$

Answer: 20, 22, 24, 26, 28, 30, 32, 34, 36

• Write ⎢44¢⎢ on the meeting strip. Provide a cup of 10 dimes, a cup of 10 nickels, and a cup of 20 pennies.

THE MEETING

calendar

• Ask your child to write the date on the calendar and meeting strip.

• Ask your child the following:

date _____ days ago, date _____ days from now

day of the week _____ days ago, day of the week _____ days from now

months of the year, _____th month, month before, month after

• Record on the meeting strip a special event and the number of days until it occurs.

weather graph

• Ask your child to color the graph and write the temperature to the nearest ten degrees in the box he/she colored.

• Ask questions about the graph.

counting

"We will practice adding and subtracting ten and one again today."

• Ask your child to select a number on the hundred number chart.

"I will say add ten, subtract ten, add one, or subtract one."

"Say the new answer and point to the number on the hundred number chart."

• Do this 6–10 times.

"What will you need to do to get back to (starting number)?"

• Ask your child to give the directions for returning to the starting number.

• Do the following 2–3 times a week:

count by 10's to 300 and backward from 300 by 10's

count by 5's to 100 and backward from 50 by 5's

say the even numbers to 100 and backward from 100

say the odd numbers to 49 and backward from 49

graph questions

• You and your child each ask a question about any of the graphs.

patterning

• Ask your child to do the following:

identify the pattern (repeating, continuing, or both)

identify the numbers to complete the pattern

read the pattern

money

• Ask your child to put the dimes, nickels, and/or pennies in the coin cup.

• Count the money in the coin cup together.

clock

• Ask your child to set the clock on the half hour or hour.

• Ask the following:

time shown on the clock

time one hour ago and time one hour from now

• Ask your child to write the digital time on the meeting strip.

• Record on the meeting strip the time an activity will occur.

number of the day

- Write three number sentences for the number of the day on the meeting strip.

ASSESSMENT

Written Assessment

"Today I would like to see what you remember from what we have been practicing."

- Give your child **Written Assessment #11**.

- Read the directions for each problem. Allow time for your child to complete that problem before continuing.

- Correct the paper, noting your child's mistakes on the **Individual Recording Form**. Review the errors with your child.

Oral Assessment

- Record your child's response(s) to the oral interview questions on the interview sheet.

THE LESSON

Creating Congruent Shapes

"A few days ago you used a geoboard to make shapes."

"Today you will learn how to make congruent shapes and designs using a geoboard."

"Congruent shapes and designs are shapes and designs that are exactly the same."

- Make the following shape on your geoboard:

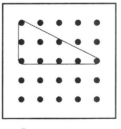

"What shape is this?"   *triangle*

"How do you know?"   *it has three sides, three angles*

- Give your child a geoboard and a geoband.

"Try to make a triangle that is exactly like mine."

- Assist your child, if necessary.

  "Put your geoboard next to mine."

  "We have congruent triangles."

  "They are exactly the same size and shape."

  "Now you will make a shape for me to copy."

- Allow time for your child to do this.

- Copy the shape on your geoboard.

- Put the geoboards next to each other.

  "Are our shapes congruent?"

  "How do you know?"

  "Now I will make a shape for you to copy."

- Make a shape on your geoboard.

  "Make a congruent shape on your geoboard."

- Allow time for your child to do this.

  "Are our shapes congruent?"

  "How do you know?"

- Repeat several times.

  "Now you will make a shape or design on the geoboard and copy it on geodot paper using a crayon."

  "Put your geoboard next to the geodot paper to show the congruent shape or design."

- Give your child **Master 2-60**.

CLASS PRACTICE

number fact practice

- Use the white fact cards to practice the last eight addition facts with your child.

- Give your child **Fact Sheet A 8.1**.

- Time your child for one minute.

- Correct the fact sheet with your child and record the score.

- Allow time for your child to complete the unfinished facts.

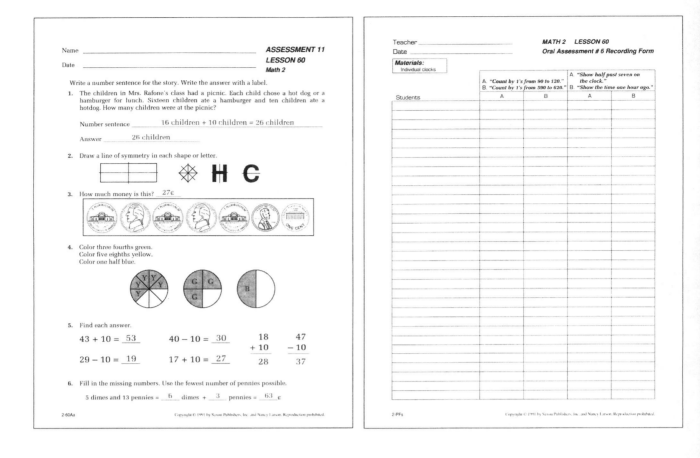

Name _____ **ASSESSMENT 11**
Date _____ **LESSON 60**
Math 2

Write a number sentence for the story. Write the answer with a label.

1. The children in Mrs. Rafone's class had a picnic. Each child chose a hot dog or a hamburger for lunch. Sixteen children ate a hamburger and ten children ate a hotdog. How many children were at the picnic?

   Number sentence _____ 16 children + 10 children = 26 children

   Answer _____ 26 children

2. Draw a line of symmetry in each shape or letter.

3. How much money is this? __27¢__

4. Color three fourths green.
   Color five eighths yellow.
   Color one half blue.

5. Find each answer.

   $43 + 10 = \underline{53}$ $40 - 10 = \underline{30}$ $\begin{array}{r} 18 \\ + 10 \\ \hline 28 \end{array}$ $\begin{array}{r} 47 \\ - 10 \\ \hline 37 \end{array}$

   $29 - 10 = \underline{19}$ $17 + 10 = \underline{27}$

6. Fill in the missing numbers. Use the fewest number of pennies possible.

   5 dimes and 13 pennies = __6__ dimes + __3__ pennies = __63__ ¢

2-60Aa

Teacher _____ **MATH 2   LESSON 60**
Date _____ *Oral Assessment # 6 Recording Form*

Materials:
Individual clocks

Students	A. "Count by 1's from 90 to 120." B. "Count by 1's from 590 to 620."		A. "Show half past seven on the clock." B. "Show the time one hour ago."	
	A	B	A	B

2-PF8

Name _____ **MASTER 2-60**
Math 2

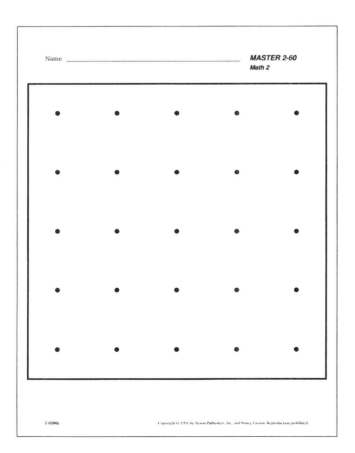

2-60Ma

Lesson 61

creating and reading a Venn diagram

THE MEETING

calendar

• Ask your child to write the date on the calendar and meeting strip.

• Ask your child the following:

date ____ days ago, date ____ days from now

day of the week ____ days ago, day of the week ____ days from now

months of the year, ____th month, month before, month after

• Record on the meeting strip a special event and the number of days until it occurs.

weather graph

• Ask your child to color the graph and write the temperature to the nearest ten degrees in the box he/she colored.

• Ask questions about the graph.

counting

"We will practice adding and subtracting ten and one again today."

- Ask your child to select a number on the hundred number chart.

  "I will say add ten, subtract ten, add one, or subtract one."

  "Say the new answer and point to the number on the hundred number chart."

- Do this 6–10 times.

  "What will you need to do to get back to (starting number)?"

- Ask your child to give the directions for returning to the starting number.

- Do the following 2–3 times a week:

  count by 10's to 300 and backward from 300 by 10's

  count by 5's to 100 and backward from 50 by 5's

  say the even numbers to 100 and backward from 100

  say the odd numbers to 49 and backward from 49

graph questions

- You and your child each ask a question about any of the graphs.

patterning

- Ask your child to do the following:

  identify the pattern (repeating, continuing, or both)

  identify the shapes to complete the pattern

  read the pattern

money

- Ask your child to put the dimes, nickels, and/or pennies in the coin cup.
- Count the money in the coin cup together.

clock

- Ask your child to set the clock on the half hour or hour.
- Ask the following:

  time shown on the clock

  time one hour ago and time one hour from now

- Ask your child to write the digital time on the meeting strip.
- Record on the meeting strip the time an activity will occur.

number of the day

- Write three number sentences for the number of the day on the meeting strip.

THE LESSON

Creating and Reading a Venn Diagram

"Today you will make a Venn diagram and answer questions about the graph."

• Open the Meeting Book to page 34.

• Show your child the graph.

"What do you think we are going to show on this graph?" vegetables that people we know like

"What will be the choices?" corn, carrots

"Today I would like to know what vegetables you and our family members, friends, and neighbors like to eat."

"We will write our initials on the graph to show which of the vegetables we like."

"If I like both carrots and corn, where will I put my initials?" where the circles intersect

"If I like only corn, where will I put my initials?" inside the corn circle but not the carrot circle

"If I like only carrots, where will I put my initials?" inside the carrot circle but not the corn circle

"If I don't like either of the vegetables, where will I put my initials?" outside the circles

• Write your initials on the graph to show which vegetables you like.

"Which vegetables do I like?"

"How do you know?"

"Now you will write your initials to show which of the vegetables you like."

"Do you like carrots?"

"Do you like corn?"

"Where will you write your initials?"

• Repeat with the choices of 5–10 family members, friends, or neighbors.

• Ask the following questions when the graph is completed:

"What can you tell me about our graph?"

• Record your child's observations on the chalkboard.

"How many people like carrots?"

"How many people like corn?"

"How many people like both carrots and corn?"

"How many people do not like either carrots or corn?"

"How many people like only carrots?"

"How many people like only corn?"

CLASS PRACTICE

number fact practice

- Use the tan, peach, and lavender fact cards to practice the subtraction facts with your child.
- Give your child **Fact Sheet S 3.4.**
- Time your child for one minute.
- Correct the fact sheet with your child and record the score.
- Allow time for your child to complete the unfinished facts.

WRITTEN PRACTICE

- Complete **Worksheet 61A** with your child.
- Complete **Worksheet 61B** with your child later in the day.

Name _____   **LESSON 61A**
                                   Math 2
Date _____

1. Joe had 4 dimes. His mother gave him 4 more dimes. Write a number sentence to show how many dimes he has now.

   Number sentence ___4 dimes + 4 dimes = 8 dimes___   Answer ___8 dimes___

   How much money is that? ___80¢___

2. Use the graph to answer the questions.

   Fruits We Like
   Bananas Grapefruit
   Amy, Ann, Sam, Steve, John, Max, Sue, Mary

   How many children's names are on this graph? ___8___

   How many children like bananas? ___7___

   How many children like only grapefruit? ___1___

   How many children like both bananas and grapefruit? ___3___

   What does Amy like? ___bananas___

3. Number the number line using the odd numbers.

   M   P
   1 3 5 7 9 11 13 15 17 19 21

   Put a point at **9**. Label it **M**. Put a point at **15**. Label it **P**.

4. Measure each line segment using inches.

   vertical line segment ___6"___

   horizontal line segment ___5"___   oblique line segment ___2"___

5. Circle the shape that is congruent to the shape on the left.

2-61Wa   Copyright © 1991 by Saxon Publishers, Inc. and Nancy Larson. Reproduction prohibited.

Name _____   **LESSON 61B**
                                   Math 2
Date _____

1. Micky had 3 dimes. His brother gave him 2 more dimes. Write a number sentence to show how many dimes he has now.

   Number sentence ___3 dimes + 2 dimes = 5 dimes___   Answer ___5 dimes___

   How much money is that? ___50¢___

2. Use the graph to answer the questions.

   Fruits We Like
   Bananas Grapefruit
   Amy, Ann, Sam, Steve, John, Max, Sue, Mary

   How many children like grapefruit? ___4___

   How many children like only bananas? ___4___

   What does Mary like? ___grapefruit___

   What does Max like? ___bananas & grapefruit___

3. Number the number line using the odd numbers.

   T   V
   1 3 5 7 9 11 13 15 17 19 21

   Put a point at **7**. Label it **T**. Put a point at **11**. Label it **V**.

4. Finish the number patterns.

   66, 67, 68, ___69___, ___70___, ___71___, ___72___, ___73___, ___74___

   20, 18, 16, ___14___, ___12___, ___10___, ___8___, ___6___, ___4___, ___2___

   5, 10, 15, ___20___, ___25___, ___30___, ___35___, ___40___, ___45___, ___50___

5. Circle the shape that is congruent to the shape on the left.

2-61Wb   Copyright © 1991 by Saxon Publishers, Inc. and Nancy Larson. Reproduction prohibited.

esson 62

reading a recipe
measuring ingredients for a recipe

---

lesson preparation

materials

recipe (from Lesson 59)

ingredients and supplies for the recipe

Fact Sheet A 8.1

in the morning

• Write the following in the pattern box on the meeting strip:

> 196,197,198, ____, ____, ____, ____, ____, ____

Answer: 196, 197, 198, 199, 200, 201, 202, 203, 204

• Write ⎡25¢⎤ on the meeting strip. Provide a cup of 10 dimes, a cup of 10 nickels, and a cup of 20 pennies.

THE MEETING

calendar

> • Ask your child to write the date on the calendar and meeting strip.

> • Ask your child the following:

>> date _____ days ago, date _____ days from now

>> day of the week _____ days ago, day of the week _____ days from now

>> months of the year, _____th month, month before, month after

> • Record on the meeting strip a special event and the number of days until it occurs.

weather graph

> • Ask your child to color the graph and write the temperature to the nearest ten degrees in the box he/she colored.

> • Ask questions about the graph.

---

counting

- Ask your child to choose a number on the hundred number chart. Ask your child to add or subtract ten or one. Repeat 6–10 times. Ask your child to give directions for returning to the starting number.

- Do the following 2–3 times a week:

    count by 10's to 300 and backward from 300 by 10's

    count by 5's to 100 and backward from 50 by 5's

    say the even numbers to 100 and backward from 50

    say the odd numbers to 49 and backward from 49

graph questions

- You and your child each ask a question about any of the graphs.

patterning

- Ask your child to do the following:

    identify the pattern (repeating, continuing, or both)

    identify the numbers to complete the pattern

    read the pattern

money

- Ask your child to put the coins in the coin cup. Count the money in the coin cup together.

- Ask your child for another way to show that amount of money. Count these coins together to check the amount.

clock

- Ask your child to set the clock on the half hour or hour.

- Ask the following:

    time shown on the clock

    time one hour ago and time one hour from now

- Ask your child to write the digital time on the meeting strip.

- Record on the meeting strip the time an activity will occur.

number of the day

- Write three number sentences for the number of the day on the meeting strip.

THE LESSON

Reading a Recipe
Measuring Ingredients for a Recipe

- Use the recipe from Lesson 59.

    "Today we will measure and mix the ingredients for our recipe."

    "Let's read our recipe together."

    "What does the first step in the directions tell us to do?"

- Continue as your child participates in the measuring and mixing.

- Complete the lesson by baking the Apple Jack Cookies. Eat, if desired.

CLASS PRACTICE

number fact practice

"Use the white fact cards to practice the oddball facts."

- Allow time for your child to do this.

- Give your child **Fact Sheet A 8.1**.

- Time your child for one minute.

- Correct the fact sheet together and record the score.

- Allow time for your child to complete the unfinished facts.

WRITTEN PRACTICE

- Complete **Worksheet 62A** with your child.

- Complete **Worksheet 62B** with your child later in the day.

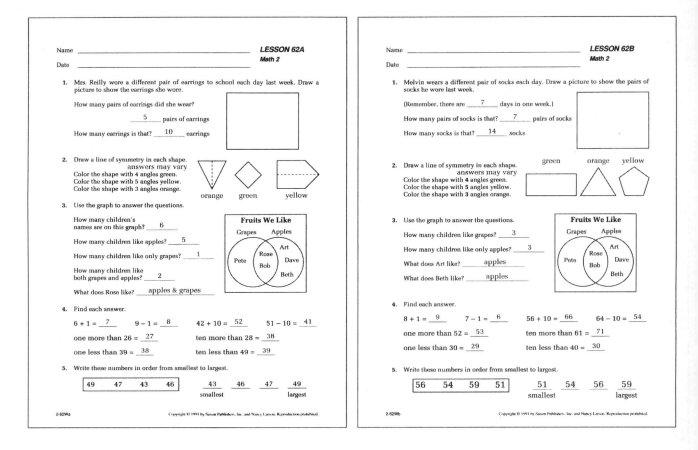

LESSON 62A
Math 2

Name _____

Date _____

1. Mrs. Reilly wore a different pair of earrings to school each day last week. Draw a picture to show the earrings she wore.

   How many pairs of earrings did she wear?

   _____5_____ pairs of earrings

   How many earrings is that? ___10___ earrings

2. Draw a line of symmetry in each shape.
   answers may vary
   Color the shape with **4** angles green.
   Color the shape with **5** angles yellow.
   Color the shape with **3** angles orange.

   orange   green   yellow

3. Use the graph to answer the questions.

   Fruits We Like

   How many children's names are on this graph? ___6___

   How many children like apples? ___5___

   How many children like only grapes? ___1___

   How many children like both grapes and apples? ___2___

   What does Rose like? ___apples & grapes___

4. Find each answer.

   6 + 1 = __7__   9 − 1 = __8__   42 + 10 = __52__   51 − 10 = __41__

   one more than 26 = __27__   ten more than 28 = __38__

   one less than 39 = __38__   ten less than 49 = __39__

5. Write these numbers in order from smallest to largest.

   | 49 | 47 | 43 | 46 |

   __43__   __46__   __47__   __49__
   smallest   largest

2-62Wa

LESSON 62B
Math 2

Name _____

Date _____

1. Melvin wears a different pair of socks each day. Draw a picture to show the pairs of socks he wore last week.

   (Remember, there are ___7___ days in one week.)

   How many pairs of socks is that? ___7___ pairs of socks

   How many socks is that? ___14___ socks

2. Draw a line of symmetry in each shape.
   answers may vary
   Color the shape with **4** angles green.
   Color the shape with **5** angles yellow.
   Color the shape with **3** angles orange.

   green   orange   yellow

3. Use the graph to answer the questions.

   Fruits We Like

   How many children like grapes? ___3___

   How many children like only apples? ___3___

   What does Art like? ___apples___

   What does Beth like? ___apples___

4. Find each answer.

   8 + 1 = __9__   7 − 1 = __6__   56 + 10 = __66__   64 − 10 = __54__

   one more than 52 = __53__   ten more than 61 = __71__

   one less than 30 = __29__   ten less than 40 = __30__

5. Write these numbers in order from smallest to largest.

   | 56 | 54 | 59 | 51 |

   __51__   __54__   __56__   __59__
   smallest   largest

2-62Wb

Lesson 63

identifying a.m. and p.m.
identifying noon and midnight
identifying dozen and half dozen

---

lesson preparation

materials

demonstration clock

1 empty egg carton

Fact Sheet A 8.2

in the morning

• Write the following in the pattern box on the meeting strip:

$$\underline{\quad}, \underline{\quad}, \underline{\quad}, 42, 43, 44, \underline{\quad}, \underline{\quad}, \underline{\quad}$$

Answer: 39, 40, 41, 42, 43, 44, 45, 46, 47

• Write ⟨89¢⟩ on the meeting strip. Provide a cup of 10 dimes, a cup of 10 nickels, and a cup of 20 pennies.

THE MEETING

calendar

- Ask your child to write the date on the calendar and meeting strip.

- Ask your child the following

    date _____ days ago, date _____ days from now

    day of the week _____ days ago, day of the week _____ days from now

    months of the year, _____th month, month before, month after

- Record on the meeting strip a special event and the number of days until it occurs.

weather graph

- Ask your child to color the graph and write the temperature to the nearest ten degrees in the box he/she colored.

- Ask questions about the graph.

counting

- Ask your child to choose a number on the hundred number chart. Ask your child to add or subtract ten or one. Repeat 6–10 times. Ask your child to give directions for returning to the starting number.

- Do the following 2–3 times a week:

    count by 10's to 300 and backward from 300 by 10's

    count by 5's to 100 and backward from 50 by 5's

    say the even numbers to 100 and backward from 50

    say the odd numbers to 49 and backward from 49

graph questions

- You and your child each ask a question about any of the graphs.

patterning

- Ask your child to do the following:

    identify the pattern (repeating, continuing, or both)

    identify the numbers to complete the pattern

    read the pattern

money

- Ask your child to put the coins in the coin cup. Count the money in the coin cup together.

- Ask your child for another way to show that amount of money. Count these coins together to check the amount.

clock

- Ask your child to set the clock on the half hour or hour.

- Ask the following:

    time shown on the clock

    time one hour ago and time one hour from now

- Ask your child to write the digital time on the meeting strip.

- Record on the meeting strip the time an activity will occur.

number of the day

- Write three number sentences for the number of the day on the meeting strip.

THE LESSON

Identifying a.m. and p.m.
Identifying Noon and Midnight

"Today you will learn how to use the a.m. and p.m. labels for time."

"You will also learn the special word we use for twelve of something."

- Set the demonstration clock so that the hour and minute hands are pointing to the twelve.

"When the clock looks like this, what time is it?"   twelve o'clock

"There are other names for this time."

"Do you know what they are?"   noon and midnight

"At midnight, both of the hands on the clock point to the twelve and it is the middle of the night."

"What are we usually doing at midnight?"

"At noon, both of the hands on the clock point to the twelve and it is the middle of the day."

"What are we usually doing at noon?"

"A new day starts at midnight."

"At noon, half of the day is gone."

"Each day there are two nine o'clocks: one in the morning and one at night."

"We need to have a way to tell people which nine o'clock we mean."

"When we begin our lessons in the morning, it is about (9:00) a.m."

"Any time before noon is called an a.m. time."

- Write "9:00 a.m." on the chalkboard.

"Midnight is 12:00 a.m."

"The p.m. time begins at noon."

"Noon is 12:00 p.m."

"We eat dinner at _____ p.m."

"Many children go to bed at 9:00 p.m."

- Write "9:00 p.m." on the chalkboard.

"Let's practice showing and writing a.m. and p.m. times."

- Show 10:00 on a large demonstration clock.

"It's morning."

"What time is it?"   10:00 a.m.

"Write this time on the chalkboard."

• Repeat with 1:00, 4:00, 10:00, and 6:00. (Alternate morning, afternoon, and evening.)

Identifying Dozen and Half Dozen

• Write the following on the chalkboard:

> hours from noon to midnight
>
> months in a year
>
> inches in a foot

"What is the same about all of these things?"

"How many hours are there from noon to midnight?"

"Let's count them on our clock to check."

• Move the hand of the demonstration clock as your child counts the hours.

"How many months are there in a year?"

"Let's count them on our birthday graph."

• Use the birthday graph in the Meeting Book.

"How many inches are in one foot?"

"Let's count the inches on a ruler."

• Hold up a ruler as your child counts the inches.

"There is a special word we can use that means twelve of something."

"Do you know what that word is?"   dozen

• Hold up an empty egg carton.

"Eggs are often sold by the dozen."

"How many eggs are in a dozen?"

"Let's count by 2's to check."

• Point to the egg cups as your child counts.

"What other things come in dozens?"   rolls, donuts

"If we have a half dozen eggs, how many eggs do we have?"   6

"Let's check this by cutting the egg carton in half."

• Cut the egg carton in half. Show your child that the two halves match.

"Let's count how many eggs are in a half dozen."

• Point to the egg cups as your child counts.

"If we have half a year, how many months is that?"

"If we have half a foot, how many inches is that?"

"What do we buy in the store that is sold by the half dozen?"   soda (pop)

CLASS PRACTICE

number fact practice

> *"Use your fact cards to practice all the addition facts."*

- Allow time for your child to do this.
- Give your child **Fact Sheet A 8.2.**
- Time your child for one minute.
- Correct the fact sheet with your child and record the score.
- Allow time for your child to complete the unfinished facts.

> *"On the back of your fact sheet, write the numbers from 160 to 192."*

WRITTEN PRACTICE

- Complete **Worksheet 63A** with your child.
- Complete **Worksheet 63B** with your child later in the day.

Name _____   **LESSON 63A**
Date _____   Math 2

1. One of these names is my mother's name. Use the clues to find my mother's name.

   It does not have six letters.
   It does not begin with a vowel.
   It is not first.   Mary (Bernice) Louise Anna Cora
   It is not fifth.
   Circle my mother's name.

2. Draw one dozen donuts.
   Your sister ate half a dozen donuts.
   Put an X on the ones she ate.

   How many donuts are left? ___6___

3. Use a red crayon to color all the shapes that are congruent to the shape on the left.

4. It is 8:30 in the morning. Is it a.m. or p.m.? ___a.m.___

   Show the time on the clock.

5. About how long is a new pencil?

   (7 inches)   15 inches   1 inch   24 inches

6. What coins could you use to make 27¢?
   answers may vary _____

2-63Wa   Copyright © 1991 by Saxon Publishers, Inc. and Nancy Larson. Reproduction prohibited.

Name _____   **LESSON 63B**
Date _____   Math 2

1. Make up your own clues for a puzzle question. Write your mother's name in one box. Write two other names in the other boxes. Write two clues.

   [   ] [   ] [   ]

   1 _____
   2 _____

2. Draw a dozen eggs.
   Your brother ate two eggs for breakfast.
   Put an X on the ones he ate.

   How many eggs are left? ___10___

3. Use a red crayon to color all the shapes that are congruent to the shape on the left.

4. It is 2:30 in the afternoon. Is it a.m. or p.m.? ___p.m.___

   Show the time on the clock.

5. About how long is an egg carton?

   2 inches   6 inches   30 inches   (12 inches)

6. What coins could you use to make 23¢?
   answers may vary _____

2-63Wb   Copyright © 1991 by Saxon Publishers, Inc. and Nancy Larson. Reproduction prohibited.

Lesson 64

adding three or more single-digit numbers

lesson preparation

materials

50 pennies

paper cup (optional)

Fact Sheet S 3.4

in the morning

• Write the following in the pattern box on the meeting strip:

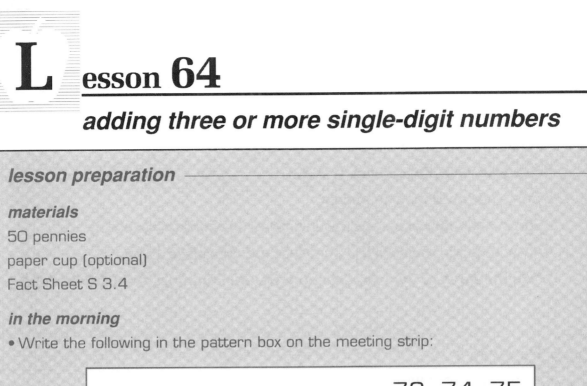

——, ——, ——, ——, ——, ——, 73, 74, 75

Answer: 67, 68, 69, 70, 71, 72, 73, 74, 75

• Write 75¢ on the meeting strip. Provide a cup of 10 dimes, a cup of 10 nickels, and a cup of 20 pennies.

THE MEETING

calendar

• Ask your child to write the date on the calendar and meeting strip.

• Ask your child the following:

date _____ days ago, date _____ days from now

day of the week _____ days ago, day of the week _____ days from now

months of the year, _____th month, month before, month after

• Record on the meeting strip a special event and the number of days until it occurs.

weather graph

• Ask your child to color the graph and write the temperature to the nearest ten degrees in the box he/she colored.

• Ask questions about the graph.

counting

- Ask your child to choose a number on the hundred number chart. Ask your child to add or subtract ten or one Repeat 6–10 times. Ask your child to give directions for returning to the starting number.

- Do the following 2–3 times a week:

    count by 10's to 300 and backward from 300 by 10's

    count by 5's to 100 and backward from 50 by 5's

    say the even numbers to 100 and backward from 50

    say the odd numbers to 49 and backward from 49

graph questions

- You and your child each ask a question about any of the graphs.

patterning

- Ask your child to do the following:

    identify the pattern (repeating, continuing, or both)

    identify the numbers to complete the pattern

    read the pattern

money

- Ask your child to put the coins in the coin cup. Count the money in the coin cup together.

- Ask your child for another way to show that amount of money. Count these coins together to check the amount.

clock

- Ask your child to set the clock on the half hour or hour.

- Ask the following:

    "It's morning (or afternoon). What time is it?"

    time one hour ago

    time one hour from now

- Ask your child to write the digital time on the meeting strip.

- Record on the meeting strip the time an activity will occur.

number of the day

- Write three number sentences for the number of the day on the meeting strip.

THE LESSON

Adding Three or More Single-Digit Numbers

"Today you will learn how to add three or more single-digit numbers."

"I have two pennies in my pocket."

- Show your child the pennies. Use a cup as a pocket, if desired.

"_____ gave me nine more pennies."

- Put nine more pennies in your pocket.

"What number sentence will we write to show what happened?"

- Ask your child to write "2 + 9" on the chalkboard.

"Then I found another penny and put it in my pocket."

- Put one more penny in your pocket.

"We can show what happened like this."

- Write "2 + 9 + 1" on the chalkboard.

"_____ gave me three more pennies."

- Put three more pennies in your pocket.

"How can we show what happened on the chalkboard problem?"

- Ask your child to write "2 + 9 + 1 + 3" on the chalkboard.
- Repeat with 8, 7, and 5 pennies.
- Write "2 + 9 + 1 + 3 + 8 + 7 + 5" on the chalkboard.

"How many pennies do you think are in my pocket now?"

- Allow time for your child to suggest answers and strategies for finding the answer.

"Let's add these numbers to find the number of pennies in my pocket."

"How do you think we will do that?"

- Allow time for your child to offer several strategies. Find the answer using each of your child's strategies.

"Let's check this by counting the pennies."

- Write the following on the chalkboard:

$$1 + 9 + 2 + 8 + 3 + 7 + 4 + 6 + 5 + 5 =$$

"This time we will pretend that I put these pennies in my pocket."

"Let's find how many pennies are in my pocket."

"Some people have a special strategy for adding a list of numbers."

"They think it is easiest to look for sums of ten first."

"This is like trading ten pennies for a dime."

"What are two numbers that have a sum of ten?" *1 + 9; 2 + 8; 3 + 7;*
4 + 6; 5 + 5

• Write the combinations on the chalkboard.

"Let's be detectives and try to find the tens in our problem."

"Do you see any numbers that have a sum of ten?"

• As your child responds, record on the chalkboard in the following manner:

"Now we can add the tens."

"What is the answer?" *50*

"Sometimes addition problems are written in a column, like this problem."

• Write the following on the chalkboard:

$$\begin{array}{r} 7 \\ 3 \\ 2 \\ 6 \\ 8 \\ +\ 1 \\ \hline \end{array}$$

"Do you see any tens?"

• Record on the chalkboard:

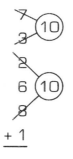

"We will circle the extra numbers."

• Record on the chalkboard:

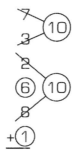

"Now we can add the circled numbers."

"What is the answer?"   27

"Now you will have a chance to try some problems."

- Write the following problems on the chalkboard one at a time. After your child completes each problem, review ways to find the sum. Crayons can be used to show groups of ten.

$$4 + 5 + 5 + 6 =$$

$$9 + 2 + 1 + 8 + 3 + 7 =$$

$$6 + 3 + 7 + 1 + 4 =$$

6	2	9
1	5	4
8	7	1
4	3	3
+ 9	8	8
	+ 2	+ 3

CLASS PRACTICE

number fact practice

"Use your fact cards to practice all the addition facts."

- Allow time for your child to do this.
- Give your child **Fact Sheet S 3.4.**
- Time your child for one minute.
- Correct the fact sheet with your child and record the score.
- Allow time for your child to complete the unfinished facts.

WRITTEN PRACTICE

- Complete **Worksheet 64A** with your child.
- Complete **Worksheet 64B** with your child later in the day.

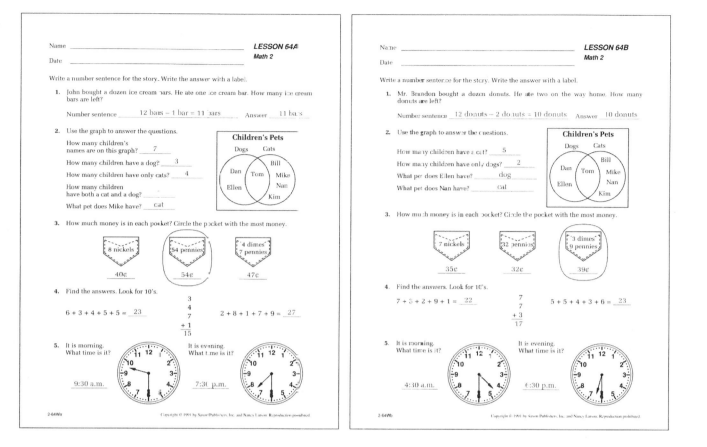

Math 2 • Lesson 64

Name _____ **LESSON 64A** / Math 2
Date _____

Write a number sentence for the story. Write the answer with a label.

1. John bought a dozen ice cream bars. He ate one ice cream bar. How many ice cream bars are left?

   Number sentence ___12 bars – 1 bar = 11 bars___   Answer ___11 bars___

2. Use the graph to answer the questions.

   Children's Pets

   How many children's names are on this graph? ___7___
   How many children have a dog? ___3___
   How many children have only cats? ___4___
   How many children have both a cat and a dog? _____
   What pet does Mike have? ___cat___

3. How much money is in each pocket? Circle the pocket with the most money.

   8 nickels — 40¢
   54 pennies — 54¢
   4 dimes 7 pennies — 47¢

4. Find the answers. Look for 10's.

   6 + 3 + 4 + 5 + 5 = ___23___

   3
   4
   7
   + 1
   ——
   15

   2 + 8 + 1 + 7 + 9 = ___27___

5. It is morning. What time is it? ___9:30 a.m.___
   It is evening. What time is it? ___7:30 p.m.___

2-64Wa

Name _____ **LESSON 64B** / Math 2
Date _____

Write a number sentence for the story. Write the answer with a label.

1. Mr. Brandon bought a dozen donuts. He ate two on the way home. How many donuts are left?

   Number sentence ___12 donuts – 2 donuts = 10 donuts___   Answer ___10 donuts___

2. Use the graph to answer the questions.

   Children's Pets

   How many children have a cat? ___5___
   How many children have only dogs? ___2___
   What pet does Ellen have? ___dog___
   What pet does Nan have? ___cat___

3. How much money is in each pocket? Circle the pocket with the most money.

   7 nickels — 35¢
   32 pennies — 32¢
   3 dimes 9 pennies — 39¢

4. Find the answers. Look for 10's.

   7 + 3 + 2 + 9 + 1 = ___22___

   7
   7
   + 3
   ——
   17

   5 + 5 + 4 + 3 + 6 = ___23___

5. It is morning. What time is it? ___4:30 a.m.___
   It is evening. What time is it? ___6:30 p.m.___

2-64Wb

Lesson 65

writing fractions using fraction notation

lesson preparation

materials

Written Assessment #12

fraction pieces (from Lesson 34)

scrap paper

small circular object for tracing

crayons

Fact Sheet A 8.2

in the morning

• Write the following in the pattern box on the meeting strip:

$$\square, \ 1, \ 3, \ \triangle, \ 5, \ 7, \ \square, \ 9, \ 11, \ \triangle, \ \underline{\quad}, \ \underline{\quad}, \ \underline{\quad}, \ \underline{\quad}$$

Answer: □, 1, 3, △, 5, 7, □, 9, 11, △, 13, 15, □, 17

• Write ⬚52¢⬚ on the meeting strip. Provide a cup of 10 dimes, a cup of 10 nickels, and a cup of 20 pennies.

THE MEETING

calendar

• Ask your child to write the date on the calendar and meeting strip.

• Ask your child the following:

date _____ days ago, date _____ days from now

day of the week _____ days ago, day of the week _____ days from now

months of the year, _____th month, month before, month after

• Record on the meeting strip a special event and the number of days until it occurs.

weather graph

• Ask your child to color the graph and write the temperature to the nearest ten degrees in the box he/she colored.

• Ask questions about the graph.

counting

• Ask your child to choose a number on the hundred number chart. Ask your child to add or subtract ten or one. Repeat 6–10 times. Ask your child to give directions for returning to the starting number.

• Do the following 2–3 times a week:

count by 10's to 300 and backward from 300 by 10's

count by 5's to 100 and backward from 50 by 5's

say the even numbers to 100 and backward from 50

say the odd numbers to 49 and backward from 49

graph questions

• You and your child each ask a question about any of the graphs.

patterning

• Ask your child to do the following:

identify the pattern (repeating, continuing, or both)

identify the numbers and shapes to complete the pattern

read the pattern

money

• Ask your child to put the coins in the coin cup. Count the money in the coin cup together.

• Ask your child for another way to show that amount of money. Count these coins together to check the amount.

clock

• Ask your child to set the clock on the half hour or hour.

• Ask the following:

"It's morning (or afternoon). What time is it?"

time one hour ago

time one hour from now

• Ask your child to write the digital time on the meeting strip.

• Record on the meeting strip the time an activity will occur.

number of the day

• Write three number sentences for the number of the day on the meeting strip.

ASSESSMENT

Written Assessment

"Today I would like to see what you remember from what we have been practicing."

- Give your child **Written Assessment #12**.

- Read the directions for each problem. Allow time for your child to complete that problem before continuing.

- Correct the paper, noting your child's mistakes on the **Individual Recording Form.** Review the errors with your child.

THE LESSON

Writing Fractions Using Fraction Notation

"Today you will learn how to write fractions using fraction notation."

- Give your child the individual fraction pieces from Lesson 34.

  "Put your fraction pieces together to make circles."

- Draw four large circles on the chalkboard.

  "We will pretend that the circles are pies."

  "What kind of pies do you think they might be?"

  "Let's pretend that you are going to eat the first pie all by yourself."

  "Point to your construction paper piece that shows how much pie you will eat."

  "We could say that you ate the whole pie."

  "Write the number 1 on this pie."

- Record this on the chalkboard pie as your child records this on the whole circle.

  "Let's pretend that you and I are going to share a pie."

  "Which construction paper pie shows how you and I could share a pie so that we will each have the same size piece?"

  "How many pieces does this pie have?"   2

  "What do we call each piece?"   one half

  "We write one half like this."

- Divide the second chalkboard circle in half and write "$\frac{1}{2}$" on each piece.

  "The bottom number of the fraction tells us how many pieces we have in the whole pie."

"Mathematicians call the bottom number the denominator."

"Write the number $\frac{1}{2}$ on each of your halves."

"Let's pretend that you will share the next pie with _____, _____, and _____."

"How many pieces will you need?" *4*

"Which color pie is divided into four pieces?"

"Show me how to cut the chalkboard pie so that we will have four equal pieces."

- Ask your child to draw lines to divide the chalkboard pie into fourths.

  "How many pieces do we have?" *4*

  "What do we call each piece?" *one fourth*

  "We wrote one half using a 1, a fraction line, and a 2 because there were two pieces."

  "How do you think we will write one fourth?"

- Record "$\frac{1}{4}$" on each piece of the chalkboard circle.

  "The bottom number of the fraction tells us how many pieces we have."

  "What do mathematicians call the bottom number?" *denominator*

  "Write the number $\frac{1}{4}$ on each of your fourths."

- Repeat with eighths.

  "I ate five eighths of the pie."

  "Use your pieces to show how much I ate."

  "We write five eighths like this."

- Write "$\frac{5}{8}$" on the chalkboard.

  "How many eighths are left for you?" *three eighths*

  "How will we write that as a fraction?"

- Ask your child to write "$\frac{3}{8}$" on the chalkboard.

- Repeat with seven eighths and three fourths.

- Draw the following on the chalkboard:

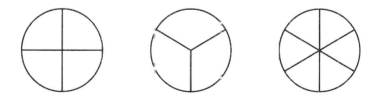

  "What fraction will we write inside each piece?"

  "How do you know?"

- Ask your child to write the fractions inside each piece.

- Shade the pies in the following way:

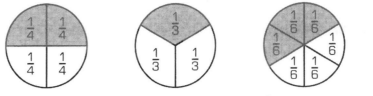

"I ate the part of each pie that is shaded."

"How much of each pie did I eat?"

- Write the fractional part of the pie that is shaded beneath each circle.

$$\frac{2}{4}, \frac{1}{3}, \frac{3}{6}, \frac{2}{8}$$

"How much of each pie is left?"

- Give your child a piece of scrap paper and a round object to trace.

"Draw three pies on this paper."

- Allow time for your child to draw the circles.

"Divide the first pie into two equal pieces."

"Divide the second pie into four equal pieces."

"Divide the third pie into eight equal pieces."

- Allow time for your child to divide the circles.

"Color three fourths of a pie green."

"Write '$\frac{3}{4}$' below the circle."

"Color six eighths of a pie yellow."

"Write '$\frac{6}{8}$' below the circle."

"Color one half of a pie blue."

"Write '$\frac{1}{2}$' below the circle."

- Save the individual fraction pieces.

CLASS PRACTICE

number fact practice

"Use your fact cards to practice all the addition facts."

- Allow time for your child to do this.

- Give your child **Fact Sheet A 8.2.**

- Time your child for one minute.

- Correct the fact sheet with your child and record the score.

- Allow time for your child to complete the unfinished facts.

WRITTEN PRACTICE

- Complete **Worksheet 65A** with your child.
- Complete **Worksheet 65B** with your child later in the day.

Name _____ **ASSESSMENT 12**
Date _____ **LESSON 65**
                                        Math 2

Write a number sentence for the story. Write the answer with a label.

1. Daniel had four nickels. Roseann gave him five more nickels. How many nickels does he have now?

   Number sentence ____4 nickels + 5 nickels = 9 nickels____

   Answer ____9 nickels____

2. Number the number line.
   Put a point at **3**. Label it **A**.
   Put a point at **6**. Label it **B**.

   0 1 2 3 4 5 6

3. How much money is this? ____47¢____

4. Circle the largest number.
   Put an X on the smallest number.
   Write the numbers in order from smallest to largest.

   47   ~~46~~   (52)   45   47   52
                           smallest   largest

5. Fill in the missing numbers on this piece of a hundred number chart.

	55	56	57	58
62	63	64	65	66	67	68
	75	76		

6. Find the answers.

   $7 - 7 =$ ___0___ $8 - 1 =$ ___7___ $12 - 6 =$ ___6___ $9 - 0 =$ ___9___

2-65Aa   Copyright © 1991 by Saxon Publishers, Inc. and Nancy Larson. Reproduction prohibited.

Name _____ **LESSON 65A**
Date _____ **Math 2**

1. Raquel had 3 dimes and 4 pennies. Her sister gave her 2 dimes.

   How many dimes does she have now? ___5___

   How many pennies does she have now? ___4___

   How much money is this? ___54¢___

2. Label each piece using a fraction.

   Color $\frac{1}{2}$ blue.

   Color $\frac{1}{4}$ red.

   Color $\frac{1}{3}$ yellow.

3. Cheryl has a dozen apples and a half dozen oranges. Draw the fruit.

4. Put these numbers in order from smallest to largest.

   23   47   50   32   23   32   47   50
                          smallest   largest

5. Show half past three in the morning.
   Write the digital time.

   ___3:30 a.m.___

6. Find the answers. Look for 10's.

   $8 + 5 + 2 + 3 + 5 =$ ___23___ $4 + 7 + 1 + 3 + 6 =$ ___21___

2-65Wa   Copyright © 1991 by Saxon Publishers, Inc. and Nancy Larson. Reproduction prohibited.

Name _____ **LESSON 65B**
Date _____ **Math 2**

1. Willie had 6 dimes and 5 pennies. His sister gave him 3 dimes.

   How many dimes does he have now? ___9___

   How many pennies does he have now? ___5___

   How much money is this? ___95¢___

2. Label each piece using a fraction.

   Color $\frac{1}{2}$ blue.

   Color $\frac{1}{4}$ red.

   Color $\frac{1}{3}$ yellow.

3. Curtis has a half dozen peaches and a dozen bananas. Draw the fruit.

4. Put these numbers in order from smallest to largest.

   35   41   17   58   17   35   41   58
                          smallest   largest

5. Show half past four in the afternoon.
   Write the digital time.

   ___4:30 p.m.___

6. Find the answers. Look for 10's.

   $3 - 5 + 9 + 5 + 1 =$ ___23___ $1 + 5 + 6 + 2 + 4 + 5 =$ ___23___

2-65Wb   Copyright © 1991 by Saxon Publishers, Inc. and Nancy Larson. Reproduction prohibited.

Lesson 66

adding two-digit numbers using dimes and pennies (part 1)

lesson preparation

materials

8 small cups

28 dimes and 30 pennies

Master 2-66

Fact Sheet A 8.2

the night before

• Put dimes and pennies in cups labeled with the following letters. Use the fewest number of dimes and pennies for each amount.

A	B	C	D	E	F	G	H
24¢	35¢	52¢	16¢	43¢	51¢	27¢	62¢

in the morning

• Write the following in the pattern box on the meeting strip:

$$\_\_\_, \_\_\_, \_\_\_, 50, 60, 70, \_\_\_, \_\_\_, \_\_\_$$

Answer: 20, 30, 40, 50, 60, 70, 80, 90, 100

• Write ⬚23¢⬚ on the meeting strip. Provide a cup of 10 dimes, a cup of 10 nickels, and a cup of 20 pennies.

THE MEETING

calendar

• Ask your child to write the date on the calendar and meeting strip.

• Ask your child the following:

  date _____ days ago, date _____ days from now

  day of the week _____ days ago, day of the week _____ days from now

  months of the year, _____th month, month before, month after

• Record on the meeting strip a special event and the number of days until it occurs.

weather graph

- Ask your child to color the graph and write the temperature to the nearest ten degrees in the box he/she colored.
- Ask questions about the graph.

counting

- Ask your child to choose a number on the hundred number chart. Ask your child to add or subtract ten or one. Repeat 6–10 times. Ask your child to give directions for returning to the starting number.
- Do the following 2–3 times a week:

    count by 10's to 300 and backward from 300 by 10's

    count by 5's to 100 and backward from 50 by 5's

    say the even numbers to 100 and backward from 50

    say the odd numbers to 49 and backward from 49

graph questions

- You and your child each ask a question about any of the graphs.

patterning

- Ask your child to do the following:

    identify the pattern (repeating, continuing, or both)

    identify the numbers to complete the pattern

    read the pattern

money

- Ask your child to put the coins in the coin cup. Count the money in the coin cup together.
- Ask your child for another way to show that amount of money. Count these coins together to check the amount.

clock

- Ask your child to set the clock on the half hour or hour.
- Ask the following:

    "It's morning (or afternoon). What time is it?"

    time one hour ago

    time one hour from now

- Ask your child to write the digital time on the meeting strip.
- Record on the meeting strip the time an activity will occur.

number of the day

- Write three number sentences for the number of the day on the meeting strip.

THE LESSON

Adding Two-Digit Numbers Using Dimes and Pennies (Part 1)

"Today you will learn how to add two-digit numbers using dimes and pennies."

- Draw the following on the chalkboard:

dimes	pennies

- Show your child cup A.

  "I have 24¢ in this cup."

  "How many dimes and pennies do I have?"   2 dimes, 4 pennies

  "I will write that at the top of my chart."

- Record the following on the chalkboard:

dimes	pennies
2	4

- Show your child cup B.

  "I have 35¢ in this cup."

  "How many dimes and pennies do I have?"   3 dimes, 5 pennies

  "How will I write that on my chart?"

- Record the following on the chalkboard:

dimes	pennies
2	4
3	5

  "Now I will put the money together in one cup."

- Pour the coins from cup B into cup A.

  "How many dimes do you think are in my cup now?"

  "How do you know?"

  "How many pennies do you think are in my cup now?"

  "How do you know?"

"Let's count to check."

- Sort and count the coins with your child.

- Record the following on the chalkboard:

	dimes	pennies
	2	4
+	3	5
	5	9

"How much money is that?" 59¢

- Write "24¢ + 35¢ = 59¢" on the chalkboard.

"I have two more cups."

- Show your child cup C.

"I have 52¢ in this cup."

"How many dimes and pennies do I have?" 5 dimes, 2 pennies

"I will write that at the top of my chart."

- Record the following on the chalkboard:

dimes	pennies
5	2

- Show your child cup D.

"I have 16¢ in this cup."

"How many dimes and pennies do I have?" 1 dime, 6 pennies

"How will I write that on my chart?"

- Record the following on the chalkboard:

dimes	pennies
5	2
1	6

"Now I will put the money together in one cup."

- Pour the coins from cup D into cup C.

"How many dimes do you think are in my cup now?"

"How do you know?"

"How many pennies do you think are in my cup now?"

"How do you know?"

"Let's count to check."

- Sort and count the coins with your child.

- Record the following on the chalkboard:

dimes	pennies
5	2
1	6
6	8

(with a "+" to the left of the second row)

"How much money is that?"

- Write "52¢ + 16¢ = 68¢" on the chalkboard.

"Now you will have a chance to add two-digit numbers."

- Give your child **Master 2-66**.

- Show your child cup E.

"I have 43¢ in cup E.

"How many dimes and pennies is that?"

"Write that in the top row of problem one."

- Demonstrate on the chalkboard.

- Show your child cup F.

"I have 51¢ in cup F."

"How many dimes and pennies is that?"

"Write that in the bottom row of problem one."

- Demonstrate on the chalkboard.

"Now I will put the money together."

- Pour the coins from cup F into cup E.

"How many dimes do I have now?"

"How many pennies do I have now?"

"Write that on your paper."

- Demonstrate on the chalkboard.

"Let's count to check."

- Count the coins with your child.

- Repeat with cups G (27¢) and H (62¢).

- Save **Master 2-66** for use in Lesson 67.

CLASS PRACTICE

number fact practice

"Use your fact cards to practice all the addition facts."

- Allow time for your child to do this.

- Give your child **Fact Sheet A 8.2.**
- Time your child for one minute.
- Correct the fact sheet with your child and record the score.
- Allow time for your child to complete the unfinished facts.

"On the back of your fact sheet write the numbers from 87 to 123."

WRITTEN PRACTICE

- Complete **Worksheet 66A** with your child.
- Complete **Worksheet 66B** with your child later in the day.

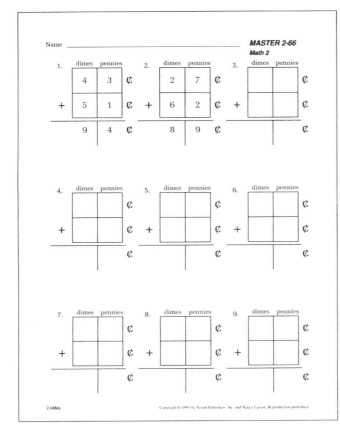

Name _____ **MASTER 2-66**
Math 2

1.
dimes	pennies	
4	3	¢
+ 5	1	¢
9	4	¢

2.
dimes	pennies	
2	7	¢
+ 6	2	¢
8	9	¢

3.
dimes	pennies	
		¢
+		¢
		¢

4.
dimes	pennies	
		¢
+		¢
		¢

5.
dimes	pennies	
		¢
+		¢
		¢

6.
dimes	pennies	
		¢
+		¢
		¢

7.
dimes	pennies	
		¢
+		¢
		¢

8.
dimes	pennies	
		¢
+		¢
		¢

9.
dimes	pennies	
		¢
+		¢
		¢

2-66Ma

Name _____ **LESSON 66A**
Math 2

Date _____

Write a number sentence for the story. Write the answer with a label.

1. On Tuesday, fifty children in Grade 2 at Washington School bought hot lunch. The other twenty-seven Grade 2 children brought their lunch from home. How many children are in Grade 2 at Washington School?

   Number sentence _____ 50 children + 27 children = 77 children

   Answer _____ 77 children

2. I have **38¢**. What coins could I have?

   answers may vary

   What is another way to make **38¢**?

   answers may vary

3. I have **63¢**. How many dimes and pennies is that?

   I have **24¢**. How many dimes and pennies is that?

   How many dimes and pennies is that altogether?

dimes	pennies
6	3
2	4
8	7

4. Put a dot inside each angle. Count the number of angles in each shape.

   ____4____ angles   ____6____ angles   ____3____ angles

5. Draw a line of symmetry in each shape in Problem 4.

6. Finish the number patterns.

   __44__ , __45__ , __46__ , 47, 48, 49, __50__ , __51__ , __52__

   196, 197, 198, __199__ , __200__ , __201__ , __202__ , __203__

2-66Wa

Name _____ **LESSON 66B**
Math 2

Date _____

Write a number sentence for the story. Write the answer with a label.

1. On Tuesday, sixty-seven children in Grade 2 at Forest School bought hot lunch. The other thirty Grade 2 children brought their lunch from home. How many children are in Grade 2 at Forest School?

   Number sentence _____ 67 children + 30 children = 97 children

   Answer _____ 97 children

2. I have **32¢**. What coins could I have?

   answers may vary

   What is another way to make **32¢**?

   answers may vary

3. I have **34¢**. How many dimes and pennies is that?

   I have **51¢**. How many dimes and pennies is that?

   How many dimes and pennies is that altogether?

dimes	pennies
3	4
5	1
8	5

4. Put a dot inside each angle. Count the number of angles in each shape.

   ____4____ angles   ____4____ angles   ____3____ angles

5. Draw a line of symmetry in each shape in Problem 4.

6. Finish the number patterns.

   __57__ , __59__ , __61__ , 63, 65, 67, __69__ , __71__ , __73__

   296, 297, 298, __299__ , __300__ , __301__ , __302__ , __303__

2-66Wb

Lesson 67

adding two-digit numbers using dimes and pennies (part 2)

---

lesson preparation

materials

empty cup

1 cup of 10 pennies and 1 cup of 10 dimes

Master 2-66 from Lesson 66

Fact Sheet A 8.2

in the morning

• Write the following in the pattern box on the meeting strip:

> 195,196,197,198, ___ , ___ , ___ , ___ , ___ , ___

Answer: 195, 196, 197, 198, 199, 200, 201, 202, 203, 204

• Write   72¢   on the meeting strip. Provide a cup of 10 dimes, a cup of 10 nickels, and a cup of 20 pennies.

THE MEETING

calendar

- • Ask your child to write the date on the calendar and meeting strip.

- • Ask your child the following:

    date ____ days ago, date ____ days from now

    day of the week ____ days ago, day of the week ____ days from now

    months of the year, ____th month, month before, month after

- • Record on the meeting strip a special event and the number of days until it occurs.

weather graph

- • Ask your child to color the graph and write the temperature to the nearest ten degrees in the box he/she colored.

- • Ask questions about the graph.

---

counting

- Ask your child to choose a number on the hundred number chart. Ask your child to add or subtract ten or one. Repeat 6–10 times. Ask your child to give directions for returning to the starting number.

- Do the following 2–3 times a week:

  count by 10's to 300 and backward from 300 by 10's

  count by 5's to 100 and backward from 50 by 5's

  say the even numbers to 100 and backward from 50

  say the odd numbers to 49 and backward from 49

graph questions

- You and your child each ask a question about any of the graphs.

patterning

- Ask your child to do the following:

  identify the pattern (repeating, continuing, or both)

  identify the numbers to complete the pattern

  read the pattern

money

- Ask your child to put the coins in the coin cup. Count the money in the coin cup together.

- Ask your child for another way to show that amount of money. Count these coins together to check the amount.

clock

- Ask your child to set the clock on the half hour or hour.

- Ask the following:

  "It's morning (or afternoon). What time is it?"

  time one hour ago

  time one hour from now

- Ask your child to write the digital time on the meeting strip.

- Record on the meeting strip the time an activity will occur.

number of the day

- Write three number sentences for the number of the day on the meeting strip.

THE LESSON

Adding Two-Digit Numbers Using Dimes and Pennies (Part 2)

- Give your child **Master 2-66** from Lesson 66.

    "What did we do yesterday?"

    "We will use money again today to help us practice adding two-digit numbers."

    "First I will put 18¢ in a cup."

    "How many dimes and pennies is that?"   *1 dime, 8 pennies*

- Put 1 dime and 8 pennies in the cup.

    "Write that at the top of problem three."

- Demonstrate on the chalkboard.

    "Now I will put 41¢ in the cup."

    "How many dimes and pennies is that?"   *4 dimes, 1 penny*

- Put 4 dimes and 1 penny in the cup.

    "Write that below the eighteen cents."

- Demonstrate on the chalkboard.

    "At the bottom of problem three, write how many dimes and pennies you think are in my cup now."

    "What did you write?"

    "How did you know?"

    "Let's count the coins to check."

    "Were you right?"

    "Let's try another problem."

- Repeat, using 34¢ and 52¢; 61¢ and 25¢; 22¢ and 36¢; 53¢ and 43¢; and 29¢ and 40¢.

    "Can you find these answers without using dimes and pennies?"

    "How?"

CLASS PRACTICE

number fact practice

    "Use your fact cards to practice all the addition facts."

- Allow time for your child to do this.

- Give your child **Fact Sheet A 8.2**.

- Time your child for one minute.

• Correct the fact sheet with your child and record the score.

• Allow time for your child to complete the unfinished facts.

WRITTEN PRACTICE

• Complete **Worksheet 67A** with your child.

• Complete **Worksheet 67B** with your child later in the day.

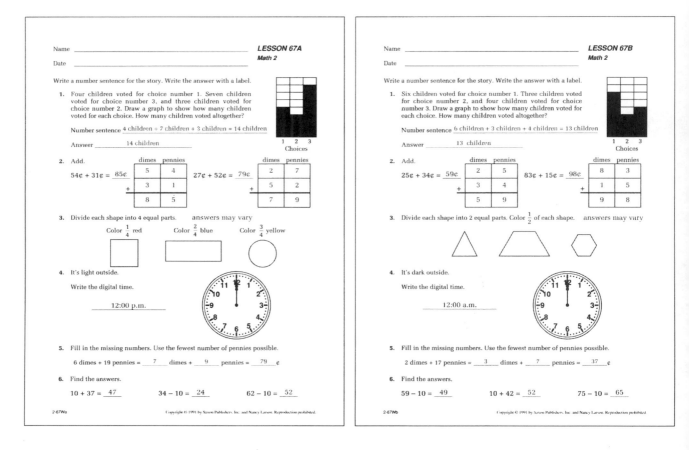

Name _____   **LESSON 67A**
                                          Math 2
Date _____

Write a number sentence for the story. Write the answer with a label.

1. Four children voted for choice number 1. Seven children voted for choice number 3, and three children voted for choice number 2. Draw a graph to show how many children voted for each choice. How many children voted altogether?

    Number sentence 4 children + 7 children + 3 children = 14 children

    Answer _____14 children_____

2. Add.

    54¢ + 31¢ = _85¢_

dimes	pennies
5	4
3	1
8	5

    27¢ + 52¢ = _79¢_

dimes	pennies
2	7
5	2
7	9

3. Divide each shape into 4 equal parts.   answers may vary

    Color $\frac{1}{4}$ red   Color $\frac{2}{4}$ blue   Color $\frac{3}{4}$ yellow

4. It's light outside.

    Write the digital time.

    _____12:00 p.m._____

5. Fill in the missing numbers. Use the fewest number of pennies possible.

    6 dimes + 19 pennies = _7_ dimes + _9_ pennies = _79_ ¢

6. Find the answers.

    10 + 37 = _47_   34 − 10 = _24_   62 − 10 = _52_

2-67Wa   Copyright © 1991 by Saxon Publishers, Inc. and Nancy Larson. Reproduction prohibited.

Name _____   **LESSON 67B**
                                          Math 2
Date _____

Write a number sentence for the story. Write the answer with a label.

1. Six children voted for choice number 1. Three children voted for choice number 2, and four children voted for choice number 3. Draw a graph to show how many children voted for each choice. How many children voted altogether?

    Number sentence 6 children + 3 children + 4 children = 13 children

    Answer _____13 children_____

2. Add.

    25¢ + 34¢ = _59¢_

dimes	pennies
2	5
3	4
5	9

    83¢ + 15¢ = _98¢_

dimes	pennies
8	3
1	5
9	8

3. Divide each shape into 2 equal parts. Color $\frac{1}{2}$ of each shape.   answers may vary

4. It's dark outside.

    Write the digital time.

    _____12:00 a.m._____

5. Fill in the missing numbers. Use the fewest number of pennies possible.

    2 dimes + 17 pennies = _3_ dimes + _7_ pennies = _37_ ¢

6. Find the answers.

    59 − 10 = _49_   10 + 42 = _52_   75 − 10 = _65_

2-67Wb   Copyright © 1991 by Saxon Publishers, Inc. and Nancy Larson. Reproduction prohibited.

L esson 68

subtracting two from a number

---

lesson preparation

materials

cup of 12 pennies

work mat

Fact Sheet S 4.0

in the morning

• Write the following in the pattern box on the meeting strip:

```
.,  . .,  . . .,  . . . .,  _____,  _____
```

Answer: ., .·., ·:·., ·:·., ·::·., ·::·.

• Write ⬛96¢⬛ on the meeting strip. Provide a cup of 10 dimes, a cup of 10 nickels, and a cup of 20 pennies.

THE MEETING

calendar

• Ask your child to write the date on the calendar and meeting strip.

• Ask your child the following:

date _____ days ago, date _____ days from now

day of the week _____ days ago, day of the week _____ days from now

months of the year, _____th month, month before, month after

• Record on the meeting strip a special event and the number of days until it occurs.

weather graph

• Ask your child to color the graph and write the temperature to the nearest ten degrees in the box he/she colored.

• Ask questions about the graph.

---

counting

- Ask your child to choose a number on the hundred number chart. Ask your child to add or subtract ten or one. Repeat 6–10 times. Ask your child to give directions for returning to the starting number.

- Do the following 2–3 times a week:

  count by 10's to 300 and backward from 300 by 10's

  count by 5's to 100 and backward from 50 by 5's

  say the even numbers to 100 and backward from 50

  say the odd numbers to 49 and backward from 49

graph questions

- You and your child each ask a question about any of the graphs.

patterning

- Ask your child to do the following:

  identify the pattern (repeating, continuing, or both)

  identify the shapes to complete the pattern

  read the pattern

money

- Ask your child to put the coins in the coin cup. Count the money in the coin cup together.

- Ask your child for another way to show that amount of money. Count these coins together to check the amount.

clock

- Ask your child to set the clock on the half hour or hour.

- Ask the following:

  "It's morning (or afternoon). What time is it?"

  time one hour ago

  time one hour from now

- Ask your child to write the digital time on the meeting strip.

- Record on the meeting strip the time an activity will occur.

number of the day

- Write three number sentences for the number of the day on the meeting strip.

THE LESSON

Subtracting Two from a Number

> *"Today you will learn the subtracting two number facts."*

- Give your child a cup of 12 pennies and a work mat.

> *"Put your pennies in groups of two."*

> *"How many groups of two do you have?"*   6

> *"Let's count the pennies."*

> *"How can we do that?"*   by 2's

- Write the even numbers in a column on the chalkboard as your child counts. (Begin at the bottom.)

> *"What do we call these numbers?"*   even numbers

> *"Put two pennies back in the cup."*

> *"How many pennies do you have now?"*   10

> *"Let's count by 2's to check."*

- Record on the chalkboard: $12 - 2 = 10$

> *"We have 10 pennies."*

> *"Take away two pennies."*

> *"How many pennies do you have now?"*   8

> *"Let's count by 2's to check."*

- Record on the chalkboard: $10 - 2 = 8$

- Repeat until all the pennies are removed from the mat.

- The chalkboard list of facts should look like the following:

$$12 - 2 = 10$$
$$10 - 2 = 8$$
$$8 - 2 = 6$$
$$6 - 2 = 4$$
$$4 - 2 = 2$$
$$2 - 2 = 0$$

> *"What happens when we subtract two from an even number?"*   the answer is the even number before it

> *"Put your pennies in groups of two again on your mat."*

> *"How many pennies do you have?"*   12

> *"Take away one penny."*

> *"How many pennies do you have now?"*   11

> *"Does every penny have a partner?"*   no

"That is why we say that 11 is an odd number."

"We have 11 pennies."

"Take away two pennies."

"How many pennies do you have now?" 9

"Let's count by 2's to check."

- Record on the chalkboard: 11 − 2 = 9
- Repeat until all the pennies are removed from the mat.
- The chalkboard list of facts should look like the following:

$$11 - 2 = 9$$
$$9 - 2 = 7$$
$$7 - 2 = 5$$
$$5 - 2 = 3$$
$$3 - 2 = 1$$

"What happens when we subtract two from an odd number?" the answer is the odd number before it

"How can we remember the answers for the subtracting two facts?" count backward by even or odd numbers

"Let's practice the subtracting two problems."

"I will say a subtracting two problem."

"Say the answer as quickly as you can."

CLASS PRACTICE

number fact practice

"Use your green fact cards to practice the subtracting two facts."

- Allow time for your child to practice independently.
- Give your child **Fact Sheet S 4.0.**
- Time your child for one minute.
- Correct the fact sheet together and record the score.
- Allow time for your child to complete the unfinished facts.

WRITTEN PRACTICE

- Complete **Worksheet 68A** with your child.
- Complete **Worksheet 68B** with your child later in the day.

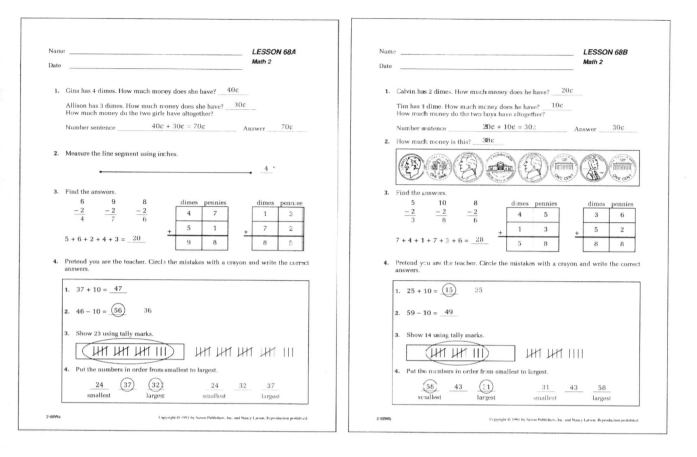

LESSON 68A
Math 2

Name _____

Date _____

1. Gina has 4 dimes. How much money does she have? __40¢__

   Allison has 3 dimes. How much money does she have? __30¢__
   How much money do the two girls have altogether?

   Number sentence ___40¢ + 30¢ = 70¢___   Answer ___70¢___

2. Measure the line segment using inches.

   ———————————————————— 4 "

3. Find the answers.

   $\begin{array}{r} 6 \\ -2 \\ \hline 4 \end{array}$ $\begin{array}{r} 9 \\ -2 \\ \hline 7 \end{array}$ $\begin{array}{r} 8 \\ -2 \\ \hline 6 \end{array}$

dimes	pennies
4	7
5	1
9	8

dimes	pennies
1	3
7	2
8	5

   5 + 6 + 2 + 4 + 3 = __20__

4. Pretend you are the teacher. Circle the mistakes with a crayon and write the correct answers.

   1. 37 + 10 = __47__

   2. 46 − 10 = (56)   36

   3. Show 23 using tally marks.

      (||||| ||||| ||||| |||) ||||| ||||| ||||| ||||| |||

   4. Put the numbers in order from smallest to largest.

      __24__ (37) (32)   __24__ __32__ __37__
      smallest largest   smallest largest

2-68Wa

LESSON 68B
Math 2

Name _____

Date _____

1. Calvin has 2 dimes. How much money does he have? __20¢__

   Tim has 1 dime. How much money does he have? __10¢__
   How much money do the two boys have altogether?

   Number sentence ___20¢ + 10¢ = 30¢___   Answer ___30¢___

2. How much money is this? __38¢__

3. Find the answers.

   $\begin{array}{r} 5 \\ -2 \\ \hline 3 \end{array}$ $\begin{array}{r} 10 \\ -2 \\ \hline 8 \end{array}$ $\begin{array}{r} 8 \\ -2 \\ \hline 6 \end{array}$

dimes	pennies
4	5
1	3
5	8

dimes	pennies
3	6
5	2
8	8

   7 + 4 + 1 + 7 + 3 + 6 = __28__

4. Pretend you are the teacher. Circle the mistakes with a crayon and write the correct answers.

   1. 25 + 10 = (15)   35

   2. 59 − 10 = __49__

   3. Show 14 using tally marks.

      (||||| ||||| |||) ||||| ||||| ||||

   4. Put the numbers in order from smallest to largest.

      (58) __43__ (31)   __31__ __43__ __58__
      smallest largest   smallest largest

2-68Wb

Lesson 69

reading a thermometer to the nearest two degrees

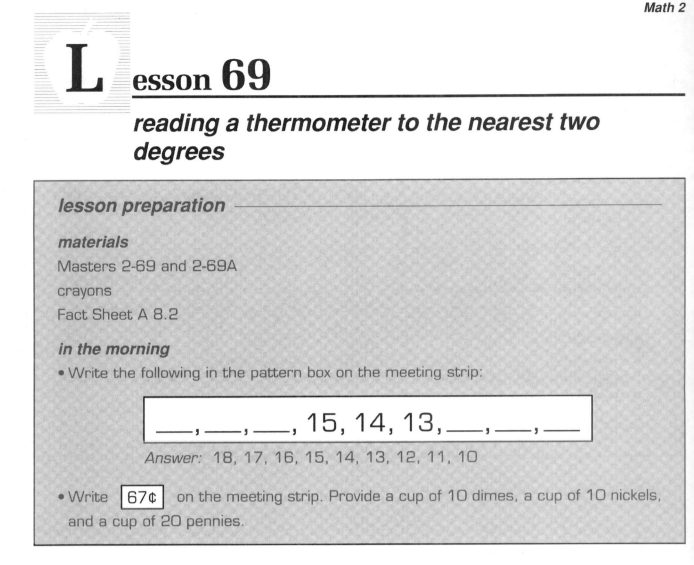

lesson preparation

materials

Masters 2-69 and 2-69A

crayons

Fact Sheet A 8.2

in the morning

• Write the following in the pattern box on the meeting strip:

___, ___, ___, 15, 14, 13, ___, ___, ___

Answer: 18, 17, 16, 15, 14, 13, 12, 11, 10

• Write [67¢] on the meeting strip. Provide a cup of 10 dimes, a cup of 10 nickels, and a cup of 20 pennies.

THE MEETING

calendar

• Ask your child to write the date on the calendar and meeting strip.

• Ask your child the following:

date _____ days ago, date _____ days from now

day of the week _____ days ago, day of the week _____ days from now

months of the year, _____th month, month before, month after

• Record on the meeting strip a special event and the number of days until it occurs.

weather graph

• Ask your child to color the graph and write the temperature to the nearest ten degrees in the box he/she colored.

• Ask questions about the graph.

counting

- Ask your child to choose a number on the hundred number chart. Ask your child to add or subtract ten or one. Repeat 6–10 times. Ask your child to give directions for returning to the starting number.

- Do the following once or twice a week:

    count by 10's to 300 and backward from 300 by 10's

    count by 5's to 100 and backward from 50 by 5's

    say the even numbers to 100 and backward from 50

    say the odd numbers to 49 and backward from 49

graph questions

- You and your child each ask a question about any of the graphs.

patterning

- Ask your child to do the following:

    identify the pattern (repeating, continuing, or both)

    identify the numbers to complete the pattern

    read the pattern

money

- Ask your child to put the coins in the coin cup. Count the money in the coin cup together.

- Ask your child for another way to show that amount of money. Count these coins together to check the amount.

clock

- Ask your child to set the clock on the half hour or hour.

- Ask the following:

    "It's morning (or afternoon). What time is it?"

    time one hour ago

    time one hour from now

- Ask your child to write the digital time on the meeting strip.

- Record on the meeting strip the time an activity will occur.

number of the day

- Write three number sentences for the number of the day on the meeting strip.

THE LESSON

Reading a Thermometer to the Nearest Two Degrees

"Each morning when we read our thermometer, we find which ten degrees the temperature is closest to."

"We use that temperature to make our graph."

"Today you will learn how to read a thermometer to the nearest two degrees."

"Many thermometers have two scales."

"One is called the Fahrenheit scale."

"The other scale is called the Celsius scale."

"Because our temperature is usually reported on TV and in the newspaper using the Fahrenheit scale, that is the scale we will use."

- Give your child **Master 2-69.**

"This is what the Fahrenheit thermometer looks like between negative 20 and 100 degrees."

"Point to the line for 0°."

"Point to the line for 30°."

"Point to the line for 70°."

"When we count the marks between the 10's, we count by 2's."

"Each mark on the thermometer means two degrees."

"Let's count together by 2's from 0 to 100 as I point to each mark on the thermometer."

- Count by 2's with your child.

"Let's begin at 20° and count backward by 2's."

"After we reach zero, we say negative two, negative four, negative six, and so forth."

- Point to the marks on the thermometer as you count backward by 2's with your child.

- Write "58°" on the chalkboard.

"How will we find 58 degrees on the thermometer?"

"To find 58 degrees, we find 50 and count by 2's to 58."

"Let's count by 2's from 50 degrees to find 58 degrees."

"Use your pencil point to show where 58 degrees is on the thermometer."

- Repeat, using 36°, 82°, 44°, 12°, 90°, and 28°.

"Water freezes at 32 degrees on the Fahrenheit scale."

"Point to 32 degrees on your thermometer."

370

"Now you will have a chance to read and shade some thermometers."

- Give your child **Master 2-69A**.

  "Let's find the temperature on the first thermometer."

  "To do this, we count up by 10's from zero and then count by 2's."

  "How high is the shading on the first thermometer?"   72°

  "We write the temperature like this."

- Record the temperature (72°F) on the chalkboard.

  "When we write the temperature, we write a small circle next to the number instead of writing the word degrees."

  "We write the capital letter F to show that we are using the Fahrenheit scale."

  "Write this temperature on the line below the thermometer."

  "Let's find the temperature that is shown on the second thermometer."

  "How will we do that?"

  "Write the temperature on the line below the thermometer."

- Repeat with the third thermometer.

  "Now you will have a chance to color in a thermometer to show the temperature."

  "The first temperature you will show is 28 degrees Fahrenheit."

  "When you color a thermometer to show the temperature, you will begin at the bottom."

  "Color up to 20 degrees and then count by 2's to find 28 degrees."

- Repeat with 46° and 74°.

- A new temperature graph will begin in January in the Meeting Book. If you are teaching this lesson in December, continue using the previous graph.

  "Tomorrow we will make a new graph for the temperature."

  "Each day you will read and graph our temperature to the nearest two degrees."

CLASS PRACTICE

number fact practice

- Use the white fact cards to practice the last eight addition facts with your child.

- Give your child **Fact Sheet A 8.2**.

- Time your child for one minute.

• Correct the fact sheet together and record the score.

• Allow time for your child to complete the unfinished facts.

WRITTEN PRACTICE

• Complete **Worksheet 69A** with your child.

• Complete **Worksheet 69B** with your child later in the day.

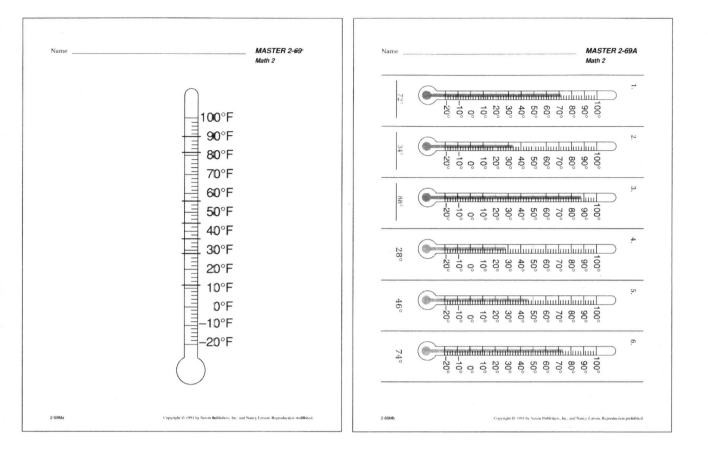

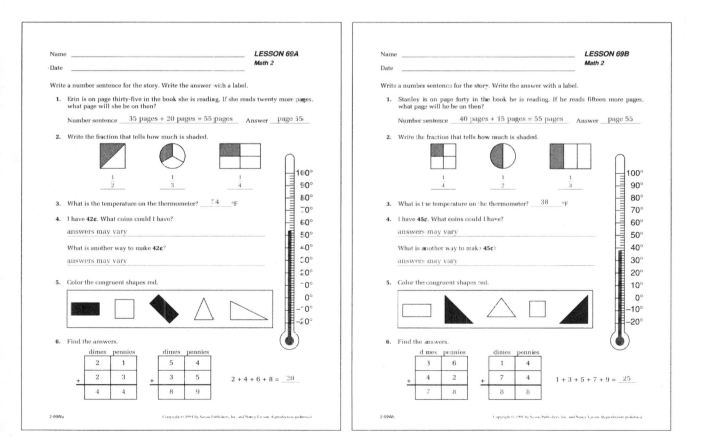

Lesson 70

identifying and creating similar shapes and designs

---

lesson preparation

materials

Written Assessment #13

Oral Assessment #7

shape pieces from Lesson 22

2 geoboards

5 geobands

crayon

Master 2-70

Teacher's Master 2-70 for the oral assessment

Fact Sheet S 4.0

in the morning

• Write the following in the pattern box on the meeting strip:

——, ——, ——, ——, ——, ——, 54, 53, 52

Answer: 60, 59, 58, 57, 56, 55, 54, 53, 52

• Write ⏐ 59¢ ⏐ on the meeting strip. Provide a cup of 10 dimes, a cup of 10 nickels, and a cup of 20 pennies.

• Beginning in January, your child will make a line graph for the daily temperature. Your child will put a dot to show the temperature to the nearest two degrees. The next day, your child will put a dot to show the temperature to the nearest two degrees and connect yesterday's dot with today's dot.

THE MEETING

calendar

• Ask your child to write the date on the calendar and meeting strip.

• Ask your child the following:

date _____ days ago, date _____ days from now

day of the week _____ days ago, day of the week _____ days from now months of the year, _____th month, month before, month after

- Record on the meeting strip a special event and the number of days until it occurs.

weather graph

- Ask your child to read today's temperature on the thermometer.

  "We will make a different graph to show our daily temperature."

  "Each morning you will put a dot above the date to show the temperature."

- Ask your child to put a dot on the line above the date to show the temperature to the nearest two degrees.

  "Let's check the dot to see if it shows _____ degrees."

- Count by 10's and 2's to check the temperature on the graph.

counting

- Ask your child to choose a number on the hundred number chart. Ask your child to add or subtract ten or one. Repeat 6–10 times. Ask your child to give directions for returning to the starting number.

- Do the following once or twice a week:

  count by 10's to 300 and backward from 300 by 10's

  count by 5's to 100 and backward from 50 by 5's

  say the even numbers to 100 and backward from 50

  say the odd numbers to 49 and backward from 49

graph questions

- You and your child each ask a question about any of the graphs.

patterning

- Ask your child to do the following:

  identify the pattern (repeating, continuing, or both)

  identify the numbers to complete the pattern

  read the pattern

money

- Ask your child to put the coins in the coin cup. Count the money in the coin cup together.

- Ask your child for another way to show that amount of money. Count these coins together to check the amount.

clock

- Ask your child to set the clock on the half hour or hour.

- Ask the following:

    "It's morning (or afternoon). What time is it?"

    time one hour ago

    time one hour from now

- Ask your child to write the digital time on the meeting strip.
- Record on the meeting strip the time an activity will occur.

number of the day

- Write three number sentences for the number of the day on the meeting strip.

ASSESSMENT

Written Assessment

"Today I would like to see what you remember from what we have been practicing."

- Give your child **Written Assessment #13**.
- Read the directions for each problem. Allow time for your child to complete that problem before continuing.
- Correct the paper, noting your child's mistakes on the **Individual Recording Form**. Review the errors with your child.

Oral Assessment

- Record your child's response(s) to the oral interview questions on the interview sheet.

THE LESSON

Identifying and Creating Similar Shapes and Designs

"A few weeks ago we used geoboards."

"What did we do with the geoboards?" worked together to make congruent shapes

"What are congruent shapes?" shapes that look exactly the same; they have the same size and shape

- Show your child the set of shape pieces from Lesson 22.

    "Find two pieces that are congruent."

    "How are they the same?" same size and shape

- Repeat several times.

- Hold up the small and large red circles.

  "These pieces are similar."

  "They are the same shape, but not the same size."

  "Find two other pieces that are similar."

  "Remember, similar pieces are the same shape but not the same size."

- Ask your child to select two pieces.

  "These are similar pieces because they are the same shape but not the same size."

- Repeat several times.

  "Today you will learn how to identify and make similar shapes and designs."

- Give your child **Master 2-70**, a geoboard, and a geoband.

  "There are small pictures of geoboards on this paper."

  "What shape do you see on the first geoboard?"

  "Let's make this shape on our geoboards."

  "Will our triangles be the same size as the triangle on the paper?"   no

  "Make a triangle on your geoboard similar to the one on the paper."

- Do this on your geoboard also.

  "Are our triangles the same?"

  "Are our triangles similar to the triangle on the paper?"

  "Why?"   they are the same shape, but not the same size

- Repeat with the next two shapes on the master.

  "Now make a design on your geoboard."

  "Use a crayon to copy it on the last geoboard picture."

CLASS PRACTICE

number fact practice

- Use the green fact cards to practice the subtracting two facts.

- Give your child **Fact Sheet S 4.0**.

- Time your child for one minute.

- Correct the fact sheet together and record the score.

- Allow time for your child to complete the unfinished facts.

Name _____

Date _____

Write a number sentence for the story. Write the answer with a label.

1. Phil had 7 dimes. He gave his brother 2 dimes. How many dimes does Phil have now?

   Number sentence _____ 7 dimes – 2 dimes = 5 dimes

   Answer ____5 dimes____   How much money is that? ___50¢

2. Darlene has 4 pairs of mittens.
   Draw the mittens.
   How many mittens did you draw? __8__

3. Show half past two on the clock.
   Write the digital time.

   | 2:30 |

   It's morning. Circle
   the correct label.   (a.m.)   p.m.

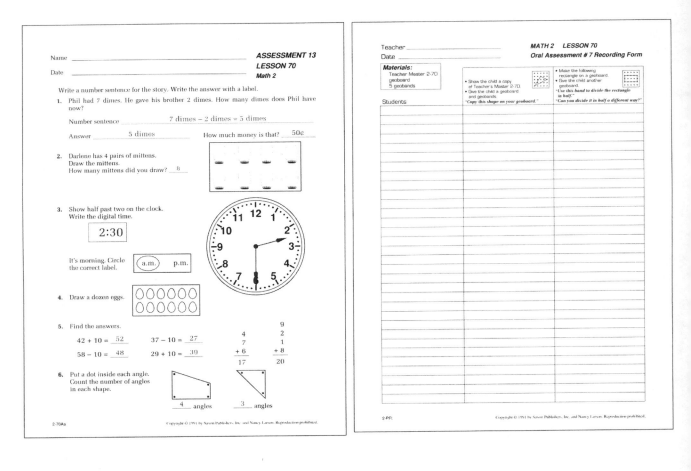

4. Draw a dozen eggs.

5. Find the answers.

   42 + 10 = __52__   37 – 10 = __27__

   58 – 10 = __48__   29 + 10 = __39__

4		9
7		2
+6		1
17		+8
		20

6. Put a dot inside each angle.
   Count the number of angles
   in each shape.

   __4__ angles   __3__ angles

Teacher _____

Date _____

Materials:
Teacher Master 2-70
geoboard
5 geobands

Students

• Show the child a copy of Teacher's Master 2-70. • Give the child a geoboard and geobands. *"Copy this shape on your geoboard."*	• Make the following rectangle on a geoboard. • Give the child another geoboard. *"Use this band to divide the rectangle in half."* *"Can you divide it in half a different way?"*

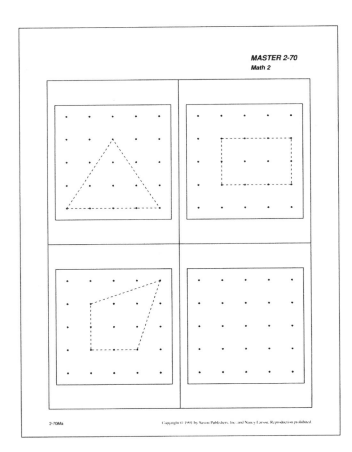

Name _____

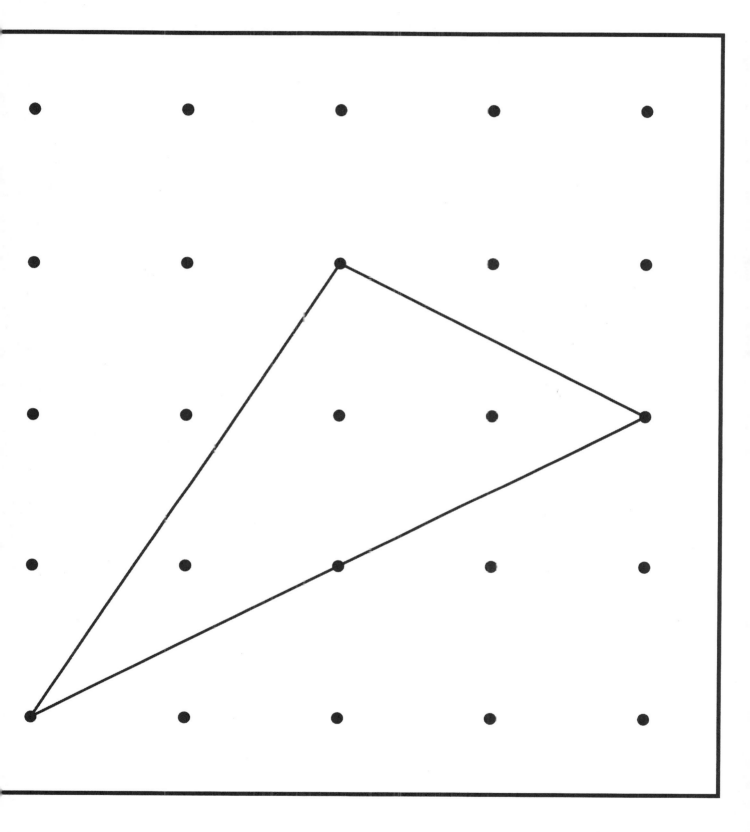

Lesson 71

adding two-digit numbers with trading using dimes and pennies (part 1)

---

lesson preparation

materials

cup of 15 pennies

cup of 10 dimes

1 piece each of yellow and white construction paper

Fact Sheet S 4.0

the night before

• Tape a piece of yellow (or other light color) construction paper and a piece of white construction paper together in the following way:

yellow	white

in the morning

• Write the following in the pattern box on the meeting strip:

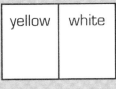

A, B, E, F, I, J, M, ___, ___, ___

Answer: A, B, E, F, I, J, M, N, Q, R

• Write 48¢ on the meeting strip. Provide a cup of 10 dimes, a cup of 10 nickels, and a cup of 20 pennies.

THE MEETING

calendar

• Ask your child to write the date on the calendar and meeting strip.

• Ask your child the following two or three times a week:

date _____ days ago, date _____ days from now

day of the week _____ days ago, day of the week _____ days from now

_____th month, month before, month after

- Record on the meeting strip a special event and the number of days until it occurs.

weather graph

"What was yesterday's temperature?"

"Do you think it is warmer or colder today?"

- Ask your child to read the thermometer and graph today's temperature to the nearest two degrees.
- Count by 10's and 2's to check the temperature on the graph.

"Connect the dot that shows yesterday's temperature to the dot that shows today's temperature."

"Is it warmer or colder than yesterday?"

counting

- Ask your child to choose a number on the hundred number chart. Ask your child to add or subtract ten or one. Repeat 6–10 times. Ask your child to give directions for returning to the starting number.
- Do the following once or twice a week:

    count by 10's to 400 and backward from 400 by 10's

    count by 5's to 100 and backward from 50 by 5's

    say the even numbers to 100 and backward from 50

    say the odd numbers to 49 and backward from 49

graph questions

- You and your child each ask a question about any of the graphs.

patterning

- Ask your child to do the following:

    identify the pattern (repeating, continuing, or both)

    identify the letters to complete the pattern

    read the pattern

money

- Ask your child to put the coins in the coin cup. Count the money in the coin cup together.
- Ask your child for another way to show that amount of money. Count these coins together to check the amount.

clock

- Ask your child to set the clock on the half hour or hour.

• Ask the following:

> *"It's (morning/afternoon/evening). What time is it?"*

> time one hour ago

> time one hour from now

• Ask your child to write the digital time on the meeting strip.

• Record on the meeting strip the time an activity will occur.

number of the day

• Write three number sentences for the number of the day on the meeting strip.

THE LESSON

Adding Two-Digit Numbers with Trading Using Dimes and Pennies (Part 1)

• Place the yellow/white mat on the table in front of you so that the yellow side is on your child's left.

> *"Today you will begin to learn how to add two-digit numbers with trading using dimes and pennies."*

• Write "24¢" on the chalkboard.

> *"How will we show 24¢ using the fewest number of dimes and pennies?"*   *2 dimes, 4 pennies*

> *"We will put the dimes on the yellow half of the mat and the pennies on the white half of the mat."*

• Put two dimes on the yellow half of the mat and four pennies on the white half of the mat. Arrange the dimes and pennies in pairs like the following:

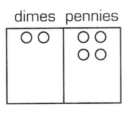

> *"Let's add 63¢."*

• Record "24¢ + 63¢" on the chalkboard.

> *"How will we show 63¢ using the fewest number of dimes and pennies?"*   *6 dimes, 3 pennies*

• Put six dimes on the yellow half of the mat and three pennies on the white half of the mat. Arrange the dimes and pennies in pairs like the following:

dimes pennies

"How much money is this altogether?"   *87¢*

• Ask your child to put the coins in the coin cup. Count the money with your child. Begin with the dimes.

• Record "24¢ + 63¢ = 87¢" on the chalkboard.

"Let's write this addition problem in a different way."

"How much money did we have to begin with?"   *24¢*

• Write "24¢" on the chalkboard.

"How much money did we add?"   *63¢*

"We can show that by writing 63¢ below the 24¢."

"We will need to be careful to write the numbers that show the number of pennies in the pennies' column and the numbers that show the number of dimes in the dimes' column."

"We will write an addition symbol in front of the bottom number."

• Write the following on the chalkboard:

$$\begin{array}{r} 24¢ \\ + \ 63¢ \\ \hline \end{array}$$

"We will add these numbers the way many people add two-digit numbers."

"Many people begin by adding the pennies first."

"How many pennies do we have altogether?"   *7*

• Record "7" on the chalkboard below the pennies' column.

"Then they add the dimes next."

"How many dimes do we have altogether?"   *8*

• Record "8" on the chalkboard below the dimes' column.

"How much money is this?"   *87¢*

"Let's try another problem."

• Repeat, using 48¢ and 21¢.

"Let's try another problem."

• Write "37¢" on the chalkboard.

"How will we show 37¢ using the fewest number of dimes and pennies?"   *3 dimes, 7 pennies*

"We will put the dimes on the yellow half of the mat and the pennies on the white half of the mat."

• Put the coins on the mat.

"Let's add 23¢."

• Record "37¢ + 23¢" on the chalkboard.

"How will we show 23¢ using the fewest number of dimes and pennies?"   *2 dimes, 3 pennies*

• Put 23¢ more on the mat.

"How much money is this?"   *60¢*

• Ask your child to put the coins in the coin cup. Count the money with your child. Begin with the dimes.

"Do we have enough pennies to trade for a dime?"   *yes*

"Let's do that."

• Trade the 10 pennies for a dime.

"What coins do we have now?"   *6 dimes, zero pennies*

"How much money is this?"   *60¢*

• Record "37¢ + 23¢ = 60¢" on the chalkboard.

"Let's write this addition problem vertically."

• Write the following on the chalkboard:

$$\begin{array}{r} 37¢ \\ +\,23¢ \\ \hline \end{array}$$

"When we write an addition problem vertically, we must be careful to write the pennies under the pennies and the dimes under the dimes."

"Let's find the answer to this problem the way many people add two-digit numbers."

"We will add the pennies first."

"Seven pennies and three pennies is how many pennies?"   *10*

"What happens when we have ten pennies?"   *we trade them for a dime*

"We will write a small '1' below the column that shows us the number of dimes."

"This shows the dime we traded for ten pennies."

• Refer to the extra dime on the mat, if necessary.

"How many pennies do we have left?"   *zero*

"We will write a '0' below the pennies' column to show that we have no pennies left."

- Lightly circle the 1 and 0 in the chalkboard problem.

  "We can see that this was ten pennies."

- Record the following on the chalkboard:

$$3\ 7\ ¢$$
$$\underline{+\ 2\ 3\ ¢}$$
$$0\ ¢$$

  "Now we will add the dimes."

  "Three dimes plus two dimes plus one dime is six dimes."

  "We will write a '6' below the dimes' column to show that we have six dimes."

- Record the following on the chalkboard:

$$3\ 7\ ¢$$
$$\underline{+\ 2\ 3\ ¢}$$
$$6\ 0\ ¢$$

  "How much money is this?"   60¢

  "Let's try another problem."

- Write "56¢" on the chalkboard.

  "How will we show 56¢ using the fewest number of dimes and pennies?"   5 dimes, 6 pennies

- Put the coins on the mat.

  "Let's add 36¢."

- Record "56¢ + 36¢" on the chalkboard.

  "How will we show 36¢ using the fewest number of dimes and pennies?"   3 dimes, 6 pennies

- Put 36¢ more on the mat.

  "How much money is this altogether?"   92¢

- Ask your child to put the coins in the coin cup. Count the money with your child. Begin with the dimes.

  "Do we have enough pennies to trade for a dime?"   yes

  "Let's do that."

- Trade the 10 pennies for a dime.

  "What coins do we have now?"   9 dimes, 2 pennies

- Record "56¢ + 36¢ = 92¢" on the chalkboard.

  "Let's write the addition problem in a different way."

  "How will we do that?"   write the problem vertically

- Record the following on the chalkboard:

$$\begin{array}{r} 56¢ \\ + 36¢ \\ \hline \end{array}$$

"When we write an addition problem vertically, we must be careful to write the pennies under the pennies and the dimes under the dimes."

"Let's find the answer to this problem the way many people add two-digit numbers."

"What will we do first?"   add the pennies

"Six pennies and six pennies is how many pennies?"   12

"What happens when we have twelve pennies?"   we trade 10 pennies for a dime

"We write a small '1' below the column that shows us the number of dimes."

"This shows that we have an extra dime to add in."

- Refer to the extra dime on the mat, if necessary.

"How many pennies do we have left?"   2

"We will write a '2' below the pennies' column to show that we have two pennies left."

"We show that like this."

- Record the following on the chalkboard:

$$\begin{array}{r} 5\ 6\ ¢ \\ + 3\ 6\ ¢ \\ {\scriptstyle 1} \\ \hline 2\ ¢ \end{array}$$

- Lightly circle the 1 and 2 in the chalkboard problem.

"This shows that we had twelve pennies when we added."

"Now we will add the dimes."

"Five dimes plus three dimes plus one dime is nine dimes."

"We will write a '9' below the dimes' column to show that we have nine dimes."

- Record the 9 below the dimes' column.

"How much money is this?"   92¢

"Let's try another problem."

- Repeat, using 17¢ and 48¢.

"Tomorrow you will have a chance to use dimes and pennies to find the answers to addition problems like these."

- Save the yellow/white mat.

CLASS PRACTICE

number fact practice

"Use your green fact cards to practice the subtracting two facts."

- Allow time for your child to practice independently.
- Give your child **Fact Sheet S 4.0.**
- Time your child for one minute.
- Correct the fact sheet with your child and record the score.
- Allow time for your child to complete the unfinished facts.

WRITTEN PRACTICE

- Complete **Worksheet 71A** with your child.
- Complete **Worksheet 71B** with your child later in the day.

Name _____ **LESSON 71A**
Date _____ Math 2

1. Mrs. Gustin put 3 pennies and 5 dimes in the coin cup.

   How much money is in the coin cup? __63¢__

   Andy took a dime out of the coin cup. How much money is in the coin cup now?

   Number sentence __63¢ − 10¢ = 53¢__   Answer __53¢__

2. Write the fraction that tells how much is shaded.

   $\frac{5}{6}$ $\frac{1}{8}$

3. Color the thermometer to show 38°F.

4. Fill in the children's names to show the ice cream flavors they like.

   ICE CREAM FLAVORS
   CHOCOLATE   VANILLA

   Amy   Chris   Sue
         Jim   Bob

   Amy likes only chocolate.
   Chris likes vanilla and chocolate.
   Sue likes only vanilla.
   Bob likes only vanilla.
   Jim likes both flavors.

   How many children like chocolate? __3__

   How many children like vanilla? __4__

5. Fill in the missing numbers on this piece of a hundred number chart.

33	34	35	36		38
43	44		46	47	48

6. Find the answers.

   67 − 10 = __57__   63 + 10 = __73__   18 − 1 = __17__

   10 less than 74 = __64__   10 more than 41 = __51__

2-71Wa   Copyright © 1991 by Saxon Publishers, Inc. and Nancy Larson. Reproduction prohibited.

Name _____ **LESSON 71B**
Date _____ Math 2

1. Mrs. Trotter put 4 pennies and 3 dimes in the coin cup.

   How much money is in the coin cup? __34¢__

   Brian took a penny out of the coin cup. How much money is in the coin cup now?

   Number sentence __34¢ − 1¢ = 33¢__   Answer __33¢__

2. Write the fraction that tells how much is shaded.

   $\frac{5}{8}$ $\frac{2}{6}$

3. Color the thermometer to show 54°F.

4. Fill in the children's names to show the ice cream flavors they like.

   ICE CREAM FLAVORS
   CHOCOLATE   STRAWBERRY

   Jim   Steve   Mike
         Mary   Lisa

   Mike likes only strawberry.
   Steve likes both flavors.
   Jim likes only chocolate.
   Mary likes both flavors.
   Lisa likes only strawberry.

   How many children like chocolate? __3__

   How many children like strawberry? __4__

5. Fill in the missing numbers on this piece of a hundred number chart.

15	16	17	18		
22	23	24	25		28	29

6. Find the answers.

   29 − 10 = __19__   49 + 10 = __59__   17 − 1 = __16__

   10 less than 53 = __43__   10 more than 27 = __37__

2-71Wb   Copyright © 1991 by Saxon Publishers, Inc. and Nancy Larson. Reproduction prohibited.

Lesson 72

adding two-digit numbers with trading using dimes and pennies (part 2)

---

lesson preparation

materials

cup of 20 pennies

cup of 10 dimes

yellow and white mat from Lesson 71

Master 2-72

Fact Sheet A 1-100

in the morning

• Write the following in the pattern box on the meeting strip:

> ## 92, 94, 96, ___, ___, ___, ___, ___, ___

Answer: 92, 94, 96, 98, 100, 102, 104, 106, 108

• Write ⟨62¢⟩ on the meeting strip. Provide a cup of 10 dimes, a cup of 10 nickels, and a cup of 20 pennies.

THE MEETING

calendar

• Ask your child to write the date on the calendar and meeting strip.

• Ask your child the following two or three times a week:

   date ____ days ago, date ____ days from now

   day of the week ____ days ago, day of the week ____ days from now

   ____th month, month before, month after

• Record on the meeting strip a special event and the number of days until it occurs.

weather graph

"What was yesterday's temperature?"

"Do you think it is warmer or colder today?"

- Ask your child to read the thermometer and graph today's temperature to the nearest two degrees.

- Count by 10's and 2's to check the temperature on the graph.

    "Connect the dot that shows yesterday's temperature to the dot that shows today's temperature."

    "Is it warmer or colder than yesterday?"

counting

- Ask your child to choose a number on the hundred number chart. Ask your child to add or subtract ten or one. Repeat 6–10 times. Ask your child to give directions for returning to the starting number.

- Do the following once or twice a week:

    count by 10's to 400 and backward from 400 by 10's

    count by 5's to 100 and backward from 50 by 5's

    say the even numbers to 100 and backward from 50

    say the odd numbers to 49 and backward from 49

graph questions

- You and your child each ask a question about any of the graphs.

patterning

- Ask your child to do the following:

    identify the pattern (repeating, continuing, or both)

    identify the numbers to complete the pattern

    read the pattern

money

- Ask your child to put the coins in the coin cup. Count the money in the coin cup together.

- Ask your child for another way to show that amount of money. Count these coins together to check the amount.

clock

- Ask your child to set the clock on the half hour or hour.

- Ask the following:

    "It's (morning/afternoon/evening). What time is it?"

    time one hour ago

    time one hour from now

- Ask your child to write the digital time on the meeting strip.

• Record on the meeting strip the time an activity will occur.

number of the day

• Write three number sentences for the number of the day on the meeting strip.

THE LESSON

Adding Two-Digit Numbers with Trading Using Dimes and Pennies (Part 2)

"Yesterday I used dimes and pennies to act out addition problems."

"Today you will learn how to use dimes and pennies to find answers for addition problems."

• Write "31¢ + 49¢" on the chalkboard.

"How will we use dimes and pennies to find this answer?"

• Ask your child to explain the steps for using dimes and pennies to find this answer. Show the addition using dimes and pennies and the yellow and white mat from Lesson 71.

"How will we write this problem vertically?"

• Ask your child to write the vertical addition problem on the chalkboard.

"How will we find the answer?" *add the pennies; trade 10 pennies for a dime, if possible; add the dimes*

"Let's find the answer together."

"What will we add first?" *pennies*

"One penny and nine pennies is how many pennies?" *10*

"Do we have enough pennies to trade for a dime?" *yes*

"How will we show this on our problem?" *write a small one at the bottom of the dimes' column and a zero below the pennies' column*

• Record this on the chalkboard problem.

"How many dimes do we have?" *8*

• Record this on the chalkboard problem.

"How much money is this?" *80¢*

"Now you will have a chance to find the answers to some addition problems using dimes and pennies."

• Give your child a cup of 10 dimes, a cup of 20 pennies, **Master 2-72**, and the yellow and white work mat.

"Read the first problem." *43¢ + 25¢*

"Let's find the answer using dimes and pennies."

"Put four dimes and three pennies on the mat."

• Allow time for your child to do this.

"Put two more dimes and five more pennies on the mat."

• Allow time for your child to do this.

"How many pennies do you have?"   8

"Do you have enough pennies to trade for a dime?"   no

"Write '8' at the bottom of the pennies' column."

"How many dimes do you have?"   6

"Write '6' at the bottom of the dimes' column."

"How much money is this?"   68¢

• Ask your child to read the second problem.

"Let's find the answer using dimes and pennies."

"What will you do first?"   put 6 dimes and 6 pennies on the mat

• Allow time for your child to do this.

"What will you do next?"   put 1 more dime and 8 more pennies on the mat

• Allow time for your child to do this.

"You will add the pennies first."

"Six pennies and eight pennies is how many pennies?"   14

"Count the pennies to make sure there are fourteen."

"Do you have enough pennies to trade for a dime?"   yes

"Trade ten pennies for a dime."

"You will need to show that you added another dime to the dimes' column."

"Write a small '1' below the other numbers in the dimes' column."

"How many pennies do you have now?"   4

"Write '4' below the pennies' column to show that you have four pennies left."

"Lightly circle the '1' and the '4.' "

"This shows the fourteen pennies you had."

• Demonstrate on the chalkboard.

$$\begin{array}{r} 6\ 6\ ¢ \\ +\ 1\ 8\ ¢ \\ \hline 4\ ¢ \end{array}$$

"How many dimes do you have altogether?"   8

"Write '8' at the bottom of the dimes' column."

"How much money is this?"   *84¢*

• Ask your child to read the third problem.

"Let's find the answer using dimes and pennies."

"What will you do first?"   *put 4 dimes and 9 pennies on the mat*

"What will you do next?"   *put 3 more dimes and 7 more pennies on the mat*

"You will add the pennies first."

"Nine pennies and seven pennies is how many pennies?"   *16*

"Do you have enough pennies to trade for a dime?"   *yes*

"Trade ten pennies for a dime."

"How will you show this on your problem?"

"Write a small '1' below the other numbers in the dimes' column to show that you traded a dime for your ten pennies."

"How many pennies do you have now?"   *6*

"Write '6' below the pennies' column to show that you have six pennies left."

• Demonstrate on the chalkboard.

$$\begin{array}{r} 4\ 9\ ¢ \\ +\ 3\ 7\ ¢ \\ {\scriptstyle 1} \\ \hline 6\ ¢ \end{array}$$

"Lightly circle the '1' and the '6.' "

"This shows the sixteen pennies you had."

"How many dimes do you have altogether?"   *8*

"Write '8' at the bottom of the dimes' column."

"How much money is this?"   *86¢*

• Ask your child to read the fourth problem.

"Let's find the answer using dimes and pennies."

"What will you do first?"   *put 4 dimes and 4 pennies on the mat*

"What will you do next?"   *put 4 more dimes and 6 more pennies on the mat*

"You will add the pennies first."

"Four pennies and six pennies is how many pennies?"   *10*

"Do you have enough pennies to trade for a dime?"   *yes*

"Trade ten pennies for a dime."

"How will you show the extra dime you traded the pennies for?"   *write a small 1 below the other numbers in the dimes' column*

"How many pennies do you have now?" zero

"What will you write below the pennies' column?" zero

- Demonstrate on the chalkboard.

$$
\begin{array}{r}
4\,4\,\cent \\
+\,4\,6\,\cent \\
\hline
0\,\cent
\end{array}
$$

"How many dimes do you have altogether?" 9

"Write '9' at the bottom of the dimes' column."

"How much money is this?" 90¢

- Repeat with the fifth and sixth problems.

CLASS PRACTICE

number fact practice

- Give your child **Fact Sheet A 1-100**.

  "There are 100 addition facts on this sheet."

  "You will have five minutes to answer as many problems as possible."

  "We will correct this fact sheet in the same way we correct the 25-problem fact sheets."

- Time your child for five minutes.

- Correct the fact sheet with your child and record the score.

- Allow time for your child to complete the unfinished facts.

WRITTEN PRACTICE

- Complete **Worksheet 72A** with your child.

- Complete **Worksheet 72B** with your child later in the day.

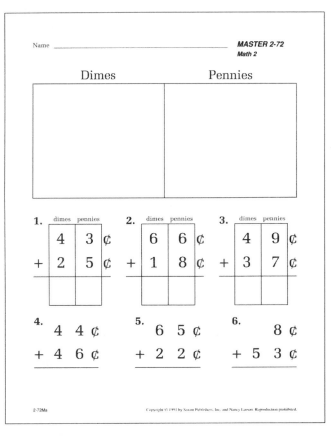

Dimes   Pennies

1.
dimes	pennies	
4	3	¢
+ 2	5	¢

2.
dimes	pennies	
6	6	¢
+ 1	8	¢

3.
dimes	pennies	
4	9	¢
+ 3	7	¢

4.
   4 4 ¢
+ 4 6 ¢

5.
   6 5 ¢
+ 2 2 ¢

6.
      8 ¢
+ 5 3 ¢

2-72Ma

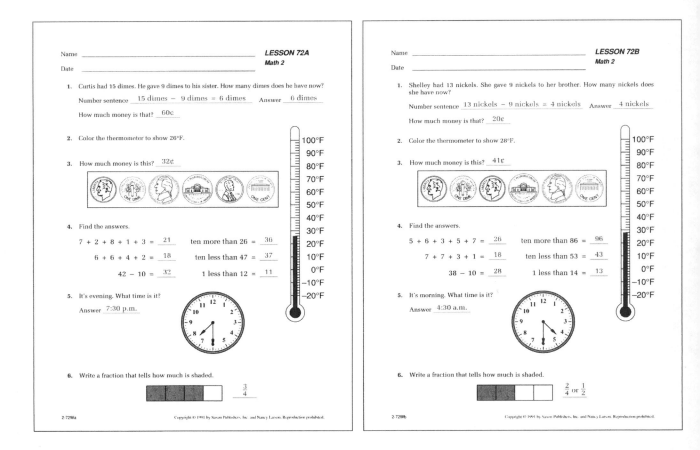

1. Curtis had 15 dimes. He gave 9 dimes to his sister. How many dimes does he have now?

   Number sentence __15 dimes − 9 dimes = 6 dimes__   Answer __6 dimes__

   How much money is that? __60¢__

2. Color the thermometer to show 26°F.

3. How much money is this? __32¢__

4. Find the answers.

   7 + 2 + 8 + 1 + 3 = __21__   ten more than 26 = __36__

       6 + 6 + 4 + 2 = __18__   ten less than 47 = __37__

            42 − 10 = __32__   1 less than 12 = __11__

5. It's evening. What time is it?

   Answer __7:30 p.m.__

6. Write a fraction that tells how much is shaded.

   $\frac{3}{4}$

2-72Wa

1. Shelley had 13 nickels. She gave 9 nickels to her brother. How many nickels does she have now?

   Number sentence __13 nickels − 9 nickels = 4 nickels__   Answer __4 nickels__

   How much money is that? __20¢__

2. Color the thermometer to show 28°F.

3. How much money is this? __41¢__

4. Find the answers.

   5 + 6 + 3 + 5 + 7 = __26__   ten more than 86 = __96__

       7 + 7 + 3 + 1 = __18__   ten less than 53 = __43__

            38 − 10 = __28__   1 less than 14 = __13__

5. It's morning. What time is it?

   Answer __4:30 a.m.__

6. Write a fraction that tells how much is shaded.

   $\frac{2}{4}$ or $\frac{1}{2}$

2-72Wb

L esson 73

subtraction facts—subtracting nine

---

lesson preparation

materials
hundred number chart
Fact Sheet S 7.0

in the morning
• Write the following in the pattern box on the meeting strip:

> a, b, C, d, e, F, ___, ___, ___, ___, ___, ___

Answer: a, b, C, d, e, F, g, h, I, j, k, L

• Write 87¢ on the meeting strip. Provide a cup of 10 dimes, a cup of 10 nickels, and a cup of 20 pennies.

THE MEETING

calendar

• Ask your child to write the date on the calendar and meeting strip.

• Ask your child the following two or three times a week:

date ____ days ago, date ____ days from now

day of the week ____ days ago, day of the week ____ days from now

____th month, month before, month after

• Record on the meeting strip a special event and the number of days until it occurs.

weather graph

• Ask your child to read and graph today's temperature to the nearest two degrees.

• Count by 10's and 2's to check the temperature on the graph.

• Ask your child to connect the dot for yesterday's temperature to the dot for today's temperature and compare the temperatures.

---

counting

- Ask your child to choose a number on the hundred number chart. Ask your child to add or subtract ten or one. Repeat 6–10 times. Ask your child to give directions for returning to the starting number.

  "Today you will learn how to count by 3's to 30."

  "We will use an AAB pattern to help us do that."

  "We can show an AAB pattern by patting our legs and clapping our hands."

  "We will pat our legs twice and then clap our hands."

  "Let's try that."

- Establish a comfortable rhythm and continue until you are patting and clapping together.

  "We will pat our legs twice as we whisper the numbers one and two."

  "Then we will clap our hands as we say 'three.'"

  "Now we will pat our legs two more times as we whisper the numbers four and five."

  "Then we will clap our hands as we say 'six.'"

  "We will keep counting until we reach 30."

  "Let's do this together slowly."

- Repeat several times.

- Do the following once or twice a week:

  count by 10's to 400 and backward from 400 by 10's

  count by 5's to 100 and backward from 50 by 5's

  say the even numbers to 100 and backward from 50

  say the odd numbers to 49 and backward from 49

graph questions

- You and your child each ask a question about any of the graphs.

patterning

- Ask your child to do the following:

  identify the pattern (repeating, continuing, or both)

  identify the letters to complete the pattern

  read the pattern

money

- Ask your child to put the coins in the coin cup. Count the money in the coin cup together.

- Ask your child for another way to show that amount of money. Count these coins together to check the amount.

clock

- Ask your child to set the clock on the half hour or hour.
- Ask the following:

  "It's (morning/afternoon/evening). What time is it?"

  time one hour ago

  time one hour from now

- Ask your child to write the digital time on the meeting strip.
- Record on the meeting strip the time an activity will occur.

number of the day

- Write three number sentences for the number of the day on the meeting strip.

THE LESSON

Subtraction Facts—Subtracting Nine

"Today you will learn how to subtract nine from a number."

"You already know the answers for 18 − 9 and 9 − 9."

"What are the answers?"

- Write the following on the chalkboard:

$$\begin{array}{cc} 18 & 9 \\ -9 & -9 \\ \hline 9 & 0 \end{array}$$

"We will use a hundred number chart to help us find the answers for the subtracting nine facts."

- Give your child a hundred number chart.
- Write the following problems on the chalkboard:

$$\begin{array}{cccccccc} 17 & 16 & 15 & 14 & 13 & 12 & 11 & 10 \\ -9 & -9 & -9 & -9 & -9 & -9 & -9 & -9 \end{array}$$

"How do you think we will use the hundred number chart to find these answers?"

"Put your finger on the number 17."

"Now count back nine spaces."

"What is the answer?"   *8*

- Fill in the answer below the chalkboard problem.
- Repeat with each problem.
- Circle the 15 number fact.

  $- 9$

  "There is a trick some people use to help them remember this answer."

  "The answer for this subtracting nine fact is the same as adding the digits one and five in the number fifteen."

  "Does this work for the other subtracting nine facts?"

  "Let's check."
- Show your child that the answer for 13 – 9 is the same as 1 + 3.

  "Let's practice the subtracting nine facts."
- Erase the chalkboard answers.

  "I will point to a number fact."

  "Try to say the answer as quickly as possible."
- Point to the number facts one at a time.
- Review the strategy for finding the answer, if necessary.

CLASS PRACTICE

number fact practice

"Use your yellow fact cards to practice the subtracting nine number facts."

- Allow time for your child to practice independently.
- Give your child **Fact Sheet S 7.0**.
- Time your child for one minute.
- Correct the fact sheet with your child and record the score.
- Allow time for your child to complete the unfinished facts.

WRITTEN PRACTICE

- Complete **Worksheet 73A** with your child.
- Complete **Worksheet 73B** with your child later in the day.

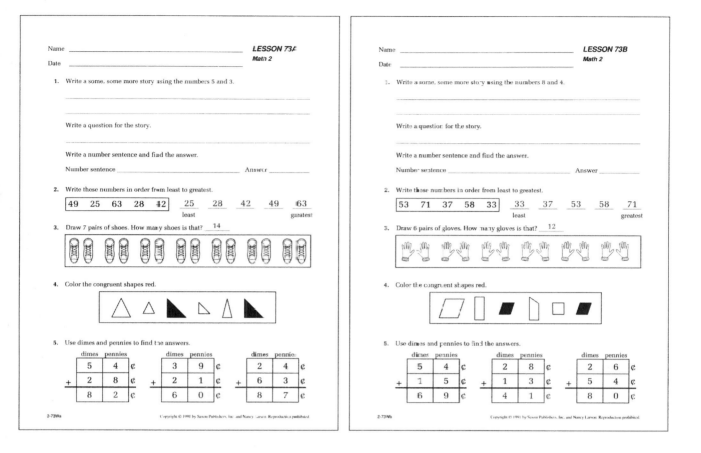

Name _____ **LESSON 73A**
Date _____ Math 2

1. Write a some, some more story using the numbers 5 and 3.

   _____

   _____

   Write a question for the story.

   _____

   Write a number sentence and find the answer.

   Number sentence _____ Answer _____

2. Write these numbers in order from least to greatest.

   | 49 | 25 | 63 | 28 | 42 | 25   28   42   49   63
   least   greatest

3. Draw 7 pairs of shoes. How many shoes is that? __14__

4. Color the congruent shapes red.

5. Use dimes and pennies to find the answers.

dimes	pennies		dimes	pennies		dimes	pennies
5	4	¢	3	9	¢	2	4	¢
+ 2	8	¢	+ 2	1	¢	+ 6	3	¢
8	2	¢	6	0	¢	8	7	¢

2-73Wa

Copyright © 1991 by Saxon Publishers, Inc. and Nancy Larson. Reproduction prohibited.

Name _____ **LESSON 73B**
Date _____ Math 2

1. Write a some, some more story using the numbers 8 and 4.

   _____

   _____

   Write a question for the story.

   _____

   Write a number sentence and find the answer.

   Number sentence _____ Answer _____

2. Write these numbers in order from least to greatest.

   | 53 | 71 | 37 | 58 | 33 | 33   37   53   58   71
   least   greatest

3. Draw 6 pairs of gloves. How many gloves is that? __12__

4. Color the congruent shapes red.

5. Use dimes and pennies to find the answers.

dimes	pennies		dimes	pennies		dimes	pennies
5	4	¢	2	8	¢	2	6	¢
+ 1	5	¢	+ 1	3	¢	+ 5	4	¢
6	9	¢	4	1	¢	8	0	¢

2-73Wb

Copyright © 1991 by Saxon Publishers, Inc. and Nancy Larson. Reproduction prohibited.

Lesson 74

measuring and drawing line segments to the nearest half inch

Math 2

---

lesson preparation

materials

Meeting Book
ruler
Master 2-74
Fact Sheet S 7.0

in the morning

• Write the following in the pattern box on the meeting strip:

___, ___, ___, 25, 30, 35, ___, ___, ___

Answer: 10, 15, 20, 25, 30, 35, 40, 45, 50

• Write [76¢] on the meeting strip. Provide a cup of 10 dimes, a cup of 10 nickels, and a cup of 20 pennies.

THE MEETING

calendar

• Ask your child to write the date on the calendar and meeting strip.

• Ask your child the following two or three times a week:

date _____ days ago, date _____ days from now

day of the week _____ days ago, day of the week _____ days from now

_____th month, month before, month after

• Record on the meeting strip a special event and the number of days until it occurs.

weather graph

• Ask your child to read and graph today's temperature to the nearest two degrees.

• Count by 10's and 2's to check the temperature on the graph.

Copyright © 1994 by Saxon Publishers, Inc. and Nancy Larson. Reproduction prohibited.

- Ask your child to connect the dot for yesterday's temperature to the dot for today's temperature and compare the temperatures.

counting

- Ask your child to choose a number on the hundred number chart. Ask your child to add or subtract ten or one. Repeat 6–10 times. Ask your child to give directions for returning to the starting number.

  "Let's use our patting and clapping pattern to help us count by 3's to 30."

  "We will pat our legs twice as we whisper the numbers one and two."

  "Then we will clap our hands as we say 'three.'"

  "Now we will pat our legs two more times as we whisper the numbers four and five."

  "Then we will clap our hands as we say 'six.'"

  "We will keep counting until we reach 30."

  "Let's do this together slowly."

- Repeat several times.

  "Now I will write the numbers we say on a counting strip in the Meeting Book as we count by 3's together."

- Record the numbers on a Meeting Book counting strip as you count by 3's to 30 with your child. Begin at the bottom of the strip and work upward.

- Do the following once or twice a week:

  count by 10's to 400 and backward from 400 by 10's

  count by 5's to 100 and backward from 50 by 5's

  say the even numbers to 100 and backward from 50

  say the odd numbers to 49 and backward from 49

graph questions

- You and your child each ask a question about any of the graphs.

patterning

- Ask your child to do the following:

  identify the pattern (repeating, continuing, or both)

  identify the numbers to complete the pattern

  read the pattern

money

- Ask your child to put the coins in the coin cup. Count the money in the coin cup together.

• Ask your child for another way to show that amount of money. Count these coins together to check the amount.

clock

• Ask your child to set the clock on the half hour or hour.

• Ask the following:

> *"It's (morning/afternoon/evening). What time is it?"*
>
> time one hour ago
>
> time one hour from now

• Ask your child to write the digital time on the meeting strip.

• Record on the meeting strip the time an activity will occur.

number of the day

• Write three number sentences for the number of the day on the meeting strip.

THE LESSON

Measuring and Drawing Line Segments to the Nearest Half Inch

"Today you will learn how to measure to the nearest half inch."

• Draw a large model of a ruler on the chalkboard with only the inch marks indicated.

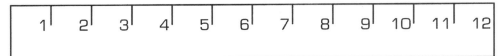

"What does this look like?"   ruler

"Is it exactly the same as your ruler?"   no, it is larger

• Hold up a regular ruler next to the line to compare.

"I drew a large picture of a ruler."

"This is a pretend 12-inch ruler."

"Where will we put the half-inch marks on this ruler?"

"Let's fill in all the half-inch marks."

• Ask your child to put marks halfway between the inches.

"Let's count by $\frac{1}{2}$'s across the chalkboard ruler."

"We will begin at the left end of the ruler."

"This is zero."

- Point to each mark as you count with your child:

$$0, \tfrac{1}{2}, 1, 1\tfrac{1}{2}, 2, 2\tfrac{1}{2}, 3, \ldots$$

- Repeat several times.

- Draw a line segment $2\tfrac{1}{2}$ inches long above the chalkboard ruler.

  "How long is this line segment?" $2\tfrac{1}{2}''$

  "We write '$2\tfrac{1}{2}$ inches' like this."

- Write "$2\tfrac{1}{2}$ inches" on the chalkboard.

  "We also can write it like this."

- Write "$2\tfrac{1}{2}''$" on the chalkboard.

- Repeat with lines 5 inches, $3\tfrac{1}{2}$ inches, and $\tfrac{1}{2}$ inch long.

- Give your child a ruler.

  "Point to the one-inch mark on your ruler."

- Repeat with 2, 3, 8, 4, and 11 inches.

  "Put your fingers on the two and the three."

  "Point to the longest line in the middle."

  "We call this $2\tfrac{1}{2}$ inches."

  "Put your fingers on the five and the six."

  "Point to the longest line in the middle."

  "We call this $5\tfrac{1}{2}$ inches."

  "Put your fingers at the left end of the ruler and the one."

  "Point to the longest line in the middle."

  "We call this $\tfrac{1}{2}$ inch."

  "Now you will use the ruler to draw and measure line segments."

- Give your child **Master 2-74**.

  "First you will draw a three-inch line segment."

  "Point to the dot next to the 3."

  "Put the left end (zero) of your ruler on the dot."

  "Carefully hold the ruler with the fingers of the hand you don't write with."

  "Put your pencil point on the dot."

  "Trace along the edge of the ruler until you come to the three."

  "Move your ruler."

  "Put a dot on the right end of the line segment."

  "Check the line segment using your ruler to make sure it is three inches long."

"Now you will use the ruler to draw a $2\frac{1}{2}$-inch line segment."

"Point to the dot next to the $2\frac{1}{2}$"."

"Put the left end (zero) of your ruler on the dot."

"Carefully hold the ruler with the fingers of the hand you don't write with."

"Put your pencil point on the dot."

"Trace along the edge of the ruler until you come to the two."

"Now continue drawing the line segment until you are halfway between the two and the three."

"Move your ruler."

"Put a dot on the right end of the line segment."

"Check the line segment using your ruler to make sure that it is $2\frac{1}{2}$ inches long."

- Repeat with $1\frac{1}{2}$, $\frac{1}{2}$, and $4\frac{1}{2}$ inches.

"Measure the line segments at the bottom of Master 2-74."

"Write the length of each line segment in the box at the left of the line segment."

CLASS PRACTICE

number fact practice

"Let's practice the subtracting nine facts together."

"Do you remember the trick for remembering the answers for the subtracting nine facts?"   the answer is the same as adding the digits in the number you are subtracting nine from

- Use the yellow fact cards to practice the subtracting nine facts with your child.

- Give your child **Fact Sheet S 7.0**.

- Time your child for one minute.

- Correct the fact sheet with your child and record the score.

- Allow time for your child to complete the unfinished facts.

WRITTEN PRACTICE

- Complete **Worksheet 74A** with your child.

- Complete **Worksheet 74B** with your child later in the day.

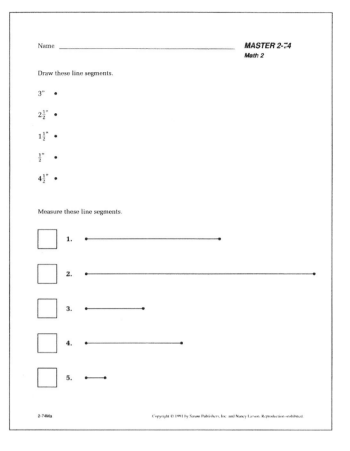

Name _____ **MASTER 2-74**
                                 Math 2

Draw these line segments.

3" •

$2\frac{1}{2}$" •

$1\frac{1}{2}$" •

$\frac{1}{2}$" •

$4\frac{1}{2}$" •

Measure these line segments.

☐ **1.** •————————————————•

☐ **2.** •——————————————————————•

☐ **3.** •————————•

☐ **4.** •——————————•

☐ **5.** •———•

2-74Ma   Copyright © 1991 by Saxon Publishers, Inc. and Nancy Larson. Reproduction prohibited.

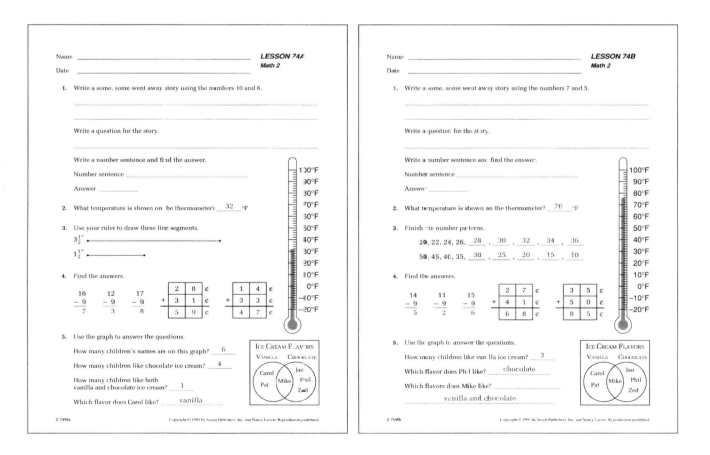

Name _____ **LESSON 74A**
Date _____ **Math 2**

1. Write a some, some went away story using the numbers 10 and 6.

    _____

    Write a question for the story.

    _____

    Write a number sentence and find the answer.

    Number sentence _____

    Answer _____

2. What temperature is shown on the thermometer? __32__ °F

3. Use your ruler to draw these line segments.

    $3\frac{1}{2}$" •————————————————•

    $1\frac{1}{2}$" •——————•

4. Find the answers.

16	12	17
−9	−9	−9
7	3	8

2	8	¢
+3	1	¢
5	9	¢

1	4	¢
+3	3	¢
4	7	¢

    Thermometer: 100°F, 90°F, 80°F, 70°F, 60°F, 50°F, 40°F, 30°F, 20°F, 10°F, 0°F, −10°F, −20°F

5. Use the graph to answer the questions.

    How many children's names are on this graph? __6__

    How many children like chocolate ice cream? __4__

    How many children like both
    vanilla and chocolate ice cream? __1__

    Which flavor does Carol like? __vanilla__

    ICE CREAM FLAVORS
    VANILLA   CHOCOLATE
    Carol   Jan
    Pat   Mike   Phil   Zed

2-74Wa   Copyright © 1991 by Saxon Publishers, Inc. and Nancy Larson. Reproduction prohibited.

Name _____ **LESSON 74B**
Date _____ **Math 2**

1. Write a some, some went away story using the numbers 7 and 5.

    _____

    Write a question for the story.

    _____

    Write a number sentence and find the answer.

    Number sentence _____

    Answer _____

2. What temperature is shown on the thermometer? __76__ °F

3. Finish the number patterns.

    20, 22, 24, 26, __28__ , __30__ , __32__ , __34__ , __36__

    50, 45, 40, 35, __30__ , __25__ , __20__ , __15__ , __10__

4. Find the answers.

14	11	15
−9	−9	−9
5	2	6

2	7	¢
+4	1	¢
6	8	¢

3	5	¢
+5	0	¢
8	5	¢

    Thermometer: 100°F, 90°F, 80°F, 70°F, 60°F, 50°F, 40°F, 30°F, 20°F, 10°F, 0°F, −10°F, −20°F

5. Use the graph to answer the questions.

    How many children like vanilla ice cream? __3__

    Which flavor does Phil like? __chocolate__

    Which flavors does Mike like?

    __vanilla and chocolate__

    ICE CREAM FLAVORS
    VANILLA   CHOCOLATE
    Carol   Jan
    Pat   Mike   Phil   Zed

2-74Wb   Copyright © 1991 by Saxon Publishers, Inc. and Nancy Larson. Reproduction prohibited.

405

Lesson 75

using the addition algorithm (part 1)

lesson preparation

materials

Written Assessment #14

cup of 20 pennies

cup of 10 dimes

yellow and white work mat from Lesson 71

Fact Sheet S 4.0

in the morning

• Write the following in the pattern box on the meeting strip:

> 13, 15, 17, ___, ___, ___, ___, ___, ___

Answer: 13, 15, 17, 19, 21, 23, 25, 27, 29

• Write 95¢ on the meeting strip. Provide a cup of 10 dimes, a cup of 10 nickels, and a cup of 20 pennies.

THE MEETING

calendar

• Ask your child to write the date on the calendar and meeting strip.

• Ask your child the following two or three times a week:

date ____ days ago, date ____ days from now

day of the week ____ days ago, day of the week ____ days from now

____th month, month before, month after

• Record on the meeting strip a special event and the number of days until it occurs.

weather graph

• Ask your child to read and graph today's temperature to the nearest two degrees.

• Count by 10's and 2's to check the temperature on the graph.

- Ask your child to connect the dot for yesterday's temperature to the dot for today's temperature and compare the temperatures.

counting

- Ask your child to choose a number on the hundred number chart. Ask your child to add or subtract ten or one. Repeat 6–10 times. Ask your child to give directions for returning to the starting number.

  "Let's use our patting and clapping pattern to help us count by 3's to 30."

- Repeat this several times.
- Do the following once or twice a week:

  count by 10's to 400 and backward from 400 by 10's

  count by 5's to 100 and backward from 50 by 5's

  say the even numbers to 100 and backward from 50

  say the odd numbers to 49 and backward from 49

graph questions

- You and your child each ask a question about any of the graphs.

patterning

- Ask your child to do the following:

  identify the pattern (repeating, continuing, or both)

  identify the numbers to complete the pattern

  read the pattern

money

- Ask your child to put the coins in the coin cup. Count the money in the coin cup together.
- Ask your child for another way to show that amount of money. Count these coins together to check the amount.

clock

- Ask your child to set the clock on the half hour or hour.
- Ask the following:

  "It's (morning/afternoon/evening). What time is it?"

  time one hour ago

  time one hour from now

- Ask your child to write the digital time on the meeting strip.
- Record on the meeting strip the time an activity will occur.

number of the day

- Write three number sentences for the number of the day on the meeting strip.

ASSESSMENT

Written Assessment

"Today I would like to see what you remember from what we have been practicing."

- Give your child **Written Assessment #14**.

- Read the directions for each problem. Allow time for your child to complete that problem before continuing.

- Correct the paper, noting your child's mistakes on the **Individual Recording Form**. Review the errors with your child.

THE LESSON

Using the Addition Algorithm (Part 1)

"Today you will learn how to add two-digit numbers."

"You will use your dimes and pennies to check your answers."

- Write the following on the chalkboard:

$$23¢ + 47¢ =$$

"Write this problem vertically on the chalkboard."

"How many dimes and pennies will we need to show these amounts of money?"   **6 dimes, 10 pennies**

- Put 6 dimes and 10 pennies on the mat.

"Should I write the amount of money we have using a 6 and a 10, like this?"

- Write "610¢" on the chalkboard below the vertical problem.

"Do we have 610¢?"

"What is wrong with what I did?"

"How much money do we really have?"   **70¢**

"What do we need to do with the coins?"   **trade 10 pennies for a dime**

- Trade the pennies for 1 dime.

"How many dimes and pennies do we have now?"   **7 dimes, zero pennies**

• Erase the 610¢.

"How will we show the addition on our chalkboard problem?"

"What will we add first?"   pennies

"How many pennies do we have altogether?"   10

"Do we have enough pennies to trade for a dime?"   yes

"How will we show that on our problem?"

"I write the '1' at the bottom of the dimes' column because it is easy for me to see how many pennies I had."

"Some people write the '1' at the top of the dimes' column."

"We write the number of extra pennies we have left at the bottom of the pennies' column."

"If there are no pennies left, we will write a zero."

• Write the following on the chalkboard:

$$
\begin{array}{r}
2\ 3\ ¢ \\
+\ 4\ 7\ ¢ \\
\underset{1}{} \\
\hline
0\ ¢
\end{array}
$$

"Now we will add the number of dimes."

"We have 70¢."

"Let's try another problem."

• Write the following on the chalkboard:

$$58¢ + 28¢ =$$

"Write this problem vertically on the chalkboard."

"How many dimes and pennies will we need to show these amounts of money?"   7 dimes, 16 pennies

• Put 7 dimes and 16 pennies on the mat.

"Should I write the amount of money we have using a 7 and a 16, like this?"

• Write "716¢" on the chalkboard below the vertical problem.

"Do we have 716¢?"

"What is wrong with what I did?"

"How much money do we really have?"   86¢

"What do we need to do with the coins?"   trade 10 pennies for 1 dime

• Trade the pennies for a dime.

"Now how many dimes and pennies do we have?"   8 dimes, 6 pennies

"How will we show the addition on our chalkboard problem?"

"What will we do first?"   add the pennies

"How many pennies do we have altogether?"   16

"Do we have enough pennies to trade for a dime?"   yes

"How will we show that on our problem?"

- Write the following on the chalkboard:

$$
\begin{array}{r}
5\ 8\ \text{¢} \\
+\ 2\ 8\ \text{¢} \\
\underline{1} \\
6\ \text{¢}
\end{array}
$$

"What will we do now?"   add the dimes

"We have 86¢."

- Write the following problems on the chalkboard:

     47¢ + 26¢   53¢ + 34¢   8¢ + 42¢   69¢ + 18¢

"Write these problems vertically on the chalkboard."

"Remember to write the pennies under the pennies and the dimes under the dimes."

- Allow time for your child to do this.

"How can we tell by looking at a problem if we will have to trade pennies for dimes?"   if there are 10 or more pennies

"What numbers will we look at to tell?"   pennies

"In which of these problems will we be able to trade ten pennies for a dime?"   first, third, fourth

"Let's find the answer for each of these problems."

"When we finish, we will use dimes and pennies to check the answers."

"Let's find the answer for the first problem."

"What will we do first?"   add the pennies

"How many pennies do we have?"   13

"Do we have enough pennies to trade for a dime?"   yes

"Write a '1' in the column with the dimes to show that we traded 10 pennies for a dime."

"How many pennies do we have left?"

"We will write the number of extra pennies we have in the pennies' column."

"Now how many dimes do we have?"   7

- Repeat with each problem.

"Now check the answers using money."

- Give your child a cup of 10 dimes, a cup of 20 pennies, and a work mat.

- Allow your child time to check his/her answers.

CLASS PRACTICE

number fact practice

- Use the green fact cards to practice the subtracting two facts with your child.

- Give your child **Fact Sheet S 4.0**.

- Time your child for one minute.

- Correct the fact sheet with your child and record the score.

- Allow time for your child to complete the unfinished facts.

WRITTEN PRACTICE

- Complete **Worksheet 75A** with your child.

- Complete **Worksheet 75B** with your child later in the day.

Name _____

Date _____

ASSESSMENT 14
LESSON 75
Math 2

1. There were a dozen cupcakes in the box. Sara, Bill, and Peter each ate one cupcake. How many cupcakes are left?

   Number sentence ___12 cup. − 3 cup. = 9 cup.___   Answer _9 cupcakes_

2. Use the graph to answer the following questions.

   How many children's names are on this graph? __6__

   How many children like peas? __3__

   How many children like squash? __5__

   How many children like both peas and squash? __2__

   Which vegetable does Jim like? ___squash___

   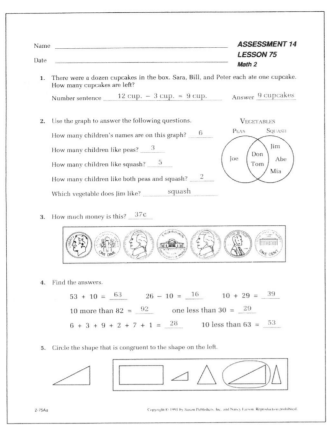

3. How much money is this? __37¢__

   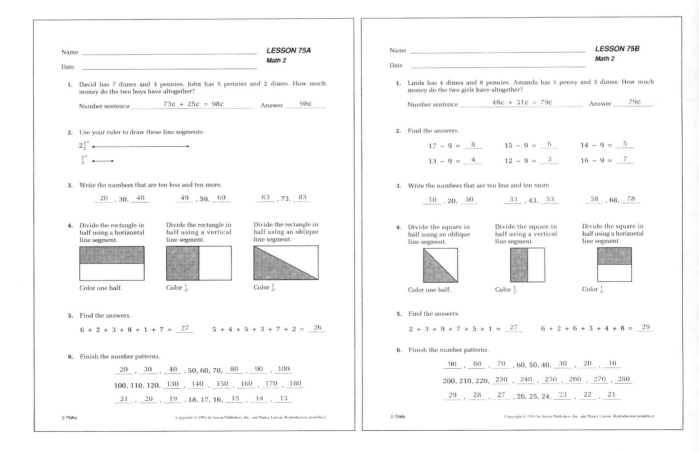

4. Find the answers.

   53 + 10 = __63__   26 − 10 = __16__   10 + 29 = __39__

   10 more than 82 = __92__   one less than 30 = __29__

   6 + 3 + 9 + 2 + 7 + 1 = __28__   10 less than 63 = __53__

5. Circle the shape that is congruent to the shape on the left.

2-75Aa

Name _____

Date _____

LESSON 75A
Math 2

1. David has 7 dimes and 3 pennies. John has 5 pennies and 2 dimes. How much money do the two boys have altogether?

   Number sentence ___73¢ + 25¢ = 98¢___   Answer ___98¢___

2. Use your ruler to draw these line segments.

   $2\frac{1}{2}$" •———————————•

   $\frac{1}{2}$" •——•

3. Write the numbers that are ten less and ten more.

   __20__, 30, __40__   __49__, 59, __69__   __63__, 73, __83__

4. Divide the rectangle in half using a horizontal line segment.

   Color one half.

   Divide the rectangle in half using a vertical line segment.

   Color $\frac{1}{2}$.

   Divide the rectangle in half using an oblique line segment.

   Color $\frac{1}{2}$.

5. Find the answers.

   6 + 2 + 3 + 8 + 1 + 7 = __27__   5 + 4 + 5 + 3 + 7 + 2 = __26__

6. Finish the number patterns.

   __20__, __30__, __40__, 50, 60, 70, __80__, __90__, __100__

   100, 110, 120, __130__, __140__, __150__, __160__, __170__, __180__

   __21__, __20__, __19__, 18, 17, 16, __15__, __14__, __13__

2-75Wa

Name _____

Date _____

LESSON 75B
Math 2

1. Linda has 4 dimes and 8 pennies. Amanda has 1 penny and 3 dimes. How much money do the two girls have altogether?

   Number sentence ___48¢ + 31¢ = 79¢___   Answer ___79¢___

2. Find the answers.

   17 − 9 = __8__   15 − 9 = __6__   14 − 9 = __5__

   13 − 9 = __4__   12 − 9 = __3__   16 − 9 = __7__

3. Write the numbers that are ten less and ten more.

   __10__, 20, __30__   __33__, 43, __53__   __58__, 68, __78__

4. Divide the square in half using an oblique line segment.

   Color one half.

   Divide the square in half using a vertical line segment.

   Color $\frac{1}{2}$.

   Divide the square in half using a horizontal line segment.

   Color $\frac{1}{2}$.

5. Find the answers.

   2 + 3 + 9 + 7 + 5 + 1 = __27__   6 + 2 + 6 + 3 + 4 + 8 = __29__

6. Finish the number patterns.

   __90__, __80__, __70__, 60, 50, 40, __30__, __20__, __10__

   200, 210, 220, __230__, __240__, __250__, __260__, __270__, __280__

   __29__, __28__, __27__, 26, 25, 24, __23__, __22__, __21__

2-75Wb

Lesson 76

using the addition algorithm (part 2)

lesson preparation

materials

Master 2-76

Fact Sheet S 4.4

in the morning

• Write the following in the pattern box on the meeting strip:

> 112, 110, 108, 106, ___, ___, ___, ___, ___, ___

Answer: 112, 110, 108, 106, 104, 102, 100, 98, 96, 94

• Write ⟨57¢⟩ on the meeting strip. Provide a cup of 10 dimes, a cup of 10 nickels, and a cup of 20 pennies.

THE MEETING

calendar

- Ask your child to write the date on the calendar and meeting strip.

- Ask your child the following two or three times a week:

    date _____ days ago, date _____ days from now

    day of the week _____ days ago, day of the week _____ days from now

    _____th month, month before, month after

- Record on the meeting strip a special event and the number of days until it occurs.

weather graph

- Ask your child to read and graph today's temperature to the nearest two degrees.

- Count by 10's and 2's to check the temperature on the graph.

- Ask your child to connect the dot for yesterday's temperature to the dot for today's temperature and compare the temperatures.

counting

- Ask your child to choose a number on the hundred number chart. Ask your child to add or subtract ten or one. Repeat 6–10 times. Ask your child to give directions for returning to the starting number.

  "Let's use our patting and clapping pattern to help us count by 3's to 30."

- Repeat this several times.

- Do the following once or twice a week:

    count by 10's to 400 and backward from 400 by 10's

    count by 5's to 100 and backward from 50 by 5's

    say the even numbers to 100 and backward from 50

    say the odd numbers to 49 and backward from 49

graph questions

- You and your child each ask a question about any of the graphs.

patterning

- Ask your child to do the following:

    identify the pattern (repeating, continuing, or both)

    identify the numbers to complete the pattern

    read the pattern

money

- Ask your child to put the coins in the coin cup. Count the money in the coin cup together.

- Ask your child for another way to show that amount of money. Count these coins together to check the amount.

clock

- Ask your child to set the clock on the half hour or hour.

- Ask the following:

    "It's (morning/afternoon/evening). What time is it?"

    time one hour ago

    time one hour from now

- Ask your child to write the digital time on the meeting strip.

- Record on the meeting strip the time an activity will occur.

number of the day

- Write three number sentences for the number of the day on the meeting strip.

THE LESSON

Using the Addition Algorithm (Part 2)

"Today you will learn how to add two-digit numbers without using dimes and pennies."

- Give your child **Master 2-76**.

  "Let's try the first problem together."

  "Write the first problem vertically on your paper."

- Allow time for your child to write the problem vertically.

  "How will you find the answer?"

  "Add the numbers together."

- Allow time for your child to do this.

  "Explain what happened in this problem."

- Ask your child to describe the steps.

  "Write the second problem vertically."

- Allow time for your child to write the problem vertically.

  "Add the numbers together."

- Allow time for your child to do this.

  "Explain what happened in this problem."

- Ask your child to describe the steps.

  "Find the answers for the other problems on this paper."

- Review the answers with your child.

CLASS PRACTICE

number fact practice

- Use the tan, peach, lavender, and green fact cards to practice the subtracting two, subtracting one, subtracting zero, subtracting half of a double, and subtracting a number from itself facts with your child.

- Give your child **Fact Sheet S 4.4**.

- Time your child for one minute.

- Correct the fact sheet with your child and record the score.

- Allow time for your child to complete the unfinished facts.

WRITTEN PRACTICE

- • Complete **Worksheet 76A** with your child.
- • Complete **Worksheet 76B** with your child later in the day.

Name _____ **MASTER 2-76**
                                   Math 2

1.   **2.**   **3.**   **4.**
12¢ + 29¢   55¢ + 26¢   73¢ + 15¢   29¢ + 21¢

+   +   +   +

5.   **6.**   **7.**
49¢ + 24¢   52¢ + 24¢   65¢ + 25¢

+   +   +
_____   _____   _____

8.   **9.**   **10.**
16¢ + 24¢   9¢ + 38¢   46¢ + 7¢

+   +   +
_____   _____   _____

2-76Ma   Copyright © 1991 by Saxon Publishers, Inc. and Nancy Larson. Reproduction prohibited.

Name _____ **LESSON 76A**
Date _____   Math 2

1. One of these is my favorite day of the week. Cross out the
days of the week that cannot be my favorite day of the week.
It is not the sixth day of the week.
It is not a weekend day.
It does not begin with a T.
It is not the day that is in the middle of the week.

What is my favorite day? ____Monday____

~~Sunday~~
Monday
~~Tuesday~~
~~Wednesday~~
~~Thursday~~
~~Friday~~
~~Saturday~~

2. Draw a shape with 4 angles. How many sides does the shape have? ___4___

answers may vary

3. Write fractions that show how much is shaded.

$\frac{3}{4}$ $\frac{5}{6}$ $\frac{1}{3}$

4. Write the numbers that are one less and one more than each number.

__11__ , 12, __13__   __39__ , 40, __41__   __78__ , 79, __80__

5. Use your ruler to measure these line segments.

____2__ ''

$4\frac{1}{2}$ ''

6. Find the answers.

39 − 10 = __29__   6 + 3 + 2 + 4 + 8 + 1 = __24__   4 7 ¢
                                                            + 1 6 ¢
10 more than 52 = __62__   10 less than 41 = __31__   6 3 ¢

2-76Wa   Copyright © 1991 by Saxon Publishers, Inc. and Nancy Larson. Reproduction prohibited.

Name _____ **LESSON 76B**
Date _____   Math 2

1. One of these is my brother's favorite day of the week. Cross out the
days of the week that cannot be my brother's favorite day of the week.
It is not the first or the third day of the week.
It does not have 6 letters.
It is not the last day of the week.
It is not the day in the middle of the week.

What is my brother's favorite day? ___Thursday___

~~Sunday~~
~~Monday~~
~~Tuesday~~
~~Wednesday~~
Thursday
~~Friday~~
~~Saturday~~

2. Draw a shape with 3 angles. How many sides does the shape have? ___3___

answers may vary

3. Write fractions that show how much is shaded.

$\frac{1}{4}$ $\frac{1}{6}$ $\frac{2}{3}$

4. Write the numbers that are one less and one more than each number.

__15__ , 16, __17__   __28__ , 29, __30__   __59__ , 60, __61__

5. What is the best estimate of the length of this paper?

5 inches   28 inches   (11 inches)   20 inches

6. Find the answers.

84 − 10 = __74__   9 + 3 + 1 + 4 + 7 = __24__   6 7 ¢
                                                        + 2 3 ¢
10 more than 72 = __82__   10 less than 63 = __53__   9 0 ¢

2-76Wb   Copyright © 1991 by Saxon Publishers, Inc. and Nancy Larson. Reproduction prohibited.

Copyright © 1994 by Saxon Publishers, Inc. and Nancy Larson. Reproduction prohibited.

Lesson 77

representing and writing mixed numbers

THE MEETING

calendar

• Ask your child to write the date on the calendar and meeting strip.

• Ask your child the following two or three times a week:

date ____ days ago, date ____ days from now

day of the week ____ days ago, day of the week ____ days from now

____th month, month before, month after

• Record on the meeting strip a special event and the number of days until it occurs.

weather graph

• Ask your child to read and graph today's temperature to the nearest two degrees.

• Count by 10's and 2's to check the temperature on the graph.

• Ask your child to connect the dot for yesterday's temperature to the dot for today's temperature and compare the temperatures.

counting

• Ask your child to choose a number on the hundred number chart. Ask your child to add or subtract ten or one. Repeat 6–10 times. Ask your child to give directions for returning to the starting number.

"Let's use our patting and clapping pattern to help us count by 3's to 30."

• Repeat this several times.

• Do the following once or twice a week:

count by 10's to 400 and backward from 400 by 10's

count by 5's to 100 and backward from 50 by 5's

say the even numbers to 100 and backward from 50

say the odd numbers to 49 and backward from 49

graph questions

• You and your child each ask a question about any of the graphs.

patterning

• Ask your child to do the following:

identify the pattern (repeating, continuing, or both)

identify the shapes to complete the pattern

read the pattern

money

• Ask your child to put the coins in the coin cup. Count the money in the coin cup together.

• Ask your child for another way to show that amount of money. Count these coins together to check the amount.

clock

• Ask your child to set the clock on the half hour or hour.

• Ask the following:

"It's (morning/afternoon/evening). What time is it?"

time one hour ago

time one hour from now

• Ask your child to write the digital time on the meeting strip.

• Record on the meeting strip the time an activity will occur.

number of the day

- Write three number sentences for the number of the day on the meeting strip.

THE LESSON

Representing and Writing Mixed Numbers

"Today you will learn how to show and write a mixed number."

"Mixed numbers look like this."

- Write the following mixed numbers on the chalkboard:

$$2\frac{1}{2} \qquad 4\frac{1}{3} \qquad 1\frac{3}{4} \qquad 3\frac{2}{3}$$

"When have we used mixed numbers before?"   *measuring and drawing line segments*

"Today you will use pattern blocks to show mixed numbers."

- Give your child a basket of pattern blocks and a piece of scrap paper.

"The yellow hexagon is one whole."

"Put three yellow hexagons on your paper."

"Cover the first hexagon using red pattern blocks."

- Allow time for your child to do this.

"How many red trapezoids did you use?"   *2*

"What fraction will we use to name the red trapezoid?" $\frac{1}{2}$

"Cover the second hexagon using blue pattern blocks."

- Allow time for your child to do this.

"How many blue parallelograms did you use?"   *3*

"What fraction will we use to name the blue parallelogram?" $\frac{1}{3}$

"Cover the third hexagon using green pattern blocks."

- Allow time for your child to do this.

"How many green triangles did you use?"   *6*

"What fraction will we use to name the green triangle?" $\frac{1}{6}$

"Hold up the pattern block that shows one third."

"Hold up the pattern block that shows one sixth."

"Hold up the pattern block that shows one half."

"Hold up the pattern block that shows one whole."

"Push your pattern blocks off your paper."

"Now I will write a mixed number on the chalkboard."

- Write "$1\frac{1}{2}$" on the chalkboard.

   "What is this number?"

   "How do you think you will show that using pattern blocks?" one yellow pattern block and one red pattern block

   "Show $1\frac{1}{2}$ using your pattern blocks."

- Write "$2\frac{1}{3}$" on the chalkboard.

   "What is this number?"

   "Show $2\frac{1}{3}$ using your pattern blocks." two yellow pattern blocks and one blue pattern block

- Write "$3\frac{1}{6}$" on the chalkboard.

   "What is this number?"

   "Show $3\frac{1}{6}$ using your pattern blocks." three yellow pattern blocks and one green pattern block

- Write "$1\frac{2}{3}$" on the chalkboard.

   "What is this number?"

   "Show $1\frac{2}{3}$ using your pattern blocks." one yellow pattern block and two blue pattern blocks

- Write "$2\frac{3}{6}$" on the chalkboard.

   "What is this number?"

   "Show $2\frac{3}{6}$ using your pattern blocks." two yellow pattern blocks and three green pattern blocks

   "Put two yellow pattern blocks and one red pattern block on your mat."

   "How will we write a mixed number to show how much this is?"

- Ask your child to describe how to write "$2\frac{1}{2}$."

   "Put one yellow pattern block and two blue pattern blocks on your mat."

   "How will we write a mixed number to show how much this is?"

- Ask your child to describe how to write "$1\frac{2}{3}$."

   "Put two yellow pattern blocks and five green pattern blocks on your mat."

   "How will we write a mixed number to show how much this is?"

- Ask your child to describe how to write "$2\frac{5}{6}$."

   "Put your pattern blocks back in the basket."

- Give your child **Master 2-77** and a crayon.

   "What do you think you will do on this master?"

   "Let's do number one together."

   "What is the mixed number we are showing?" $2\frac{1}{2}$

"Quickly color the first two squares."

"How will you color one half of a square?"

"Use your pencil to divide the next square in half."

"Now color one half of the square."

"What is the mixed number you will show in problem two?" $3\frac{1}{2}$

"What will you do first?"   color 3 circles

"Quickly color the first three circles."

"What will you do next?"   divide a circle in half using a pencil and color one half

"Use your pencil to divide the next circle in half."

"Now color one half of the circle."

- Repeat with problems 3 and 4.

"What will you do for problem five?"   write a mixed number to show how much is shaded

- Ask your child to read the directions.

"How do you think you will do this?"

"How many whole squares are shaded?"   5

"Write that on the line."

"Now you will write a fraction to show how much of the last square is shaded."

"What fraction will you write for this?" $\frac{1}{4}$

"Write that next to the whole number."

- Repeat with the last three problems. $3\frac{1}{3}$; $4\frac{2}{3}$; $2\frac{3}{4}$

CLASS PRACTICE

number fact practice

- Use the yellow fact cards to practice the subtracting nine facts with your child.

- Give your child **Fact Sheet S 7.0.**

- Time your child for one minute.

- Correct the fact sheet with your child and record the score.

- Allow time for your child to complete the unfinished facts.

WRITTEN PRACTICE

• Complete **Worksheet 77A** with your child.

• Complete **Worksheet 77B** with your child later in the day.

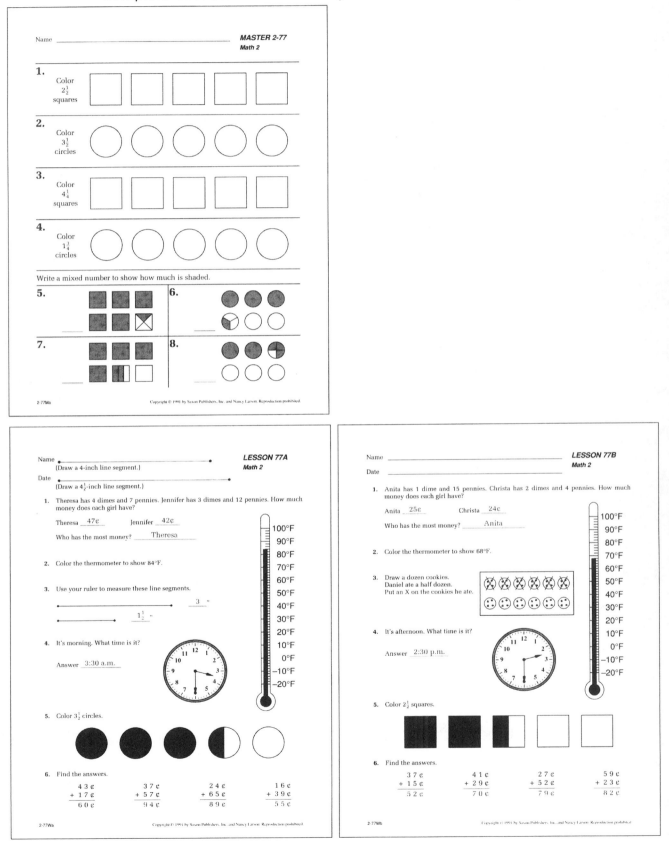

Lesson 78

ordering three-digit numbers

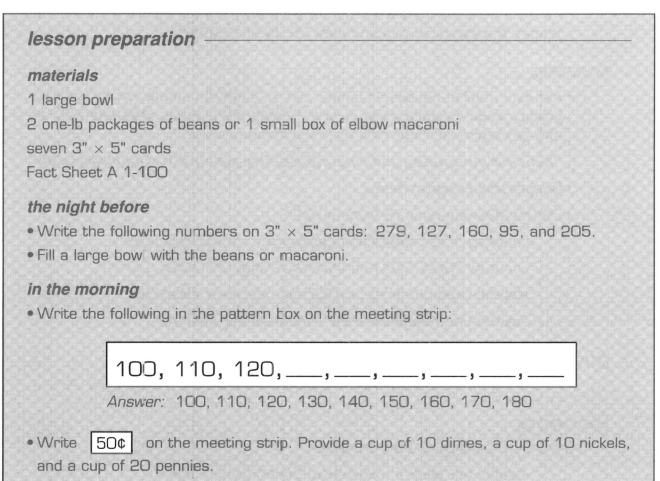

lesson preparation

materials

1 large bowl

2 one-lb packages of beans or 1 small box of elbow macaroni

seven 3" × 5" cards

Fact Sheet A 1-100

the night before

• Write the following numbers on 3" × 5" cards: 279, 127, 160, 95, and 205.

• Fill a large bowl with the beans or macaroni.

in the morning

• Write the following in the pattern box on the meeting strip:

```
100, 110, 120, ____, ____, ____, ____, ____, ____
```

Answer: 100, 110, 120, 130, 140, 150, 160, 170, 180

• Write 50¢ on the meeting strip. Provide a cup of 10 dimes, a cup of 10 nickels, and a cup of 20 pennies.

THE MEETING

calendar

• Ask your child to write the date on the calendar and meeting strip.

• Ask your child the following two or three times a week:

date _____ days ago, date _____ days from now

day of the week _____ days ago, day of the week _____ days from now

_____th month, month before, month after

• Record on the meeting strip a special event and the number of days until it occurs.

weather graph

- Ask your child to read and graph today's temperature to the nearest two degrees.
- Count by 10's and 2's to check the temperature on the graph.
- Ask your child to connect the dot for yesterday's temperature to the dot for today's temperature and compare the temperatures.

counting

- Ask your child to choose a number on the hundred number chart. Ask your child to add or subtract ten or one. Repeat 6–10 times. Ask your child to give directions for returning to the starting number.

    "Let's use our patting and clapping pattern to help us count by 3's to 30."

- Repeat this several times.
- Do the following once or twice a week:

    count by 10's to 400 and backward from 400 by 10's

    count by 5's to 100 and backward from 50 by 5's

    say the even numbers to 100 and backward from 50

    say the odd numbers to 49 and backward from 49

graph questions

- You and your child each ask a question about any of the graphs.

patterning

- Ask your child to do the following:

    identify the pattern (repeating, continuing, or both)

    identify the numbers to complete the pattern

    read the pattern

money

- Ask your child to put the coins in the coin cup. Count the money in the coin cup together.
- Ask your child for another way to show that amount of money. Count these coins together to check the amount.

clock

- Ask your child to set the clock on the half hour or hour.
- Ask the following:

    "It's (morning/afternoon/evening). What time is it?"

    time one hour ago

    time one hour from now

- Ask your child to write the digital time on the meeting strip.
- Record on the meeting strip the time an activity will occur.

number of the day

- Write three number sentences for the number of the day on the meeting strip.

THE LESSON

Ordering Three-Digit Numbers

"Today you will learn how to order three-digit numbers."

"I have a bowl of beans (macaroni)."

"Let's guess how many beans are in this bowl."

- Give your child a 3" × 5" card.

"Write your guess on this card."

"I will write my guess on another card."

"Tomorrow we will check our guesses by counting the beans in the bowl."

- Put the number cards on the table in the following order:

279   127   160   95   205

"Let's pretend that these were guesses for the number of beans in the bowl."

"Put these guesses in order from least to greatest."

- Allow time for your child to order the number cards.

"Put the card with your estimate in the correct place."

"Put the card with my estimate in the correct place."

"Now let's read the number cards from left to right."

"Are the numbers in order from least to greatest?"

"Now you will practice putting other numbers in order from least to greatest."

- Write the following numbers on the chalkboard:

257   194   85   148   239

____ ____ ____ ____ ____

"Write the number that shows the smallest amount on the first line."

"This number is called the least."

"Cross out that number in the top row."

"Now write the number that shows the second smallest amount on the next line."

"Cross out that number in the top row."

- Repeat with the other numbers.

"Let's read the numbers from least to greatest."

- Write the following numbers on the chalkboard:

121   293   206   57   138

____   ____   ____   ____   ____

"Put these numbers in order from least to greatest."

- Allow time for your child to do this.

- Repeat with different numbers, if necessary.

CLASS PRACTICE

number fact practice

- Give your child **Fact Sheet A 1-100**.

"There are 100 addition facts on this sheet."

"You will have five minutes to answer as many problems as possible."

"We will correct this fact sheet in the same way we correct the 25-problem fact sheets."

- Time your child for five minutes.

- Correct the fact sheet with your child and record the score.

- Allow time for your child to complete the unfinished facts.

WRITTEN PRACTICE

- Complete **Worksheet 78A** with your child.

- Complete **Worksheet 78B** with your child later in the day.

LESSON 78A
Math 2

Name _____ (Draw a 3-inch line segment.)

Date _____ (Draw a 3½-inch line segment.)

1. Ryan and David each ate 5 marshmallows. Michael ate 2 marshmallows. How many marshmallows did the 3 boys eat altogether?

   Number sentence ___5 mm's + 5 mm's + 2 mm's = 12 mm's___

   Answer _12 marshmallows_

2. Write these numbers in order from least to greatest.

321	114	259	170

   ___114___ ___170___ ___259___ ___321___
   least   greatest

3. One of these is my favorite number.
   Cross out the numbers that cannot be my favorite number.
   It has 2 digits.
   It is between 80 and 100.
   It is not an odd number.
   What is my favorite number? __98__

   98
   ~~32~~
   ~~127~~
   ~~84~~
   ~~9~~

4. Write the examples vertically. Find the answers.

   74¢ + 9¢
7	4	¢
+	9	¢
8	3	¢

   17¢ + 43¢
1	7	¢
+ 4	8	¢
6	5	¢

   44¢ + 36¢
4	4	¢
+ 3	6	¢
8	0	¢

   82¢ + 16¢
8	2	¢
+ 1	6	¢
9	8	¢

5. Fill in the missing numbers in the number patterns.

   _20_ , _25_ , _30_ , 35, 40, 45, _50_ , _55_ , _60_

   15, 17, 19, _21_ , _23_ , _25_ , _27_ , _29_ , _31_

2-78Wa

Copyright © 1991 by Saxon Publishers, Inc. and Nancy Larson. Reproduction prohibited.

LESSON 78B
Math 2

Name _____

Date _____

1. Joe and Robert each ate 3 pieces of watermelon. Collin ate 4 pieces of watermelon. How many pieces of watermelon did the 3 boys eat altogether?

   Number sentence ___3 pieces + 3 pieces + 4 pieces = 10 pieces___

   Answer ___10 pieces___

2. Write these numbers in order from least to greatest.

148	220	171	205

   ___148___ ___171___ ___205___ ___220___
   least   greatest

3. One of these is my brother's favorite number.
   Cross out the numbers that cannot be his favorite number.
   It does not have 3 digits.
   It is between 10 and 30.
   It is not an odd number.
   What is my brother's favorite number? __24__

   ~~148~~
   24
   ~~15~~
   ~~36~~
   ~~32~~

4. Write the examples vertically. Find the answers.

   51¢ + 16¢
5	1	¢
+ 1	6	¢
6	7	¢

   7¢ + 37¢
	7	¢
+ 3	7	¢
4	4	¢

   55¢ + 29¢
5	5	¢
+ 2	9	¢
8	4	¢

   14¢ + 36¢
1	4	¢
+ 3	6	¢
5	0	¢

5. Fill in the missing numbers in the number patterns.

   70, 65, 60, _55_ , _50_ , _45_ , _40_ , _35_ , _30_

   12, 14, 16, _18_ , _20_ , _22_ , _24_ , _26_ , _28_

2-78Wb

Copyright © 1991 by Saxon Publishers, Inc. and Nancy Larson. Reproduction prohibited.

Lesson 79

representing three-digit numbers pictorially

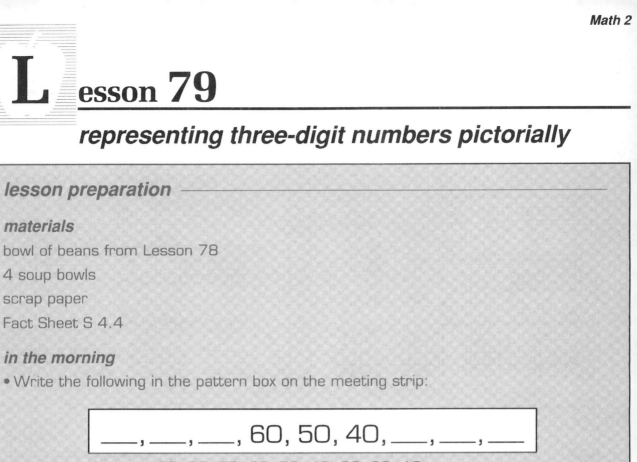

lesson preparation

materials

bowl of beans from Lesson 78

4 soup bowls

scrap paper

Fact Sheet S 4.4

in the morning

• Write the following in the pattern box on the meeting strip:

___, ___, ___, 60, 50, 40, ___, ___, ___

Answer: 90, 80, 70, 60, 50, 40, 30, 20, 10

• Write ⎡83¢⎤ on the meeting strip. Provide a cup of 10 dimes, a cup of 10 nickels, and a cup of 20 pennies.

THE MEETING

calendar

• Ask your child to write the date on the calendar and meeting strip.

• Ask your child the following two or three times a week:

date _____ days ago, date _____ days from now

day of the week _____ days ago, day of the week _____ days from now

_____th month, month before, month after

• Record on the meeting strip a special event and the number of days until it occurs.

weather graph

• Ask your child to read and graph today's temperature to the nearest two degrees.

• Count by 10's and 2's to check the temperature on the graph.

• Ask your child to connect the dot for yesterday's temperature to the dot for today's temperature and compare the temperatures.

counting

- Ask your child to choose a number on the hundred number chart. Ask your child to add or subtract ten or one. Repeat 6–10 times. Ask your child to give directions for returning to the starting number.

  "Let's use our patting and clapping pattern to help us count by 3's to 30."

- Repeat this several times.

- Do the following once or twice a week:

  count by 10's to 400 and backward from 400 by 10's

  count by 5's to 100 and backward from 50 by 5's

  say the even numbers to 100 and backward from 50

  say the odd numbers to 49 and backward from 49

graph questions

- You and your child each ask a question about any of the graphs.

patterning

- Ask your child to do the following:

  identify the pattern (repeating, continuing, or both)

  identify the numbers to complete the pattern

  read the pattern

money

- Ask your child to put the coins in the coin cup. Count the money in the coin cup together.

- Ask your child for another way to show that amount of money. Count these coins together to check the amount.

clock

- Ask your child to set the clock on the half hour or hour.

- Ask the following:

  "It's (morning/afternoon/evening). What time is it?"

  time one hour ago

  time one hour from now

- Ask your child to write the digital time on the meeting strip.

- Record on the meeting strip the time an activity will occur.

number of the day

- Write three number sentences for the number of the day on the meeting strip.

THE LESSON

Representing Three-Digit Numbers Pictorially

"Today you will learn how to show a three-digit number using a picture."

- Show your child the bowl of beans.

  "Yesterday we estimated the number of beans in this bowl."

  "Today we will check our guesses by counting the beans in the bowl."

  "How do you think we could do that?"

  "Let's work together to make groups of ten beans."

  "Then we can count by 10's to find the number of beans in the bowl."

  "Whenever we have 100 beans, we will put them in a soup bowl."

- As your child counts by 10's, put 10 groups of 10 in a bowl. Leave the extra tens and ones on the demonstration table.

  "We have _____ hundreds, _____ tens, and _____ ones."

- Point to the groups of beans as you name each group.

  "We write that like this."

- Write the amount on the chalkboard.

  "How close were our estimates from yesterday?"

- Write the following above the digits:

  hundreds   tens   ones

  "How will we use the bowls, groups of ten, and ones to show one hundred twenty-five?"

- Ask your child to show this number of beans.

  "How will we write this amount?"

- Write "125" below the "hundreds, tens, ones" heading on the chalkboard.
- Repeat with 160, 53, and 214.
- Write "246" on the chalkboard.

  "How will we show this amount using the hundreds, tens, and ones?"

- Ask your child to show the amount.

  "Let's count to check."

- Repeat with 105, 230, and 74.
- Give your child a piece of scrap paper.

  "Now we will draw pictures to show the amount for a number."

  "If we want to show hundreds, we will draw a large square like this."

"Each square will stand for 100 beans."

• Draw the following on the chalkboard:

"If we want to show tens, we will draw a tower like this."

• Draw the following on the chalkboard:

"If we want to show ones, we will draw a small square like this."

• Draw the following on the chalkboard:

• Write "416" on the chalkboard.

"Write four hundred sixteen on your paper."

"Draw a picture for this number."

• Allow time for your child to draw the picture.

"How did you draw the picture?"

• Draw the picture on the chalkboard as your child describes the large squares, tower, and small squares used.

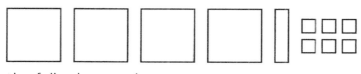

• Repeat with the following numbers:

    135   302   260   57

Class Practice

number fact practice

• Use the tan, peach, lavender, and green fact cards to practice the subtracting two, subtracting one, subtracting zero, subtracting half of a double, and subtracting a number from itself facts with your child.

• Give your child **Fact Sheet S 4.4.**

• Time your child for one minute.

• Correct the fact sheet with your child and record the score.

• Allow time for your child to complete the unfinished facts.

WRITTEN PRACTICE

- Complete **Worksheet 79A** with your child.
- Complete **Worksheet 79B** with your child later in the day.

Name ●━━━━━━━━━●
(Draw a 2-inch line segment.)

Date
(Draw a $2\frac{1}{2}$-inch line segment.)

LESSON 79A
Math 2

1. There are 17 children in Room 12. Nine children went to the nurse's office to have their eyes checked. How many children are in Room 12 now?

   Number sentence _____17 children − 9 children = 8 children_____

   Answer _8 children_

2. Write these numbers in order from least to greatest.

   | 291 | 67 | 134 | 178 |

   _67_   _134_   _178_   _291_
   least   greatest

3. Color the thermometer to show 44°F.

4. Draw a picture to show 251. (Use ☐ for 100, ☐ for 10, ☐ for 1.)

5. Color $2\frac{3}{4}$ circles.

6. Write the examples vertically. Find the answers.

   43¢ + 28¢

4	3	¢
+ 2	8	¢
7	1	¢

   31¢ + 9¢

3	1	¢
+	9	¢
4	0	¢

   52¢ + 13¢

5	2	¢
+ 1	3	¢
6	5	¢

2-79Wa

Name _____

Date _____

LESSON 79B
Math 2

1. There are 15 children in Room 17. Nine children left to go to the library. How many children are in Room 17 now?

   Number sentence _____15 children − 9 children = 6 children_____

   Answer _6 children_

2. Write these numbers in order from least to greatest.

   | 152 | 212 | 73 | 243 |

   _73_   _152_   _212_   _243_
   least   greatest

3. Color the thermometer to show 56°F.

4. Draw a picture to show 143. (Use ☐ for 100, ☐ for 10, ☐ for 1.)

5. Color $1\frac{1}{4}$ squares.

6. Write the examples vertically. Find the answers.

   62¢ + 27¢

6	2	¢
+ 2	7	¢
8	9	¢

   17¢ + 45¢

1	7	¢
+ 4	5	¢
6	2	¢

   8¢ + 86¢

	8	¢
+ 8	6	¢
9	4	¢

2-79Wb

Lesson 80

identifying and creating overlapping geometric designs

---

lesson preparation

materials

Written Assessment #15

Oral Assessment #8

basket of color tiles of mixed colors

Master 2-80

geoboard

geobands

Fact Sheet S 7.0

in the morning

• Write the following in the pattern box on the meeting strip:

> 6, 16, 26, 36, ____, ____, ____, ____, ____, ____

Answer: 6, 16, 26, 36, 46, 56, 66, 76, 86, 96

• Write $\boxed{69¢}$ on the meeting strip. Provide a cup of 10 dimes, a cup of 10 nickels, and a cup of 20 pennies.

THE MEETING

calendar

• Ask your child to write the date on the calendar and meeting strip.

• Ask your child the following two or three times a week:

    date _____ days ago, date _____ days from now

    day of the week _____ days ago   day of the week _____ days from now

    _____th month, month before, month after

• Record on the meeting strip a special event and the number of days until it occurs.

---

weather graph

- Ask your child to read and graph today's temperature to the nearest two degrees.
- Count by 10's and 2's to check the temperature on the graph.
- Ask your child to connect the dot for yesterday's temperature to the dot for today's temperature and compare the temperatures.

counting

- Ask your child to choose a number on the hundred number chart. Ask your child to add or subtract ten or one. Repeat 6–10 times. Ask your child to give directions for returning to the starting number.

  "Let's use our patting and clapping pattern to help us count by 3's to 30."

- Repeat this several times.
- Do the following once or twice a week:

  count by 10's to 400 and backward from 400 by 10's

  count by 5's to 100 and backward from 50 by 5's

  say the even numbers to 100 and backward from 50

  say the odd numbers to 49 and backward from 49

graph questions

- You and your child each ask a question about any of the graphs.

patterning

- Ask your child to do the following:

  identify the pattern (repeating, continuing, or both)

  identify the numbers to complete the pattern

  read the pattern

money

- Ask your child to put the coins in the coin cup. Count the money in the coin cup together.
- Ask your child for another way to show that amount of money. Count these coins together to check the amount.

clock

- Ask your child to set the clock on the half hour or hour.
- Ask the following:

  "It's (morning/afternoon/evening). What time is it?"

  time one hour ago

time one hour from now

- Ask your child to write the digital time on the meeting strip.
- Record on the meeting strip the time an activity will occur.

number of the day

- Write three number sentences for the number of the day on the meeting strip.

ASSESSMENT

Written Assessment

"Today I would like to see what you remember from what we have been practicing."

- Give your child **Written Assessment #15**.
- Read the directions for each problem. Allow time for your child to complete that problem before continuing.
- Correct the paper, noting your child's mistakes on the **Individual Recording Form**. Review the errors with your child.

Oral Assessment

- Record your child's response(s) to the oral interview questions on the interview sheet.

THE LESSON

Identifying and Creating Overlapping Geometric Designs

"Today you will learn how to identify and create overlapping geometric shapes."

- Give your child **Master 2-80**, a geoboard, and an assortment of geobands.

"Copy the first design on your geoboard."

- Allow time for your child to do this.

"Does your design look like the one on the paper?"

"How do you know?"

"Is your geoboard design the same size as the design on the paper?"

"What do we call designs that are the same shape but not the same size?"   similar designs or shapes

"How many triangles do you see in this design?"   3 (2 medium, 1 large)

"Carefully take the geobands off the geoboard."

"Copy the second design on your geoboard."

- Allow time for your child to do this.

"Does your design look like the one on the paper?"

"How do you know?"

"Is your geoboard design the same size as the design on the paper?"

"What do we call designs that are the same shape but not the same size?"   similar designs or shapes

"How many squares do you see in this design?"   3 (1 small, 1 medium, 1 large)

"Carefully take the geobands off the geoboard."

"Copy the third design on your geoboard."

- Allow time for your child to do this.

"Does your design look like the one on the paper?"

"How do you know?"

"How many triangles do you see in this design?"   6 (2 small, 3 medium, 1 large)

"Carefully take the geobands off the geoboard."

"Make an overlapping design on your geoboard using either triangles or squares."

"When you finish, copy the design in the fourth square."

CLASS PRACTICE

number fact practice

- Use the yellow fact cards to practice the subtracting nine facts with your child.
- Give your child **Fact Sheet S 7.0.**
- Time your child for one minute.
- Correct the fact sheet with your child and record the score.
- Allow time for your child to complete the unfinished facts.

Left Page

Teacher _____ **MATH 2 LESSON 80**
Date _____ Oral Assessment #8 Recording Form

Materials:
basket of color tiles of mixed colors

"Take a double handful of color tiles."
A "Make a graph to show the colors of the tiles you picked."
 • Allow time for the child to make the graph.
B "Tell me about your graph."
C "How many more _____ tiles are there than _____ tiles?"

Students	A. Graphs with 1–1 Correspondence	B. Describes the Graph	C. Compares Columns

2-80La   Copyright © 1991 by Saxon Publishers, Inc. and Nancy Larson. Reproduction prohibited.

Right Page

Name _____ **ASSESSMENT 15**
Date _____ **LESSON 80**
 Math 2

1. Ryan has 5 dimes and 7 pennies. How much money does he have? __57¢__

   Daniel has 3 dimes and 2 pennies. How much money does he have? __32¢__

   How much money do the two boys have altogether?

   Number sentence __57¢ + 32¢ = 89¢__   Answer __89¢__

2. Write these numbers in order from least to greatest.

   | 48 | 27 | 25 | 43 | 39 |

   __25__   __27__   __39__   __43__   __48__
   least   greatest

3. Write the fractions that show what part is shaded.

   $\frac{3}{4}$ $\frac{1}{2}$

4. Draw a dozen donuts. Color a half dozen brown to show that they are chocolate.

   B B B B B B

   How many donuts are chocolate? __6__

   How many donuts are not chocolate? __6__

5. Add.

   $6 + 4 + 7 + 2 + 3 + 1 =$ __23__

   $\begin{array}{r} 24¢ \\ + 53¢ \\ \hline 77¢ \end{array}$ $\begin{array}{r} 16¢ \\ + 72¢ \\ \hline 88¢ \end{array}$

2-80Aa   Copyright © 1991 by Saxon Publishers, Inc. and Nancy Larson. Reproduction prohibited.

Bottom Page

Name _____ **MASTER 2-80**
 Math 2

How many squares and triangles can you find in each design?

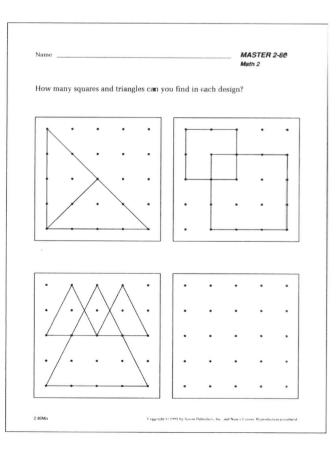

2-80Ma   Copyright © 1991 by Saxon Publishers, Inc. and Nancy Larson. Reproduction prohibited.

Lesson 81

writing a three-digit number for a model

---

lesson preparation —————————————————————

materials

scrap paper

Fact Sheet S 4.4

in the morning

• Write the following in the pattern box on the meeting strip:

> 37, 39, 41, ——, ——, ——, ——, ——, ——

Answer: 37, 39, 41, 43, 45, 47, 49, 51, 53

• Write 85¢ on the meeting strip. Provide a cup of 10 dimes, a cup of 10 nickels, and a cup of 20 pennies.

THE MEETING

calendar

- Ask your child to write the date on the calendar and meeting strip.

- Ask your child the following two or three times a week:

    date ____ days ago, date ____ days from now

    day of the week ____ days ago, day of the week ____ days from now

    ____th month, month before, month after

- Record on the meeting strip a special event and the number of days until it occurs.

weather graph

- Ask your child to read and graph today's temperature to the nearest two degrees.

- Count by 10's and 2's to check the temperature on the graph.

- Ask your child to connect the dot for yesterday's temperature to the dot for today's temperature and compare the temperatures.

counting

- Ask your child to choose a number on the hundred number chart. Ask your child to add or subtract ten or one. Repeat 6–10 times. Ask your child to give directions for returning to the starting number.

   "Let's use our patting and clapping pattern to help us count by 3's to 30."

- Repeat this several times.

- Do the following once or twice a week:

   count by 10's to 400 and backward from 400 by 10's

   count by 5's to 100 and backward from 50 by 5's

   say the even numbers to 100 and backward from 50

   say the odd numbers to 49 and backward from 49

graph questions

- You and your child each ask a question about any of the graphs.

patterning

- Ask your child to do the following:

   identify the pattern (repeating, continuing, or both)

   identify the numbers to complete the pattern

   read the pattern

money

- Ask your child to put the coins in the coin cup. Count the money in the coin cup together.

- Ask your child for another way to show that amount of money. Count these coins together to check the amount.

clock

- Ask your child to set the clock on the half hour or hour.

- Ask the following:

   "It's (morning/afternoon/evening). What time is it?"

   time one hour ago

   time one hour from now

- Ask your child to write the digital time on the meeting strip.

- Record on the meeting strip the time an activity will occur.

number of the day

- Write three number sentences for the number of the day on the meeting strip.

THE LESSON

Writing a Three-Digit Number for a Model

    "A few days ago you learned how to draw pictures for three-digit numbers."

    "Today you will learn how to write a three-digit number for a picture."

- Give your child a piece of scrap paper.
- Write "247" on the chalkboard.

    "Draw a picture for this number."

- Allow time for your child to do this.

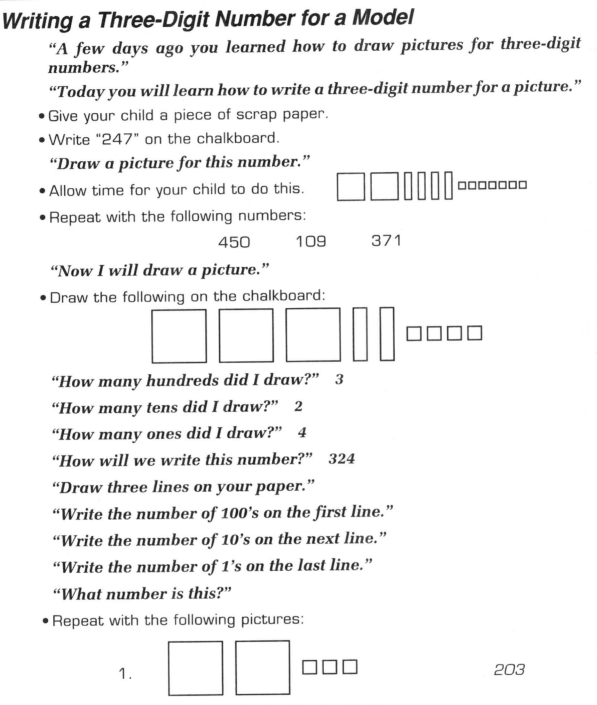

- Repeat with the following numbers:

           450   109   371

    "Now I will draw a picture."

- Draw the following on the chalkboard:

    "How many hundreds did I draw?"   *3*

    "How many tens did I draw?"   *2*

    "How many ones did I draw?"   *4*

    "How will we write this number?"   *324*

    "Draw three lines on your paper."

    "Write the number of 100's on the first line."

    "Write the number of 10's on the next line."

    "Write the number of 1's on the last line."

    "What number is this?"

- Repeat with the following pictures:

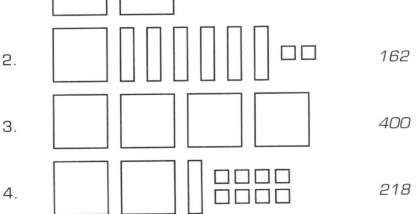

1.   *203*

2.   *162*

3.   *400*

4.   *218*

5. ☐ ☐

101

CLASS PRACTICE

number fact practice

- Use the tan, peach, lavender, and green fact cards to practice the subtracting two, subtracting one, subtracting zero, subtracting half of a double, and subtracting a number from itself facts with your child.
- Give your child **Fact Sheet S 4.4**.
- Time your child for one minute.
- Correct the fact sheet with your child and record the score.
- Allow time for your child to complete the unfinished facts.

WRITTEN PRACTICE

- Complete **Worksheet 81A** with your child.
- Complete **Worksheet 81B** with your child later in the day.

Name _____
(Draw a 3-inch line segment.)
Date _____
(Draw a 3½-inch line segment.)

LESSON 81A
Math 2

1. Draw 5 apples. Divide each apple in half.

How many pieces of apple do you have? ___10___

2. What number does this picture show? ___312___

| 100 | 100 | 100 | 10 | 1 | 1 |

3. Write these numbers in order from least to greatest.

| 249 434 193 216 180 | ___180___ ___193___ ___216___ ___249___ ___434___ |
| | least   greatest |

4. I have 3 dimes, 2 nickels, and 4 pennies. Draw the coins.

(D)(D)(D)(N)(N)(P)(P)(P)(P)

How much money is this? ___44¢___

5. Find the answers.

46¢ + 27¢

	4	6	¢
+	2	7	¢
	7	3	¢

35¢ + 55¢

	3	5	¢
+	5	5	¢
	9	0	¢

21¢ + 48¢

	2	1	¢
+	4	8	¢
	6	9	¢

2-81Wa

Copyright © 1994 by Saxon Publishers, Inc. and Nancy Larson. Reproduction prohibited.

Name _____
Date _____

LESSON 81B
Math 2

1. Draw 4 bananas. Divide each banana in half.

How many pieces of banana do you have? ___8___

2. What number does this picture show? ___214___

| 100 | 100 | 10 | 1 | 1 | 1 | 1 |

3. Write these numbers in order from least to greatest.

| 273 192 154 337 230 | ___154___ ___192___ ___230___ ___273___ ___337___ |
| | least   greatest |

4. I have 5 dimes, 3 nickels, and 1 penny. Draw the coins.

(D)(D)(D)(D)(D)(N)(N)(N)(P)

How much money is this? ___66¢___

5. Find the answers.

53¢ + 36¢

	5	3	¢
+	3	6	¢
	8	9	¢

29¢ + 17¢

	2	9	¢
+	1	7	¢
	4	6	¢

48¢ + 22¢

	4	8	¢
+	2	2	¢
	7	0	¢

2-81Wb

Copyright © 1994 by Saxon Publishers, Inc. and Nancy Larson. Reproduction prohibited.

Lesson 82

identifying and writing addition and subtraction fact families

lesson preparation

materials

Master 2-82

Fact Sheet A 1-100

in the morning

• Write the following in the pattern box on the meeting strip:

▨, △, ▣, ◭, ▢, ◭, ____, ____, ____, ____

Answer: ▣, ◭, ▣, ◭, ▣, ◭, ▣, ◭, ▣, ◭

• Write 41¢ on the meeting strip. Provide a cup of 10 dimes, a cup of 10 nickels, and a cup of 20 pennies.

THE MEETING

calendar

• Ask your child to write the date on the calendar and meeting strip.

• Ask your child the following two or three times a week:

date _____ days ago, date _____ days from now

day of the week _____ days ago, day of the week _____ days from now

_____th month, month before, month after

• Record on the meeting strip a special event and the number of days until it occurs.

weather graph

• Ask your child to read and graph today's temperature to the nearest two degrees.

• Count by 10's and 2's to check the temperature on the graph.

- Ask your child to connect the dot for yesterday's temperature to the dot for today's temperature and compare the temperatures.

counting

- Ask your child to choose a number on the hundred number chart. Ask your child to add or subtract ten or one. Repeat 6–10 times. Ask your child to give directions for returning to the starting number.

  "Let's use our patting and clapping pattern to help us count by 3's to 30."

- Repeat this several times.

- Do the following once or twice a week:

  count by 10's to 400 and backward from 400 by 10's

  count by 5's to 100 and backward from 50 by 5's

  say the even numbers to 100 and backward from 50

  say the odd numbers to 49 and backward from 49

graph questions

- You and your child each ask a question about any of the graphs.

patterning

- Ask your child to do the following:

  identify the pattern (repeating, continuing, or both)

  identify the shapes to complete the pattern

  read the pattern

money

- Ask your child to put the coins in the coin cup. Count the money in the coin cup together.

- Ask your child for another way to show that amount of money. Count these coins together to check the amount.

clock

- Ask your child to set the clock on the half hour or hour.

- Ask the following:

  "It's (morning/afternoon/evening). What time is it?"

  time one hour ago

  time one hour from now

- Ask your child to write the digital time on the meeting strip.

- Record on the meeting strip the time an activity will occur.

number of the day

- Write three number sentences for the number of the day on the meeting strip.

THE LESSON

Identifying and Writing Addition and Subtraction Fact Families

"Today you will learn how to write addition and subtraction fact families."

- Write the numbers 5, 1, and 6 on the chalkboard.

5   1   6

"What do you notice about these three numbers?"

- Allow your child to offer as many observations as possible.

"What number sentences can we write using only these three numbers?"

- Write the number sentences on the chalkboard as your child names them.

5 + 1 = 6   1 + 5 = 6   6 – 1 = 5   6 – 5 = 1

"These number sentences make a fact family."

"Let's write these fact family number sentences as vertical problems."

"How will we do that?"

- Write each of the following on the chalkboard:

$$
\begin{array}{cccc}
5 & 1 & 6 & 6 \\
+1 & +5 & -1 & -5 \\
\hline
6 & 6 & 5 & 1
\end{array}
$$

- Write the numbers 2, 7, and 9 on the chalkboard.

"What fact family number sentences can we write for these three numbers?"

- Write the number sentences on the chalkboard as your child names them.

2 + 7 = 9   7 + 2 = 9   9 – 2 = 7   9 – 7 = 2

"What do you notice about the numbers in these fact family number sentences?" the greatest number is always on the right or the left; the other two numbers switch places

"Let's write these fact family number sentences as vertical problems."

"How will we do that?"

- Write each of the following on the chalkboard:

$$
\begin{array}{cccc}
2 & 7 & 9 & 9 \\
+7 & +2 & -2 & -7 \\
\hline
9 & 9 & 7 & 2
\end{array}
$$

"What do you notice about the numbers in these fact family problems?" the greatest number is always on the top or the bottom; the other two numbers switch places

- Write the numbers 3 and 9 on the chalkboard.

"We always need three numbers to make a fact family."

"What is a third number that we can use for this fact family?" 6 or 12

"Let's use twelve as the third number."

- Write "12" on the chalkboard next to the other two numbers.

"How will we write a fact family for these three numbers?"

- Write the number sentences on the chalkboard as your child names them.

$$3 + 9 = 12 \qquad 9 + 3 = 12 \qquad 12 - 3 = 9 \qquad 12 - 9 = 3$$

"What do you notice about the numbers in these fact family number sentences?" the greatest number is always on the right or the left; the other two numbers switch places

"Let's write these fact family number sentences as vertical problems."

"How will we do that?"

- Write each of the following on the chalkboard:

$$
\begin{array}{cccc}
3 & 9 & 12 & 12 \\
+9 & +3 & -9 & -3 \\
\hline
12 & 12 & 3 & 9
\end{array}
$$

"What do you notice about the numbers in these fact family problems?" the greatest number is always on the top or the bottom; the other two numbers switch places

"Now let's use six as the third number."

- Write the fact families in the same way as above using 3, 6, and 9.
- Give your child **Master 2-82**.

"What numbers will we use for the fact family in problem one?" 4, 5, 9

"What is one number sentence you can write using the numbers four, nine, and five?"

"Write this number sentence on the first line."

"What is another number sentence that you can write using the numbers four, nine, and five?"

"Write this number sentence on the next line."

- Repeat with the next two number sentences.

"Now you will write these fact family number sentences as vertical problems."

- Allow time for your child to do this.
- Repeat with the second problem.

"There is a missing number in each of the last two problems."

"What numbers could you use for the third problem?"   7 or 11

"Choose one of these numbers and write the fact family."

- Repeat with the last problem.

"Tomorrow you will learn how we can use fact families to help us find the answers for subtraction problems."

CLASS PRACTICE

number fact practice

- Give your child **Fact Sheet A 1-100.**

  "There are 100 addition facts on this sheet."

  "You will have five minutes to answer as many problems as possible."

  "We will correct this fact sheet in the same way we correct the 25-problem fact sheets."

- Time your child for five minutes.

- Correct the fact sheet with your child and record the score.

- Allow time for your child to complete the unfinished facts.

WRITTEN PRACTICE

- Complete **Worksheet 82A** with your child.

- Complete **Worksheet 82B** with your child later in the day.

Name _____ MASTER 2-82
                              Math 2

| 4 | 9 | 5 | | 8 | 6 | 2 |

_____
_____
_____
_____
_____

| 9 | 2 | ☐ | | 8 | 9 | ☐ |

_____
_____
_____
_____
_____

2-82Ma   Copyright © 1991 by Saxon Publishers, Inc. and Nancy Larson. Reproduction prohibited.

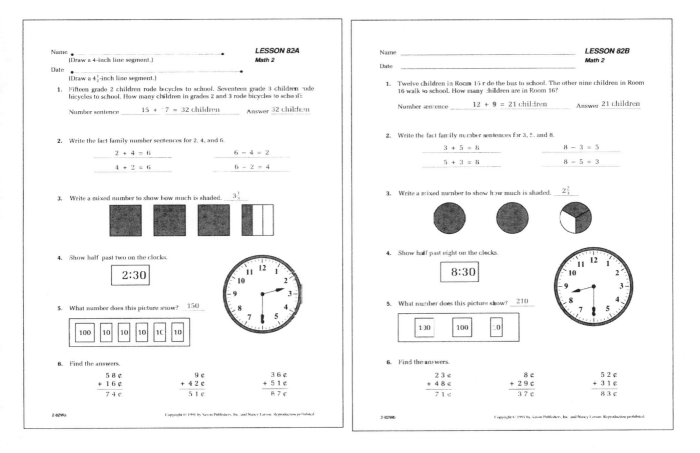

Name _____ LESSON 82A
     (Draw a 4-inch line segment.)   Math 2
Date _____
     (Draw a 4½-inch line segment.)

1. Fifteen grade 2 children rode bicycles to school. Seventeen grade 3 children rode bicycles to school. How many children in grades 2 and 3 rode bicycles to school?

   Number sentence ___15 + 17 = 32 children___   Answer _32 children_

2. Write the fact family number sentences for 2, 4, and 6.

   2 + 4 = 6   6 − 4 = 2
   4 + 2 = 6   6 − 2 = 4

3. Write a mixed number to show how much is shaded. 3¼

4. Show half past two on the clocks.

   2:30

5. What number does this picture show? ___150

   | 100 | 10 | 10 | 10 | 10 | 10 |

6. Find the answers.

   58¢   9¢   36¢
  + 16¢   + 42¢   + 51¢
   74¢   51¢   87¢

2-82Wa   Copyright © 1991 by Saxon Publishers, Inc. and Nancy Larson. Reproduction prohibited.

Name _____ LESSON 82B
                              Math 2
Date _____

1. Twelve children in Room 16 ride the bus to school. The other nine children in Room 16 walk to school. How many children are in Room 16?

   Number sentence ___12 + 9 = 21 children___   Answer _21 children_

2. Write the fact family number sentences for 3, 5, and 8.

   3 + 5 = 8   8 − 3 = 5
   5 + 3 = 8   8 − 5 = 3

3. Write a mixed number to show how much is shaded. ___2⅔

4. Show half past eight on the clocks.

   8:30

5. What number does this picture show? ___210

   | 100 | 100 | 10 |

6. Find the answers.

   23¢   8¢   52¢
  + 48¢   + 29¢   + 31¢
   71¢   37¢   83¢

2-82Wb   Copyright © 1991 by Saxon Publishers, Inc. and Nancy Larson. Reproduction prohibited.

Lesson 83

subtraction facts—differences of 1, 2, and 9

lesson preparation

materials

scrap paper

Fact Sheet S 7.4

in the morning

• Write the following in the pattern box on the meeting strip:

80, 75, 70, 65, ___, ___, ___, ___, ___, ___

Answer: 80, 75, 70, 65, 60, 55, 50, 45, 40, 35

• Write 73¢ on the meeting strip. Provide a cup of 10 dimes, a cup of 10 nickels, and a cup of 20 pennies.

THE MEETING

calendar

• Ask your child to write the date on the calendar and meeting strip.

• Ask your child the following two or three times a week:

  date ____ days ago, date ____ days from now

  day of the week ____ days ago, day of the week ____ days from now

  ____th month, month before, month after

• Record on the meeting strip a special event and the number of days until it occurs.

weather graph

• Ask your child to read and graph today's temperature to the nearest two degrees.

• Count by 10's and 2's to check the temperature on the graph.

• Ask your child to connect the dot for yesterday's temperature to the dot for today's temperature and compare the temperatures.

counting

- Ask your child to choose a number on the hundred number chart. Ask your child to add or subtract ten or one. Repeat 6–10 times. Ask your child to give directions for returning to the starting number.

  "Let's use our patting and clapping pattern to help us count by 3's to 30."

- Repeat this several times.

- Do the following once or twice a week:

  count by 10's to 400 and backward from 400 by 10's

  count by 5's to 100 and backward from 50 by 5's

  say the even numbers to 100 and backward from 50

  say the odd numbers to 49 and backward from 49

graph questions

- You and your child each ask a question about any of the graphs.

patterning

- Ask your child to do the following:

  identify the pattern (repeating continuing, or both)

  identify the numbers to complete the pattern

  read the pattern

money

- Ask your child to put the coins in the coin cup. Count the money in the coin cup together.

- Ask your child for another way to show that amount of money. Count these coins together to check the amount.

clock

- Ask your child to set the clock on the half hour or hour.

- Ask the following:

  "It's (morning/afternoon/evening). What time is it?"

  time one hour ago

  time one hour from now

- Ask your child to write the digital time on the meeting strip.

- Record on the meeting strip the time an activity will occur.

number of the day

- Write three number sentences for the number of the day on the meeting strip.

fact practice

"Today we will add something new to The Meeting."

"Each day I will choose three fact family numbers."

"Then you will write the fact family number sentences for these three numbers."

"Let's try that."

- Write "4, 5, 9" on the chalkboard.

"Let's write four number sentences using only these numbers."

"What number sentences could we write?"

- Ask your child to write the number sentences on the chalkboard.

- Continue until the four number sentences are listed:

$$4 + 5 = 9; \quad 5 + 4 = 9; \quad 9 - 5 = 4; \quad 9 - 4 = 5$$

"These number sentences make a fact family."

THE LESSON

Subtraction Facts—Differences of 1, 2, and 9

"Yesterday you learned how to write fact families for three numbers."

"Today you will use fact families to help you learn the answers to some new subtraction problems."

"These subtraction problems all will have answers of one, two, or nine."

- Write the numbers 1, 5, and 6 on the chalkboard.

- Give your child a piece of scrap paper.

"Write the fact family number sentences for one, five, and six."

- Allow time for your child to write the number sentences.

"What number sentences did you write?"

- Write the number sentences on the chalkboard.

- Circle 6 − 5 = 1.

"What is another way we can write this problem?"

- Record the following on the chalkboard:

$$\begin{array}{r} 6 \\ -\ 5 \\ \hline 1 \end{array}$$

"The answer in a subtraction problem is called the difference."

"The difference in this subtraction problem is one."

"Let's write some more subtraction problems that have a difference of one."

- Ask your child to suggest problems.
- Record the problems on the chalkboard. Check the answers using pennies, if necessary.

"What do you notice about the top and the middle numbers in these problems?" first number is one more than the second number

- Encourage your child to offer observations about the numbers.
- Write the numbers 7, 2, and 5 on the chalkboard.

"Write the fact family number sentences for seven, two, and five."

- Allow time for your child to write the number sentences.

"What number sentences did you write?"

- Write the number sentences on the chalkboard.
- Circle 7 – 5 = 2.

"What is another way we can write this problem?"

- Record the following on the chalkboard:

$$\begin{array}{r} 7 \\ -\ 5 \\ \hline 2 \end{array}$$

"The answer in a subtraction problem is called the difference."

"The difference in this subtraction problem is two."

"Let's write some more subtraction problems that have a difference of two."

- Ask your child to suggest problems.
- Record the problems on the chalkboard. Check the answers using pennies, if necessary.

"What do you notice about the top and the middle numbers in these problems?" first number is two more than the second number

- Encourage your child to offer observations about the numbers.
- Write the numbers 13, 4, and 9 on the chalkboard.

"Write the fact family number sentences for thirteen, four, and nine."

- Allow time for your child to write the number sentences.

"What number sentences did you write?"

- Write the number sentences on the chalkboard.
- Circle 13 – 4 = 9.

"What is another way we can write this problem?"

- Record the following on the chalkboard:

$$\begin{array}{r} 13 \\ -\ 4 \\ \hline 9 \end{array}$$

"What do we call the answer in a subtraction problem?" the difference

"Let's write some more subtraction problems that have a difference of nine."

- Ask your child to suggest problems.

- Record the problems on the chalkboard. Check the answers using pennies, if necessary.

"What do you notice about the top and the middle numbers in these problems?"

- Encourage your child to offer observations about the numbers.

CLASS PRACTICE

number fact practice

- Use the yellow, peach, and green fact cards to practice the subtraction facts with your child.

- Give your child **Fact Sheet S 7.4.**

- Time your child for one minute.

- Correct the fact sheet with your child and record the score.

- Allow time for your child to complete the unfinished facts.

WRITTEN PRACTICE

- Complete **Worksheet 83A** with your child.

- Complete **Worksheet 83B** with your child later in the day.

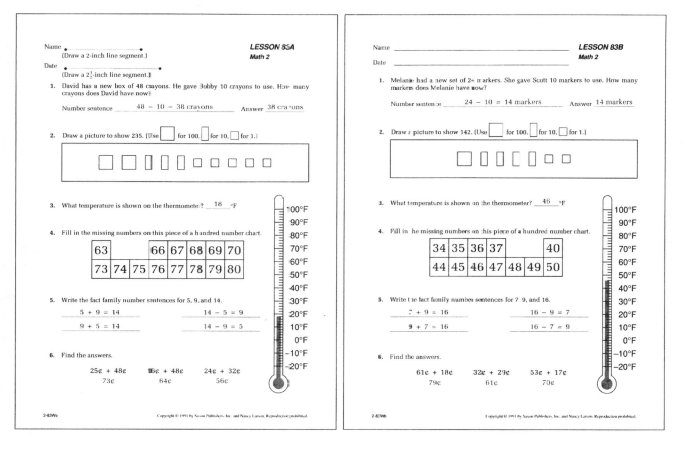

Name _____
(Draw a 2-inch line segment.)

Date _____
(Draw a 2½-inch line segment.)

LESSON 83A
Math 2

1. David has a new box of 48 crayons. He gave Bobby 10 crayons to use. How many crayons does David have now?

    Number sentence ___48 – 10 = 38 crayons___   Answer ___38 crayons___

2. Draw a picture to show 235. (Use ☐ for 100, ☐ for 10, ☐ for 1.)

3. What temperature is shown on the thermometer? ___18___ °F

4. Fill in the missing numbers on this piece of a hundred number chart.

63		66	67	68	69	70
73	74	75	76	77	78	79	80

5. Write the fact family number sentences for 5, 9, and 14.

    $5 + 9 = 14$ $14 - 5 = 9$
    $9 + 5 = 14$ $14 - 9 = 5$

6. Find the answers.

    $25¢ + 48¢$ $16¢ + 48¢$ $24¢ + 32¢$
        73¢   64¢   56¢

2-83Wa

Name _____

Date _____

LESSON 83B
Math 2

1. Melanie had a new set of 24 markers. She gave Scott 10 markers to use. How many markers does Melanie have now?

    Number sentence ___24 – 10 = 14 markers___   Answer ___14 markers___

2. Draw a picture to show 142. (Use ☐ for 100, ☐ for 10, ☐ for 1.)

3. What temperature is shown on the thermometer? ___46___ °F

4. Fill in the missing numbers on this piece of a hundred number chart.

34	35	36	37			40
44	45	46	47	48	49	50

5. Write the fact family number sentences for 7, 9, and 16.

    $7 + 9 = 16$ $16 - 9 = 7$
    $9 + 7 = 16$ $16 - 7 = 9$

6. Find the answers.

    $61¢ + 18¢$ $32¢ + 29¢$ $53¢ + 17¢$
        79¢   61¢   70¢

2-83Wb

Lesson 84

telling and showing time to five-minute intervals

lesson preparation

materials

demonstration clock

Master 2-84

Fact Sheet S 7.4

in the morning

• Write the following in the pattern box on the meeting strip:

> 24, 26, 28, ___, ___, ___, ___, ___, ___
>
> *Answer:* 24, 26, 28, 30, 32, 34, 36, 38, 40

• Write   98¢   on the meeting strip. Provide a cup of 10 dimes, a cup of 10 nickels, and a cup of 20 pennies.

THE MEETING

calendar

- Ask your child to write the date on the calendar and meeting strip.

- Ask your child the following two or three times a week:

    date _____ days ago, date _____ days from now

    day of the week _____ days ago, day of the week _____ days from now

    _____th month, month before, month after

- Record on the meeting strip a special event and the number of days until it occurs.

weather graph

- Ask your child to read and graph today's temperature to the nearest two degrees.

- Count by 10's and 2's to check the temperature on the graph.

- Ask your child to connect the dot for yesterday's temperature to the dot for today's temperature and compare the temperatures.

counting

- Ask your child to choose a number on the hundred number chart. Ask your child to add or subtract ten or one. Repeat 6–10 times. Ask your child to give directions for returning to the starting number.

    "Let's use our patting and clapping pattern to help us count by 3's to 30."

- Repeat this several times.

- Do the following once or twice a week:

    count by 10's to 400 and backward from 400 by 10's

    count by 5's to 100 and backward from 50 by 5's

    say the even numbers to 100 and backward from 50

    say the odd numbers to 49 and backward from 49

graph questions

- You and your child each ask a question about any of the graphs.

patterning

- Ask your child to do the following

    identify the pattern (repeating, continuing, or both)

    identify the numbers to complete the pattern

    read the pattern

money

- Ask your child to put the coins in the coin cup. Count the money in the coin cup together.

- Ask your child for another way to show that amount of money. Count these coins together to check the amount.

clock

- Ask your child to set the clock on the half hour or hour.

- Ask the following:

    "It's (morning/afternoon/evening). What time is it?"

    time one hour ago

    time one hour from now

- Ask your child to write the digital time on the meeting strip.

- Record on the meeting strip the time an activity will occur.

number of the day

- Write three number sentences for the number of the day on the meeting strip.

fact practice

- Write three fact family numbers (e.g., 2, 7, 9) on the chalkboard.
- Allow time for your child to write the four fact family number sentences on the chalkboard.

THE LESSON

Telling and Showing Time to Five-Minute Intervals

"Today you will learn how to read and show time to five-minute intervals."

- Draw a large circle on the chalkboard.

"We will pretend that this is a clock."

"What do I need to do to make this circle look like the clock on the wall (demonstration clock)?"   number the clock face; draw the hands

"First I will find the center of the clock."

- Put a dot in the center of the chalkboard circle.

"This is where the hands of the clock start."

"Now I will put the numbers on the clock."

"I will begin by putting a line or mark at the top and bottom of the circle."

"Then I will put a line or mark in the middle of the right side and then in the middle of the left side."

"What numbers will I write next to these marks?"   12, 6, 3, 9

"Now I draw marks for the other numbers."

"What numbers will I write between the 12 and the 3? . . . the 3 and the 6? . . . the 6 and the 9? . . . the 9 and the 12?"

- Stress the importance of careful spacing.

"What do these numbers tell us?"   the hour

"Which hand on the clock shows the hour?"   hour hand or short hand

"The long hand on the clock is called the minute hand."

"This tells us the number of minutes past the hour."

"Each small line or dot on the clock shows one minute."

"How many lines (dots) do you see between the twelve and the one on the demonstration clock?"   4

"When we get to the one, we are at the fifth line."

"This shows five minutes because it is the fifth mark."

"When we get to the two, we are at the tenth line."

"This shows ten minutes because it is the tenth mark."

- Continue around the clock face.

- Number the clock in the following way:

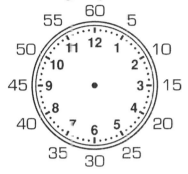

- Set the demonstration clock at 9:00.

  "What time is it?"

  "Which hand tells us the hour?"

  "Now I will move the minute hand."

- Move the minute hand to the one.

  "How many minutes past nine is it?"   5

  "Let's count to check."

  "We say that it is five minutes past nine o'clock or 9:05."

- Write "9:05" on the chalkboard.

- Move the minute hand to two.

  "How many minutes past nine is it now?"   10

  "Let's count to check."

  "We say that it is ten minutes past nine o'clock or 9:10."

- Write "9:10" on the chalkboard.

- Move the minute hand to three. Repeat above.

- Repeat around the clock face.

- Circle each of the numbers around the outside of the clock.

  "What do you notice about these numbers?"   same as counting by 5's

  "We can use counting by 5's to help us tell the time."

- Set the clock at 2:40.

  "Let's count by 5's to find the number of minutes past two."

- Point to the numbers on the clock face as you count by 5's with your child.

  "We write the time like this."

- Record "2:40" on the chalkboard.

"We can read the time as two forty."

"When we read the minutes first, we say forty minutes past two."

- Repeat with 7:25, 4:55, 12:20, and 4:05.
- Give your child the demonstration clock.

  "Show 3:10 on the clock."

- Write "3:10" on the chalkboard.

  "Where does the hour hand point?"

  "Where does the minute hand point?"

  "What is another way to read this time?"   *ten minutes past three*

- Repeat with 6:15, 8:50, 11:05, and 12:30.
- Give your child **Master 2-84.**

  "Look at the top of the paper."

  "What do you think you will do?"   *write the times shown on the clocks*

  "Look at the bottom of the paper."

  "What do you think you will do?"   *draw the hands on the clocks*

CLASS PRACTICE

number fact practice

- Use the yellow, peach, and green fact sheets to practice the subtraction facts with your child.
- Give your child **Fact Sheet S 7.4.**
- Time your child for one minute.
- Correct the fact sheet with your child and record the score.
- Allow time for your child to complete the unfinished facts.

WRITTEN PRACTICE

- Complete **Worksheet 84A** with your child.
- Complete **Worksheet 84B** with your child later in the day.

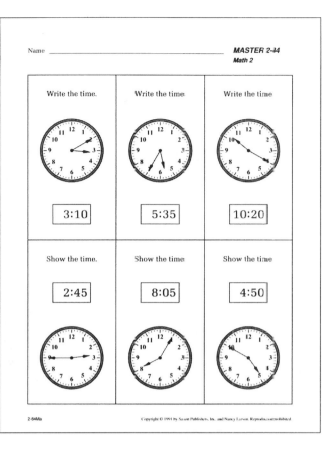

Name _____ **MASTER 2-84**
Math 2

Write the time.	Write the time.	Write the time
(clock)	(clock)	(clock)
3:10	5:35	10:20

Show the time.	Show the time.	Show the time
2:45	8:05	4:50
(clock)	(clock)	(clock)

2-84Ma   Copyright © 1991 by Saxon Publishers, Inc. and Nancy Larson. Reproduction prohibited.

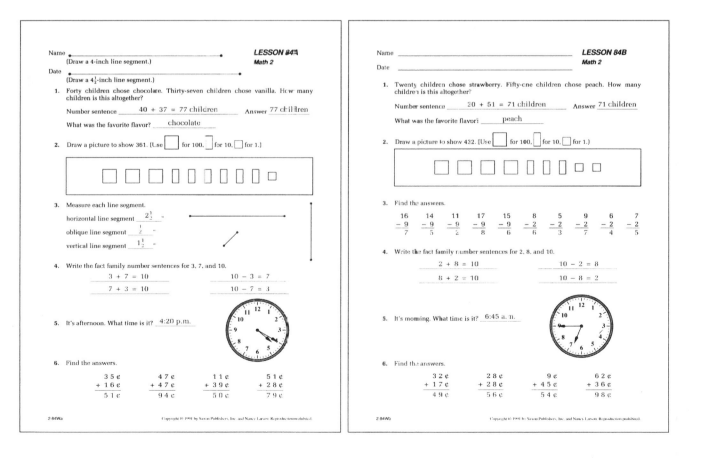

Name •————————————• **LESSON 84A**
(Draw a 4-inch line segment.)   **Math 2**
Date •———————————————•
(Draw a 4½-inch line segment.)

1. Forty children chose chocolate. Thirty-seven children chose vanilla. How many children is this altogether?

   Number sentence ___40 + 37 = 77 children___   Answer _77 children_

   What was the favorite flavor? ___chocolate___

2. Draw a picture to show 361. (Use ☐ for 100, ☐ for 10, ☐ for 1.)

3. Measure each line segment.

   horizontal line segment ___2½___
   oblique line segment ___½___
   vertical line segment ___1½___

4. Write the fact family number sentences for 3, 7, and 10.

   3 + 7 = 10   10 − 3 = 7
   7 + 3 = 10   10 − 7 = 3

5. It's afternoon. What time is it? ___4:20 p.m.___

6. Find the answers.

3 5 ¢	4 7 ¢	1 1 ¢	5 1 ¢
+ 1 6 ¢	+ 4 7 ¢	+ 3 9 ¢	+ 2 8 ¢
5 1 ¢	9 4 ¢	5 0 ¢	7 9 ¢

2-84Wa   Copyright © 1991 by Saxon Publishers, Inc. and Nancy Larson. Reproduction prohibited.

Name _____ **LESSON 84B**
Date _____   **Math 2**

1. Twenty children chose strawberry. Fifty-one children chose peach. How many children is this altogether?

   Number sentence ___20 + 51 = 71 children___   Answer _71 children_

   What was the favorite flavor? ___peach___

2. Draw a picture to show 432. (Use ☐ for 100, ☐ for 10, ☐ for 1.)

3. Find the answers.

16	14	11	17	15	8	5	9	6	7
− 9	− 9	− 9	− 9	− 9	− 2	− 2	− 2	− 2	− 2
7	5	2	8	6	6	3	7	4	5

4. Write the fact family number sentences for 2, 8, and 10.

   2 + 8 = 10   10 − 2 = 8
   8 + 2 = 10   10 − 8 = 2

5. It's morning. What time is it? ___6:45 a.m.___

6. Find the answers.

3 2 ¢	2 8 ¢	9 ¢	6 2 ¢
+ 1 7 ¢	+ 2 8 ¢	+ 4 5 ¢	+ 3 6 ¢
4 9 ¢	5 6 ¢	5 4 ¢	9 8 ¢

2-84Wb   Copyright © 1991 by Saxon Publishers, Inc. and Nancy Larson. Reproduction prohibited.

Lesson 85

adding three two-digit numbers with a sum less than 100

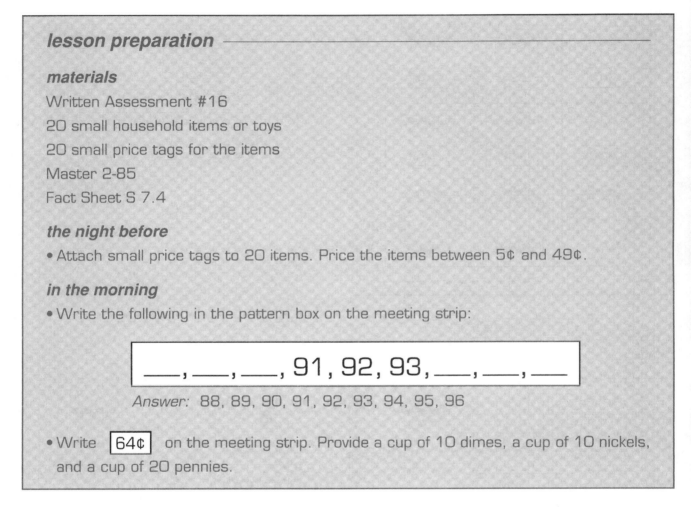

lesson preparation

materials

Written Assessment #16

20 small household items or toys

20 small price tags for the items

Master 2-85

Fact Sheet S 7.4

the night before

• Attach small price tags to 20 items. Price the items between 5¢ and 49¢.

in the morning

• Write the following in the pattern box on the meeting strip:

___, ___, ___, 91, 92, 93, ___, ___, ___

Answer: 88, 89, 90, 91, 92, 93, 94, 95, 96

• Write 64¢ on the meeting strip. Provide a cup of 10 dimes, a cup of 10 nickels, and a cup of 20 pennies.

THE MEETING

calendar

• Ask your child to write the date on the calendar and meeting strip.

• Ask your child the following two or three times a week:

  date _____ days ago, date _____ days from now

  day of the week _____ days ago, day of the week _____ days from now

  _____th month, month before, month after

• Record on the meeting strip a special event and the number of days until it occurs.

weather graph

- Ask your child to read and graph today's temperature to the nearest two degrees.

- Count by 10's and 2's to check the temperature on the graph.

- Ask your child to connect the dot for yesterday's temperature to the dot for today's temperature and compare the temperatures.

counting

- Ask your child to choose a number on the hundred number chart. Ask your child to add or subtract ten or one. Repeat 6–10 times. Ask your child to give directions for returning to the starting number.

    "Let's use our patting and clapping pattern to help us count by 3's to 30."

- Repeat this several times.

- Do the following once or twice a week:

    count by 10's to 400 and backward from 400 by 10's

    count by 5's to 100 and backward from 50 by 5's

    say the even numbers to 100 and backward from 50

    say the odd numbers to 49 and backward from 49

graph questions

- You and your child each ask a question about any of the graphs.

patterning

- Ask your child to do the following:

    identify the pattern (repeating, continuing, or both)

    identify the numbers to complete the pattern

    read the pattern

money

- Ask your child to put the coins in the coin cup. Count the money in the coin cup together.

- Ask your child for another way to show that amount of money. Count these coins together to check the amount.

clock

- Set the clock to a five-minute interval.

- Ask the following:

    "It's (morning/afternoon/evening). What time is it?"

    time one hour ago

    time one hour from now

- Ask your child to write the digital time on the meeting strip.
- Record on the meeting strip the time an activity will occur.

number of the day

- Write three number sentences for the number of the day on the meeting strip.

fact practice

- Write three fact family numbers (e.g., 2, 7, 9) on the chalkboard.
- Allow time for your child to write the four fact family number sentences on the chalkboard.

ASSESSMENT

Written Assessment

"Today I would like to see what you remember from what we have been practicing."

- Pass out **Written Assessment #16**.
- Read the directions for each problem. Allow time for your child to complete each problem before continuing.
- Correct the paper, noting your child's mistakes on the **Individual Recording Form**. Review the errors with your child.

THE LESSON

Adding Three Two-Digit Numbers with a Sum Less Than 100

"We have been adding two-digit numbers."

"Today you will learn how to add three two-digit numbers."

"Let's pretend that you are the owner of a store."

"We will pretend that I am the customer."

"I am going to come to your store to buy things."

- Show your child the 20 household items or toys with price tags.
- Give your child **Master 2-85**.

"These are small sales slips."

"The first items I would like to buy are _____ and _____."

- Choose two items.

"What is the cost of each of these items?"

- Write the amounts on the chalkboard horizontally.

  "How will we find the total cost of these two things?"   add

  "Use the first sales slip to find the amount of money I owe."

- Allow time for your child to do this.

- Check the answer with coins, if desired or necessary.

- Repeat with two other combinations of items. Make sure at least two out of the three problems require regrouping.

- Choose three items with a total value of less than one dollar.

  "This time I would like to buy _____ , _____ , and _____ ."

  "How will you write that on the sales slip?"

  "Add to find the amount of money I owe."

- Correct the problem together.

- Repeat two more times using other combinations.

CLASS PRACTICE

number fact practice

- Use the yellow, peach, and green fact cards to practice the subtraction facts with your child.

- Give your child **Fact Sheet S 7.4.**

- Time your child for one minute.

- Correct the fact sheet with your child and record the score.

- Allow time for your child to complete the unfinished facts.

WRITTEN PRACTICE

- Complete **Worksheet 85A** with your child.

- Complete **Worksheet 85B** with your child later in the day.

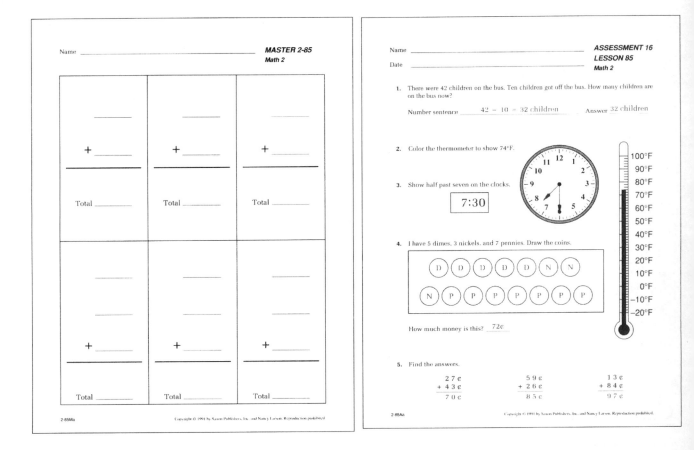

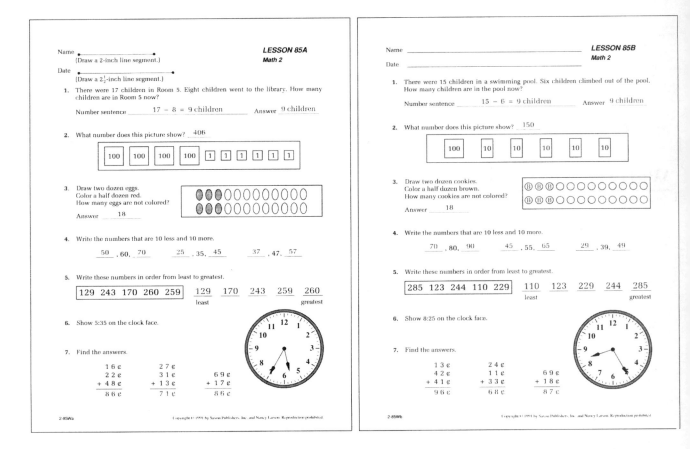

L esson 86

estimating and counting large collections grouping by 10's and 100's

---

lesson preparation

materials

5–7 containers of objects (see lesson)

10 small and 4 large paper cups

piece of paper

Master 2-86

Fact Sheet S 7.4

the night before

- Use a large collection of 100–400 similar objects for each container. Possibilities include cereal, screws, paper clips, buttons, puzzle pieces, baseball cards, elastic bands, safety pins, etc.

- Label the containers with letters. Display containers filled with the objects to be counted.

in the morning

- Write the following in the pattern box on the meeting strip:

> 396, 397, 398, ____, ____, ____, ____, ____, ____

Answer: 396, 397, 398, 399, 400, 401, 402, 403, 404

- Write 55¢ on the meeting strip. Provide a cup of 10 dimes, a cup of 10 nickels, and a cup of 20 pennies.

THE MEETING

calendar

- Ask your child to write the date on the calendar and meeting strip.

- Ask your child the following two or three times a week:

date _____ days ago, date _____ days from now

day of the week _____ days ago, day of the week _____ days from now

_____th month, month before, month after

- Record on the meeting strip a special event and the number of days until it occurs.

weather graph

- Ask your child to read and graph today's temperature to the nearest two degrees.
- Count by 10's and 2's to check the temperature on the graph.
- Ask your child to connect the dot for yesterday's temperature to the dot for today's temperature and compare the temperatures.

counting

- Ask your child to choose a number on the hundred number chart. Ask your child to add or subtract ten or one. Repeat 6–10 times. Ask your child to give directions for returning to the starting number.

  "Let's use our patting and clapping pattern to help us count by 3's to 30."

- Repeat this several times.
- Do the following once or twice a week:

  count by 10's to 400 and backward from 400 by 10's

  count by 5's to 100 and backward from 50 by 5's

  say the even numbers to 100 and backward from 50

  say the odd numbers to 49 and backward from 49

graph questions

- You and your child each ask a question about any of the graphs.

patterning

- Ask your child to do the following:

  identify the pattern (repeating, continuing, or both)

  identify the numbers to complete the pattern

  read the pattern

money

- Ask your child to put the coins in the coin cup. Count the money in the coin cup together.
- Ask your child for another way to show that amount of money. Count these coins together to check the amount.

clock

- Set the clock to a five-minute interval.
- Ask the following:

"It's (morning/afternoon/evening). What time is it?"

time one hour ago

time one hour from now

- Ask your child to write the digital time on the meeting strip.
- Record on the meeting strip the time an activity will occur.

number of the day

- Write three number sentences for the number of the day on the meeting strip.

fact practice

- Write three fact family numbers (e.g., 2, 7, 9) on the chalkboard.
- Allow time for your child to write the four fact family number sentences on the chalkboard.

THE LESSON

Estimating and Counting Large Collections Grouping by 10's and 100's

"Today you will learn how to estimate and count a large collection of objects."

- Display the containers for your child to see

"I filled these containers with different objects."

"What objects did I use?"

"Which container do you think has the most objects?"

"Which container do you think has the fewest objects?"

"Let's estimate how many objects are in each container."

- Give your child **Master 2-86**.

"What does the word 'estimate' mean?"   to make a sensible guess

"Look closely at each container and make your estimate."

"Write your estimate next to the letter of the container."

"I'll write my estimates on another piece of paper."

- Write your estimates on another piece of paper.

"Now we will work together to count the objects in container A."

- Hold up a small and a large cup.

"We will use the small cups for groups of ten."

"We will use the large cups for ten groups of ten."

"We will put the objects from ten small cups into a large cup just like we put ten groups of ten beans in a bowl."

"How many objects will be in the large cup?"   *100*

"When we finish counting the objects in container A, you will draw a picture of the cups we used at the bottom of your paper."

"You will draw the cups like this."

- Draw the following on the chalkboard.

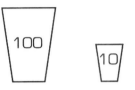

- Count the objects in container A with your child.

"How many _____ were in container A?"

"Let's count the cups to check."

"Write this number in the last column."

"Were our estimates good estimates?"

"Why?"

- Return the objects to container A and repeat with the other containers. Do not draw the pictures for containers B–G.

CLASS PRACTICE

number fact practice

- Use the yellow, peach, and green fact cards to practice the subtraction facts with your child.
- Give your child **Fact Sheet S 7.4**.
- Time your child for one minute.
- Correct the fact sheet with your child and record the score.
- Allow time for your child to complete the unfinished facts.

WRITTEN PRACTICE

- Complete **Worksheet 86A** with your child.
- Complete **Worksheet 86B** with your child later in the day.

Name _____ **MASTER 2-86**
                                    Mata 2

Container	Estimate	Actual
A		
B		
C		
D		
E		
F		
G		

Picture for container _____ .

2-86Ma   Copyright © 1991 by Saxon Publishers, Inc. and Nancy Larson. Reproduction prohibited.

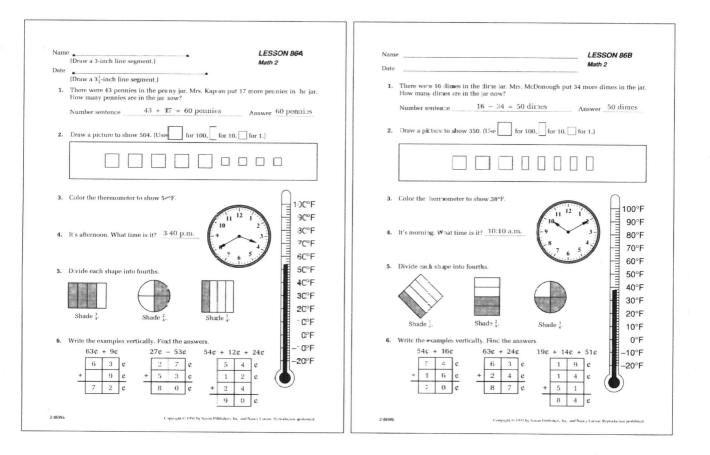

Name _____ **LESSON 86A**
(Draw a 3-inch line segment.)   **Math 2**
Date _____
(Draw a 3½-inch line segment.)

1. There were 43 pennies in the penny jar. Mrs. Kaplan put 17 more pennies in the jar. How many pennies are in the jar now?

   Number sentence _____ 43 + 17 = 60 pennies _____   Answer 60 pennies

2. Draw a picture to show 504. (Use ☐ for 100, ☐ for 10, ☐ for 1.)

3. Color the thermometer to show 54°F.

4. It's afternoon. What time is it? __3:40 p.m.__

5. Divide each shape into fourths.

   Shade 3/4.   Shade 2/4.   Shade 1/4.

6. Write the examples vertically. Find the answers.

   63¢ + 9¢   27¢ − 53¢   54¢ + 12¢ + 24¢

	6	3	¢
+		9	¢
	7	2	¢

	2	7	¢
+	5	3	¢
	8	0	¢

	5	4	¢
	1	2	¢
+	2	4	¢
	9	0	¢

2-86Wa   Copyright © 1991 by Saxon Publishers, Inc. and Nancy Larson. Reproduction prohibited.

Name _____ **LESSON 86B**
                                    Math 2
Date _____

1. There were 16 dimes in the dime jar. Mrs. McDonough put 34 more dimes in the jar. How many dimes are in the jar now?

   Number sentence _____ 16 − 34 = 50 dimes _____   Answer 50 dimes

2. Draw a picture to show 350. (Use ☐ for 100, ☐ for 10, ☐ for 1.)

3. Color the thermometer to show 38°F.

4. It's morning. What time is it? __10:10 a.m.__

5. Divide each shape into fourths.

   Shade 1/4.   Shade 2/4.   Shade 3/4.

6. Write the examples vertically. Find the answers.

   54¢ + 16¢   63¢ + 24¢   19¢ + 14¢ + 51¢

	5	4	¢
+	1	6	¢
	7	0	¢

	6	3	¢
+	2	4	¢
	8	7	¢

	1	9	¢
	1	4	¢
+	5	1	¢
	8	4	¢

2-86Wb   Copyright © 1991 by Saxon Publishers, Inc. and Nancy Larson. Reproduction prohibited.

esson 87

subtraction facts—subtracting a number from ten

---

lesson preparation

materials

demonstration clock
Fact Sheet S 6.0

in the morning

• Write the following in the pattern box on the meeting strip:

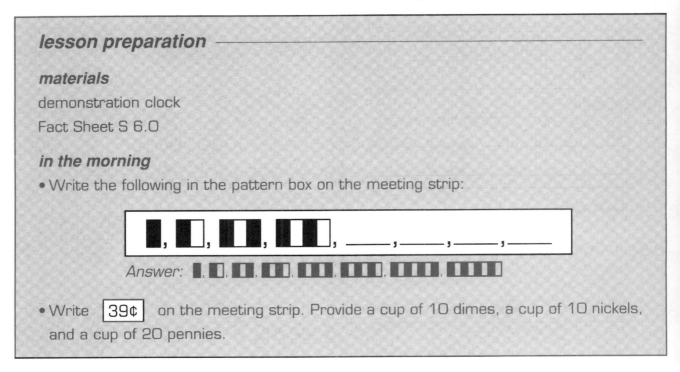

Answer:

• Write ⟨39¢⟩ on the meeting strip. Provide a cup of 10 dimes, a cup of 10 nickels, and a cup of 20 pennies.

THE MEETING

calendar

• Ask your child to write the date on the calendar and meeting strip.

• Ask your child the following two or three times a week:

date _____ days ago, date _____ days from now

day of the week _____ days ago, day of the week _____ days from now

_____th month, month before, month after

• Record on the meeting strip a special event and the number of days until it occurs.

weather graph

• Ask your child to read and graph today's temperature to the nearest two degrees.

• Count by 10's and 2's to check the temperature on the graph.

• Ask your child to connect the dot for yesterday's temperature to the dot for today's temperature and compare the temperatures.

counting

- Ask your child to choose a number on the hundred number chart. Ask your child to add or subtract ten or one. Repeat 6–10 times. Ask your child to give directions for returning to the starting number.

  "Let's use our patting and clapping pattern to help us count by 3's to 30."

- Repeat this several times.

- Do the following once or twice a week:

  count by 10's to 400 and backward from 400 by 10's

  count by 5's to 100 and backward from 50 by 5's

  say the even numbers to 100 and backward from 50

  say the odd numbers to 49 and backward from 49

graph questions

- You and your child each ask a question about any of the graphs.

patterning

- Ask your child to do the following:

  identify the pattern (repeating, continuing, or both)

  identify the shapes to complete the pattern

  read the pattern

money

- Ask your child to put the coins in the coin cup. Count the money in the coin cup together.

- Ask your child for another way to show that amount of money. Count these coins together to check the amount.

clock

- Set the clock to a five-minute interval.

- Ask the following:

  "It's (morning/afternoon/evening). What time is it?"

  time one hour ago

  time one hour from now

- Ask your child to write the digital time on the meeting strip.

- Record on the meeting strip the time an activity will occur.

number of the day

- Write three number sentences for the number of the day on the meeting strip.

fact practice

- Write three fact family numbers (e.g., 2, 7, 9) on the chalkboard.
- Allow time for your child to write the four fact family number sentences on the chalkboard.

THE LESSON

Subtraction Facts—Subtracting a Number from Ten

"Today you will learn the subtracting a number from ten facts."

"What are two numbers that have a sum of ten?"

- Write the combinations on the chalkboard as your child names them.

$$0 + 10 = 10 \qquad 10 + 0 = 10$$
$$1 + 9 = 10 \qquad 9 + 1 = 10$$
$$2 + 8 = 10 \qquad 8 + 2 = 10$$
$$3 + 7 = 10 \qquad 7 + 3 = 10$$
$$4 + 6 = 10 \qquad 6 + 4 = 10$$
$$5 + 5 = 10$$

"We can use these sums of ten to help us find the answer when we subtract a number from ten."

- Write "10 − 7 = _____" on the chalkboard.

"If we want to know what ten minus seven is, we can think of what number plus seven equals ten."

"What is the answer?"

- Record "10 − 7 = 3."

- Repeat with several facts.

"We can always use addition to help us find a subtraction answer."

- Write the following on the chalkboard:

$$\begin{array}{ccccccccc} 10 & 10 & 10 & 10 & 10 & 10 & 10 & 10 & 10 \\ \underline{-9} & \underline{-3} & \underline{-8} & \underline{-5} & \underline{-1} & \underline{-6} & \underline{-4} & \underline{-2} & \underline{-7} \end{array}$$

"To find the answer for ten minus nine, we can think 'some number plus nine is ten.' "

- Point to each part of the problem as you say the numbers. Work from the bottom up.

"What is the answer?"

- Record the answer on the chalkboard.

- Repeat with the other problems.

"Let's practice the subtracting a number from ten facts."

- Erase the chalkboard answers.

"I will point to a number fact."

"Say the answer as quickly as possible."

- Point to the number facts one at a time.

CLASS PRACTICE

number fact practice

- Give your child **Fact Sheet S 6.0.**
- Time your child for one minute.
- Correct the fact sheet with your child and record the score.
- Allow time for your child to complete the unfinished facts.

telling time to five-minute intervals

- Give your child a demonstration clock.
- Write "4:20" on the chalkboard.

"Show 4:20 on your clock."

"Where does the hour hand point?"

"Where does the minute hand point?"

- Repeat with 7:55, 2:10, 6:25, 11:40, and 3:05.

WRITTEN PRACTICE

- Complete **Worksheet 87A** with your child.
- Complete **Worksheet 87B** with your child later in the day.

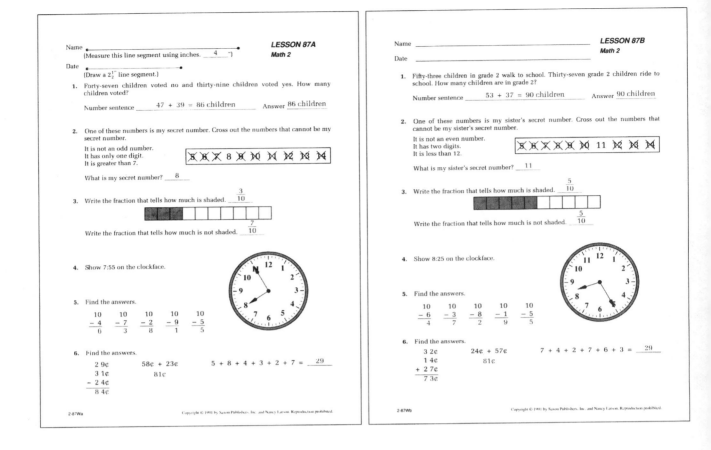

Name _____ **LESSON 87A**
(Measure this line segment using inches. ___4___ ") **Math 2**

Date _____
(Draw a 2½" line segment.)

1. Forty-seven children voted no and thirty-nine children voted yes. How many children voted?

   Number sentence ___47 + 39 = 86 children___ Answer __86 children__

2. One of these numbers is my secret number. Cross out the numbers that cannot be my secret number.

   It is not an odd number.
   It has only one digit.
   It is greater than 7.

   ~~5~~ ~~6~~ ~~7~~ 8 ~~9~~ ~~10~~ ~~11~~ ~~12~~ ~~13~~ ~~14~~

   What is my secret number? __8__

3. Write the fraction that tells how much is shaded. ___3/10___

   Write the fraction that tells how much is not shaded. ___7/10___

4. Show 7:55 on the clockface.

5. Find the answers.

10	10	10	10	10
− 4	− 7	− 2	− 9	− 5
6	3	8	1	5

6. Find the answers.

   29¢
   31¢
   − 24¢
   84¢

   58¢ + 23¢
   81¢

   5 + 8 + 4 + 3 + 2 + 7 = ___29___

2-87Wa

Name _____ **LESSON 87B**

Date _____ **Math 2**

1. Fifty-three children in grade 2 walk to school. Thirty-seven grade 2 children ride to school. How many children are in grade 2?

   Number sentence ___53 + 37 = 90 children___ Answer __90 children__

2. One of these numbers is my sister's secret number. Cross out the numbers that cannot be my sister's secret number.

   It is not an even number.
   It has two digits.
   It is less than 12.

   ~~5~~ ~~6~~ ~~7~~ ~~8~~ ~~9~~ ~~10~~ 11 ~~12~~ ~~13~~ ~~14~~

   What is my sister's secret number? __11__

3. Write the fraction that tells how much is shaded. ___5/10___

   Write the fraction that tells how much is not shaded. ___5/10___

4. Show 8:25 on the clockface.

5. Find the answers.

10	10	10	10	10
− 6	− 3	− 8	− 1	− 5
4	7	2	9	5

6. Find the answers.

   32¢
   14¢
   + 27¢
   73¢

   24¢ + 57¢
   81¢

   7 + 4 + 2 + 7 + 6 + 3 = ___29___

2-87Wb

Lesson 88

creating a bar graph with a scale of two

lesson preparation

materials

2 cups of 20 pennies

Master 2-88

crayons

Fact Sheet S 6.0

in the morning

• Write the following in the pattern box on the meeting strip:

——:——, ——, ——, ——, ——, 22, 24, 26, 28

Answer: 10, 12, 14, 16, 18, 20, 22, 24, 26, 28

• Write | 43¢ | on the meeting strip. Provide a cup of 10 dimes, a cup of 10 nickels, and a cup of 20 pennies.

THE MEETING

calendar

• Ask your child to write the date on the calendar and meeting strip.

• Ask your child the following two or three times a week:

   date ____ days ago, date ____ days from now

   day of the week ____ days ago, day of the week ____ days from now

   ____th month, month before month after

• Record on the meeting strip a special event and the number of days until it occurs.

weather graph

• Ask your child to read and graph today's temperature to the nearest two degrees.

• Count by 10's and 2's to check the temperature on the graph.

• Ask your child to connect the dot for yesterday's temperature to the dot for today's temperature and compare the temperatures.

counting

- Ask your child to choose a number on the hundred number chart. Ask your child to add or subtract ten or one. Repeat 6–10 times. Ask your child to give directions for returning to the starting number.

  "Let's use our patting and clapping pattern to help us count by 3's to 30."

- Repeat this several times.

- Do the following once or twice a week:

  count by 10's to 400 and backward from 400 by 10's

  count by 5's to 100 and backward from 50 by 5's

  say the even numbers to 100 and backward from 50

  say the odd numbers to 49 and backward from 49

graph questions

- You and your child each ask a question about any of the graphs.

patterning

- Ask your child to do the following:

  identify the pattern (repeating, continuing, or both)

  identify the numbers to complete the pattern

  read the pattern

money

- Ask your child to put the coins in the coin cup. Count the money in the coin cup together.

- Ask your child for another way to show that amount of money. Count these coins together to check the amount.

clock

- Set the clock to a five-minute interval.

- Ask the following:

  "It's (morning/afternoon/evening). What time is it?"

  time one hour ago

  time one hour from now

- Ask your child to write the digital time on the meeting strip.

- Record on the meeting strip the time an activity will occur.

number of the day

- Write three number sentences for the number of the day on the meeting strip.

fact practice

- Write three fact family numbers (e.g., 2, 7, 9) on the chalkboard.
- Allow time for your child to write the four fact family number sentences on the chalkboard.

THE LESSON

Creating a Bar Graph with a Scale of Two

"Today you will learn how to draw a bar graph with a scale of two."

"We will use pennies to make a graph."

"When we looked at pennies before, we noticed that there was a date on each coin."

"Do you know what that date means?"   the year the penny was minted

"The mint date of a penny is just like a birth date of the penny."

"It tells us the year the penny was made."

- Hold up a cup of 20 pennies.

"There are 20 pennies in this cup."

"Do you think that more of these pennies were made before or after you were born?"

"Let's check to see."

"Let's tally and make a graph to show the mint dates of the pennies."

- Give your child the cup of pennies.
- Draw the following on the chalkboard:

Before 1965 1965-1969 1970-1974 1975-1979 1980-1984 1985-1989 1990-now

"Take a penny out of the cup."

"What is the mint date of the penny?"

"Where will I draw a tally mark to show that?"

- Ask your child to identify where to draw the tally mark.
- Repeat for all the pennies.

"Let's show this information on a graph."

- Draw **Master 2-88** on the chalkboard.

"How many pennies were minted before 1965?"

"How will we show that on our graph?"

- Shade the graph to show this information.
- Repeat with each column on the graph.

"How many pennies were minted between 1985 and 1989?"

"How many pennies were minted between 1975 and 1979?"

"Between what years were most of the pennies minted?"

"Between what years were the fewest of the pennies minted?"

"How many more pennies were minted between _____ and _____ than between _____ and _____?"

- Repeat with several columns.

"Let's circle the column that has the pennies that are just about the same age as you are."

- Circle the dates at the bottom of the appropriate column.

"Where are the older pennies on the graph?"   on the left

"Where are the younger pennies on the graph?"   on the right

"Are more pennies older or younger than you?"

"Now you will make your own graph to show the mint dates of 20 different pennies."

"Do you think your graph will look the same as the one we made together?"

- Ask your child to explain why he/she thinks the graphs will or will not look the same.
- Give your child **Master 2-88** and a cup of 20 pennies.

"Tally and draw a graph to show the mint dates of the pennies in your cup."

"Do this just like we did when we worked together."

"Use your crayons to color your graph."

"When you finish, answer the questions about your graph."

CLASS PRACTICE

number fact practice

- Use the blue fact cards to practice the subtracting a number from ten facts with your child.
- Give your child **Fact Sheet S 6.0.**
- Time your child for one minute.
- Correct the fact sheet with your child and record the score.
- Allow time for your child to complete the unfinished facts.

WRITTEN PRACTICE

- • Complete **Worksheet 88A** with your child.

- • Complete **Worksheet 88B** with your child later in the day.

Name _____ MASTER 2-88
Math 2

Before 1965 1965-1969 1970-1974 1975-1979 1980-1984 1985-1989 1990-now

20
18
16
14
12
10
8
6
4
2
0

Before 1965 1965-1969 1970-1974 1975-1979 1980-1984 1985-1989 1990-now

How many pennies were minted in the years 1980 through 1984? _____

How many pennies were minted in the years 1985 through 1989? _____

When were most of your pennies minted? _____

2-88Ma

Name _____ **LESSON 88A**
(Measure this line segment using inches. ___3___ ") Math 2

Date _____
(Draw a $1\frac{1}{2}$" line segment.)

1. There are 83 children in grade 2 at Haley School. Ten second graders were absent on Monday. How many grade 2 children were in school?

   Number sentence ___83 – 10 = 73 children___   Answer _73 children_

2. Shelly has a half dozen dimes and a dozen pennies.

   How many dimes is this? __6__   How much money is that? _60¢_

   How many pennies is this? __12__   How much money is that? _12¢_

   How much money does Shelly have? _72¢_

3. This is a tally to show how many children chose each color.

yellow	|||| ||||
purple	|||| ||
pink	||||

   Colors Children Chose

   Shade the graph to show the colors the children chose.

4. It's morning. What time is it? _7:45 a.m._

   It's evening. What time is it? _8:05 p.m._

5. Find the answers.

   $$\begin{array}{r} 26¢ \\ + 46¢ \\ \hline 72¢ \end{array}$$ $$\begin{array}{r} 37¢ \\ + 42¢ \\ \hline 79¢ \end{array}$$ $$\begin{array}{r} 16¢ \\ 43¢ \\ + 28¢ \\ \hline 87¢ \end{array}$$

2-88Wa

Name _____ **LESSON 88B**
Math 2

Date _____

1. The label on the bag says that there are 72 pieces of candy in the bag. Mindy ate 10 pieces. How many pieces are left?

   Number sentence ___72 – 10 = 62 pieces___   Answer _62 pieces_

2. Curtis has nine dimes and a half dozen pennies.

   How many dimes is this? __9__   How much money is that? _90¢_

   How many pennies is this? __6__   How much money is that? _6¢_

   How much money does Curtis have? _96¢_

3. This is a tally to show how many children chose each color.

red	|||| |||
blue	|||| ||||	
green	|||

   Colors Children Chose

   Shade the graph to show the colors the children chose.

4. It's morning. What time is it? _4:35 a.m._

   It's evening. What time is it? _3:50 p.m._

5. Find the answers.

   $$\begin{array}{r} 27¢ \\ + 53¢ \\ \hline 80¢ \end{array}$$ $$\begin{array}{r} 16¢ \\ + 51¢ \\ \hline 67¢ \end{array}$$ $$\begin{array}{r} 13¢ \\ 38¢ \\ + 37¢ \\ \hline 88¢ \end{array}$$

2-88Wb

Lesson 89

writing number sentences to show equal groups
multiplying by ten

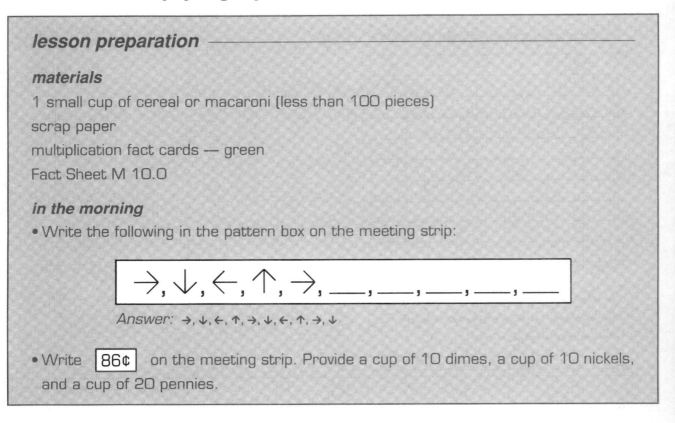

lesson preparation

materials

1 small cup of cereal or macaroni (less than 100 pieces)

scrap paper

multiplication fact cards — green

Fact Sheet M 10.0

in the morning

• Write the following in the pattern box on the meeting strip:

$$\rightarrow, \downarrow, \leftarrow, \uparrow, \rightarrow, \underline{\quad}, \underline{\quad}, \underline{\quad}, \underline{\quad}, \underline{\quad}$$

Answer: $\rightarrow, \downarrow, \leftarrow, \uparrow, \rightarrow, \downarrow, \leftarrow, \uparrow, \rightarrow, \downarrow$

• Write 86¢ on the meeting strip. Provide a cup of 10 dimes, a cup of 10 nickels, and a cup of 20 pennies.

THE MEETING

calendar

• Ask your child to write the date on the calendar and meeting strip.

• Ask your child the following two or three times a week:

  date _____ days ago, date _____ days from now

  day of the week _____ days ago, day of the week _____ days from now

  _____th month, month before, month after

• Record on the meeting strip a special event and the number of days until it occurs.

weather graph

• Ask your child to read and graph today's temperature to the nearest two degrees.

• Count by 10's and 2's to check the temperature on the graph.

- Ask your child to connect the dot for yesterday's temperature to the dot for today's temperature and compare the temperatures.

counting

- Ask your child to choose a number on the hundred number chart. Ask your child to add or subtract ten or one. Repeat 6–10 times. Ask your child to give directions for returning to the starting number.

    "Let's use our patting and clapping pattern to help us count by 3's to 30."

- Repeat this several times.

- Do the following once or twice a week:

    count by 10's to 400 and backward from 400 by 10's

    count by 5's to 100 and backward from 50 by 5's

    say the even numbers to 100 and backward from 50

    say the odd numbers to 49 and backward from 49

graph questions

- You and your child each ask a question about any of the graphs.

patterning

- Ask your child to do the following:

    identify the pattern (repeating, continuing, or both)

    identify the shapes to complete the pattern

    read the pattern

money

- Ask your child to put the coins in the coin cup. Count the money in the coin cup together.

- Ask your child for another way to show that amount of money. Count these coins together to check the amount.

clock

- Set the clock to a five-minute interval.

- Ask the following:

    "It's (morning/afternoon/evening). What time is it?"

    time one hour ago

    time one hour from now

- Ask your child to write the digital time on the meeting strip.

- Record on the meeting strip the time an activity will occur.

number of the day

- Write three number sentences for the number of the day on the meeting strip.

fact practice

- Write three fact family numbers (e.g., 2, 7, 9) on the chalkboard.

- Allow time for your child to write the four fact family number sentences on the chalkboard.

THE LESSON

Writing Number Sentences to Show Equal Groups Multiplying by Ten

"Today you will learn how to write equal groups number sentences."

"You also will learn how to multiply by ten."

"I will give you a cup of cereal (macaroni) to count."

"Put the pieces of cereal in groups of ten on your paper as you count them."

"Put the extra pieces of cereal in the cup."

- Give your child a small cup of cereal (macaroni) and a piece of scrap paper. Allow time for your child to count and group the cereal (macaroni).

"How many groups of ten do you have?"

- Record on the chalkboard "____ groups of ten equals ____."

"Let's count by 10's to find this answer."

- Record the answer next to the phrase.

"There is another way mathematicians write this problem."

- Write "____ × 10," filling in the appropriate number of groups.

"Writing the × is a short way of writing 'groups of.' "

"Put the cereal in the cup."

- Write the following problems on the chalkboard:

   9 groups of 10   6 groups of 10   12 groups of 10

"Write these problems the short way on your paper." **9 × 10, 6 × 10, 12 × 10**

"How can we find out how many pieces of cereal will be in nine groups of ten?" **count by 10's**

"How many pieces of cereal is that?" **90**

- Record "9 × 10 = 90" on the chalkboard.

- Repeat with 6 × 10 and 12 × 10.

- Write the following on the chalkboard:

$$3 \times 10 = \qquad 7 \times 10 = \qquad 4 \times 10 = \qquad 1 \times 10 =$$
$$5 \times 10 = \qquad 10 \times 10 = \qquad 8 \times 10 = \qquad 0 \times 10 =$$

"How will we read the first problem?"

- This should be read "three groups of ten equals."

"How many pieces of cereal are there in three groups of ten?"   *30*

- Fill in the answer.

- Repeat with each problem.

"What we did today is called multiplication."

"You have learned how to multiply by ten."

"We can write these problems another way."

- Write the following on the chalkboard:

$$
\begin{array}{ccccccccccc}
10 & 10 & 10 & 10 & 10 & 10 & 10 & 10 & 10 & 10 & 10 \\
\times 0 & \times 1 & \times 2 & \times 3 & \times 4 & \times 5 & \times 6 & \times 7 & \times 8 & \times 9 & \times 10
\end{array}
$$

"Which number in each problem do you think tells us the number of groups now?"   *bottom number*

"What is each answer?"

"Do you see a pattern?"

- Give your child the green multiplication fact cards. Use a separate plastic bag to store these cards.

"Use the fact cards to practice the multiplying by ten facts."

CLASS PRACTICE

number fact practice

- Give your child **Fact Sheet M 10.0.**

- Time your child for one minute.

- Correct the fact sheet with your child and record the score.

- Allow time for your child to complete the unfinished facts.

WRITTEN PRACTICE

- •Complete **Worksheet 89A** with your child.
- •Complete **Worksheet 89B** with your child later in the day.

Name _____

(Draw a 4½" line segment.)

LESSON 89A
Math 2

Date _____

(Measure this line segment using inches. ___3___ ")

1. Blake bought a marker for 48¢ and an eraser for 37¢. How much money did he spend?

   Number sentence ___48¢ + 37¢ = 85¢___   Answer ___85¢___

2. Write these examples using the multiplication symbol.

   5 groups of 10   3 groups of 10   9 groups of 10
   ___5 × 10___   ___3 × 10___   ___9 × 10___

3. Draw a picture to show 2 hundreds + 3 ones.

   What number does the picture show? ___203___

4. Use the graph to answer the questions.

   How many children like hiking? ___6___

   How many children like skating? ___4___

   How many children like only hiking? ___4___

   What does Curt like? ___both___

   Sports Children Like
   Hiking   Skating
   Phil
   May Bob Jill
   Leah Curt Skip
   Sarah

5. Shade the thermometer to show 74°F.

6. Find each product.

   3 × 10 = ___30___   8 × 10 = ___80___   6 × 10 = ___60___   0 × 10 = ___0___

7. How much money is this? ___56¢___

2-89Wa   Copyright © 1991 by Saxon Publishers, Inc. and Nancy Larson. Reproduction prohibited.

Name _____

LESSON 89B
Math 2

Date _____

1. Alexis bought a note pad for 64¢ and a pencil for 19¢. How much money did she spend?

   Number sentence ___64¢ + 19¢ = 83¢___   Answer ___83¢___

2. Write these examples using the multiplication symbol.

   4 groups of 10   8 groups of 10   2 groups of 10
   ___4 × 10___   ___8 × 10___   ___2 × 10___

3. Draw a picture to show 3 hundreds + 2 tens + 7 ones.

   What number does the picture show? ___327___

4. Use the graph to answer the questions.

   How many children like soccer? ___3___

   How many children like swimming? ___5___

   How many children like only soccer? ___2___

   What does Anna like? ___swimming___

   Sports Children Like
   Soccer   Swimming
          Eric
   Carl   Anna
   Kris Jan Joel
          Jim

5. Shade the thermometer to show 48°F.

6. Find each product.

   2 × 10 = ___20___   7 × 10 = ___70___   9 × 10 = ___90___   1 × 10 = ___10___

7. How much money is this? ___37¢___

2-89Wb   Copyright © 1991 by Saxon Publishers, Inc. and Nancy Larson. Reproduction prohibited.

esson 90

covering designs with tangram pieces

---

lesson preparation ──────────────────────────────────

materials

Written Assessment #17

Oral Assessment #9

1 set of tangrams (or Master 2-90A)

Masters 2-90B and 2-90C

thermometer

Fact Sheet S 6.0

the night before

• If a tangram set is not available, paste Master 2-90A on tagboard or construction paper and cut apart the pieces.

in the morning

• Write the following in the pattern box on the meeting strip:

┌───┐
│ ___, ___, ___,73, 75, 77,___, ___, ___   │
└───┘

Answer: 67, 69, 71, 73, 75, 77, 79, 81, 83

• Write │ 80¢ │ on the meeting strip. Provide a cup of 10 dimes, a cup of 10 nickels, and a cup of 20 pennies.

THE MEETING

calendar

• Ask your child to write the date on the calendar and meeting strip.

• Ask your child the following two or three times a week:

    date _____ days ago, date _____ days from now

    day of the week _____ days ago, day of the week _____ days from now

    _____th month, month before, month after

• Record on the meeting strip a special event and the number of days until it occurs.

---

weather graph

- Ask your child to read and graph today's temperature to the nearest two degrees.

- Count by 10's and 2's to check the temperature on the graph.

- Ask your child to connect the dot for yesterday's temperature to the dot for today's temperature and compare the temperatures.

counting

- Ask your child to choose a number on the hundred number chart. Ask your child to add or subtract ten or one. Repeat 6–10 times. Ask your child to give directions for returning to the starting number.

    "Let's use our patting and clapping pattern to help us count by 3's to 30."

- Repeat this several times.

- Do the following once or twice a week:

    count by 10's to 400 and backward from 400 by 10's

    count by 5's to 100 and backward from 50 by 5's

    say the even numbers to 100 and backward from 50

    say the odd numbers to 49 and backward from 49

graph questions

- You and your child each ask a question about any of the graphs.

patterning

- Ask your child to do the following:

    identify the pattern (repeating, continuing, or both)

    identify the numbers to complete the pattern

    read the pattern

money

- Ask your child to put the coins in the coin cup. Count the money in the coin cup together.

- Ask your child for another way to show that amount of money. Count these coins together to check the amount.

clock

- Set the clock to a five-minute interval.

- Ask the following:

    "It's (morning/afternoon/evening). What time is it?"

    time one hour ago

time one hour from now

- Ask your child to write the digital time on the meeting strip.

- Record on the meeting strip the time an activity will occur.

number of the day

- Write three number sentences for the number of the day on the meeting strip.

fact practice

- Write three fact family numbers (e.g., 2, 7, 9) on the chalkboard.

- Allow time for your child to write the four fact family number sentences on the chalkboard.

ASSESSMENT

Written Assessment

"Today I would like to see what you remember from what we have been practicing."

- Give your child **Written Assessment #17**.

- Read the directions for each problem. Allow time for your child to complete that problem before continuing.

- Correct the paper, noting your child's mistakes on the **Individual Recording Form**. Review the errors with your child.

Oral Assessment

- Record your child's response(s) to the oral interview questions on the interview sheet.

THE LESSON

Covering Designs with Tangram Pieces

"Today you will learn how to cover designs using tangram pieces."

"A tangram is a square that has been cut into seven pieces."

"We can use the pieces to make designs or pictures."

"I will give you a set of tangram pieces to use."

- Give your child a set of tangram pieces.

"Make a design or picture using your tangram pieces."

- Allow time for your child to explore the pieces and to make a design or picture.

  "Sort your tangram pieces."

- Allow time for your child to sort the pieces.

  "How did you sort your pieces?"

- Ask your child to describe how he/she sorted the tangram pieces.

  "Are any of the tangram pieces the same size?"   yes

  "Which ones?"   2 large triangles and 2 small triangles

  "What do we call the shapes that are not triangles?"   square and parallelogram

  "Now you will use the two small triangles to try to cover the other pieces."

  "Which piece do you think the two small triangles will cover?"

  "Try to cover the square using the two small triangles."

- Allow time for your child to cover the square.

  "Try to cover the parallelogram using the two small triangles."

- Allow time for your child to cover the parallelogram.

  "Try to cover the medium triangle using the two small triangles."

- Allow time for your child to cover the triangle.

  "Now try to use your pieces to cover the large triangle."

- Allow time for your child to cover the large triangle.

  "What pieces did you use?"

- Ask your child to name the pieces used.

  "Try to cover the large triangle using different pieces."

- Allow time for your child to do this.

  "What pieces did you use?"

- Ask your child to name the pieces used.

  "Now you will have a chance to cover designs using the tangram pieces."

- Give your child **Master 2-90B.**

  "All of these shapes can be covered using one set of tangram pieces."

  "Try to cover all of these shapes using your tangram pieces."

- Give your child **Master 2-90C.**

  "Cover these shapes one at a time using your tangram pieces."

- The tangram set pieces will be used again in Lesson 110.

CLASS PRACTICE

number fact practice

- Use the blue fact cards to practice the subtracting a number from ten facts with your child.

- Give your child **Fact Sheet S 6.0.**

- Time your child for one minute.

- Correct the fact sheet with your child and record the score.

- Allow time for your child to complete the unfinished facts.

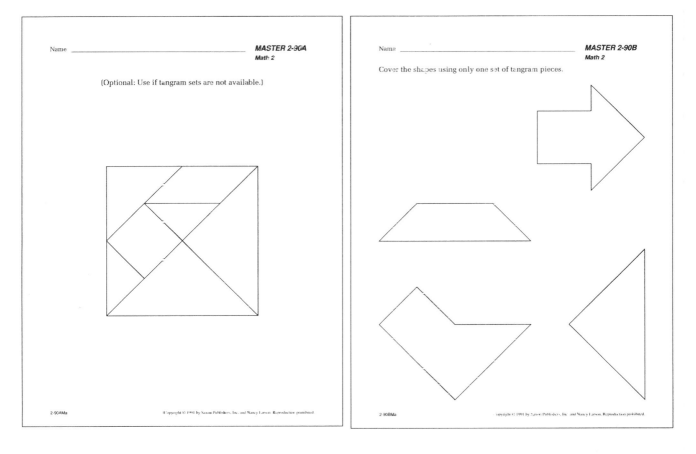

Name _____ **MASTER 2-90A**
Math 2

(Optional: Use if tangram sets are not available.)

2-90AMa   © Copyright © 1994 by Saxon Publishers, Inc. and Nancy Larson. Reproduction prohibited.

Name _____ **MASTER 2-90B**
Math 2

Cover the shapes using only one set of tangram pieces.

2-90BMa   copyright © 1994 by Saxon Publishers, Inc. and Nancy Larson. Reproduction prohibited.

Name _____ **MASTER 2-90C**
Math 2

Cover this shape using only the triangles.

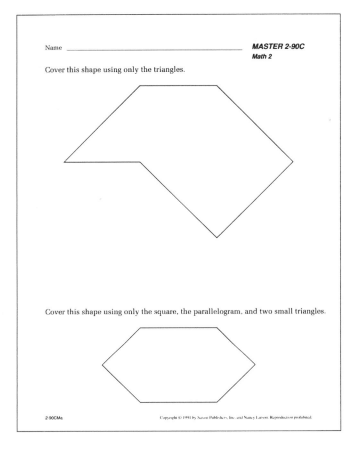

Cover this shape using only the square, the parallelogram, and two small triangles.

2-90CMa   Copyright © 1994 by Saxon Publishers, Inc. and Nancy Larson. Reproduction prohibited.

Teacher _____ **MATH 2   LESSON 90**
Date _____ Oral Assessment # 9 Recording Form

Materials: Thermometer	"What are the even numbers?" • Stop child's counting at 20. "What are the odd numbers?" • Stop child's counting at 19.	• Show the child the thermometer. "What is the temperature?" • Accept answers to the nearest 2°
Students		

2-90La   Copyright © 1994 by Saxon Publishers, Inc. and Nancy Larson. Reproduction prohibited.

Name _____ **ASSESSMENT 17**
LESSON 90
Date _____ **Math 2**

1. Amanda bought a ruler for 27¢ and an eraser for 46¢. How much money did she spend?

   Number sentence _____ 27¢ + 46¢ = 73¢ _____   Answer ____ 73¢ ____

2. One of these names is the name of my dog. Cross out the names that cannot be my dog's name.

   It is not the name that is last.   ~~Lady~~ Duffer ~~Rover~~ ~~Spot~~ ~~Ebony~~
   It does not have 4 letters.
   It is not the name in the middle.

   What is my dog's name? ____ Duffer ____

3. What temperature does the thermometer show? __ 58 __ °F

4. Write fractions that tell how much is shaded.

   [shaded bar diagram] $\frac{5}{10}$ [shaded grid] $\frac{3}{6}$

5. Measure the line segment using inches. __ $5\frac{1}{2}$ __ inches

   Draw a $2\frac{1}{2}''$ line segment.

6. Find the answers.

56¢	68¢	24¢	42¢ + 19¢
+ 32¢	+ 17¢	+ 36¢	
88¢	85¢	60¢	

	4	2	¢
+	1	9	¢
	6	1	¢

2-90Aa   Copyright © 1994 by Saxon Publishers, Inc. and Nancy Larson. Reproduction prohibited.

L esson 91

writing numbers in expanded form

lesson preparation

materials

demonstration clock

Fact Sheet M 10.0

in the morning

• Write the following in the pattern box on the meeting strip:

| ___, ___, ___, 45, 50, 55, ___, ___, ___ |

Answer: 30, 35, 40, 45, 50, 55, 60, 65, 70

• Write 33¢ on the meeting strip. Provide a cup of 10 dimes, a cup of 10 nickels, and a cup of 20 pennies.

THE MEETING

calendar

• Ask your child to write the date on the calendar and meeting strip.

• Ask your child the following two or three times a week:

   date _____ days ago, date _____ days from now

   day of the week _____ days ago, day of the week _____ days from now

   _____th month, month before, month after

• Record on the meeting strip a special event and the number of days until it occurs.

weather graph

• Ask your child to read and graph today's temperature to the nearest two degrees.

• Count by 10's and 2's to check the temperature on the graph.

• Ask your child to connect the dot for yesterday's temperature to the dot for today's temperature and compare the temperatures.

counting

- Ask your child to choose a number on the hundred number chart. Ask your child to add or subtract ten or one. Repeat 6–10 times. Ask your child to give directions for returning to the starting number.

  "Let's use our patting and clapping pattern to help us count by 3's to 30."

- Repeat this several times.

- Do the following once or twice a week:

  count by 10's to 400 and backward from 400 by 10's

  count by 5's to 100 and backward from 50 by 5's

  say the even numbers to 100 and backward from 50

  say the odd numbers to 49 and backward from 49

graph questions

- You and your child each ask a question about any of the graphs.

patterning

- Ask your child to do the following:

  identify the pattern (repeating, continuing, or both)

  identify the numbers to complete the pattern

  read the pattern

money

- Ask your child to put the coins in the coin cup. Count the money in the coin cup together.

- Ask your child for another way to show that amount of money. Count these coins together to check the amount.

clock

- Set the clock to a five-minute interval.

- Ask the following:

  "It's (morning/afternoon/evening). What time is it?"

  time one hour ago

  time one hour from now

- Ask your child to write the digital time on the meeting strip.

- Record on the meeting strip the time an activity will occur.

number of the day

- Write three number sentences for the number of the day on the meeting strip.

fact practice

- Write three fact family numbers (e.g., 2, 7, 9) on the chalkboard.
- Allow time for your child to write the four fact family number sentences on the chalkboard.

THE LESSON

Writing Numbers in Expanded Form

"Today you will learn how to write two- and three-digit numbers in expanded form."

"When we counted large collections, we grouped by hundreds, tens, and ones."

- Draw the following on the chalkboard:

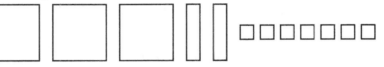

"What number will we write to show what I have drawn?" 327

"We can write that number in expanded form by writing three hundred plus twenty plus seven."

- Write the following on the chalkboard below the picture:

$$300 + 20 + 7 = 327$$

- Draw the following on the chalkboard:

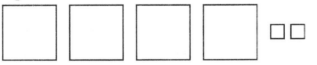

"What number will we write to show what I have drawn?" 402

"We can write that number in expanded form by writing four hundred plus two."

- Write the following on the chalkboard below the picture:

$$400 + 2 = 402$$

- Draw the following on the chalkboard:

"What number will we write to show what I have drawn?" 150

"We can write that number in expanded form by writing one hundred plus fifty."

- Write the following on the chalkboard below the picture:

$$100 + 50 = 150$$

"How do you think we will write 582 in expanded form?"

- Record "582 = 500 + 80 + 2" on the chalkboard.

"How do you think we will write 913 in expanded form?"

- Record "913 = 900 + 10 + 3" on the chalkboard.

"How do you think we will write 460 in expanded form?"

- Record "460 = 400 + 60" on the chalkboard.

"How do you think we will write 209 in expanded form?"

- Record "209 = 200 + 9" on the chalkboard.

"How do you think we will write 57 in expanded form?"

- Record "57 = 50 + 7" on the chalkboard.

"Which digit tells us the number of hundreds?"   *third digit from the right*

"Which digit tells us the number of tens?"   *second digit from the right*

"Which digit tells us the number of ones?"   *digit on the right*

- Write "481" on the chalkboard.

"Write this number in expanded form."

- Ask your child to write the expanded form on the chalkboard.

- Repeat with the following numbers: 340, 865, 106, 83.

- Write the following on the chalkboard: 500 + 20 + 7 =

"Now I wrote a number in expanded form."

"Write the number for this expanded form."

- Ask your child to write the number on the chalkboard.

- Repeat with the following:

$$300 + 80 + 6 = \qquad 600 + 8 = \qquad 900 + 60 =$$

CLASS PRACTICE

telling time to five-minute intervals

"Now we will practice showing time to five-minute intervals."

- Give your child a demonstration clock.

- Write "1:05" on the chalkboard.

"Show 1:05 on your clock."

"Where does the hour hand point?"

"Where does the minute hand point?"

- Repeat with 9:30, 12:15, 5:50, 12:35, and 8:45.

number fact practice

- Use the green multiplication fact cards to practice the multiplying by ten facts with your child.

- Give your child **Fact Sheet M 10.0**.

- Time your child for one minute.

- Correct the fact sheet with your child and record the score.

- Allow time for your child to complete the unfinished facts.

WRITTEN PRACTICE

- Complete **Worksheet 91A** with your child.

- Complete **Worksheet 91B** with your child later in the day.

Name _____   **LESSON 91A**
[Draw a 4" line segment.]   **Math 2**
Date _____
[Draw a 2½" line segment.]

1. Marsha counted the tiles and put them in groups of 10. When she finished she counted 3 groups of tiles. Draw a picture to show the tiles.

   Marsha has ___3___ groups of ___10___ tiles.

   How many tiles does she have? ___30___

2. Circle the largest and the smallest numbers.

   (71)   47   (24)   38

   Find their sum. ___95___

   Find the sum of the two numbers that are not circled. ___85___

3. Write 437 in expanded form. _____400 + 30 + 7_____

   Write the number for 200 + 60 + 7. ___267___

   Write the number for 300 + 8. ___308___

4. Find the answers.

   10 less than 61 = ___51___   10 – 7 = ___3___   10 – 4 = ___6___

   13 – 9 = ___4___   7 × 10 = ___70___   4 × 10 = ___40___

5. Color the shapes that are congruent to the oval on the left.

2-91Wa

Copyright © 1991 by Saxon Publishers, Inc. and Nancy Larson. Reproduction prohibited.

Name _____   **LESSON 91B**
                                         Math 2
Date _____

1. Stephen counted the cubes and put them in groups of 10. When he finished, he counted 5 groups of cubes. Draw a picture to show the cubes.

   Stephen has ___5___ groups of ___10___ cubes.

   How many cubes does he have? ___50___

2. Circle the largest and the smallest numbers.

   37   (26)   (54)   47

   Find their sum. ___80___

   Find the sum of the two numbers that are not circled. ___84___

3. Write 163 in expanded form. _____100 + 60 + 3_____

   Write the number for 300 + 20 + 9. ___329___

   Write the number for 400 + 70. ___470___

4. Find the answers.

   1 less than 40 = ___39___   10 – 6 = ___4___   10 – 3 = ___7___

   12 – 9 = ___3___   5 × 10 = ___50___   3 × 10 = ___30___

5. Color the shapes that are congruent to the rectangle on the left.

2-91Wb

Copyright © 1991 by Saxon Publishers, Inc. and Nancy Larson. Reproduction prohibited.

Lesson 92

subtraction facts—subtracting using the doubles plus one facts

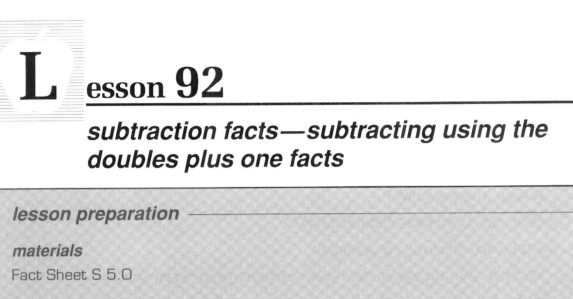

lesson preparation ───────────────────────────

materials

Fact Sheet S 5.0

in the morning

• Write the following in the pattern box on the meeting strip:

> 11, ___, 13, ___, 15, ___, 17, ___, 19, ___

Answer: 11, 12, 13, 14, 15, 16, 17, 18, 19, 20

• Write 91¢ on the meeting strip. Provide a cup of 10 dimes, a cup of 10 nickels, and a cup of 20 pennies.

THE MEETING

calendar

• Ask your child to write the date on the calendar and meeting strip.

• Ask your child the following two or three times a week:

> date _____ days ago, date _____ days from now
>
> day of the week _____ days ago, day of the week _____ days from now
>
> _____th month, month before, month after

• Record on the meeting strip a special event and the number of days until it occurs.

weather graph

• Ask your child to read and graph today's temperature to the nearest two degrees.

• Count by 10's and 2's to check the temperature on the graph.

• Ask your child to connect the dot for yesterday's temperature to the dot for today's temperature and compare the temperatures.

counting

- Ask your child to choose a number on the hundred number chart. Ask your child to add or subtract ten or one. Repeat 6–10 times. Ask your child to give directions for returning to the starting number.

  "Let's use our patting and clapping pattern to help us count by 3's to 30."

- Repeat this several times

- Do the following once or twice a week:

  count by 10's to 400 and backward from 400 by 10's

  count by 5's to 100 and backward from 50 by 5's

  say the even numbers to 100 and backward from 50

  say the odd numbers to 49 and backward from 49

graph questions

- You and your child each ask a question about any of the graphs.

patterning

- Ask your child to do the following:

  identify the pattern (repeating, continuing, or both)

  identify the numbers to complete the pattern

  read the pattern

money

- Ask your child to put the coins in the coin cup. Count the money in the coin cup together.

- Ask your child for another way to show that amount of money. Count these coins together to check the amount.

clock

- Set the clock to a five-minute interval.

- Ask the following:

  "It's (morning/afternoon/evening). What time is it?"

  time one hour ago

  time one hour from now

- Ask your child to write the digital time on the meeting strip.

- Record on the meeting strip the time an activity will occur.

number of the day

- Write three number sentences for the number of the day on the meeting strip.

fact practice

- Write three fact family numbers (e.g., 2, 7, 9) on the chalkboard.

- Allow time for your child to write the four fact family number sentences on the chalkboard.

THE LESSON

Subtraction Facts—Subtracting Using the Doubles Plus One Facts

"Each morning we have been practicing fact families."

"Today you will learn how to find the answer for subtraction problems by using the doubles plus one addition facts."

"What is an example of a doubles plus one fact?"

- Write the combinations on the chalkboard as your child names them.

0 + 1 = 1	1 + 0 = 1
1 + 2 = 3	2 + 1 = 3
2 + 3 = 5	3 + 2 = 5
3 + 4 = 7	4 + 3 = 7
4 + 5 = 9	5 + 4 = 9
5 + 6 = 11	6 + 5 = 11
6 + 7 = 13	7 + 6 = 13
7 + 8 = 15	8 + 7 = 15
8 + 9 = 17	9 + 8 = 17

"Let's write the two subtraction problems that go with each pair of addition problems."

- Write the subtraction facts on the chalkboard as your child names them.

0 + 1 = 1	1 + 0 = 1	1 − 0 = 1	1 − 1 = 0
1 + 2 = 3	2 + 1 = 3	3 − 1 = 2	3 − 2 = 1
2 + 3 = 5	3 + 2 = 5	5 − 2 = 3	5 − 3 = 2
3 + 4 = 7	4 + 3 = 7	7 − 3 = 4	7 − 4 = 3
4 + 5 = 9	5 + 4 = 9	9 − 4 = 5	9 − 5 = 4
5 + 6 = 11	6 + 5 = 11	11 − 5 = 6	11 − 6 = 5
6 + 7 = 13	7 + 6 = 13	13 − 6 = 7	13 − 7 = 6
7 + 8 = 15	8 + 7 = 15	15 − 7 = 8	15 − 8 = 7
8 + 9 = 17	9 + 8 = 17	17 − 8 = 9	17 − 9 = 8

"We can always use addition to help us find a subtraction answer."

- Write the following on the chalkboard:

$$\begin{array}{cccccccccccccc} 7 & 17 & 11 & 13 & 9 & 15 & 17 & 5 & 15 & 9 & 11 & 7 & 13 \\ -3 & -8 & -5 & -6 & -4 & -7 & -9 & -3 & -8 & -5 & -6 & -4 & -7 \end{array}$$

"To find the answer for seven minus three, we can think 'some number plus three is seven.' "

- Point to each part of the problem as you say the numbers. Work from the bottom up.

  "What is the answer?"   *4*

- Record the answer below the problem.

- Repeat with the other problems.

  "Let's practice the subtracting using the doubles plus one facts."

- Erase the answers for the vertical problems.

  "I will point to a number fact."

  "Say the answer as quickly as possible."

- Point to the number facts one at a time.

- Review the strategy for finding the answer, if necessary.

  "Take out your pink fact cards."

  "Practice saying the subtraction facts to yourself."

  "Check the back of each card to see if the answer is correct."

- Allow time for your child to practice independently.

CLASS PRACTICE

number fact practice

- Give your child **Fact Sheet S 5.0**.
- Time your child for one minute.
- Correct the fact sheet with your child and record the score.
- Allow time for your child to complete the unfinished facts.

WRITTEN PRACTICE

- Complete **Worksheet 92A** with your child.
- Complete **Worksheet 92B** with your child later in the day.

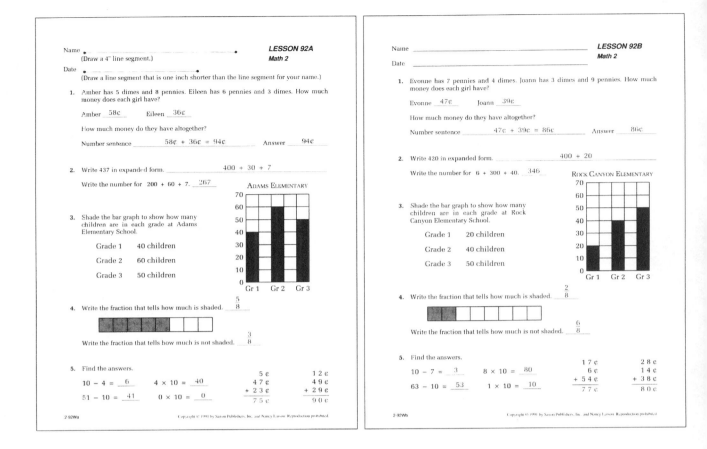

Lesson 93

writing money amounts using $ signs and ¢ symbols

---

lesson preparation ————————————————————————

materials

advertisements containing $ and ¢ symbols (from newspapers, etc.)

scrap paper

Fact Sheet S 5.0

the night before

• Cut out advertisements containing dollar signs and cent symbols from newspaper flyers.

in the morning

• Write the following in the pattern box on the meeting strip:

> 250, 260, 270, ___, ___, ___, ___, ___, ___

Answer: 250, 260, 270, 280, 290, 300, 310, 320, 330

• Write 79¢ on the meeting strip. Provide a cup of 10 dimes, a cup of 10 nickels, and a cup of 20 pennies.

THE MEETING

calendar

> • Ask your child to write the date on the calendar and meeting strip.
>
> • Ask your child the following two or three times a week:
>
>> date _____ days ago, date _____ days from now
>>
>> day of the week _____ days ago, day of the week _____ days from now
>>
>> _____th month, month before, month after
>
> • Record on the meeting strip a special event and the number of days until it occurs.

weather graph

> • Ask your child to read and graph today's temperature to the nearest two degrees.

- Count by 10's and 2's to check the temperature on the graph.
- Ask your child to connect the dot for yesterday's temperature to the dot for today's temperature and compare the temperatures.

counting

- Ask your child to choose a number on the hundred number chart. Ask your child to add or subtract ten or one. Repeat 6–10 times. Ask your child to give directions for returning to the starting number.

    "Let's use our patting and clapping pattern to help us count by 3's to 30."

- Repeat this several times.
- Do the following once or twice a week:

    count by 10's to 400 and backward from 400 by 10's

    count by 5's to 100 and backward from 50 by 5's

    say the even numbers to 100 and backward from 50

    say the odd numbers to 49 and backward from 49

graph questions

- You and your child each ask a question about any of the graphs.

patterning

- Ask your child to do the following:

    identify the pattern (repeating, continuing, or both)

    identify the numbers to complete the pattern

    read the pattern

money

- Ask your child to put the coins in the coin cup. Count the money in the coin cup together.
- Ask your child for another way to show that amount of money. Count these coins together to check the amount.

clock

- Set the clock to a five-minute interval.
- Ask the following:

    "It's (morning/afternoon/evening). What time is it?"

    time one hour ago

    time one hour from now

- Ask your child to write the digital time on the meeting strip.
- Record on the meeting strip the time an activity will occur.

number of the day

- Write three number sentences for the number of the day on the meeting strip.

fact practice

- Write three fact family numbers (e.g., 2, 7, 9) on the chalkboard.

- Allow time for your child to write the four fact family number sentences on the chalkboard.

THE LESSON

Writing Money Amounts Using $ Signs and ¢ Symbols

- Show your child examples of ads with dollar signs and cent symbols.

  "These are some ways that money amounts are written in the newspaper."

  "What do you notice about the way money amounts are written?" they are written with a cent symbol or with a dollar sign

  "Today you will learn how to write money amounts using a dollar sign or a cent symbol."

  "When something costs less than one dollar, we usually write the amount using a cent symbol."

  "Let's pretend that the cost of a pencil is 38 cents."

  "How will we write that?"

- Write "38¢" on the chalkboard.

  "When something costs more than one dollar, we write the amount a different way."

  "We write the amount using a dollar sign."

  "Let's write two dollars and fifty cents."

  "First we will write a dollar sign."

  "When we write a dollar sign, we write an S with a vertical line through it."

- Demonstrate on the chalkboard.

  "We write a two next to the dollar sign to show two dollars."

- Demonstrate on the chalkboard.

  "Now we use a decimal point to separate the dollars from the cents."

  "We write '50' after the decimal point to show fifty cents."

- Demonstrate on the chalkboard.

"*We never use a dollar sign and a cent symbol in the same problem.*"

"*Now we will practice writing some money amounts using the dollar sign.*"

"*Let's write three dollars and twenty-five cents.*"

"*What will we write first?*"   dollar sign

- Demonstrate on the chalkboard.

"*What will we write next?*"   three

"*What will we use to separate the dollars and the cents?*"   decimal point

"*How will we write twenty-five cents?*"

"*Now you will have a chance to write money amounts on the chalkboard.*"

"*Write four dollars and eighty cents on the chalkboard.*"

- Repeat with $5.91, $2.10, and $6.33.

"*Whenever we have less than ten cents, we must always use a zero in front of the number of cents.*"

"*When we write four dollars and five cents, we write it like this.*"

- Write "$4.05" on the chalkboard.

"*The zero means that I do not have any dimes.*"

"*We can also write amounts less than one dollar using a dollar sign.*"

"*Let's write sixty-three cents using a dollar sign.*"

"*Do we have any dollars?*"   no

"*What will we write to show that we don't have any dollars?*"   0

"*What will we write next?*"   decimal point

"*How will we write sixty-three cents?*"

- Write "$0.63" on the chalkboard.

"*How will we write sixty-three cents using a cent symbol?*"

- Write "63¢" on the chalkboard.

"*When we use a cent symbol, we do not use a decimal point.*"

"*Now you will have a chance to practice writing some money amounts.*"

- Give your child a piece of scrap paper.

"*First you will practice making a dollar sign.*"

"*How do we make a dollar sign?*"

"*Make five dollar signs on your paper.*"

- Allow time for your child to do this.

"*Now you will write money amounts on your paper using the dollar sign.*"

"Write two dollars and seventy-five cents."

• Allow time for your child to do this.

• Repeat with the following amounts:

   "Three dollars and forty-one cents."

   "Four dollars and twelve cents."

   "One dollar and eight cents."

   "Eight dollars and ninety cents."

   "Twenty-six dollars."

   "Nine dollars and fifteen cents."

   "Fifty-eight cents."

CLASS PRACTICE

number fact practice

• Use the pink fact cards to practice the subtracting using the doubles plus one facts with your child.

• Give your child **Fact Sheet S 5.0**.

• Time your child for one minute.

• Correct the fact sheet with your child and record the score.

• Allow time for your child to complete the unfinished facts.

WRITTEN PRACTICE

• Complete **Worksheet 93A** with your child.

• Complete **Worksheet 93B** with your child later in the day.

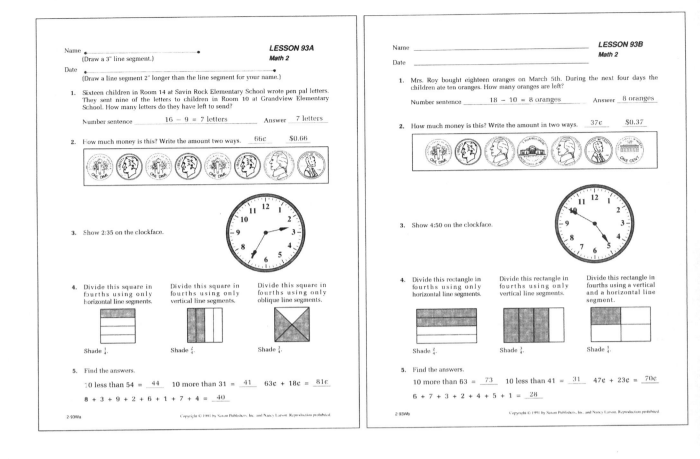

LESSON 93A
Math 2

Name _____ (Draw a 3" line segment.)

Date _____
(Draw a line segment 2" longer than the line segment for your name.)

1. Sixteen children in Room 14 at Savin Rock Elementary School wrote pen pal letters. They sent nine of the letters to children in Room 10 at Grandview Elementary School. How many letters do they have left to send?

    Number sentence ___16 − 9 = 7 letters___   Answer ___7 letters___

2. How much money is this? Write the amount two ways.   66¢ $0.66

3. Show 2:35 on the clockface.

4. Divide this square in fourths using only horizontal line segments.

    Shade ¼.

    Divide this square in fourths using only vertical line segments.

    Shade 2/4.

    Divide this square in fourths using only oblique line segments.

    Shade ¾.

5. Find the answers.

    10 less than 54 = ___44___   10 more than 31 = ___41___   63¢ + 18¢ = ___81¢___

    8 + 3 + 9 + 2 + 6 + 1 + 7 + 4 = ___40___

2-93Wa

LESSON 93B
Math 2

Name _____

Date _____

1. Mrs. Roy bought eighteen oranges on March 5th. During the next four days the children ate ten oranges. How many oranges are left?

    Number sentence ___18 − 10 = 8 oranges___   Answer ___8 oranges___

2. How much money is this? Write the amount in two ways.   37¢ $0.37

3. Show 4:50 on the clockface.

4. Divide this rectangle in fourths using only horizontal line segments.

    Shade 2/4.

    Divide this rectangle in fourths using only vertical line segments.

    Shade ¾.

    Divide this rectangle in fourths using a vertical and a horizontal line segment.

    Shade ¼.

5. Find the answers.

    10 more than 63 = ___73___   10 less than 41 = ___31___   47¢ + 23¢ = ___70¢___

    6 + 7 + 3 + 2 + 4 + 5 + 1 = ___28___

2-93Wb

L esson 94

measuring height in feet and inches

---

lesson preparation

materials

ruler

tape

step stool

Fact Sheet S 6.4

in the morning

• Write the following in the pattern box on the meeting strip:

⊕, ⊕, ⊖, ⊕, ⊕, ___, ___, ___, ___

Answer: ⊖,⊕,⊕,⊕,⊕,⊕,⊕ ⊕,⊕

• Write 99¢ on the meeting strip. Provide a cup of 10 dimes, a cup of 10 nickels, and a cup of 20 pennies.

THE MEETING

calendar

• Ask your child to write the date on the calendar and meeting strip.

• Ask your child the following two or three times a week:

   date _____ days ago, date _____ days from now

   day of the week _____ days ago, day of the week _____ days from now

   _____th month, month before, month after

• Record on the meeting strip a special event and the number of days until it occurs.

weather graph

• Ask your child to read and graph today's temperature to the nearest two degrees.

• Count by 10's and 2's to check the temperature on the graph.

- Ask your child to connect the dot for yesterday's temperature to the dot for today's temperature and compare the temperatures.

counting

- Ask your child to choose a number on the hundred number chart. Ask your child to add or subtract ten or one. Repeat 6–10 times. Ask your child to give directions for returning to the starting number.

"Let's use our patting and clapping pattern to help us count by 3's to 30."

- Repeat this several times.
- Do the following once or twice a week:

count by 10's to 400 and backward from 400 by 10's

count by 5's to 100 and backward from 50 by 5's

say the even numbers to 100 and backward from 50

say the odd numbers to 49 and backward from 49

graph questions

- You and your child each ask a question about any of the graphs.

patterning

- Ask your child to do the following:

identify the pattern (repeating, continuing, or both)

identify the shapes to complete the pattern

read the pattern

money

- Ask your child to put the coins in the coin cup. Count the money in the coin cup together.
- Ask your child for another way to show that amount of money. Count these coins together to check the amount.

"Write this amount on the chalkboard using a dollar sign."

clock

- Set the clock to a five-minute interval.
- Ask the following:

"It's (morning/afternoon/evening). What time is it?"

time one hour ago

time one hour from now

- Ask your child to write the digital time on the meeting strip.
- Record on the meeting strip the time an activity will occur.

number of the day

- Write three number sentences for the number of the day on the meeting strip.

fact practice

- Write three fact family numbers (e.g., 2, 7, 9) on the chalkboard.
- Allow time for your child to write the four fact family number sentences on the chalkboard.

THE LESSON

Measuring Height in Feet and Inches

"When you go to the doctor's office, what does the nurse usually do before you see the doctor?" checks height and weight

"How does the nurse do that?"

"Today you will learn how to measure height in feet and inches."

"When we measure our height, do we tell people that we are 50 miles tall or 60 pounds tall?"

"What unit of measure do we use?" inches or feet and inches

"We can tell people our height in different ways."

"One way we tell them our height is to tell them how tall we are in feet and inches."

"Let's find your height from the bottom of your feet to the top of your head."

"About how many feet and inches tall do you think you are?"

"What are some ways we could measure your height?"

- Allow your child to suggest ways to measure his/her height.

"It is easier to measure the wall than it is to measure a person."

- Use the door frame or a convenient wall.

"Stand with your back and heels against the wall."

"Now I will put a ruler on the top of your head."

"I will put a piece of tape to mark where the ruler meets the wall."

- Do this.

"Now you can move away from the wall."

"We will measure your height by measuring from the floor to the tape."

"We will work together to do this."

"As I measure one foot, put your finger at the end of the ruler."

"Then I will move the ruler and measure another foot."

"You will move your finger to the end of the ruler and count another foot."

"We will be careful to keep the ruler straight."

- Measure your child's height.

"How many full feet tall are you?"

- Record "_____ feet" on the chalkboard.

"Are there any inches left over?"

"Let's count them."

- Record "_____ feet, _____ inches" on the chalkboard.

"Was your estimate close?"

"Let's try to estimate how tall I am in feet and extra inches."

"About how many feet and inches tall do you think I am?"

- Write the estimate on the chalkboard.

"We are trying to find my height from the bottom of my feet to the top of my head."

"How are we going to measure my height?"

- Ask your child to restate the steps as you work together to measure your height. Measure your height against the same door frame or wall. Provide a stool for your child to stand on, if necessary.

"How many full feet tall am I?"

- Record "_____ feet" on the chalkboard.

"Are there any inches left over?"

"Let's count them."

- Record "_____ feet, _____ inches" on the chalkboard.

"Was your estimate close?"

"Who was taller?"

"How many feet and inches taller?"

CLASS PRACTICE

number fact practice

- Use the tan, peach, lavender, green, pink, and blue fact cards to practice the subtraction facts with your child.

- Give your child **Fact Sheet S 6.4.**

- Time your child for one minute.

• Correct the fact sheet with your child and record the score.

• Allow time for your child to complete the unfinished facts.

WRITTEN PRACTICE

• Complete **Worksheet 94A** with your child.

• Complete **Worksheet 94B** with your child later in the day.

Name _____

(Measure this line segment. __4__ ")

Date _____

(Measure this line segment. __½__ ")

LESSON 94A
Math 2

1. Mary bought a package of candy. There are 10 candies in a package. Mary ate 6 candies. How many candies are left?

    Number sentence ___10 − 6 = 4 candies___   Answer ___4 candies___

2. Put these numbers in order from least to greatest.

    | 29 | 43 | 27 | 48 | 19 |

    __19__   __27__   __29__   __43__   __48__
    least   greatest

    Add the least and the greatest numbers. __67__

    Add the other three numbers __99__

3. About how tall is your teacher? _____

4. Brian has 3 dimes and 16 pennies. How much money is this?

    Write the amount two ways.   __46¢__   __$0.46__

5. My favorite time of day is 8:40 p.m. Show that time on the clock.

    Is it morning or evening? __evening__

6. Write the fact family number sentences for 2, 7, and 9.

    | 2 + 7 = 9 | 9 − 2 = 7 |
    | 7 + 2 = 9 | 9 − 7 = 2 |

2-94Wa

Name _____

Date _____

LESSON 94B
Math 2

1. Silvia bought a package of pencils. There are 10 pencils in a package. Silvia gave 4 pencils to her sister. How many pencils does she have left?

    Number sentence ___10 − 4 = 6 pencils___   Answer ___6 pencils___

2. Put these numbers in order from least to greatest.

    | 38 | 24 | 33 | 18 | 26 |

    __18__   __24__   __26__   __33__   __38__
    least   greatest

    Add the least and the greatest numbers. __56__

    Add the other three numbers. __83__

3. About how long is your bed? _____

4. Evan has 5 dimes and 14 pennies. How much money is this?

    Write the amount two ways.   __64¢__   __$0.64__

5. What is your favorite time of day?

    _____

    Show that time on the clock.

6. Write the fact family number sentences for 1, 8, and 9.

    | 1 + 8 = 9 | 9 − 1 = 8 |
    | 8 + 1 = 9 | 9 − 8 = 1 |

2-94Wb

Lesson 95

adding two-digit numbers with a sum greater than 100

---

lesson preparation

materials

Written Assessment #18

Fact Sheet S 5.0

in the morning

• Write the following in the pattern box on the meeting strip:

> 51, 61, 71, ____, ____, ____, ____, ____, ____

Answer: 51, 61, 71, 81, 91, 101, 111, 121, 131

• Write [71¢] on the meeting strip. Provide a cup of 10 dimes, a cup of 10 nickels, and a cup of 20 pennies.

THE MEETING

calendar

- Ask your child to write the date on the calendar and meeting strip.

- Ask your child the following two or three times a week:

    date ____ days ago, date ____ days from now

    day of the week ____ days ago, day of the week ____ days from now

    ____th month, month before, month after

- Record on the meeting strip a special event and the number of days until it occurs.

weather graph

- Ask your child to read and graph today's temperature to the nearest two degrees.

- Count by 10's and 2's to check the temperature on the graph.

- Ask your child to connect the dot for yesterday's temperature to the dot for today's temperature and compare the temperatures.

---

counting

- Ask your child to choose a number on the hundred number chart. Ask your child to add or subtract ten or one. Repeat 6–10 times. Ask your child to give directions for returning to the starting number.

  "Let's use our patting and clapping pattern to help us count by 3's to 30."

- Repeat this several times.

- Do the following once or twice a week:

  count by 10's to 400 and backward from 400 by 10's

  count by 5's to 100 and backward from 50 by 5's

  say the even numbers to 100 and backward from 50

  say the odd numbers to 49 and backward from 49

graph questions

- You and your child each ask a question about any of the graphs.

patterning

- Ask your child to do the following

  identify the pattern (repeating, continuing, or both)

  identify the numbers to complete the pattern

  read the pattern

money

- Ask your child to put the coins in the coin cup. Count the money in the coin cup together.

- Ask your child for another way to show that amount of money. Count these coins together to check the amount.

  "Write this amount on the chalkboard using a dollar sign."

clock

- Set the clock to a five-minute interval.

- Ask the following:

  "It's (morning/afternoon/evening). What time is it?"

  time one hour ago

  time one hour from now

- Ask your child to write the digital time on the meeting strip.

- Record on the meeting strip the time an activity will occur.

number of the day

- Write three number sentences for the number of the day on the meeting strip.

fact practice

- Write three fact family numbers (e.g., 2, 7, 9) on the chalkboard.

- Allow time for your child to write the four fact family number sentences on the chalkboard.

ASSESSMENT

Written Assessment

"Today I would like to see what you remember from what we have been practicing."

- Give your child **Written Assessment #18.**

- Read the directions for each problem. Allow time for your child to complete each problem before continuing.

- Correct the paper, noting your child's mistakes on the **Individual Recording Form**. Review the errors with your child.

THE LESSON

Adding Two-Digit Numbers with a Sum Greater Than 100

"Today you will learn how to add two-digit numbers that have a sum greater than 100."

"Tell me a two-digit number."

- Write the number on the chalkboard.

"What number could we add to this number so the sum will be greater than 100?"

- Write the suggested number below the first number.

"Why do you think the sum will be greater than 100?"

- Encourage your child to justify his/her answer.

"Let's try it to see."

- Do the addition with your child.

"Were you right?"

"Let's try another problem."

"Tell me a two-digit number."

- Write the number on the chalkboard.

  "Tell me a different two-digit number."

- Write the number below the first number.

  "Do you think the sum will be greater than 100?"

  "How do you know?"

- Encourage your child to justify his/her answer.

  "Let's try it to see."

- Do the addition with your child.

  "Were you right?"

- Repeat one or two more times.

  "Now you will have a chance to try some problems."

- Write the following numbers on the chalkboard:

  45   28   71   84   35   59   66   92

  "Pick two numbers you think will have a sum greater than 100."

- Write the numbers vertically on the chalkboard.

  "Why do you think the sum of these numbers will be greater than 100?"

- Ask your child to add the numbers.

  "What is the sum?"

  "Is the sum greater than 100?"

  "Pick two other numbers you think will have a sum greater than 100."

- Repeat the above steps using these two numbers.

- Repeat until as many numbers as possible are used.

CLASS PRACTICE

number fact practice

- Use the pink fact cards to practice the subtracting using the doubles plus one facts with your child.

- Give your child **Fact Sheet S 5.0**.

- Time your child for one minute.

- Correct the fact sheet with your child and record the score.

- Allow time for your child to complete the unfinished facts.

WRITTEN PRACTICE

- Complete **Worksheet 95A** with your child.
- Complete **Worksheet 95B** with your child later in the day.

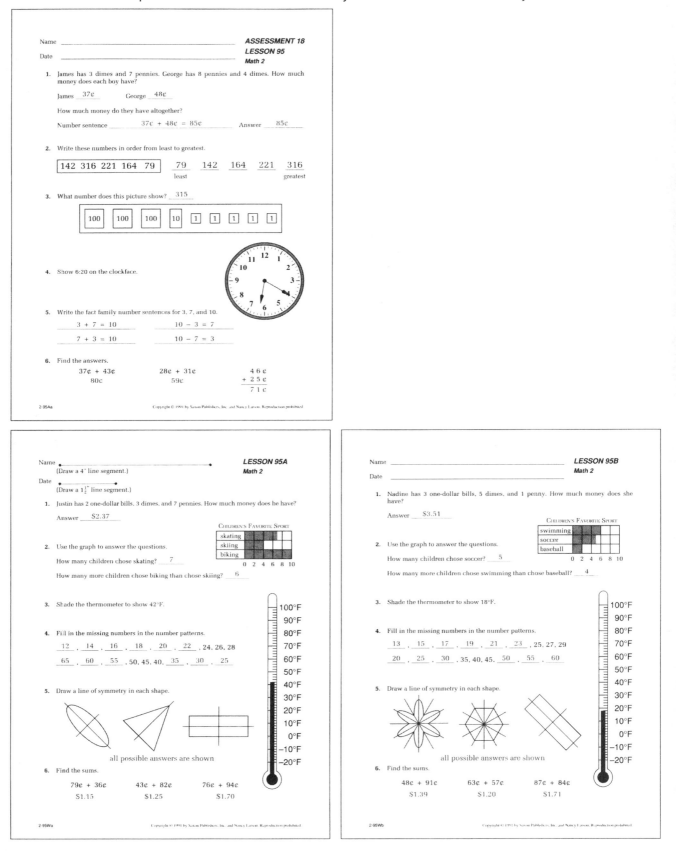

Name _____ **ASSESSMENT 18**
Date _____ **LESSON 95**
                                   Math 2

1. James has 3 dimes and 7 pennies. George has 8 pennies and 4 dimes. How much money does each boy have?

   James __37¢__   George __48¢__

   How much money do they have altogether?

   Number sentence _____37¢ + 48¢ = 85¢_____   Answer __85¢__

2. Write these numbers in order from least to greatest.

   | 142 316 221 164 79 | __79__   __142__   __164__   __221__   __316__
                             least   greatest

3. What number does this picture show? __315__

   | 100 | 100 | 100 | 10 | 1 | 1 | 1 | 1 | 1 |

4. Show 6:20 on the clockface.

5. Write the fact family number sentences for 3, 7, and 10.

   __3 + 7 = 10__   __10 − 3 = 7__
   __7 + 3 = 10__   __10 − 7 = 3__

6. Find the answers.

   37¢ + 43¢   28¢ + 31¢   4 6 ¢
      80¢   59¢   + 2 5 ¢
                                      7 1 ¢

2-95Aa   Copyright © 1991 by Saxon Publishers, Inc. and Nancy Larson. Reproduction prohibited.

Name •_____•   **LESSON 95A**
(Draw a 4" line segment.)   Math 2
Date _____
(Draw a 1½" line segment.)

1. Justin has 2 one-dollar bills, 3 dimes, and 7 pennies. How much money does he have?

   Answer __$2.37__

   CHILDREN'S FAVORITE SPORT

skating	
skiing	
biking	
   0 2 4 6 8 10

2. Use the graph to answer the questions.

   How many children chose skating? __7__

   How many more children chose biking than chose skiing? __6__

3. Shade the thermometer to show 42°F.

4. Fill in the missing numbers in the number patterns.

   __12__ , __14__ , __16__ , 18 , 20 , __22__ , 24, 26, 28

   __65__ , __60__ , __55__ , 50, 45, 40, __35__ , __30__ , 25

5. Draw a line of symmetry in each shape.

   all possible answers are shown

6. Find the sums.

   79¢ + 36¢   43¢ + 82¢   76¢ + 94¢
     $1.15 $1.25 $1.70

2-95Wa   Copyright © 1991 by Saxon Publishers, Inc. and Nancy Larson. Reproduction prohibited.

Name _____   **LESSON 95B**
                                    Math 2
Date _____

1. Nadine has 3 one-dollar bills, 5 dimes, and 1 penny. How much money does she have?

   Answer __$3.51__

   CHILDREN'S FAVORITE SPORT

swimming	
soccer	
baseball	
   0 2 4 6 8 10

2. Use the graph to answer the questions.

   How many children chose soccer? __5__

   How many more children chose swimming than chose baseball? __4__

3. Shade the thermometer to show 18°F.

4. Fill in the missing numbers in the number patterns.

   __13__ , __15__ , __17__ , __19__ , __21__ , __23__ , 25, 27, 29

   __20__ , __25__ , __30__ , 35, 40, 45, __50__ , __55__ , __60__

5. Draw a line of symmetry in each shape.

   all possible answers are shown

6. Find the sums.

   48¢ + 91¢   63¢ + 57¢   87¢ + 84¢
     $1.39 $1.20 $1.71

2-95Wb   Copyright © 1991 by Saxon Publishers, Inc. and Nancy Larson. Reproduction prohibited.

Lesson 96

finding one half of a set of an even number of objects

---

lesson preparation

materials

piece of paper

20 small hard candies (color tiles) in a bag

Master 2-96

Fact Sheet A 1-100

in the morning

• Write the following in the pattern box on the meeting strip:

> 4, x, 8, x, 12, x, ___, ___, ___, ___, ___, ___

Answer: 4, x, 8, x, 12, x, 16, x, 20, x, 24, x

• Write 94¢ on the meeting strip. Provide a cup of 10 dimes, a cup of 10 nickels, and a cup of 20 pennies.

THE MEETING

calendar

• Ask your child to write the date on the calendar and meeting strip.

• Ask your child the following two or three times a week:

> date _____ days ago, date _____ days from now
>
> day of the week _____ days ago, day of the week _____ days from now
>
> _____th month, month before, month after

• Record on the meeting strip a special event and the number of days until it occurs.

weather graph

• Ask your child to read and graph today's temperature to the nearest two degrees.

• Count by 10's and 2's to check the temperature on the graph.

- Ask your child to connect the dot for yesterday's temperature to the dot for today's temperature and compare the temperatures.

counting

- Ask your child to choose a number on the hundred number chart. Ask your child to add or subtract ten or one. Repeat 6–10 times. Ask your child to give directions for returning to the starting number.

"Let's use our patting and clapping pattern to help us count by 3's to 30."

- Repeat this several times.

- Do the following once or twice a week:

count by 10's to 400 and backward from 400 by 10's

count by 5's to 100 and backward from 50 by 5's

say the even numbers to 100 and backward from 50

say the odd numbers to 49 and backward from 49

graph questions

- You and your child each ask a question about any of the graphs.

patterning

- Ask your child to do the following:

identify the pattern (repeating, continuing, or both)

identify the numbers and letters to complete the pattern

read the pattern

money

- Ask your child to put the coins in the coin cup. Count the money in the coin cup together.

- Ask your child for another way to show that amount of money. Count these coins together to check the amount.

"Write this amount on the chalkboard using a dollar sign."

clock

- Set the clock to a five-minute interval.

- Ask the following:

"It's (morning/afternoon/evening). What time is it?"

time one hour ago

time one hour from now

- Ask your child to write the digital time on the meeting strip.

- Record on the meeting strip the time an activity will occur.

number of the day

- Write three number sentences for the number of the day on the meeting strip.

fact practice

- Write three fact family numbers (e.g., 2, 7, 9) on the chalkboard.

- Allow time for your child to write the four fact family number sentences on the chalkboard.

THE LESSON

Finding One Half of a Set of an Even Number of Objects

"Today you will learn how to divide a set of objects in half."

- Give your child a piece of paper.

  "You will use this paper as a work mat."

  "Fold the paper in half."

  "How many equal parts do you have?"   2

  "What is each part called?"   one half

  "Each half is the same size."

  "When we divide something in half, both parts must be the same."

- Give your child a bag of 20 hard candies (color tiles).

  "Put twelve candies on the table."

  "I would like to share these candies with you so we will each have half of the candies."

  "How can I do that?"

- Use a strategy your child suggests to divide the candies.

  "How many candies did you get?"   6

  "How many candies did I get?"   6

  "Is this fair?"

  "When we divide something in half, each of the two parts must be the same."

  "Half of twelve is six."

  "Some people use the doubles facts to help them find one half of a set of objects."

  "How do you think they do that?"

- Write "6 + 6 = 12" on the chalkboard.

"Now put ten candies on the table."

"Let's pretend that you are sharing these ten candies with me."

"Put half of the ten candies on one side of the fold of the paper and the other half of the candies on the other side of the fold."

- Allow time for your child to divide the candies.

"How many candies are on each half?"　5

"How do you know this is fair?"　both sides are the same

"Five doubled is ten."

"What is one half of ten?"　5

"We can write that like this."

- Write the following on the chalkboard:

one half of 10 is 5

"Now you will show how to share 14 candies with me."

"How many more candies will you need to take out of the bag?"　4

"Put half of the 14 candies on one side of the fold of your paper and the other half of the candies on the other side of the fold."

- Allow time for your child to divide the tiles.

"How many candies are on each half?"　7

"How do you know this is fair?"　both sides are the same

"Seven doubled is fourteen."

"What is one half of 14?"　7

- Write the following on the chalkboard:

one half of 14 is 7

- Repeat with 6, 16, and 20 candies.

"Now let's try some problems without using the candies."

- Give your child **Master 2-96**.

"I will tell you a number of candies."

"Pretend that you are sharing them with me."

"You will give me half and you will keep half."

"Draw a picture to show how many candies you and I will each have."

- Use the following amounts:

8 candies　　18 candies　　12 candies

- When your child finishes, continue.

"What is half of 8?"　4

"What is half of 18?"　9

"What is half of 12?" **6**

- Save **Master 2-96**, candies, and the folded paper for use in Lesson 97.

CLASS PRACTICE

number fact practice

- Give your child **Fact Sheet A 1-100.**

- Time your child for five minutes.

- Correct the fact sheet with your child and record the score.

- Allow time for your child to complete the unfinished facts.

WRITTEN PRACTICE

- Complete **Worksheet 96A** with your child.

- Complete **Worksheet 96B** with your child later in the day.

Name _____   **MASTER 2-96**
           Math 2

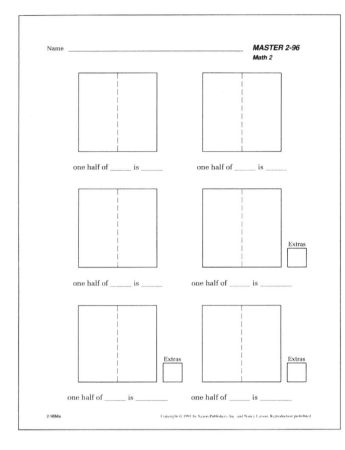

one half of _____ is _____   one half of _____ is _____

one half of _____ is _____   one half of _____ is _____

one half of _____ is _____   one half of _____ is _____

2-96Ma   Copyright © 1991 by Saxon Publishers, Inc. and Nancy Larson. Reproduction prohibited.

Name •_____•   **LESSON 96A**
(Draw a $4\frac{1}{2}$" line segment.)   Math 2
Date •_____•
(Draw a 3" line segment.)

1. The kindergarten children made a graph to show the shoes and sneakers they were wearing. They counted 10 shoes and 14 sneakers. Draw a picture to show the shoes and sneakers. Circle the pairs.

    How many children are in the kindergarten class? __12__

2. What number does this picture show? __215__

   | 100 | 100 | 10 | 1 | 1 | 1 | 1 | 1 |

   Write the number in expanded form. ___200 + 10 + 5___

3. I have 3 dimes, 4 nickels, and 6 pennies. Draw the coins. How much money is this? Write the amount two ways.

   __56¢__   __$0.56__   D D D N N N N   P P P P P P

4. Show how two children will share ten books equally.   ☐ = 1 book

   How many books will each child have? __5__

5. I can see the stars. What time is it?   one half of 10 is __5__

   Answer __1:20 a.m.__

2-96Wa   Copyright © 1991 by Saxon Publishers, Inc. and Nancy Larson. Reproduction prohibited.

Name _____   **LESSON 96B**
          Math 2
Date _____

1. The first grade children made a graph to show the shoes and sneakers they were wearing. They counted 8 shoes and 12 sneakers. Draw a picture to show the shoes and sneakers. Circle the pairs.

   How many children are in the first grade class? __10__

2. What number does this picture show? __327__

   | 100 | 100 | 100 | 10 | 10 | 1 | 1 | 1 | 1 | 1 | 1 | 1 |

   Write the number in expanded form. ___300 + 20 + 7___

3. I have 4 dimes, 3 nickels, and 6 pennies. Draw the coins. How much money is this? Write the amount two ways.

   __61¢__   __$0.61__   D D D D N N N   P P P P P P

4. Show how two children will share six books equally.   ☐ = 1 book

   How many books will each child have? __3__

5. The sun is shining. What time is it?   one half of 6 is __3__

   Answer __11:25 a.m.__

2-96Wb   Copyright © 1991 by Saxon Publishers, Inc. and Nancy Larson. Reproduction prohibited.

Lesson 97

finding one half of a set of an odd number of objects

THE MEETING

calendar

• Ask your child to write the date on the calendar and meeting strip.

• Ask your child the following two or three times a week:

date _____ days ago date _____ days from now

day of the week _____ days ago, day of the week _____ days from now

_____th month, month before, month after

• Record on the meeting strip a special event and the number of days until it occurs.

weather graph

• Ask your child to read and graph today's temperature to the nearest two degrees.

• Count by 10's and 2's to check the temperature on the graph.

• Ask your child to connect the dot for yesterday's temperature to the dot for today's temperature and compare the temperatures.

counting

• Ask your child to choose a number on the hundred number chart. Ask your child to add or subtract ten or one. Repeat 6–10 times. Ask your child to give directions for returning to the starting number.

"Let's use our patting and clapping pattern to help us count by 3's to 30."

• Repeat this several times.

• Do the following once or twice a week:

count by 10's to 400 and backward from 400 by 10's

count by 5's to 100 and backward from 50 by 5's

say the even numbers to 100 and backward from 50

say the odd numbers to 49 and backward from 49

graph questions

• You and your child each ask a question about any of the graphs.

patterning

• Ask your child to do the following:

identify the pattern (repeating, continuing, or both)

identify the numbers to complete the pattern

read the pattern

money

• Ask your child to put the coins in the coin cup. Count the money in the coin cup together.

• Ask your child for another way to show that amount of money. Count these coins together to check the amount.

"Write this amount on the chalkboard using a dollar sign."

clock

• Set the clock to a five-minute interval.

• Ask the following:

"It's (morning/afternoon/evening). What time is it?"

time one hour ago

time one hour from now

• Ask your child to write the digital time on the meeting strip.

• Record on the meeting strip the time an activity will occur.

number of the day

- Write three number sentences for the number of the day on the meeting strip.

fact practice

- Write three fact family numbers (e.g., 2, 7, 9) on the chalkboard.

- Allow time for your child to write the four fact family number sentences on the chalkboard.

THE LESSON

Finding One Half of a Set of an Odd Number of Objects

"Yesterday you used candies to help you find one half of a set of objects."

"What was half of ten?"   5

"What other numbers of candies did you find half of?"   6, 8, 12, 14, 16, 18, 20

- Write the numbers on the chalkboard.

"What types of numbers are these?"   even numbers

"How many candies were in each half?"

- Write the following on the chalkboard:

6	8	10	12	14	16	18	20
3 3	4 4	5 5	6 6	7 7	8 8	9 9	10 10

"Today you will learn how to find half of an odd number of objects."

- Give your child the folded paper from Lesson 96.

"We will use this paper as a work mat."

"How many equal parts do you have?"   2

"What is each part called?"   one half

"Each half is the same size."

"When we divide something in half, both parts must be the same."

- Give your child the bag of candies.

"Put seven candies on the table."

"I would like to share these candies with you so we will each have half of the candies."

"How can I do that?"

- Use a strategy your child suggests to divide the candies.

"How many candies did you get?"   3

"How many candies did I get?"   3

"Is this fair?"

"What about the extra candy?"   it is left over or remaining; because it is a hard candy, it cannot be divided in half

"When we divide something in half, each of the two parts must be the same."

"Half of seven is three with one left over or remaining."

"Put nine hard candies on the table."

"Let's pretend that you are sharing these candies with me."

"Put half of the nine candies on one side of the fold of the paper and the other half of the candies on the other side of the fold."

• Allow time for your child to divide the tiles.

"What happened?"   there is one tile left over

"How many candies are on each half?"   4

"How do you know this is fair?"   both sides are the same

"Four and four is eight and one more is nine."

"What is one half of nine?"   4 with one left over

"We can write that this way."

• Write the following on the chalkboard:

one half of 9 is 4 R1

"We write a capital R before the number of left over or remaining candies."

"Now we will show how to share 15 candies with a friend."

"How many more candies will you need to take out of the bag?"   6

"Put half of the 15 candies on one side of the fold of your paper and the other half of the candies on the other side of the fold."

• Allow time for your child to divide the tiles.

"What happened?"   there is one tile left over

"How many candies are on each half?"   7

"I will draw a picture to show that."

• Draw the following on the chalkboard:

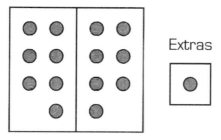

Extras

"Seven and seven is fourteen and one more is fifteen."

"What is one half of 15?" 7 with one candy left over

• Write the following on the chalkboard below the picture:

one half of 15 is 7 R1

"We write a capital R before the number of left over or remaining candies."

• Repeat with 5, 13, and 19 candies.

"Now let's try some problems without using the candies."

• Give your child **Master 2-96** from Lesson 96.

"I will tell you a number of candies."

"Pretend that you are sharing them with me."

"You will give me half and you will keep half."

"Draw a picture to show how many candies you and I will each have."

"If there is an extra candy left over, draw the extra candy in the box on the right."

• Use the following amounts:

11 candies   17 candies   9 candies

• When your child finishes, continue.

"What is half of 11?" 5 R1

"What is half of 17?" 8 R1

"What is half of 9?" 4 R1

"What numbers of candies did you find half of today?" 5, 7, 9, 11, 13, 15, 17, 19

"What types of numbers are these?" odd numbers

"What happened each time you found half of an odd number of candies?" there was one candy left over

Class Practice

number fact practice

- Use the pink fact cards to practice the subtracting using the doubles plus one facts with your child.
- Give your child **Fact Sheet S 5.0.**
- Time your child for one minute.
- Correct the fact sheet with your child and record the score.
- Allow time for your child to complete the unfinished facts.

Written Practice

- Complete **Worksheet 97A** with your child.
- Complete **Worksheet 97B** with your child later in the day.

Name _____ (Draw a $3\frac{1}{2}$" line segment.)

LESSON 97A
Math 2

Date _____ (Draw a 3" line segment.)

1. There are 27 children in Room 6, 22 children in Room 7, and 25 children in Room 8. How many children are in the three classes?

   Number sentence ___ $27 + 22 + 25 = 74$ children ___ Answer 74 children

2. Twelve children chose vanilla ice cream, nine children chose chocolate ice cream, and five children chose strawberry ice cream.

   Shade the graph to show the ice cream flavors the children chose.

   ICE CREAM FLAVORS
   chocolate
   vanilla
   strawberry
   0 2 4 6 8 10 12 14

   How many more children chose vanilla than chose chocolate? ___ 3

3. Write 806 in expanded form. ___ $800 + 6$

   Draw a picture to show this amount. (Use ☐ for 100, ☐ for 10, ☐ for 1.)

4. Show how 2 children will share 11 pennies equally.

   ○ = 1 penny

   How many pennies will each child have? ___ 5

   How many extra pennies are there? ___ 1

   one half of 11 is 5 R1

5. Find the answers.

   $7 \times 10 =$ ___ 70 $4 \times 10 =$ ___ 40 $41 - 10 =$ ___ 31

   $8 + 6 + 5 + 2 + 3 + 5 + 3 + 4 + 5 + 1 + 2 =$ ___ 44

Name _____

LESSON 97B
Math 2

Date _____

1. There are 18 children in Room 9, 23 children in Room 10, and 25 children in Room 11. How many children are in the three classes?

   Number sentence ___ $18 + 23 + 25 = 66$ children ___ Answer 66 children

2. Seven children chose a banana, ten children chose an apple, and eleven children chose an orange.

   Shade the graph to show the fruits the children chose.

   FRUITS
   orange
   banana
   apple
   0 2 4 6 8 10 12 14

   How many more children chose an orange than chose an apple? ___ 1

3. Write 350 in expanded form. ___ $300 + 50$

   Draw a picture to show this amount. (Use ☐ for 100, ☐ for 10, ☐ for 1.)

4. Show how 2 children will share 9 pennies equally.

   ○ = 1 penny

   How many pennies will each child have? ___ 4

   How many extra pennies are there? ___ 1

   one half of 9 is 4 R1

5. Find the answers.

   $3 \times 10 =$ ___ 30 $6 \times 10 =$ ___ 60 $24 - 10 =$ ___ 14

   $4 + 5 + 2 + 3 + 7 + 9 + 5 + 1 + 1 + 4 =$ ___ 41

L esson 98

counting quarters

THE MEETING

calendar

• Ask your child to write the date on the calendar and meeting strip.

• Ask your child the following two or three times a week:

date _____ days ago, date _____ days from now

day of the week _____ days ago, day of the week _____ days from now

_____th month, month before, month after

• Record on the meeting strip a special event and the number of days until it occurs.

weather graph

• Ask your child to read and graph today's temperature to the nearest two degrees.

• Count by 10's and 2's to check the temperature on the graph.

- Ask your child to connect the dot for yesterday's temperature to the dot for today's temperature and compare the temperatures.

counting

- Ask your child to choose a number on the hundred number chart. Ask your child to add or subtract ten or one. Repeat 6–10 times. Ask your child to give directions for returning to the starting number.

"Let's use our patting and clapping pattern to help us count by 3's to 30."

- Repeat this several times.

- Do the following once or twice a week:

    count by 10's to 400 and backward from 400 by 10's

    count by 5's to 100 and backward from 50 by 5's

    say the even numbers to 100 and backward from 50

    say the odd numbers to 49 and backward from 49

graph questions

- You and your child each ask a question about any of the graphs.

patterning

- Ask your child to do the following:

    identify the pattern (repeating, continuing, or both)

    identify the letters to complete the pattern

    read the pattern

money

- Ask your child to put the coins in the coin cup. Count the money in the coin cup together.

- Ask your child for another way to show that amount of money. Count these coins together to check the amount.

"Write this amount on the chalkboard using a dollar sign."

clock

- Set the clock to a five-minute interval.

- Ask the following:

    "It's (morning/afternoon/evening). What time is it?"

    time one hour ago

    time one hour from now

- Ask your child to write the digital time on the meeting strip.

- Record on the meeting strip the time an activity will occur.

number of the day

- Write three number sentences for the number of the day on the meeting strip.

fact practice

- Write three fact family numbers (e.g., 2, 7, 9) on the chalkboard.
- Allow time for your child to write the four fact family number sentences on the chalkboard.

THE LESSON

Counting Quarters

"Today you will learn how to count and show money amounts using a new coin."

- Hold a one-dollar bill in one hand and a quarter in the other. Show your child the amount of money in each hand.

"If I told you that you could have the money I have in one of my hands, which hand would you choose?"

"Why?"

- Hold up a quarter.

"What is the name of this coin?" quarter

"Today you will learn how to count quarters."

- Put 2 quarters in one hand and a dollar in the other.

"If I told you that you could have the money I have in one of my hands, which hand would you choose now?"

"Why?"

- Put 3 quarters in one hand and a dollar in the other.

"If I told you that you could have the money I have in one of my hands, which hand would you choose now?"

"Why?"

- Put 4 quarters in one hand and a dollar in the other.

"If I told you that you could have the money I have in one of my hands, which hand would you choose now?"

"Why?"

- Put 5 quarters in one hand and a dollar in the other hand.

"If I told you that you could have the money I have in one of my hands, which hand would you choose now?"

"Why?"

"When we count quarters, what do we count by?"   25's

"Let's count by 25's to one dollar as I hold up quarters."

- Hold up one quarter at a time as you count by 25's to 100 with your child.

"How many quarters are there in one dollar?"   4

- Write "4 quarters = $1.00" on the chalkboard.

"When we say one quarter, it means the same as one fourth."

- Draw a rectangle like the following on the chalkboard:

"How will we divide this rectangle into four equal parts?"

- Ask your child to divide the rectangle into fourths.

"What do we call each part?"   one fourth

"If this was a one-dollar bill, each part would be worth one quarter."

- Write "25¢" inside each quarter.

25¢	25¢
25¢	25¢

"What is the value of one fourth of a dollar?"   25¢

"How many quarters is that?"   1

"What is the value of two fourths of a dollar?"   50¢

"How many quarters is that?"   2

"What is the value of three fourths of a dollar?"   75¢

"How many quarters is that?"   3

"What is the value of four fourths of a dollar?" $1.00

"How many quarters is that?"   4

"We will make a counting strip to help us count by 25's."

"Let's count by 25's slowly as I write the numbers on a counting strip in the Meeting Book."

- Write the numbers from 0 to 300 as you count by 25's together.

300
275
250
225
200
175
150
125
100
75
50
25
0

- Give your child a cup of 12 quarters.

  "Put the quarters in a row on the table."

  "Point to each quarter as we count by 25's to see how much money this is." $3.00

- Count the quarters with your child.

  "Put three quarters in the cup."

  "How many quarters are in front of you now?" 9

  "Let's count by 25's to see how much money this is." $2.25

- Repeat, using 5, 8, 10, 2, 9, and 4 quarters.

  "Show $1.50 using quarters."

  "How many quarters did you use?" 6

- Repeat with 75¢, $2.25, and $1.75.

  "Put the quarters in the cup."

CLASS PRACTICE

number fact practice

- Use the tan, peach, lavender, green, pink, and blue fact cards to practice the subtraction facts with your child.

- Give your child **Fact Sheet S 6.4**.

- Time your child for one minute.

- Correct the fact sheet with your child and record the score.

- Allow time for your child to complete the unfinished facts.

WRITTEN PRACTICE

- Complete **Worksheet 98A** with your child.
- Complete **Worksheet 98B** with your child later in the day.

Name ●————————————————●   **LESSON 98A**
     (Draw a line segment 1" shorter than the date line segment.)   **Math 2**
Date ●————————————————————●

1. Ellen has 5 dimes and 8 pennies. Kay has 2 dimes and 14 pennies. How much money does each girl have?

   Ellen __58¢__   Kay __34¢__   Who has the most money? __Ellen__
   Write a number sentence to show how to find how much money the girls have altogether.

   Number sentence ____58¢ + 34¢ = 92¢____   Answer ___92¢___

   Show two ways to write how much money they have altogether. __92¢__ __$0.92__

2. Show how the children will share the markers equally.

   8 markers ▭   15 markers ▭

   one half of 8 is __4__   one half of 15 is __7 R1__

3. Two fact family number sentences are 6 + 7 = 13 and 13 − 6 = 7. Write the other two fact family number sentences.

   ____7 + 6 = 13____   ____13 − 7 = 6____

4. How much money is this? __$1.50__

5. Find the answers.

1 6 ¢	4 9 ¢	5 × 10 = __50__	10 − 8 = __2__
2 5 ¢	+ 3 3 ¢		
+ 3 9 ¢	————	15 − 7 = __8__	17 − 9 = __8__
————	8 2 ¢		
8 0 ¢			

2-98Ws   Copyright © 1994 by Saxon Publishers, Inc. and Nancy Larson. Reproduction prohibited.

Name _____   **LESSON 98B**
      Math 2
Date _____

1. Andrew has 2 dimes and 17 pennies. Danny has 3 dimes and 5 pennies. How much money does each boy have?

   Andrew __37¢__   Danny __35¢__   Who has the most money? __Andrew__
   Write a number sentence to show how to find how much money the boys have altogether.

   Number sentence ____37¢ + 35¢ = 72¢____   Answer ___72¢___

   Show two ways to write how much money they have altogether. __72¢__ __$0.72__

2. Show how the children will share the markers equally.

   12 markers ▭   5 markers ▭

   one half of 12 is __6__   one half of 5 is __2 R1__

3. Two fact family number sentences are 8 + 9 = 17 and 17 − 9 = 8. Write the other two fact family number sentences.

   ____9 + 8 = 17____   ____17 − 8 = 9____

4. How much money is this? __$1.25__

5. Find the answers.

2 3 ¢	6 5	9 × 10 = __90__	10 − 6 = __4__
4 8 ¢	+ 7 3		
+ 1 7 ¢	————	13 − 6 = __7__	17 − 8 = __9__
————	1 3 8		
8 8 ¢			

2-98Wb   Copyright © 1994 by Saxon Publishers, Inc. and Nancy Larson. Reproduction prohibited.

Lesson 99

multiplying by one
multiplying by one hundred

---

lesson preparation

materials

multiplication fact cards — beige
demonstration clock
Fact Sheet M 10.0

the night before

• Separate the beige multiplication fact cards.

in the morning

• Write the following in the pattern box on the meeting strip:

> 20, 18, 16, 14, ___, ___, ___, ___, ___, ___

Answer: 20, 18, 16, 14, 12, 10, 8, 6, 4, 2

• Write 54¢ on the meeting strip. Provide a cup of 6 quarters, a cup of 10 dimes, a cup of 10 nickels, and a cup of 20 pennies.

THE MEETING

calendar

• Ask your child to write the date on the calendar and meeting strip.

• Ask your child the following two or three times a week:

> date _____ days ago, date _____ days from now
>
> day of the week _____ days ago, day of the week _____ days from now
>
> _____th month, month before, month after

• Record on the meeting strip a special event and the number of days until it occurs.

weather graph

• Ask your child to read and graph today's temperature to the nearest two degrees.

- Count by 10's and 2's to check the temperature on the graph.
- Ask your child to connect the dot for yesterday's temperature to the dot for today's temperature and compare the temperatures.

counting

- Count by 25's to 300 and backward from 300 by 25's.
- Count by 3's to 30 and backward from 30 by 3's.
- Do the following once a week:

  count by 10's to 400 and backward from 400 by 10's

  count by 5's to 100 and backward from 50 by 5's

  say the even numbers to 100 and backward from 50

  say the odd numbers to 49 and backward from 49

graph questions

- You and your child each ask a question about any of the graphs.

patterning

- Ask your child to do the following:

  identify the pattern (repeating, continuing, or both)

  identify the numbers to complete the pattern

  read the pattern

money

- Ask your child to put the coins in the coin cup. Count the money in the coin cup together.
- Ask your child for another way to show that amount of money. Count these coins together to check the amount.

  "Write this amount on the chalkboard using a dollar sign."

clock

- Set the clock to a five-minute interval.
- Ask the following:

  "It's (morning/afternoon/evening). What time is it?"

  time one hour ago

  time one hour from now

- Ask your child to write the digital time on the meeting strip.
- Record on the meeting strip the time an activity will occur.

number of the day

- Write three number sentences for the number of the day on the meeting strip

fact practice

- Write three fact family numbers (e.g., 2, 7, 9) on the chalkboard.

- Allow time for your child to write the four fact family number sentences on the chalkboard.

THE LESSON

Multiplying by One
Multiplying by One Hundred

"A few days ago you learned how to multiply by ten."

- Write the following on the chalkboard:

  3 groups of 10 =   4 groups of 10 =   9 groups of 10 =

  "How much will each of these equal?"

  "Write these problems using the multiplication sign."

- Ask your child to write the multiplication problems on the chalkboard.

  "What happens when we multiply a number by ten?"   we add a zero

  "Today you will learn how to multiply by one and by one hundred."

- Write the following on the chalkboard:

  6 groups of 1 =   4 groups of 1 =   9 groups of 1 =

  "How much will each of these equal?"

  "Write these problems using the multiplication sign."

- Ask your child to write the multiplication problems on the chalkboard.

- Write the following on the chalkboard:

  $3 \times 1 =$ $7 \times 1 =$ $4 \times 1 =$ $10 \times 1 =$

  $5 \times 1 =$ $\times 1 =$ $8 \times 1 =$ $\times 1 =$

  "Read the first problem."

- It should be read "3 groups of 1 each."

  "What is the answer?"

- Fill in the answer.

- Repeat with each problem.

  "We can write these problems another way."

- Write the following on the chalkboard:

$$\begin{array}{ccccccccccc} 1 & 1 & 1 & 1 & 1 & 1 & 1 & 1 & 1 & 1 & 1 \\ \underline{\times\,0} & \underline{\times\,1} & \underline{\times\,2} & \underline{\times\,3} & \underline{\times\,4} & \underline{\times\,5} & \underline{\times\,6} & \underline{\times\,7} & \underline{\times\,8} & \underline{\times\,9} & \underline{\times\,10} \end{array}$$

"Which number do you think tells us the number of groups now?" bottom number

"What is each answer?"

- Fill in the answers.

"Do you see a pattern?"

"What happens when we multiply by one?"

- Write the following on the chalkboard:

6 groups of 100 =   4 groups of 100 =   9 groups of 100 =

"How much will each of these equal?"

"Write these problems using the multiplication sign."

- Ask your child to write the multiplication problems on the chalkboard.

- Write the following on the chalkboard:

$3 \times 100 =$ $7 \times 100 =$ $4 \times 100 =$ $10 \times 100 =$

$5 \times 100 =$ $1 \times 100 =$ $8 \times 100 =$ $0 \times 100 =$

"Read the first problem."

- It should be read "3 groups of 100 each."

"What is the answer?"

- Fill in the answer.

- Repeat with each problem.

"What happens when we multiply a number by 100?"   answer is the same number with two zeros

"Today I will give you multiplying by one fact cards."

- Give your child the beige multiplication fact cards.

"Read the problems and say the answers to yourself."

CLASS PRACTICE

number fact practice

- Give your child **Fact Sheet M 10.0.**

- Time your child for one minute.

- Correct the fact sheet with your child and record the score.
- Allow time for your child to complete the unfinished facts.

"On the back of the fact sheet, write the numbers from 485 to 523."

telling time to five-minute intervals

"Now you will practice showing time to five-minute intervals."

- Give your child a demonstration clock.
- Write "5:40" on the chalkboard.

"Show 5:40 on your clock."

"Where does the hour hand point?"

"Where does the minute hand point?"

- Repeat with 9:35, 2:25, 12:10, 2:05, and 6:55.

WRITTEN PRACTICE

- Complete **Worksheet 99A** with your child.
- Complete **Worksheet 99B** with your child later in the day.

Name _____ **LESSON 99A**
(Draw a 4" line segment.) **Math 2**
Date
(Draw a line segment 2" shorter than the line segment for your name.)

1. Fourteen children were in the gym. Twenty-five children from Room 2 joined them. Ten minutes later fifteen children from Room 6 arrived. How many children are in the gym now?

Number sentence __14 + 25 + 15 = 54 children__ Answer 54 children

2. Show how the children will share the markers equally.

9 markers ▭   14 markers ▭

one half of 9 is 4 R1   one half of 14 is 7

3. What would be a good estimate of the height of a desk in your classroom?

5 feet 1 inch   1 foot 2 inches   (2 feet 1 inch)   12 feet 2 inches

4. What fractional part of each shape is shaded? $\frac{1}{4}$ $\frac{2}{3}$

5. How much money is this? Write the amount two ways. 75¢ $0.75

6. Find the products.

$9 \times 1 =$ __9__ $7 \times 10 =$ __70__ $5 \times 100 =$ __500__ $14 \times 1 =$ __4__

7. Find the answers.

$88 + 34 = 122$ $37 + 92 = 129$ $\begin{array}{r}15\\23\\+47\\\hline85\end{array}$

2-99Wa

Name _____ **LESSON 99B**
Math 2
Date

1. Eighteen children were in the lunch room. Twenty-two children from Room 7 arrived. Five minutes later, thirteen children from Room 21 joined them. How many children are in the lunch room now?

Number sentence __18 + 22 + 13 = 53 children__ Answer 53 children

2. Show how 2 children will share the markers equally.

10 markers ▭   7 markers ▭

one half of 10 is 5   one half of 7 is 3 R1

3. What would be a good estimate of the height of a car?

(5 feet 1 inch)   1 foot 2 inches   2 feet 1 inch   12 feet 2 inches

4. What fractional part of each shape is shaded? $\frac{3}{4}$ $\frac{1}{2}$

5. How much money is this? Write the amount two ways. 50¢ $0.50

6. Find the products.

$2 \times 1 =$ __2__ $4 \times 10 =$ __40__ $3 \times 100 =$ __300__ $18 \times 1 =$ __18__

7. Find the answers.

$56 + 63 = 19$ $29 + 38 = 67$ $\begin{array}{r}25\\32\\+8\\\hline65\end{array}$

2-99Wb

Lesson 100

finding area using 1" color tiles

lesson preparation

materials

Written Assessment #19

Oral Assessment #10

20 color tiles

Master 2-100

10 dimes and 20 pennies

scrap paper

Fact Sheet S 6.4

in the morning

• Write the following in the pattern box on the meeting strip:

1, 2, 5, 6, 9, 10, ___, ___, ___, ___, ___, ___

Answer: 1, 2, 5, 6, 9, 10, 13, 14, 17, 18, 21, 22

• Write ⬚82¢ on the meeting strip. Provide a cup of 6 quarters, a cup of 10 dimes, a cup of 10 nickels, and a cup of 20 pennies.

THE MEETING

calendar

• Ask your child to write the date on the calendar and meeting strip.

• Ask your child the following two or three times a week:

date _____ days ago, date _____ days from now

day of the week _____ days ago, day of the week _____ days from now

_____th month, month before, month after

• Record on the meeting strip a special event and the number of days until it occurs.

weather graph

• Ask your child to read and graph today's temperature to the nearest two degrees.

- Count by 10's and 2's to check the temperature on the graph.
- Ask your child to connect the dot for yesterday's temperature to the dot for today's temperature and compare the temperatures.

counting

- Count by 25's to 300 and backward from 300 by 25's.
- Count by 3's to 30 and backward from 30 by 3's.
- Do the following once a week:

    count by 10's to 400 and backward from 400 by 10's

    count by 5's to 100 and backward from 50 by 5's

    say the even numbers to 100 and backward from 50

    say the odd numbers to 49 and backward from 49

graph questions

- You and your child each ask a question about any of the graphs.

patterning

- Ask your child to do the following:

    identify the pattern (repeating, continuing, or both)

    identify the numbers to complete the pattern

    read the pattern

money

- Ask your child to put the coins in the coin cup. Count the money in the coin cup together.
- Ask your child for another way to show that amount of money. Count these coins together to check the amount.

    "Write this amount on the chalkboard using a dollar sign."

clock

- Set the clock to a five-minute interval.
- Ask the following:

    "It's (morning/afternoon/evening). What time is it?"

    time one hour ago

    time one hour from now

- Ask your child to write the digital time on the meeting strip.
- Record on the meeting strip the time an activity will occur.

number of the day

- Write three number sentences for the number of the day on the meeting strip.

fact practice

- Write three fact family numbers (e.g., 2, 7, 9) on the chalkboard.
- Allow time for your child to write the four fact family number sentences on the chalkboard.

ASSESSMENT

Written Assessment

"Today I would like to see what you remember from what we have been practicing."

- Give your child **Written Assessment #19.**
- Read the directions for each problem. Allow time for your child to complete that problem before continuing.
- Correct the paper, noting your child's mistakes on the **Individual Recording Form.** Review the errors with your child.

Oral Assessment

- Record your child's response(s) to the oral interview questions on the interview sheet.

THE LESSON

Finding Area Using 1" Color Tiles

"Today you will learn how to use one-inch square color tiles to find the area of a shape."

- Give your child **Master 2-100.**

"You will use color tiles to cover the four shapes on this paper."

"Which shape do you think you will need the most color tiles to cover?"

"Which shape do you think you will need the fewest color tiles to cover?"

- Write your child's predictions on the chalkboard.

"Now you will use color tiles to cover each of the shapes on the paper."

- Give your child a bag of 20 color tiles.

"Use your 1" tiles to cover shape A."

- Allow time for your child to do this.

  "How many color tiles did you use to cover shape A?"

  "You will record the number of 1" color tiles you used to cover shape A on the chart at the bottom right-hand corner of the paper."

- Repeat with shapes B, C, and D.

  "Which shape has the largest area?"   *B*

  "How many color tiles did you use to cover it?"   *12*

  "Which shape has the smallest area?"   *A*

  "How many color tiles did you use to cover it?"   *8*

  "What is the area of shape C?"   *10 color tiles*

  "What is the area of shape D?"   *9 color tiles*

  "Turn over your paper."

  "Make a square using four color tiles."

- Allow time for your child to do this.

  "Make a rectangle using six color tiles."

- Allow time for your child to do this.

  "Can we make a square using six color tiles?"   *no*

- If desired, ask your child to make rectangles using 14 and 20 tiles.

CLASS PRACTICE

number fact practice

- Use the tan, peach, lavender, green, pink, and blue fact cards to practice the subtraction facts with your child.
- Give your child **Fact Sheet S 6.4**.
- Time your child for one minute.
- Correct the fact sheet with your child and record the score.
- Allow time for your child to complete the unfinished facts.

Name _____ **MASTER 2-100**

Math 2

Use 1" color tiles to cover each shape.

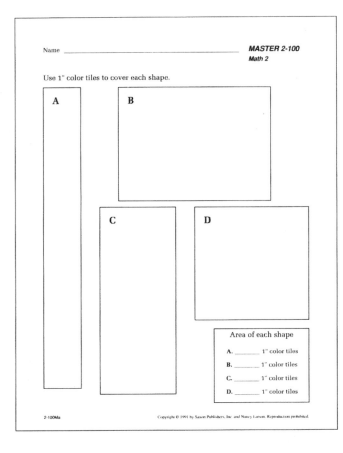

A

B

C

D

Area of each shape

A. _____ 1" color tiles

B. _____ 1" color tiles

C. _____ 1" color tiles

D. _____ 1" color tiles

2-100Ma   Copyright © 1991 by Saxon Publishers, Inc. and Nancy Larson. Reproduction prohibited.

Teacher _____

Date _____

MATH 2   LESSON 100

Oral Assessment # 10 Recording Form

Materials:
10 dimes
20 pennies
scrap paper

Students

	"Show me 36¢." • Hand the child 47¢ *"How much money did I give you?"* *"Show that using the fewest number of pennies."*		• Write 36¢ + 47¢ on a piece of scrap paper. *"Show how to find the answer for this example."* *"Explain each step."*		
	Counts Money	Trades Pennies for Dimes	Sets Up Example	Adds Correctly	Explains Steps (References Money)

2-PFw   Copyright © 1991 by Saxon Publishers, Inc. and Nancy Larson. Reproduction prohibited.

Name _____

Date _____

ASSESSMENT 19

LESSON 100

Math 2

1. Sam has 18 stickers, Cedric has 27 stickers, and Tony has 32 stickers. How many stickers do the three boys have altogether?

Number sentence ___18 + 27 + 32 = 77 stickers___   Answer _77 stickers_

2. Write these numbers in expanded form.

421 = ___400 + 20 + 1___   307 = ___300 + 7___

3. Find the answers.

$3 \times 10 =$ _30_ $8 \times 10 =$ _80_ $10 - 7 =$ _3_ $34 - 10 =$ _24_

$7 \times 10 =$ _70_ $5 \times 10 =$ _50_ $10 - 4 =$ _6_ $49 + 10 =$ _59_

4. Shade the graph to show that there are 20 children in Room 10.

How many children
are in Room 9? _16_

How many children
are in Room 8? _19_

NUMBER OF CHILDREN IN EACH CLASSROOM

Room 8

Room 9

Room 10

0 2 4 6 8 10 12 14 16 18 20 22

How many more children are in Room 10 than in Room 9? _4_

5. It's afternoon. What time is it? _4:45 p.m._

6. Find the answers.

7 4 ¢	2 3 ¢	3 9 ¢
+ 1 7 ¢	4 2 ¢	1 9 ¢
9 1 ¢	+ 1 5 ¢	+ 3 4 ¢
	8 0 ¢	9 2 ¢

2-100Aa

L esson 101

subtraction facts—last sixteen facts

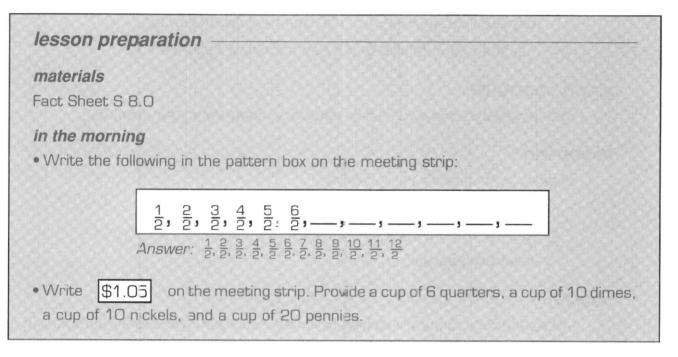

lesson preparation

materials

Fact Sheet S 8.0

in the morning

• Write the following in the pattern box on the meeting strip:

$$\frac{1}{2}, \quad \frac{2}{2}, \quad \frac{3}{2}, \quad \frac{4}{2}, \quad \frac{5}{2}, \quad \frac{6}{2}, \quad \text{\_\_\_}, \quad \text{\_\_\_}, \quad \text{\_\_\_}, \quad \text{\_\_\_}, \quad \text{\_\_\_}, \quad \text{\_\_\_}$$

Answer: $\frac{1}{2}, \frac{2}{2}, \frac{3}{2}, \frac{4}{2}, \frac{5}{2}, \frac{6}{2}, \frac{7}{2}, \frac{8}{2}, \frac{9}{2}, \frac{10}{2}, \frac{11}{2}, \frac{12}{2}$

• Write $\boxed{\$1.05}$ on the meeting strip. Provide a cup of 6 quarters, a cup of 10 dimes, a cup of 10 nickels, and a cup of 20 pennies.

THE MEETING

calendar

• Ask your child to write the date on the calendar and meeting strip.

• Ask your child the following two or three times a week:

date ____ days ago, date ____ days from now

day of the week ____ days ago, day of the week ____ days from now

____th month, month before, month after

• Record on the meeting strip a special event and the number of days until it occurs.

weather graph

• Ask your child to read and graph today's temperature to the nearest two degrees.

• Count by 10's and 2's to check the temperature on the graph.

• Ask your child to connect the dot for yesterday's temperature to the dot for today's temperature and compare the temperatures.

counting

• Count by 25's to 300 and backward from 300 by 25's.

- Count by 3's to 30 and backward from 30 by 3's.
- Do the following once a week:

    count by 10's to 400 and backward from 400 by 10's

    count by 5's to 100 and backward from 50 by 5's

    say the even numbers to 100 and backward from 50

    say the odd numbers to 49 and backward from 49

graph questions

- You and your child each ask a question about any of the graphs.

patterning

- Ask your child to do the following:

    identify the pattern (repeating, continuing, or both)

    identify the numbers to complete the pattern

    read the pattern

money

- Ask your child to put the coins in the coin cup. Count the money in the coin cup together.
- Ask your child for another way to show that amount of money. Count these coins together to check the amount.

clock

- Set the clock to a five-minute interval.
- Ask the following:

    "It's (morning/afternoon/evening). What time is it?"

    time one hour ago

    time one hour from now

- Ask your child to write the digital time on the meeting strip.
- Record on the meeting strip the time an activity will occur.

number of the day

- Write three number sentences for the number of the day on the meeting strip.

fact practice

- Write three fact family numbers (e.g., 2, 7, 9) on the chalkboard.
- Allow time for your child to write the four fact family number sentences on the chalkboard.

THE LESSON

Subtraction Facts—Last Sixteen Facts

"Today you will learn the last sixteen subtraction facts."

"We will use the addition facts we call the oddballs to help us."

- Write the following on the chalkboard:

$$
\begin{array}{cccccccc}
5 & 6 & 8 & 3 & 7 & 8 & 7 & 8 \\
+3 & +3 & +3 & +5 & +4 & +4 & +5 & +5 \\
\end{array}
$$

"What are these answers?"

- Record the answers on the chalkboard.

"We can use these addition facts to help us learn the last sixteen subtraction facts."

"What two fact family subtraction facts can we write for 5 + 3 = 8?"

- Write the following on the chalkboard:

$$
\begin{array}{cc}
8 & 8 \\
-3 & -5 \\
\end{array}
$$

"What are the answers?"

- Write the answers below the problems.

"We can check subtraction answers by adding up."

- Check the answers with your child.

"What two fact family subtraction facts can we write for 6 + 3 = 9?"

- Write the following on the chalkboard:

$$
\begin{array}{cc}
9 & 9 \\
-3 & -6 \\
\end{array}
$$

"What are the answers?"

- Write the answers below the problems.

"We can check subtraction answers by adding up."

- Repeat with each of the other addition problems.

"Some people think it is easier to remember subtraction answers by thinking of addition."

"Let's practice our new subtraction facts together."

- Erase the chalkboard answers.

"I will point to a number fact."

"Say the answer as quickly as possible."

• Point to the number facts one at a time.

CLASS PRACTICE

number fact practice

- • Use the white fact cards to practice the last sixteen subtraction facts with your child.
- • Give your child **Fact Sheet S 8.0.**
- • Time your child for one minute.
- • Correct the fact sheet with your child and record the score.
- • Allow time for your child to complete the unfinished facts.

WRITTEN PRACTICE

- • Complete **Worksheet 101A** with your child.
- • Complete **Worksheet 101B** with your child later in the day.

Name _____   **LESSON 101A**
    (Draw a 3½" line segment.)   **Math 2**
Date _____
    (Draw a 2½" line segment.)

1. Catherine has 164 baseball cards. Steve has 247 baseball cards. Susan has 187 baseball cards, and Carl has 128 baseball cards.

    Who has the most cards? __Steve__   Who has the fewest cards? __Carl__

    Write the names of the children in order from the one who has the most cards to the one who has the fewest cards.

    __Steve__   __Susan__   __Catherine__   __Carl__

2. Draw a picture to show three hundred twenty-one. (Use ☐ for 100, ☐ for 10, and ☐ for 1.)

    [☐ ☐ ☐ ☐ ☐]

    Write this number in expanded form. __300 + 20 + 1__
    Circle the number that shows three hundred twenty-one.

    3002001   (321)   30021   3021

3. I have 9 quarters. Draw the quarters.

    Ⓠ Ⓠ Ⓠ Ⓠ Ⓠ Ⓠ Ⓠ Ⓠ Ⓠ

    How much money do I have? __$2.25__

4. Shade the thermometer to show 24°F.

5. It's morning. What time is it? __10:40 a.m.__

6. Find the products.

    6 × 10 = __60__   8 × 1 = __8__   7 × 100 = __700__

2-101Wa   Copyright © 1994 by Saxon Publishers, Inc. and Nancy Larson. Reproduction prohibited.

Name _____   **LESSON 101B**
                                 Math 2
Date _____

1. Michael has 259 stamps in his stamp collection. Vera has 145 stamps, Crystal has 232 stamps, and Selby has 95 stamps.

    Who has the most stamps? __Michael__   Who has the fewest stamps? __Selby__

    Write the names of the children in order from the one who has the most stamps to the one who has the fewest stamps.

    __Michael__   __Crystal__   __Vera__   __Selby__

2. Draw a picture to show two hundred fifty-three. (Use ☐ for 100, ☐ for 10, and ☐ for 1.)

    [☐ ☐ ☐ ☐ ☐ ☐ ☐ ☐]

    Write this number in expanded form. __200 + 50 + 3__
    Circle the number that shows two hundred fifty-three.

    20053   2053   200503   (253)

3. I have 7 quarters. Draw the quarters.

    Ⓠ Ⓠ Ⓠ Ⓠ Ⓠ Ⓠ Ⓠ

    How much money do I have? __$1.75__

4. Shade the thermometer to show 86°F.

5. It's afternoon. What time is it? __5:25 p.m.__

6. Find the products.

    9 × 10 = __90__   3 × 1 = __3__   6 × 100 = __600__

2-101Wb   Copyright © 1994 by Saxon Publishers, Inc. and Nancy Larson. Reproduction prohibited.

Lesson 102

using comparison symbols (>, <, and =)

THE MEETING

calendar

• Ask your child to write the date on the calendar and meeting strip.

• Ask your child the following two or three times a week:

date ____ days ago, date ____ days from now

day of the week ____ days ago, day of the week ____ days from now

____th month, month before, month after

• Record on the meeting strip a special event and the number of days until it occurs.

weather graph

• Ask your child to read and graph today's temperature to the nearest two degrees.

- Count by 10's and 2's to check the temperature on the graph.
- Ask your child to connect the dot for yesterday's temperature to the dot for today's temperature and compare the temperatures.

counting

- Count by 25's to 300 and backward from 300 by 25's.
- Count by 3's to 30 and backward from 30 by 3's.
- Do the following once a week:

   count by 10's to 400 and backward from 400 by 10's

   count by 5's to 100 and backward from 50 by 5's

   say the even numbers to 100 and backward from 50

   say the odd numbers to 49 and backward from 49

graph questions

- You and your child each ask a question about any of the graphs.

patterning

- Ask your child to do the following:

   identify the pattern (repeating, continuing, or both)

   identify the numbers to complete the pattern

   read the pattern

money

- Ask your child to put the coins in the coin cup. Count the money in the coin cup together.
- Ask your child for another way to show that amount of money. Count these coins together to check the amount.

clock

- Set the clock to a five-minute interval.
- Ask the following:

   "It's (morning/afternoon/evening). What time is it?"

   time one hour ago

   time one hour from now

- Ask your child to write the digital time on the meeting strip.
- Record on the meeting strip the time an activity will occur.

number of the day

- Write three number sentences for the number of the day on the meeting strip.

fact practice

- Write three fact family numbers (e.g., 2, 7, 9) on the chalkboard.
- Allow time for your child to write the four fact family number sentences on the chalkboard.

THE LESSON

Using Comparison Symbols (>, <, and =)

"We have been using the equal sign for our number sentences."

"Today you will learn how to use a new symbol that mathematicians use for number sentences."

- Give your child a piece of paper and a bag of 10 color tiles.

"Fold your paper in half."

"Put half of your tiles in a pile on the left side and the other half of the tiles in a pile on the right side."

- Allow time for your child to do this.

"How many tiles did you put on each side of your paper?"

- Record the numbers on the chalkboard.

"Which pile of tiles is taller?" they are the same

"We can say that five equals five."

"Mathematicians write that like this."

- Write "5 = 5" on the chalkboard.

"Move some tiles to the pile on the left side of the paper."

- Allow time for your child to do this.

"How many tiles are on the left side?"

"How many tiles are on the right side?"

"Which pile of tiles is taller?" pile on the left

"We can say that the pile on the left has a greater number of tiles than the pile on the right."

"Mathematicians write that like this."

- Write "(#) > (#)" on the chalkboard.
- Point to the numbers and the symbol as you say the following:

"We can say that (_____) is greater than (_____)."

"The open or bigger side of the symbol is next to the greater number."

"Some people think that this symbol looks like an alligator's mouth."

"The open mouth of the alligator eats the greater number."

551

"Now make the pile of tiles on the right side taller."

- Allow time for your child to do this.

"How many tiles are on the left side?"

"How many tiles are on the right side?"

"Which pile of tiles is taller?"   pile on the right

"We can say that the pile on the right has a greater number of tiles than the pile on the left."

"Mathematicians write that like this."

- Write "(#) < (#)" on the chalkboard.

- Point to the numbers and the symbol as you say the following (begin on the right):

"We can say that (_____) is greater than (_____)."

"The open or bigger side of the symbol is next to the greater number."

"We can remember that by remembering that the open mouth eats the greater number."

- Give your child the square with the > symbol written on one side and the = sign on the other side.

"Use some of your tiles and the equal sign to make a true number sentence."

"How many tiles did you use on each side?"

"Now use your tiles and the greater than symbol to make a true number sentence."

"How many tiles did you use on each side?"

- Repeat several times.

- Record the combinations on the chalkboard.

"What do you notice about the numbers next to the open side of the symbol?"   they are the greater numbers

"Now turn your greater than symbol around."

"Use your tiles to make a true number sentence."

"How many tiles did you use on each side?"

- Repeat several times.

- Record the combinations on the chalkboard.

"What do you notice about the numbers now?"   the numbers on the right are greater

"Let's practice drawing this new symbol."

"When we draw a greater than symbol, we draw an arrow without a stick."

- Demonstrate on the chalkboard.

  "Draw five 'greater than' symbols on your paper."

- Allow time for your child to do this.

  "Now we will draw a 'greater than' symbol that points the other way."

- Demonstrate on the chalkboard.

  "Draw five symbols on your paper."

- Write the following on the chalkboard:

$$2 \; \square \; 5$$

  "Write the correct symbol between the numbers."

- Ask your child to write the symbol on the chalkboard.

- Repeat with the following:

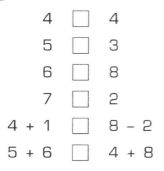

$$4 \; \square \; 4$$
$$5 \; \square \; 3$$
$$6 \; \square \; 8$$
$$7 \; \square \; 2$$
$$4 + 1 \; \square \; 8 - 2$$
$$5 + 6 \; \square \; 4 + 8$$

- Write the following on the chalkboard:

$$7 > \bigcirc$$

  "Fill in a number that will make the number sentence true."

- Repeat with the following:

$$\bigcirc < 4$$
$$5 > \bigcirc$$
$$\bigcirc > 8$$
$$7 < \bigcirc$$
$$4 + \bigcirc = 8 + 4$$

Class Practice

number fact practice

- Use the white fact cards to practice the last sixteen subtraction facts with your child.

- Give your child **Fact Sheet S 8.0**.

- Time your child for one minute.

- Correct the fact sheet with your child and record the score.

- Allow time for your child to complete the unfinished facts.

WRITTEN PRACTICE

- Complete **Worksheet 102A** with your child.

- Complete **Worksheet 102B** with your child later in the day.

Name ●━━━━━━━━━━━━━●
(Draw a $2\frac{1}{2}$" line segment.)

LESSON 102A
Math 2

Date ━━━━━━━━━━━━━━●
(Draw a line segment 1" longer than the line segment for your name.)

1. The cost of a small notebook is 67¢. The cost of a pencil is 23¢. How much money does Stephen need to buy a small notebook and a pencil?

Number sentence ____67¢ + 23¢ = 90¢____ Answer ___90¢___

Draw the coins he could use to buy the notebook and pencil.

(Q) (Q) (Q) (D) (N)

2. Show how to divide the pennies in half.

18 pennies Ⓟ

Extras

17 pennies Ⓟ

Extras Ⓟ

one half of 18 is ___9___

one half of 17 is _8 R1_

3. Fill in the correct comparison symbol (>, <, or =).

7 [<] 9   17 [>] 6   100 [<] 200

4. Find the answers.

```
  84        25
+ 34      + 17
-----     -----
 118        42
```

8 × 100 = _800_

2 × 10 = _20_

```
 51
 19
+23
----
 93
```

```
 3
 7
 2
 4
 8
+5
---
29
```

5. Fill in the missing numbers in the number patterns.

296, 297, 298, _299_, _300_, _301_, _302_, _303_, _304_

_130_, _140_, _150_, 160, 170, 180, _190_, _200_, _210_

2-102Wa

Name _____

LESSON 102B
Math 2

Date _____

1. The cost of a marker is 74¢. The cost of a small eraser is 19¢. How much money does DeAnna need to buy a marker and a small eraser?

Number sentence ____74¢ + 19¢ = 93¢____ Answer ___93¢___

Draw the coins she could use to buy the marker and eraser.

(Q) (Q) (Q) (D) (N) (P) (P) (P)

2. Show how to divide the pennies in half.

16 pennies Ⓟ

Extras

11 pennies Ⓟ

Extras Ⓟ

one half of 16 is ___8___

one half of 11 is _5 R1_

3. Fill in the correct comparison symbol (>, <, or =).

12 [>] 7   15 [<] 18   230 [>] 179

4. Find the answers.

```
  74        36
+ 85      + 39
-----     -----
 159        75
```

7 × 100 = _700_

3 × 10 = _30_

```
 46
 23
+16
----
 85
```

```
 3
 1
 7
 8
 6
+2
---
27
```

5. Fill in the missing numbers in the number patterns.

305, 304, 303, _302_, _301_, _300_, _299_, _298_, _297_

_70_, _80_, _90_, _100_, _110_, _120_, 130, 140, 150

2-102Wb

Lesson 103

identifying geometric solids (cone, cube, sphere, cylinder, rectangular solid, and pyramid)

---

lesson preparation

materials

set of 6 geometric solids (cone, cube, sphere, cylinder, rectangular solid, pyramid) or objects from your home with these shapes

6 pieces of 1" × 4" construction paper

tape

Master 2-103

Fact Sheet S 5.4

in the morning

• Write the following in the pattern box on the meeting strip:

| _____, _____, _____, _____, _____, _____, 197, 196, 195 |

Answer: 203, 202, 201, 200, 199, 198, 197, 196, 195

• Write $0.97   on the meeting strip. Provide a cup of 6 quarters, a cup of 10 dimes, a cup of 10 nickels, and a cup of 20 pennies.

THE MEETING

calendar

- Ask your child to write the date on the calendar and meeting strip.

- Ask your child the following two or three times a week:

    date _____ days ago, date _____ days from now

    day of the week _____ days ago, day of the week _____ days from now

    _____th month, month before, month after

- Record on the meeting strip a special event and the number of days until it occurs.

weather graph

- Ask your child to read and graph today's temperature to the nearest two degrees.

- Count by 10's and 2's to check the temperature on the graph.
- Ask your child to connect the dot for yesterday's temperature to the dot for today's temperature and compare the temperatures.

counting

- Count by 25's to 300 and backward from 300 by 25's.
- Count by 3's to 30 and backward from 30 by 3's.
- Do the following once a week:

    count by 10's to 400 and backward from 400 by 10's

    count by 5's to 100 and backward from 50 by 5's

    say the even numbers to 100 and backward from 50

    say the odd numbers to 49 and backward from 49

graph questions

- You and your child each ask a question about any of the graphs.

patterning

- Ask your child to do the following:

    identify the pattern (repeating, continuing, or both)

    identify the numbers to complete the pattern

    read the pattern

money

- Ask your child to put the coins in the coin cup. Count the money in the coin cup together.
- Ask your child for another way to show that amount of money. Count these coins together to check the amount.

clock

- Set the clock to a five-minute interval.
- Ask the following:

    "It's (morning/afternoon/evening). What time is it?"

    time one hour ago

    time one hour from now

- Ask your child to write the digital time on the meeting strip.
- Record on the meeting strip the time an activity will occur.

number of the day

- Write three number sentences for the number of the day on the meeting strip.

fact practice

- Write three fact family numbers (e.g., 2, 7, 9) on the chalkboard.

- Allow time for your child to write the four fact family number sentences on the chalkboard.

THE LESSON

Identifying Geometric Solids (Cone, Cube, Sphere, Cylinder, Rectangular Solid, and Pyramid)

- Display models of the 6 geometric solids (cone, cube, sphere, cylinder, rectangular solid, and pyramid).

  "Each of these geometric solids has a special name."

  "Today you will learn their special names."

  "Do you know the special name for any of these solids?"

- If your child correctly identifies a solid, write the name on a tag and tape it to the geometric solid.

- Identify the shapes your child does not know and write the names on tags.

- Tape the tags to the geometric solids.

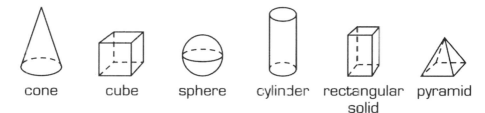

| cone | cube | sphere | cylinder | rectangular solid | pyramid |

  "Find two geometric solids that are the same in some way."

  "How are they the same?"

  "How are they different?"

- Repeat several times.

  "Now you will try to find objects in our house that have the same shapes as these geometric solids."

  "Try to find at least one object for each solid."

  "Which solid do you think will be the easiest to find?"

  "Which solid do you think will be the hardest to find?"

- Give your child **Master 2-103**.

  "Write the name or draw a picture of the objects you find on this recording sheet."

- Assist your child, if necessary.

"What did you find that has the shape of a cone? . . . cube? . . . sphere? . . . cylinder? . . . rectangular solid? . . . pyramid?"

"Which solid shape was the easiest to find?"

"Which solid shape was the hardest to find?"

CLASS PRACTICE

number fact practice

- Use the fact cards to practice all the subtraction facts with your child.
- Give your child **Fact Sheet S 5.4.**
- Time your child for one minute.
- Correct the fact sheet with your child and record the score.
- Allow time for your child to complete the unfinished facts.

WRITTEN PRACTICE

- Complete **Worksheet 103A** with your child.
- Complete **Worksheet 103B** with your child later in the day.

Name _____ MASTER 2-103
Math 2

Cone	Cube	Sphere
Cylinder	Rectangular Solid	Pyramid

2-103Ma   Copyright © 1991 by Saxon Publishers, Inc. and Nancy Larson. Reproduction prohibited.

Name _____ ● **LESSON 103A**
[Draw a 4" line segment.]   Math 2
Date _____
[Draw a 1½" line segment.]

1. There were a dozen pairs of shoes in Roseann's closet. Her dog chewed one shoe. Draw a picture of the unchewed shoes Roseann has left.

How many unchewed shoes does Roseann have left? __23__

2. About how long is a new pencil?

   1 foot 2 inches   3 inches   18 inches   (7 inches)

3. I have 6 dimes, 4 nickels, and 7 pennies. Draw the coins.

   D D D D D D N N
   N N P P P P P P P

   How much money do I have? __87¢__

4. Fill in a number to make each number sentence true. some answers may vary

   [ ] > 7   6 + 4 = [4] + 6   5 > [ ]

5. How many cubes did the children find? __13__
   How many spheres did the children find? __10__
   How many more cubes than spheres did they find? __3__

   Geometric Solids Children Found
   0 2 4 6 8 10 12 14 16

6. Circle the name of this geometric solid.

   cone   rectangular solid   (pyramid)   cylinder

2-103Wa   Copyright © 1991 by Saxon Publishers, Inc. and Nancy Larson. Reproduction prohibited.

Name _____ **LESSON 103B**
Math 2
Date _____

1. Martha had a half dozen pairs of earrings. She lost one earring. Draw a picture of the earrings she has left.

   How many earrings does she have left? __11__

2. About how long is a telephone?

   2 feet   (8 inches)   2 pounds   3 inches

3. I have 2 dimes, 6 nickels, and 3 pennies. Draw the coins.

   D D N N N N N N
   P P P P P P P P

   How much money do I have? __53¢__

4. Fill in a number to make each number sentence true. some answers may vary

   4 > [ ]   8 + 9 = 9 + [8]   [ ] > 8

5. How many cones did the children find? __11__
   How many pyramids did the children find? __6__
   How many more cones than pyramids did they find? __5__

   Geometric Solids Children Found
   0 2 4 6 8 10 12 14 16

6. Circle the name of this geometric solid.

   cone   rectangular solid   pyramid   (cylinder)

2-103Wb   Copyright © 1991 by Saxon Publishers, Inc. and Nancy Larson. Reproduction prohibited.

Lesson 104

adding three two-digit numbers with a sum greater than 100

---

lesson preparation

materials

twenty-five 3" × 5" cards or pieces of construction paper

piece of paper

Master 2-104

Fact Sheet A 2-100

the night before

• Write the following money amounts on the 3" × 5" cards using a marker:

12¢	15¢	17¢	18¢	23¢	24¢	26¢	27¢	29¢	31¢
35¢	36¢	38¢	39¢	42¢	43¢	47¢	49¢	52¢	54¢
56¢	57¢	60¢	64¢	66¢					

• Mix the cards.

in the morning

• Write the following in the pattern box on the meeting strip:

> 2, B, 4, D, 6, F, ___, ___, ___, ___, ___, ___

Answer: 2, B, 4, D, 6, F, 8, H, 10, J, 12, L

• Write $1.35 on the meeting strip. Provide a cup of 6 quarters, a cup of 10 dimes, a cup of 10 nickels, and a cup of 20 pennies.

THE MEETING

calendar

• Ask your child to write the date on the calendar and meeting strip.

• Ask your child the following two or three times a week:

date _____ days ago, date _____ days from now

day of the week _____ days ago, day of the week _____ days from now

_____th month, month before, month after

- Record on the meeting strip a special event and the number of days until it occurs.

weather graph

- Ask your child to read and graph today's temperature to the nearest two degrees.
- Count by 10's and 2's to check the temperature on the graph.
- Ask your child to connect the dot for yesterday's temperature to the dot for today's temperature and compare the temperatures.

counting

- Count by 25's to 300 and backward from 300 by 25's.
- Count by 3's to 30 and backward from 30 by 3's.
- Do the following once a week:

    count by 10's to 400 and backward from 400 by 10's

    count by 5's to 100 and backward from 50 by 5's

    say the even numbers to 100 and backward from 50

    say the odd numbers to 49 and backward from 49

graph questions

- You and your child each ask a question about any of the graphs.

patterning

- Ask your child to do the following:

    identify the pattern (repeating, continuing, or both)

    identify the numbers and letters to complete the pattern

    read the pattern

money

- Ask your child to put the coins in the coin cup. Count the money in the coin cup together.
- Ask your child for another way to show that amount of money. Count these coins together to check the amount.

clock

- Set the clock to a five-minute interval.
- Ask the following:

    "It's (morning/afternoon/evening). What time is it?"

    time one hour ago

    time one hour from now

- Ask your child to write the digital time on the meeting strip.
- Record on the meeting strip the time an activity will occur.

number of the day

- Write three number sentences for the number of the day on the meeting strip.

fact practice

- Write three fact family numbers (e.g., 2, 7, 9) on the chalkboard.
- Allow time for your child to write the four fact family number sentences on the chalkboard.

THE LESSON

Adding Three Two-Digit Numbers with a Sum Greater Than 100

"Today you will learn how to add three two-digit numbers."

- Write the following on the chalkboard:

37¢   25¢   18¢

"How can we find out how much money this is altogether?"

"How should we write these numbers so we can add them together?"

- Write the following on the chalkboard:

37¢
25¢
+ 18¢

"What will we add first?"   pennies

"Let's start at the top of the column and add the number of pennies."

"What is seven pennies plus five pennies?"

"What is twelve pennies plus eight pennies?"

"Will we write 20 pennies below the pennies' column?"   no

"Why not?"   because if we have more than 10 pennies, we trade them for dimes

"Can we trade 20 pennies for dimes?"   yes

"How many dimes will we have?"   2

"Do we have any pennies left over?"   no

"We show that like this."

- Write the following on the chalkboard:

$$37¢$$
$$25¢$$
$$+ \ 18¢$$
$$\overline{0}$$

"Let's start at the top of the tens' column and add the number of dimes."

"What is three dimes plus two dimes plus one dime plus two dimes?"   *8 dimes*

- Write the 8 below the 10s' column.

"Now we will play a game called 'Add It Up.' "

- Give your child **Master 2-104**.

"I have written some two-digit numbers on cards."

- Show your child the cards.

"I will mix the cards and give you three cards and I will take three cards."

"You will write your three numbers in the space marked 'Round 1' on your recording sheet."

"Then you will add your numbers together as we did on the board."

"I will add my three numbers on a piece of paper."

"When we finish, we will check our answers."

"If your answer is correct, we will pretend that you will win that amount of money."

"If my answer is correct, we will pretend that I will win that amount of money."

"We will play three rounds."

"The person with the most money at the end of the game is the winner."

- Play the game.
- Check the answers together.

"Now we will add the amounts we each won at the bottom of our paper."

"How much money did you win?"

"How much money did I win?"

"Who won the most money?"

Class Practice

number fact practice

- Give your child **Fact Sheet A 2-100**.

- Time your child for five minutes.

- Correct the fact sheet with your child and record the score.

- Allow time for your child to complete the unfinished facts.

Written Practice

- Complete **Worksheet 104A** with your child.

- Complete **Worksheet 104B** with your child later in the day.

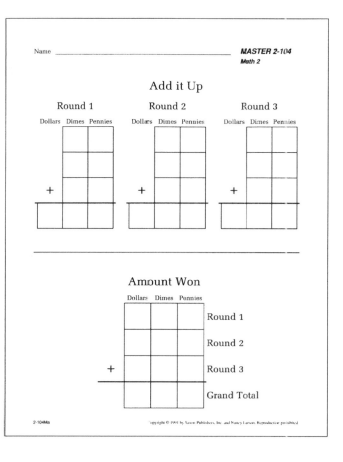

Add it Up

Round 1   Round 2   Round 3

Dollars | Dimes | Pennies

Amount Won

Dollars | Dimes | Pennies

Round 1
Round 2
Round 3
Grand Total

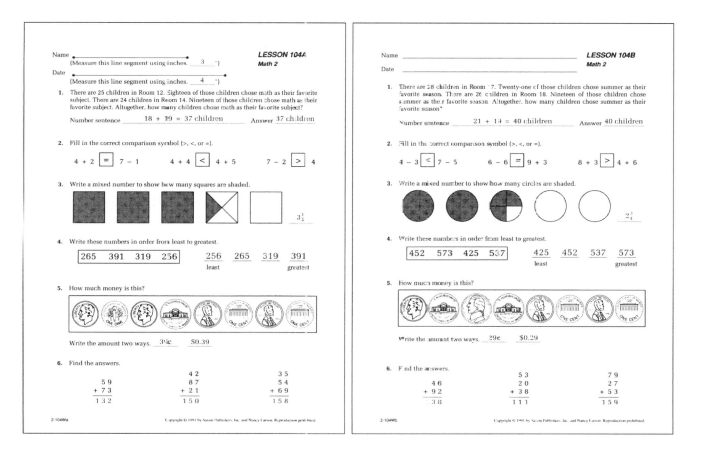

LESSON 104A
Math 2

Name _____
(Measure this line segment using inches. ___3___")

Date _____
(Measure this line segment using inches. ___4___")

1. There are 25 children in Room 12. Eighteen of those children chose math as their favorite subject. There are 24 children in Room 14. Nineteen of those children chose math as their favorite subject. Altogether, how many children chose math as their favorite subject?

   Number sentence ___18 + 19 = 37 children___   Answer _37 children_

2. Fill in the correct comparison symbol (>, <, or =).

   $4 + 2 \boxed{=} 7 - 1$ $4 + 4 \boxed{<} 4 + 5$ $7 - 2 \boxed{>} 4$

3. Write a mixed number to show how many squares are shaded.

   $3\frac{1}{4}$

4. Write these numbers in order from least to greatest.

   | 265 | 391 | 319 | 256 | ___256___ ___265___ ___319___ ___391___

   least   greatest

5. How much money is this?

   Write the amount two ways. ___39¢___   ___$0.39___

6. Find the answers.

	42	35
59	87	54
+ 73	+ 21	+ 69
132	150	158

LESSON 104B
Math 2

Name _____

Date _____

1. There are 28 children in Room 17. Twenty-one of those children chose summer as their favorite season. There are 26 children in Room 18. Nineteen of those children chose summer as their favorite season. Altogether, how many children chose summer as their favorite season?

   Number sentence ___21 + 19 = 40 children___   Answer _40 children_

2. Fill in the correct comparison symbol (>, <, or =).

   $4 - 3 \boxed{<} 7 - 5$ $6 - 6 \boxed{=} 9 + 3$ $8 + 3 \boxed{>} 4 + 6$

3. Write a mixed number to show how many circles are shaded.

   $2\frac{1}{4}$

4. Write these numbers in order from least to greatest.

   | 452 | 573 | 425 | 537 | ___425___ ___452___ ___537___ ___573___

   least   greatest

5. How much money is this?

   Write the amount two ways. ___29¢___   ___$0.29___

6. Find the answers.

	53	79
46	20	27
+ 92	+ 38	+ 53
138	111	159

Lesson **105**

measuring and drawing line segments using centimeters

THE MEETING

calendar

• Ask your child to write the date on the calendar and meeting strip.

• Ask your child the following two or three times a week:

date _____ days ago, date _____ days from now

day of the week _____ days ago, day of the week _____ days from now

_____th month, month before, month after

• Record on the meeting strip a special event and the number of days until it occurs.

weather graph

• Ask your child to read and graph today's temperature to the nearest two degrees.

• Count by 10's and 2's to check the temperature on the graph.

- Ask your child to connect the dot for yesterday's temperature to the dot for today's temperature and compare the temperatures.

counting

- Count by 25's to 300 and backward from 300 by 25's.
- Count by 3's to 30 and backward from 30 by 3's.
- Do the following once a week:

    count by 10's to 400 and backward from 400 by 10's

    count by 5's to 100 and backward from 50 by 5's

    say the even numbers to 100 and backward from 50

    say the odd numbers to 49 and backward from 49

graph questions

- You and your child each ask a question about any of the graphs.

patterning

- Ask your child to do the following:

    identify the pattern (repeating, continuing, or both)

    identify the numbers to complete the pattern

    read the pattern

money

- Ask your child to put the coins in the coin cup. Count the money in the coin cup together.
- Ask your child for another way to show that amount of money. Count these coins together to check the amount.

clock

- Set the clock to a five-minute interval.
- Ask the following:

    "It's (morning/afternoon/evening). What time is it?"

    time one hour ago

    time one hour from now

- Ask your child to write the digital time on the meeting strip.
- Record on the meeting strip the time an activity will occur.

number of the day

- Write three number sentences for the number of the day on the meeting strip.

fact practice

- Write three fact family numbers (e.g., 2, 7, 9) on the chalkboard.
- Allow time for your child to write the four fact family number sentences on the chalkboard.

ASSESSMENT

Written Assessment

"Today I would like to see what you remember from what we have been practicing."

- Give your child **Written Assessment #20**.
- Read the directions for each problem. Allow time for your child to complete each problem before continuing.
- Correct the paper, noting your child's mistakes on the **Individual Recording Form.** Review the errors with your child.

THE LESSON

Measuring and Drawing Line Segments Using Centimeters

"We have been measuring and drawing line segments to the nearest inch and half inch."

"Today you will learn how to measure and draw line segments using centimeters."

- Give your child a centimeter/inch ruler and a piece of paper ($8\frac{1}{2}$" × 11").
- Draw the following on the chalkboard:

"Let's fold the paper so it will look like the picture on the chalkboard."

"You will do that using only two folds."

"How do you think you will do that?"

"Fold your paper in half like this."

- Demonstrate with another piece of paper.

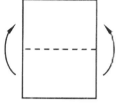

"*Now fold your paper in half again like this.*"

- Demonstrate.

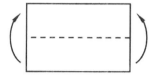

"*Open your paper.*"

"*How many equal pieces do you have?*"   4

"*What do we call each piece?*"   one fourth

"*Now you will practice drawing line segments.*"

"*You will draw one line segment in each fourth of the paper.*"

"*In the first rectangle, carefully draw a four-inch line segment.*"

"*Remember to put a small dot at each endpoint.*"

- Allow time for your child to do this.

"*Now you will use the other side of your ruler to measure the line segment.*"

"*These units are called centimeters.*"

"*About how many centimeters long is your ruler?*"   30 centimeters

"*Now you will measure the line segment using centimeters.*"

"*There is usually a line near the beginning of the ruler that shows you where to begin measuring.*"

"*Put this line on the first endpoint.*"

"*Look along your ruler until you come to the other endpoint.*"

"*About how many centimeters long is the line segment?*"   10 cm

"*Write '10 centimeters' below the line segment.*"

"*When we abbreviate centimeter, we write 'cm' without a period.*"

- Demonstrate on the chalkboard.

"*Draw a line segment 15 centimeters long in the next rectangle.*"

"*Write '15 centimeters' below the line segment.*"

- Allow time for your child to draw and label the line segment.

"*Now you will use the inch side of the ruler to measure the line segment.*"

"About how many inches long is the line segment?"   6 inches

"Now you will practice drawing and measuring some more line segments."

"Draw a five-centimeter line segment in the third rectangle."

"Write '5 centimeters' below the line segment."

"Now measure the line segment using inches."

"About how many inches long is the line segment?"   2 inches

"Draw a 20-centimeter line segment in the last rectangle."

"Write '20 centimeters' below the line segment."

"Now measure the line segment using inches."

"About how many inches long is the line segment?"   8 inches

"Draw a 13-centimeter line segment on the back of your paper."

"Write '13 centimeters' below the line segment."

"Measure the line segment using inches."

"About how many inches long is the line segment?"   5 inches

"Draw a 5 1/2-inch line segment in the next rectangle."

"Measure the line segment using centimeters."

"About how many centimeters long is the line segment?"   14 centimeters

"What do you notice about centimeters?"

CLASS PRACTICE

number fact practice

- Use the white fact cards to practice the last sixteen subtraction facts with your child.
- Give your child **Fact Sheet S 8.0**.
- Time your child for one minute.
- Correct the fact sheet with your child and record the score.
- Allow time for your child to complete the unfinished facts.

WRITTEN PRACTICE

- Complete **Worksheet 105A** with your child.
- Complete **Worksheet 105B** with your child later in the day.

Name _____
(Draw a 4-inch line segment.)

Date _____
(Draw a 1½" line segment.)

ASSESSMENT 20
LESSON 105
Math 2

1. Fred has 14 baseball cards. Show how he will share them equally with his brother.

14 cards ☐

extras

one half of 14 is __7__

How many baseball cards will each boy have? __7__

2. About how tall are you? __answers may vary__

3. Measure these line segments using inches.

1½"

5½"

4. I have 2 dimes, 3 nickels, and 7 pennies. Draw the coins. How much money is this?

(D) (D) (N) (N) (N) (P)
(P) (P) (P) (P) (P) (P)

Write the amount two ways. __42¢__   __$0.42__

5. Color 2¼ circles.

6. Find the answers.

$$\begin{array}{r} 65 \\ +48 \\ \hline 113 \end{array}\qquad \begin{array}{r} 57 \\ +92 \\ \hline 149 \end{array}\qquad \begin{array}{r} 14¢ \\ 23¢ \\ +36¢ \\ \hline 73¢ \end{array}\qquad \begin{array}{l} 62¢ + 18¢ \\ 80¢ \end{array}$$

Name _____
(Draw a 9 cm line segment.)

Date _____
(Measure this line segment to the nearest centimeter. __8__ cm)

LESSON 105A
Math 2

1. Erica went to a party at 11:00 a.m. She left when the party was over at 2:00 p.m.

How long was she at the party? __3 hours__

2. Draw a picture to show one hundred thirteen. (Use ☐ for 100, ☐ for 10, and ☐ for 1.)

Write the number in expanded form. __100 + 10 + 3__
Circle the number that shows one hundred thirteen.

10013   (113)   1013   100103

3. Color the cone yellow.
Color the pyramid blue.
Color the rectangular solid red.

4. I have 10 quarters. Draw the coins. How much money do I have? __$2.50__

(Q)(Q)(Q)(Q)(Q)(Q)(Q)(Q)(Q)(Q)

5. Fill in the missing numbers.

$1 \times \boxed{9} = 9$ $\boxed{2} \times 10 = 20$ $3 \times \boxed{100} = 300$

6. Find the answers.

64 + 73   68 + 7 + 13   29 + 73 + 21
137   88   123

Name _____

Date _____

LESSON 105B
Math 2

1. Luis went to visit his grandfather at 4:00 p.m. He left for home at 8:00 p.m.

How long was he at his grandfather's? __4 hours__

2. Draw a picture to show three hundred twenty-four. (Use ☐ for 100, ☐ for 10, and ☐ for 1.)

Write the number in expanded form. __300 + 20 + 4__
Circle the number that shows three hundred twenty-four.

3204   (324)   30024   300204

3. Color the sphere orange.
Color the cylinder green.
Color the cube purple.

4. I have 8 quarters. Draw the coins. How much money do I have? __$2.00__

(Q)(Q)(Q)(Q)(Q)(Q)(Q)(Q)

5. Fill in the missing numbers.

$4 \times \boxed{10} = 40$ $\boxed{9} \times 100 = 900$ $1 \times \boxed{8} = 8$

6. Find the answers.

84 + 65   24 + 8 + 56   73 + 46 + 31
149   88   150

Lesson 106

multiplying by five

lesson preparation

materials

elbow macaroni

small cup

8 nickels

multiplication fact cards — yellow

Fact Sheet M 14.0

the night before

• Put 45–55 pieces of macaroni in a small cup.

in the morning

• Write the following in the pattern box on the meeting strip:

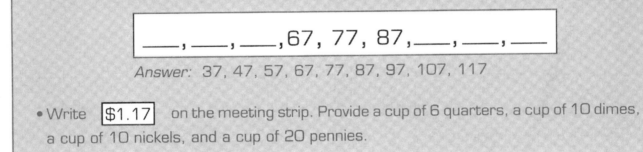

____, ____, ____, 67, 77, 87, ____, ____, ____

Answer: 37, 47, 57, 67, 77, 87, 97, 107, 117

• Write $1.17 on the meeting strip. Provide a cup of 6 quarters, a cup of 10 dimes, a cup of 10 nickels, and a cup of 20 pennies.

THE MEETING

calendar

• Ask your child to write the date on the calendar and meeting strip.

• Ask your child the following two or three times a week:

    date _____ days ago, date _____ days from now

    day of the week _____ days ago, day of the week _____ days from now

    _____th month, month before, month after

• Record on the meeting strip a special event and the number of days until it occurs.

weather graph

- Ask your child to read and graph today's temperature to the nearest two degrees.
- Count by 10's and 2's to check the temperature on the graph.
- Ask your child to connect the dot for yesterday's temperature to the dot for today's temperature and compare the temperatures.

counting

- Count by 25's to 300 and backward from 300 by 25's.
- Count by 3's to 30 and backward from 30 by 3's.
- Do the following once a week:

    count by 10's to 400 and backward from 400 by 10's

    count by 5's to 100 and backward from 50 by 5's

    say the even numbers to 100 and backward from 50

    say the odd numbers to 49 and backward from 49

graph questions

- You and your child each ask a question about any of the graphs.

patterning

- Ask your child to do the following:

    identify the pattern (repeating, continuing, or both)

    identify the numbers to complete the pattern

    read the pattern

money

- Ask your child to put the coins in the coin cup. Count the money in the coin cup together.
- Ask your child for another way to show that amount of money. Count these coins together to check the amount.

clock

- Set the clock to a five-minute interval.
- Ask the following:

    "It's (morning/afternoon/evening). What time is it?"

    time one hour ago

    time one hour from now

- Ask your child to write the digital time on the meeting strip.
- Record on the meeting strip the time an activity will occur.

number of the day

- Write three number sentences for the number of the day on the meeting strip.

fact practice

- Write three fact family numbers (e.g., 2, 7, 9) on the chalkboard.

- Allow time for your child to write the four fact family number sentences on the chalkboard.

THE LESSON

Multiplying by Five

"Today you will learn how to multiply by five."

"I am going to give you a small cup of macaroni."

"You will put the pieces of macaroni in groups of five."

"Make as many groups of five as possible."

"Put the extra macaroni back in the cup."

- Give your child a cup of macaroni.

"How many groups of five do you have?"

- Record on the chalkboard "_____ groups of five."

"Let's count by 5's to find this answer."

- Record the answer next to the problem.

"What is a shorter way of writing '_____ groups of five'?"

- Write "_____ × 5" next to the problem. (Fill in the appropriate number of groups.)

"Writing the 'x' is a short way of writing 'groups of.' "

- Write the following on the chalkboard:

$$0 \times 5 = \qquad 6 \times 5 =$$
$$1 \times 5 = \qquad 7 \times 5 =$$
$$2 \times 5 = \qquad 8 \times 5 =$$
$$3 \times 5 = \qquad 9 \times 5 =$$
$$4 \times 5 = \qquad 10 \times 5 =$$
$$5 \times 5 =$$

"Let's fill in these answers."

"How many pieces of macaroni are in one group of five?"   5

"How many pieces of macaroni are in two groups of five?"   10

"How many pieces of macaroni are in three groups of five?"   15

"How many pieces of macaroni are in four groups of five?"   20

"How many pieces of macaroni are in five groups of five?"   25

- Record each answer on the chalkboard.

- Repeat with 6, 7, 8, 9, and 10 groups of five.

"If we didn't have any groups of five, how many pieces of macaroni would we have?"   zero

"Do you see a pattern in the answers?"   it is like counting by 5's

"We can write these problems another way."

- Write the following on the chalkboard

$$\begin{array}{cccccccccccc} 5 & 5 & 5 & 5 & 5 & 5 & 5 & 5 & 5 & 5 & 5 \\ \times\,0 & \times\,1 & \times\,2 & \times\,3 & \times\,4 & \times\,5 & \times\,6 & \times\,7 & \times\,8 & \times\,9 & \times\,10 \end{array}$$

"Let's read these problems together as we say the answers."

- Read the problems as "zero groups of five equal zero," "one group of five equals five," etc.

"Which number do you think tells us the number of groups now?"   bottom number

"Use the macaroni to show two groups of five."

"Put the rest of your macaroni in the cup."

- Allow time for your child to do this.

"How many pieces of macaroni did you use?"   10

"We can write a number sentence to show what we did like this."

- Write "2 × 5 = 10" on the chalkboard.

"This means two groups of five."

"Leave this macaroni on the table."

"Now show five groups of two on your mat."

- Allow time for your child to do this.

"How many pieces of macaroni did you use to show five groups of two?"   10

"We can write a number sentence to show what we did like this."

- Write "5 × 2 = 10" below "2 × 5 = 10" on the chalkboard.

"This means five groups of two."

"What do you notice about these number sentences?"   the 2 and 5 changed places; the total number of pieces of macaroni used in each problem is 10

"Show three groups of five."

- Allow time for your child to do this.

  "How many pieces of macaroni did you use?"   15

  "We can write a number sentence to show what we did like this."

- Write "3 × 5 = 15" on the chalkboard.

  "This means three groups of five."

  "Leave this on the table."

  "Now show five groups of three on the table."

- Allow time for your child to do this.

  "How many pieces of macaroni did you use to show five groups of three?"   15

  "We can write a number sentence to show what we did like this."

- Write "5 × 3 = 15" below "3 × 5 = 15" on the chalkboard.

  "This means five groups of three."

  "What do you notice about these number sentences?"   the 3 and 5 changed places; the total number of pieces of macaroni used in each problem is 15

- Write   2 below   5.
         × 5   × 2

  "What do we know about this answer?"   it is the same as 5 × 2

- Write   3 below   5.
         × 5   × 3

  "What do we know about this answer?"   it is the same as 5 × 3

  "We will call these problems 'switcharounds.' "

  "We can switch the numbers around when we multiply and the answers will be the same."

  "Mathematicians call this the commutative property, but we will call these problems 'switcharounds.' "

- Write   5 and   9 × 5 =   on the chalkboard.
         × 7

  "What are these answers?"

  "How will we write a switcharound problem for each of these problems?"

- Ask your child to write the switcharound problems on the chalkboard.

  "Carefully slide the macaroni back into your cup."

  "When do we use counting by 5's?"   counting nickels, telling time

- Give your child 3 nickels.

  "What did I give you?"   3 nickels

"What is each nickel worth?" 5 cents

"How much money is this?" 15 cents

"We can write a multiplication number sentence to show how much money I gave you."

"I gave you three groups of five cents."

- Write "3 × 5¢ = 15¢" on the chalkboard.

- Give your child three more nickels.

"How many nickels do you have?" 6

"What is each nickel worth?" 5 cents

"Let's write a multiplication number sentence to show how much money you have now."

"I gave you six groups of five cents."

- Write "6 × 5¢ = _____" on the chalkboard.

"How much money is this?"

- Write "6 × 5¢ = 30¢" on the chalkboard.

"Let's count by 5's to check."

- Give your child two more nickels.

"How many nickels do you have?" 8

"What is each nickel worth?" 5 cents

"How will we write a multiplication number sentence to show how much money you have now?"

- Write "8 × 5¢ = 40¢" on the chalkboard.

"Let's count by 5's to check."

"Let's practice the multiplying by five facts."

- Erase the chalkboard answers for the vertical number facts.

"I will point to a number fact."

"Say the answer as quickly as possible."

- Point to the number facts one at a time.

"I will give you the multiplying by five fact cards."

- Give your child the yellow multiplication fact cards.

"Match the switcharound facts."

"When you finish, practice saying the answers to yourself."

CLASS PRACTICE

number fact practice

- Give your child **Fact Sheet M 14.0.**
- Time your child for one minute.
- Correct the fact sheet with your child and record the score.
- Allow time for your child to complete the unfinished facts.

WRITTEN PRACTICE

- Complete **Worksheet 106A** with your child.
- Complete **Worksheet 106B** with your child later in the day.

Name _____ **LESSON 106A**
(Draw a 7 cm line segment.) **Math 2**

Date _____
(Measure this line segment using centimeters. ___6___ cm)

1. Leah said that whenever she adds two odd numbers the answer is always an even number. Add 3 pairs of odd numbers to see if she is right. answers may vary

    ☐ + ☐ = ☐   ☐ + ☐ = ☐   ☐ + ☐ = ☐
    odd   odd ___   odd   odd ___   odd   odd ___

2. Show how to share the markers equally.

    13 markers ✏   extras ☐

    8 markers ✏   extras ☐

    one half of 13 is _6 R1_   one half of 8 is __4__

3. Use the clues to write the children's names on the Venn diagram.
    Sue has only a dog.
    Mary has a cat and a dog.
    Peter has only a cat.
    Mark has both pets.
    Sam has only a cat.

    Children's Pets
    Dogs   Cats
    Sue Mary Peter
         Mark Sam

    How many children have a cat? __4__

    How many children have only a dog? __1__

4. Find the products.

5	5	5	5	5	5	5	5	5	5
×2	×7	×5	×1	×8	×10	×5	×0	×4	×9	×6
10	35	25	5	40	50	15	0	20	45	30

5. Find the answers.

    25¢   95   63 + 97   16 + 86
    37¢   92   160   102
    +18¢   +21
    ‾‾‾‾   ‾‾‾
    80¢   208

2-106Wa   Copyright © 1994 by Saxon Publishers, Inc. and Nancy Larson. Reproduction prohibited.

Name _____ **LESSON 106B**
Math 2

Date _____

1. Martell said that whenever he adds two even numbers the answer is always an even number. Add 3 pairs of even numbers to see if he is right. answers may vary

    ☐ + ☐ = ☐   ☐ + ☐ = ☐   ☐ + ☐ = ☐
    even   even ___   even   even ___   even   even ___

2. Show how to share the pencils equally.

    12 pencils ✏   extras ☐

    15 pencils ✏   extras ☐

    one half of 12 is __6__   one half of 15 is _7 R1_

3. Use the clues to write the children's names on the Venn diagram.
    Bob has fish and birds.
    Tom has only a bird.
    Carol has both pets.
    Tim has only a bird.
    Frank has only fish for pets.
    Karen has only a bird.

    Children's Pets
    Fish   Birds
    Frank Bob Karen
           Carol Tom
                  Tim

    How many children have birds? __5__

    How many children have only fish? __1__

4. Fill in the missing numbers.

5	5	5	5	5	5	5	5
×3	×4	×8	×0	×5	×10	×1	×9
15	20	40	0	25	50	5	45

5. Find the answers.

    46¢   84   86 + 34   17 + 97
    19¢   73   120   114
    +25¢   +61
    ‾‾‾‾   ‾‾‾
    90¢   218

2-106Wb   Copyright © 1994 by Saxon Publishers, Inc. and Nancy Larson. Reproduction prohibited.

L esson **107**

subtracting two-digit numbers using dimes and pennies (part 1)

---

lesson preparation

materials

1 cup of 10 dimes

1 cup of 20 pennies

yellow/white mat from Lesson 71

Fact Sheet M 14.0

in the morning

• Write the following in the pattern box on the meeting strip:

$$\frac{1}{4}, \frac{2}{4}, \frac{3}{4}, \frac{4}{4}, \frac{5}{4}, \frac{6}{4}, \underline{\quad}, \underline{\quad}, \underline{\quad}, \underline{\quad}, \underline{\quad}, \underline{\quad}$$

Answer: $\frac{1}{4}, \frac{2}{4}, \frac{3}{4}, \frac{4}{4}, \frac{5}{4}, \frac{6}{4}, \frac{7}{4}, \frac{8}{4}, \frac{9}{4}, \frac{10}{4}, \frac{11}{4}, \frac{12}{4}$

• Write $\boxed{\$1.55}$ on the meeting strip. Provide a cup of 6 quarters, a cup of 10 dimes, a cup of 10 nickels, and a cup of 20 pennies.

THE MEETING

calendar

• Ask your child to write the date on the calendar and meeting strip.

• Ask your child the following two or three times a week:

date _____ days ago, date _____ days from now

day of the week _____ days ago, day of the week _____ days from now

_____th month, month before, month after

• Record on the meeting strip a special event and the number of days until it occurs.

weather graph

• Ask your child to read and graph today's temperature to the nearest two degrees.

• Count by 10's and 2's to check the temperature on the graph.

- Ask your child to connect the dot for yesterday's temperature to the dot for today's temperature and compare the temperatures.

counting

- Count by 25's to 300 and backward from 300 by 25's.
- Count by 3's to 30 and backward from 30 by 3's.
- Do the following once a week:

    count by 10's to 400 and backward from 400 by 10's

    count by 5's to 100 and backward from 50 by 5's

    say the even numbers to 100 and backward from 50

    say the odd numbers to 49 and backward from 49

graph questions

- You and your child each ask a question about any of the graphs.

patterning

- Ask your child to do the following:

    identify the pattern (repeating, continuing, or both)

    identify the numbers to complete the pattern

    read the pattern

money

- Ask your child to put the coins in the coin cup. Count the money in the coin cup together.
- Ask your child for another way to show that amount of money. Count these coins together to check the amount.

clock

- Set the clock to a five-minute interval.
- Ask the following:

    "It's (morning/afternoon/evening). What time is it?"

    time one hour ago

    time one hour from now

- Ask your child to write the digital time on the meeting strip.
- Record on the meeting strip the time an activity will occur.

number of the day

- Write three number sentences for the number of the day on the meeting strip.

fact practice

- Write three fact family numbers (e.g., 2, 7, 9) on the chalkboard.
- Allow time for your child to write the four fact family number sentences on the chalkboard.

THE LESSON

Subtracting Two-Digit Numbers Using Dimes and Pennies (Part 1)

"We have been adding two-digit numbers."

"Today you will begin to learn how to subtract two-digit numbers using dimes and pennies."

- Write the following problems on the chalkboard or chart paper:

$$75¢ - 23¢ = \qquad 50¢ - 23¢ =$$
$$87¢ - 54¢ = \qquad 92¢ - 36¢ =$$

"What kind of problems are these?"   some, some went away

"What is another way to write the first problem?"

- Ask your child to write the problem vertically on the chalkboard.

"What happens in a some, some went away problem?"   there are some and some go away

"How much money does this problem tell us to begin with?"   75¢

"Let's show 75¢ using only dimes and pennies."

"We will use the fewest number of pennies possible."

"How many dimes and pennies will we use for 75¢?"   7 dimes, 5 pennies

- Put the dimes and pennies on the yellow and white mat from Lesson 71.

"What does the problem tell us to do now?"   take away 23¢

- Write the following on the chalkboard:

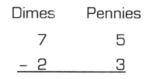

"We will take away the pennies first."

"How many pennies does our problem tell us to take away?"   3

"I will take away three pennies and put them in the penny cup."

- Remove 3 pennies.

   "How many pennies do we have now?"   2

- Record "2" under the pennies' column.

   "How many dimes does our problem tell us to take away?"   2

   "I will take away two dimes and put them in the dime cup."

- Remove 2 dimes.

   "How many dimes do we have now?"   5

- Record "5" under the dimes' column.

   "How much money do we have left?"   52¢

- Put the coins back in the coin cups.

- Point to 50¢ – 23¢.

   "What is another way to write this problem?"

- Ask your child to write the following problem vertically on the chalkboard:

$$\begin{array}{r} 50¢ \\ -\ 23¢ \\ \hline \end{array}$$

   "How much money does this problem tell us to begin with?"   50¢

   "We will make 50¢ using only dimes and pennies."

   "We will use the fewest number of pennies possible."

   "How many dimes and pennies will we use to show 50¢?"   5 dimes

- Put the dimes on the mat.

   "What does the problem tell us to do now?"   take away 23¢

- Write the following on the chalkboard:

Dimes	Pennies
5	0
– 2	3

   "We will take away the pennies first."

   "How many pennies does our problem tell us to take away?"   3

   "Can we take away three pennies and put them in the cup?"   no

   "Why not?"   we don't have any pennies on the mat

   "What could we do so we will have some pennies to use?"   trade a dime for 10 pennies

- Trade a dime for 10 pennies.

   "How many dimes do we have now?"   4

"How many pennies do we have now?"   10

"Do we still have 50¢?"   yes

"Whenever we trade a dime for pennies, we must show this on our problem."

"This is how we will show what we did."

- Record the following on the chalkboard. **Do not** cross out the zero and write "10" above the zero.

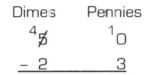

Dimes   Pennies

⁴⁄₅   ¹0

− 2   3

"Can we take away three pennies now?"   yes

- Remove 3 pennies.

  "How many pennies do we have left?"   7

- Record "7" under the pennies' column.

  "How many dimes does our problem tell us to take away?"   2

- Remove 2 dimes.

  "How many dimes do we have now?"   2

- Record "2" under the dimes' column.

  "How much money do we have left?"   27¢

- Put the coins back in the coin cups.

- Point to 87¢ − 54¢.

  "What is another way to write this problem?"

- Ask your child to write the problem vertically on the chalkboard.

      87¢
    − 54¢

"How much money does this problem tell us to begin with?"   87¢

"How many dimes and pennies will we use to show 87¢?"   8 dimes, 7 pennies

- Put the dimes and pennies on the mat.

  "What does the problem tell us to do?"   take away 54¢

- Write the following on the chalkboard:

  Dimes   Pennies

  8   7

  − 5   4

"We will take away the pennies first."

"How many pennies does our problem tell us to take away?"   4

"Can we take away four pennies and put them in the cup?"   yes

- Remove 4 pennies.

"How many pennies do we have left?"   3

- Record "3" under the pennies' column.

"How many dimes does our problem tell us to take away?"   5

- Remove 5 dimes.

"How many dimes do we have now?"   3

- Record "3" under the dimes' column.

"How much money do we have left?"   33¢

- Put the coins back in the coin cups.

- Point to 92¢ – 36¢.

"What is another way to write this problem?"

- Ask your child to write the following problem vertically on the chalkboard:

$$
\begin{array}{r}
92¢ \\
- \ 36¢ \\
\hline
\end{array}
$$

"How much money does this problem tell us to begin with?"   92¢

"How many dimes and pennies will we use to show 92¢?"   9 dimes, 2 pennies

- Put the dimes and pennies on the mat.

"What does the problem tell us to do?"   take away 36¢

- Write the following on the chalkboard or chart paper:

Dimes	Pennies
9	2
– 3	6

"We will take away the pennies first."

"How many pennies does our problem tell us to take away?"   6

"Can we take away six pennies and put them in the cup?"   no

"Why not?"   we only have 2 pennies

"What could we do so we will have enough pennies?"   trade a dime for 10 pennies

"Show me how to do that."

- Ask your child to trade a dime for 10 pennies.

"How many dimes do we have now?" 8

"How many pennies do we have now?" 12

"Do we still have 92¢?"

"Whenever we trade a dime for pennies, we must show this on our problem."

"This is how we will show what we did."

- Write the following on the chalkboard. **Do not** cross out the 2 and write "12" above the 2.

$$
\begin{array}{cc}
\text{Dimes} & \text{Pennies} \\
{}^{8}\cancel{9} & {}^{1}2 \\
\underline{-\ 3} & \underline{\quad\ 6}
\end{array}
$$

"Can we take away six pennies now?" yes

- Remove 6 pennies.

"How many pennies do we have left?" 6

- Record "6" under the pennies' column.

"How many dimes does our problem tell us to take away?" 3

- Remove 3 dimes.

"How many dimes do we have now?" 5

- Record "5" under the dimes' column.

"How much money do we have left?" 56¢

"Tomorrow you will have a chance to act out subtraction problems using dimes and pennies."

- Save the yellow/white mat.

CLASS PRACTICE

number fact practice

- Use the yellow multiplication fact cards to practice the multiplying by five facts with your child.
- Give your child **Fact Sheet M 14.0**.
- Time your child for one minute.
- Correct the fact sheet with your child and record the score.
- Allow time for your child to complete the unfinished facts.

WRITTEN PRACTICE

- Complete **Worksheet 107A** with your child.
- Complete **Worksheet 107B** with your child later in the day.

Name _____ **LESSON 107A**
(Draw an 11 cm line segment.) **Math 2**

Date _____
(Measure this line segment using centimeters. __7__ cm)

1. Ahmad put the peanuts in groups of 10. When he finished, he counted four groups of peanuts. Draw a picture and write a number sentence to show how many peanuts he has.

   __4__ groups of __10__ peanuts. Number sentence __4 × 10 = 40__

   How many peanuts does he have? __40__

2. It's morning.
   What time is it? __6:50 a.m.__

3. Color the cubes red.
   Color the cylinders yellow.
   Color the pyramids blue.
   Shade the graph to show the number of cylinders, cubes, and pyramids.

	0 1 2 3 4 5 6 7 8 9
Cubes	
Cylinders	
Pyramids	

4. Find the answers.

   $7 \times 5 = $ __35__ $3 \times 10 = $ __30__ $24 + 97 = $ __121__   7 8
   $4 \times 5 = $ __20__ $52 - 10 = $ __42__ $6 \times 100 = $ __600__   2 4
   $2 + 6 + 3 + 7 + 4 = $ __22__   + 3 2
                                                                           1 3 4

2-107Wa   Copyright © 1991 by Saxon Publishers, Inc. and Nancy Larson. Reproduction prohibited.

Name _____ **LESSON 107B**
  Math 2

Date _____

1. Justin put the paper clips in groups of 10. When he finished, he counted six groups of paper clips. Draw a picture and write a number sentence to show how many paper clips he has.

   __6__ groups of __10__ paper clips. Number sentence __6 × 10 = 60__

   How many paper clips does he have? __60__

2. It's afternoon.
   What time is it? __2:35 p.m.__

3. Color the spheres yellow.
   Color the rectangular solids blue.
   Color the cones red.
   Shade the graph to show the number of spheres, rectangular solids, and cones.

	0 1 2 3 4 5 6 7 8 9
Spheres	
Rect. Sol.	
Cones	

4. Find the answers.

   $9 \times 5 = $ __45__ $7 \times 10 = $ __70__ $75 + 47 = $ __122__   8 4
   $6 \times 5 = $ __30__ $63 - 10 = $ __53__ $8 \times 100 = $ __800__   9 3
   $4 + 2 + 9 + 3 + 1 + 6 = $ __25__   + 1 6
                                                                           1 9 3

2-107Wb   Copyright © 1991 by Saxon Publishers, Inc. and Nancy Larson. Reproduction prohibited.

586

L esson 108

subtracting two-digit numbers using dimes and pennies (part 2)

lesson preparation

materials

yellow/white mat from Lesson 71

1 cup of 10 dimes

1 cup of 20 pennies

Fact Sheet S 8.1

in the morning

• Write the following in the pattern box on the meeting strip:

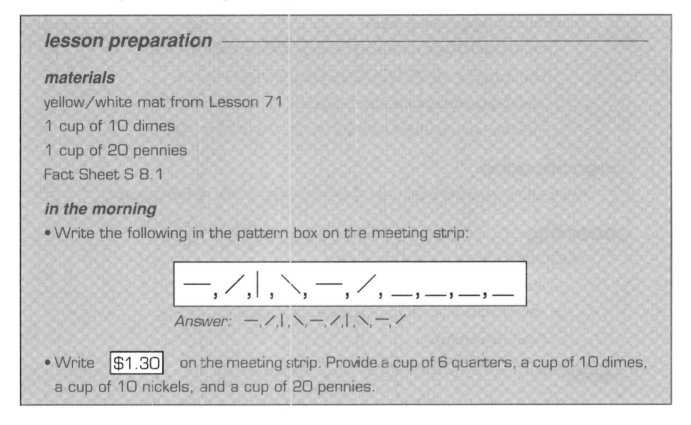

Answer: —,╱,│,╲,—,╱,│,╲,—,╱

• Write $1.30 on the meeting strip. Provide a cup of 6 quarters, a cup of 10 dimes, a cup of 10 nickels, and a cup of 20 pennies.

THE MEETING

calendar

• Ask your child to write the date on the calendar and meeting strip.

• Ask your child the following two or three times a week:

  date _____ days ago, date _____ days from now

  day of the week _____ days ago, day of the week _____ days from now

  _____th month, month before, month after

• Record on the meeting strip a special event and the number of days until it occurs.

weather graph

• Ask your child to read and graph today's temperature to the nearest two degrees.

• Count by 10's and 2's to check the temperature on the graph.

• Ask your child to connect the dot for yesterday's temperature to the dot for today's temperature and compare the temperatures.

counting

• Count by 25's to 300 and backward from 300 by 25's.

• Count by 3's to 30 and backward from 30 by 3's.

• Do the following once a week:

count by 10's to 400 and backward from 400 by 10's

count by 5's to 100 and backward from 50 by 5's

say the even numbers to 100 and backward from 50

say the odd numbers to 49 and backward from 49

graph questions

• You and your child each ask a question about any of the graphs.

patterning

• Ask your child to do the following:

identify the pattern (repeating, continuing, or both)

identify the shapes to complete the pattern

read the pattern

money

• Ask your child to put the coins in the coin cup. Count the money in the coin cup together.

• Ask your child for another way to show that amount of money. Count these coins together to check the amount.

clock

• Set the clock to a five-minute interval.

• Ask the following:

"It's (morning/afternoon/evening). What time is it?"

time one hour ago

time one hour from now

• Ask your child to write the digital time on the meeting strip.

• Record on the meeting strip the time an activity will occur.

number of the day

• Write three number sentences for the number of the day on the meeting strip.

fact practice

- Write three fact family numbers (e.g., 2, 7, 9) on the chalkboard.
- Allow time for your child to write the four fact family number sentences on the chalkboard.

THE LESSON

Subtracting Two-Digit Numbers Using Dimes and Pennies (Part 2)

"Today you will continue to learn how to subtract two-digit numbers using dimes and pennies."

- Write the following problem on the chalkboard:

$$45¢ - 31¢ =$$

"What is another way to write this problem?"

- Ask your child to write the following problem vertically on the chalkboard:

$$\begin{array}{r} 45¢ \\ -\ 31¢ \\ \hline \end{array}$$

"Now you will use the dimes and pennies to find this answer."

- Give your child the yellow/white mat, a cup of 10 dimes, and a cup of 20 pennies.

"What coins will you put on the left side of your paper?" dimes

"What coins will you put on the right side of your paper?" pennies

"How much money does this problem tell us to begin with?" 45¢

"Show 45¢ using the fewest number of dimes and pennies."

"How many dimes and pennies did you use?" 4 dimes, 5 pennies

"What does the problem tell us to do now?" take away 31¢

- Write the following on the chalkboard:

Dimes	Pennies
4	5
- 3	1

"We will begin with the pennies."

"How many pennies does our problem tell us to take away?" 1

"Take away one penny and put it in the penny cup."

"How many pennies do you have now?" 4

- Record "4" under the pennies' column.

"How many dimes does our problem tell us to take away?"   3

"Take away three dimes and put them in the dime cup."

"How many dimes do you have now?"   1

- Record "1" under the dimes' column.

"How much money do you have left?"   14¢

"Put the coins in your coin cups."

- Write the following problem on the chalkboard:

60¢ – 26¢ =

"How much money does the problem tell us to begin with?"   60¢

"Show 60¢ using the fewest number of dimes and pennies."

"What coins did you use?"   6 dimes

"What does the problem tell us to do now?"   take away 26¢

- Write the following on the chalkboard:

Dimes	Pennies
6	0
– 2	6

"We will begin with the pennies."

"How many pennies does our problem tell us to take away?"   6

"Can you take away six pennies and put them in the cup?"   no

"Why not?"   we don't have any pennies

"What could you do so that you can take away six pennies?"   trade a dime for ten pennies

"Trade a dime for ten pennies."

- Allow time for your child to do this.

"How many dimes do you have now?"   5

"How many pennies do you have now?"   10

"Do you still have 60¢?"

"Whenever we trade a dime for pennies, we must show this on our problem."

"This is how we show what we did."

- Record the following on the chalkboard:

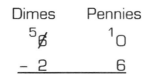

Dimes	Pennies
⁵6̸	¹0
– 2	6

"Can you take away six pennies now?"   yes

"Put six pennies in your cup."

"How many pennies do you have left?"   4

• Record "4" under the pennies' column.

"How many dimes does our problem tell us to take away?"   2

"Take away two dimes and put them in your cup."

"How many dimes do you have now?"   3

• Record "3" under the dimes' column.

"How much money do you have left?"   34¢

"Put the coins in your coin cups."

• Write the following problem on the chalkboard:

$$74¢ - 43¢ =$$

"How much money does this problem tell us to begin with?"   74¢

"How many dimes and pennies will you use to show 74¢?"   7 dimes, 4 pennies

"Put the coins on your paper."

"What does the problem tell us to do now?"   take away 43¢

• Write the following on the chalkboard:

Dimes	Pennies
7	4
− 4	3

"What will you do first?"   take away 3 pennies

"Do you have enough pennies that you can take away three pennies?"   yes

"Take away three pennies and put them in your cup."

"How many pennies do you have now?"   1

• Record "1" under the pennies' column.

"What will you do next?"   take away 4 dimes

"Take away four dimes and put them in your cup."

"How many dimes do you have now?"   3

• Record "3" under the dimes' column.

"How much money do you have left?"   31¢

"Put the coins in your coin cups."

• Write the following problem on the chalkboard:

$$52¢ - 38¢ =$$

"How much money does this problem tell us to begin with?"   52¢

"Show this on your paper."

"What coins did you use?"   5 dimes, 2 pennies

"What does this problem tell us to do now?"   take away 38¢

- Write the following on the chalkboard:

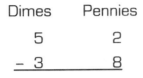

"We will begin with the pennies."

"How many pennies does our problem tell us to take away?"   8

"Can you take away eight pennies and put them in the cup?"   no

"Why not?"   we only have 2 pennies

"What could you do so you will have enough pennies to use?"   trade a dime for 10 pennies

"Put a dime in the dime cup and take ten pennies out of the penny cup."

- Allow time for your child to do this.

"How many dimes do you have now?"   4

"How many pennies do you have now?"   12

"Do you still have 52¢?"   yes

"Whenever we trade a dime for pennies, we must show this on our problem."

"This is how we show what we did."

- Write the following on the chalkboard:

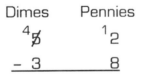

"Can you take away eight pennies now?"   yes

"Put eight pennies in the cup."

"How many pennies do you have left?"   4

- Record "4" under the pennies' column.

"How many dimes does our problem tell us to take away?"   3

"Take away three dimes and put them in your cup."

"How many dimes do you have now?"   1

- Record "1" under the dimes' column.

"How much money do you have left?" 14¢

"Tomorrow you will learn how to write subtraction problems on paper."

"You will use dimes and pennies to check the answers."

- Save the yellow/white mat.

CLASS PRACTICE

number fact practice

- Use the white fact cards to practice the last sixteen subtraction facts with your child.

- Give your child **Fact Sheet S 8.1.**

- Time your child for one minute.

- Correct the fact sheet with your child and record the score.

- Allow time for your child to complete the unfinished facts.

WRITTEN PRACTICE

- Complete **Worksheet 108A** with your child.

- Complete **Worksheet 108B** with your child later in the day.

Name _____ **LESSON 108A**
(Draw an 11 cm line segment.) Math 2
Date _____
(Measure this line segment using centimeters. _10_ cm)

1. The grade 2 children at Emerson School collected cans for recycling. During the first week they collected 37 cans, during the second week they collected 88 cans, and during the third week they collected 96 cans. How many cans did they collect during the first two weeks?

   Number sentence ___37 + 88 = 125 cans___ Answer _125 cans_

   During which week were the most cans collected? ___third___

2. Draw a picture to show three hundred fifteen. (Use ☐ for 100, ☐ for 10, and ☐ for 1.)

   Write this number in expanded form. ___300 + 10 + 5___

   Write three hundred fifteen using digits. _315_

3. Show how to share the pattern blocks equally.

   13 green pattern blocks △   16 orange pattern blocks ☐

   one half of 13 is _6 R1_   one half of 16 is _8_

4. Circle the ones that are the same as 56¢.

   6 dimes 5 pennies   (56 pennies)
   (4 dimes 16 pennies)   (6 pennies 5 dimes)

5. Find the answers.

   4 × 5 = _20_   6 × 100 = _600_   59 + 41 = 100   38 − 31 = 129

   9 × 5 = _45_   0 × 5 = _0_

2-108Wa   Copyright © 1991 by Saxon Publishers, Inc. and Nancy Larson. Reproduction prohibited.

Name _____ **LESSON 108B**
Math 2
Date _____

1. The grade 2 children in Mrs. Haller's class and Mrs. Carroll's class collected bottles for recycling. During the first week they collected 53 bottles, during the second week they collected 49 bottles, and during the third week they collected 32 bottles. How many bottles did they collect during the first two weeks?

   Number sentence ___63 + 49 = 112 bottles___ Answer _112 bottles_

   During which week were the most bottles collected? ___first___

2. Draw a picture to show two hundred sixteen. (Use ☐ for 100, ☐ for 10, and ☐ for 1.)

   Write this number in expanded form. ___200 + 10 + 6___

   Write two hundred sixteen using digits. _216_

3. Show how to share the pattern blocks equally.

   17 green pattern blocks △   14 orange pattern blocks ☐

   one half of 17 is _8 R1_   one half of 14 is _7_

4. Circle the ones that are the same as 42¢.

   (3 dimes 12 pennies)   (2 pennies 4 dimes)
   2 dimes 4 pennies   42 dimes

5. Find the answers.

   4 × 5 = _30_   9 × 100 = _900_   37 + 63 = 100   72 + 83 = 155

   4 × 5 = _40_   0 × 10 = _0_

2-108Wb   Copyright © 1991 by Saxon Publishers, Inc. and Nancy Larson. Reproduction prohibited.

Lesson 109

subtracting two-digit numbers (part 1)

---

lesson preparation

materials

1 cup of 10 dimes

1 cup of 20 pennies

yellow/white mat from Lesson 71

scrap paper

Fact Sheet S 8.1

in the morning

• Write the following in the pattern box on the meeting strip:

> 2, 7, 12, 17, 22, ____, ____, ____, ____, ____

Answer: 2, 7, 12, 17, 22, 27, 32, 37, 42, 47

• Write $1.32 on the meeting strip. Provide a cup of 6 quarters, a cup of 10 dimes, a cup of 10 nickels, and a cup of 20 pennies.

THE MEETING

calendar

• Ask your child to write the date on the calendar and meeting strip.

• Ask your child the following two or three times a week:

  date _____ days ago, date _____ days from now

  day of the week _____ days ago, day of the week _____ days from now

  _____th month, month before, month after

• Record on the meeting strip a special event and the number of days until it occurs.

weather graph

• Ask your child to read and graph today's temperature to the nearest two degrees.

• Count by 10's and 2's to check the temperature on the graph.

- Ask your child to connect the dot for yesterday's temperature to the dot for today's temperature and compare the temperatures.

counting

- Count by 25's to 300 and backward from 300 by 25's.
- Count by 3's to 30 and backward from 30 by 3's.
- Do the following once a week:

  count by 10's to 400 and backward from 400 by 10's

  count by 5's to 100 and backward from 50 by 5's

  say the even numbers to 100 and backward from 50

  say the odd numbers to 49 and backward from 49

graph questions

- You and your child each ask a question about any of the graphs.

patterning

- Ask your child to do the following:

  identify the pattern (repeating, continuing, or both)

  identify the numbers to complete the pattern

  read the pattern

money

- Ask your child to put the coins in the coin cup. Count the money in the coin cup together.
- Ask your child for another way to show that amount of money. Count these coins together to check the amount.

clock

- Set the clock to a five-minute interval
- Ask the following:

  "It's (morning/afternoon/evening). What time is it?"

  time one hour ago

  time one hour from now

- Ask your child to write the digital time on the meeting strip.
- Record on the meeting strip the time an activity will occur.

number of the day

- Write three number sentences for the number of the day on the meeting strip.

fact practice

- Write three fact family numbers (e.g., 2, 7, 9) on the chalkboard.
- Allow time for your child to write the four fact family number sentences on the chalkboard.

THE LESSON

Subtracting Two-Digit Numbers (Part 1)

"Yesterday you practiced subtracting two-digit numbers using dimes and pennies."

"Today you will learn how to record the steps for two-digit subtraction problems on paper."

- Give your child a cup of 10 dimes, a cup of 20 pennies, the yellow/white mat, and a piece of scrap paper.
- Write the following story on the chalkboard:

    Jamelle has 52¢.

    She only has dimes and pennies.

    She spent 35¢ at lunch.

"What type of story is this?"   some, some went away

"How much money did Jamelle have to begin with?"   52¢

"Show 52¢ using the fewest number of dimes and pennies."

- Allow time for your child to do this.

"What coins did you use?"   5 dimes, 2 pennies

"What happened in this story?"   she spent 35¢ at lunch

"How will you show that using dimes and pennies?"   take away 5 pennies and 3 dimes

"Do you have enough pennies that you can take away five?"   no

"What will you need to do?"   trade 1 dime for 10 pennies

"Show that with your coins."

- Allow time for your child to do this.

"How many dimes and how many pennies do you have now?"   4 dimes, 12 pennies

"How much money is this?"   52¢

"What will you do next?"   take away 5 pennies

"Do that."

"What will you do now?"   take away 3 dimes

"Do that."

"How much money is left?" 17¢

"What number sentence can we write for this story?" 52¢ − 35¢ = 17¢

- Write the number sentence horizontally on the chalkboard.

"Put the coins back in the cup."

"Now we will find the answer for 52¢ minus 35¢ without using dimes and pennies."

- Write the following on the chalkboard:

$$52¢$$
$$- 35¢$$

"This problem tells us that we have 52¢ and we are taking away 35¢."

"Write this problem on your paper."

"Make sure the pennies are in the pennies' column and the dimes are in the dimes' column."

- Allow time for your child to write the problem.

"What did you do first when you used dimes and pennies to find the answer?" checked to make sure that there were enough pennies to subtract 5 pennies

"Did you have enough pennies?" no

"What did you do next?" traded 1 dime for 10 pennies

"How will we show this on our problem?" cross out the 5, write a 4 above the 5, write a small 1 in front of the 2

- Record this on the chalkboard in the following way:

$$^{4}\cancel{5} \ ^{1}2¢$$
$$- \ 3 \ \ 5¢$$

"Show this on the problem on your paper."

- Allow time for your child to do this.

"What did you do next?" took away 5 pennies from the 12 pennies and 3 dimes from the 4 dimes

"How will we show this on our problem?" write the 7 below the pennies' column and the 1 below the dimes' column

- Record the following on the chalkboard:

$$^{4}\cancel{5} \ ^{1}2¢$$
$$- \ 3 \ \ 5¢$$
$$\overline{ 1 \ \ 7¢}$$

"Show this on your problem."

"Let's try another story."

• Write the following story on the chalkboard:

> Steve has 64¢.
>
> He only has dimes and pennies.
>
> He spent 37¢ at the school store.

"What type of story is this?"   some, some went away

"How much money does Steve have to begin with?"   64¢

"Show this using the fewest number of dimes and pennies."

• Allow time for your child to do this.

"What coins did you use?"   6 dimes, 4 pennies

"What happened in this story?"   he spent 37¢ at the school store

"How will you show that using dimes and pennies?"   take away 7 pennies and 3 dimes

"Do you have enough pennies that you can take away seven?"   no

"What should you do?"   trade 1 dime for 10 pennies

"Show that with your coins."

• Allow time for your child to do this.

"How many dimes and pennies do you have now?"   5 dimes, 14 pennies

"How much money is this?"   64¢

"What will you do now?"   take away 7 pennies

"Do that."

"What will you do next?"   take away 3 dimes

"Do that."

"How much money is left?"   27¢

"What number sentence can we write for this story?"   64¢ − 37¢ = 27¢

• Write the number sentence horizontally on the chalkboard.

"Put the coins back in the cups."

"Now we will find the answer for 64¢ minus 37¢ without using dimes and pennies."

• Write the following on the chalkboard:

$$\begin{array}{r} 64¢ \\ -\ 37¢ \\ \hline \end{array}$$

"This problem tells us that we have 64¢ and we are taking away 37¢."

"Write this problem on your paper."

• Allow time for your child to write the problem.

"What did you do first when you used dimes and pennies?" tried to subtract the pennies but we didn't have enough pennies

"What did you do next?" traded 1 dime for 10 pennies

"How many dimes and pennies did you have then?" 5 dimes, 14 pennies

"How will we show this on our problem?" cross out the 6, write a 5 above the 6, write a small 1 in front of the 4

- Record this on the chalkboard in the following way:

$$\begin{array}{r} ^5\cancel{6}4¢ \\ -\ 3\ \ 7¢ \\ \hline \end{array}$$

"Show this on the problem on your paper."

- Allow time for your child to do this.

"What did you do next?" took away 7 pennies from the 14 pennies and 3 dimes from the 5 dimes

- Record the following on the chalkboard:

$$\begin{array}{r} ^5\cancel{6}\ ^14¢ \\ -\ 3\ \ 7¢ \\ \hline 2\ \ 7¢ \end{array}$$

"Show this on your problem."

"Let's try another story."

- Write the following story on the chalkboard:

  Carl has 78¢.

  He only has dimes and pennies.

  He gave 32¢ to Eric.

"What type of story is this?" some, some went away

"Use your coins to show how much money Carl has to begin with."

- Allow time for your child to do this.

"What coins did you use?" 7 dimes, 8 pennies

"What happened in this story?" Carl gave Eric 32¢

"How will you show that using dimes and pennies?" take away 2 pennies and 3 dimes

"Do you have enough pennies to do that?" yes

"Do you need to trade a dime for ten pennies?" no

"Why not?" we have enough pennies to use

"Take away two pennies."

"What will you do next?" take away 3 dimes

"Do that."

"How much money is left?"   *46¢*

"What number sentence can we write for this story?"   *78¢ − 32¢ = 46¢*

- Write the number sentence horizontally on the chalkboard.

"Put the coins back in the cups."

"Now we will show how to find the answer for 78¢ minus 32¢ without using dimes and pennies."

"Write the problem vertically on your paper."

- Write the problem vertically on the chalkboard:

$$
\begin{array}{r}
78¢ \\
- \ 32¢ \\
\hline
\end{array}
$$

"This problem tells us that we have 78¢ and we are taking away 32¢."

"What will we do first?"   *check to see if we have enough pennies*

"Do we have enough pennies?"   *yes*

"What will we do next?"   *subtract 2 pennies from 8 pennies and 3 dimes from 7 dimes*

"Show that on your problem."

- Record the following on the chalkboard:

$$
\begin{array}{r}
78¢ \\
- \ 32¢ \\
\hline
46¢
\end{array}
$$

- Write the following problem on the chalkboard:

$$
\begin{array}{r}
82¢ \\
- \ 59¢ \\
\hline
\end{array}
$$

"Find the answer using dimes and pennies."

- Allow time for your child to do this.

"How much money is left?"   *23¢*

"What did you do to find this answer?"

"Show the subtraction steps on your paper."

- Allow time for your child to record the steps and the answer.

- Ask your child to explain the steps.

- Repeat with the following problems. Write the problems on the chalkboard one at a time. Include additional problems, if desired.

$$
\begin{array}{r}
71¢ \\
- \ 42¢ \\
\hline
\end{array}
\qquad
\begin{array}{r}
96¢ \\
- \ 31¢ \\
\hline
\end{array}
$$

CLASS PRACTICE

number fact practice

- Use the white fact cards to practice the last sixteen subtraction facts with your child.
- Give your child **Fact Sheet S 8.1.**
- Time your child for one minute.
- Correct the fact sheet with your child and record the score.
- Allow time for your child to complete the unfinished facts.

WRITTEN PRACTICE

- Complete **Worksheet 109A** with your child.
- Complete **Worksheet 109B** with your child later in the day.

Name _____
(Draw a 9 cm line segment.)

LESSON 109A
Math 2

Date _____
(Measure this line segment using centimeters. __8__ cm)

1. The children in Room 6 collected 16 quarters. They spent 9 quarters for new markers for the classroom. Draw a picture and write a number sentence to show what happened.

Number sentence _____ 16 − 9 = 7 quarters _____

How many quarters do they have left? __7__   How much money is that? $1.75

2. Write the fact family number sentences for 4, 5, and 9.

$4 + 5 = 9$ $9 − 4 = 5$
$5 + 4 = 9$ $9 − 5 = 4$

3. Show 94°F on the thermometer.

4. Find each answer.

$$5\ 2¢ \quad\quad 3\ 6¢ \quad\quad 5\ 8¢$$
$$-\ 2\ 4¢ \quad -\ 2\ 3¢ \quad +\ 3\ 6¢$$
$$2\ 8¢ \quad\quad 1\ 3¢ \quad\quad 9\ 4¢$$

5. Write four hundred twenty using digits. __420__

Write this number in expanded form. ___ 400 + 20 ___

6. Draw a line of symmetry in each shape. Color one side. answers may vary

What fractional part of each shape did you color? $\frac{1}{2}$

Thermometer: 100°F, 90°F, 80°F, 70°F, 60°F, 50°F, 40°F, 30°F, 20°F, 10°F, 0°F, −10°F, −20°F

2-109Wa

Name _____

LESSON 109B
Math 2

Date _____

1. Sharon had 5 quarters. Her sister gave her nine more quarters. Draw a picture and write a number sentence to show what happened.

Number sentence _____ 5 + 9 = 14 quarters _____

How many quarters does Sharon have now? __14__

How much money is that? $3.50

2. Write the fact family number sentences for 6, 7, and 13.

$6 + 7 = 13$ $13 − 6 = 7$
$7 + 6 = 13$ $13 − 7 = 6$

3. Show 26°F on the thermometer.

4. Find each answer.

$$6\ 3¢ \quad\quad 7\ 5¢ \quad\quad 4\ 7¢$$
$$-\ 4\ 7¢ \quad -\ 6\ 1¢ \quad +\ 1\ 6¢$$
$$1\ 6¢ \quad\quad 1\ 4¢ \quad\quad 6\ 3¢$$

5. Write five hundred seven using digits. __507__

Write this number in expanded form. ___ 500 + 7 ___

6. Draw a line of symmetry in each shape. Color one side. answers may vary

What fractional part of each shape did you color? $\frac{1}{2}$

Thermometer: 100°F, 90°F, 80°F, 70°F, 60°F, 50°F, 40°F, 30°F, 20°F, 10°F, 0°F, −10°F, −20°F

2-109Wb

Lesson 110

covering the same design in different ways using tangram pieces

---

lesson preparation

materials

Written Assessment #21

Oral Assessment #11

1 set of tangrams

Master 2-110

demonstration clock

individual clock

Fact Sheet S 5.4

in the morning

• Write the following in the pattern box on the meeting strip:

> ___ , ___ , ___ , 145, 135, 125, ___ , ___ , ___

Answer: 175, 165, 155, 145, 135, 125, 115, 105, 95

• Write $1.24 on the meeting strip. Provide a cup of 6 quarters, a cup of 10 dimes, a cup of 10 nickels, and a cup of 20 pennies.

THE MEETING

calendar

• Ask your child to write the date on the calendar and meeting strip.

• Ask your child the following two or three times a week:

  date _____ days ago, date _____ days from now

  day of the week _____ days ago, day of the week _____ days from now

  _____th month, month before, month after

• Record on the meeting strip a special event and the number of days until it occurs.

weather graph

- Ask your child to read and graph today's temperature to the nearest two degrees.
- Count by 10's and 2's to check the temperature on the graph.
- Ask your child to connect the dot for yesterday's temperature to the dot for today's temperature and compare the temperatures.

counting

- Count by 25's to 300 and backward from 300 by 25's.
- Count by 3's to 30 and backward from 30 by 3's.
- Do the following once a week:

      count by 10's to 400 and backward from 400 by 10's

      count by 5's to 100 and backward from 50 by 5's

      say the even numbers to 100 and backward from 50

      say the odd numbers to 49 and backward from 49

graph questions

- You and your child each ask a question about any of the graphs.

patterning

- Ask your child to do the following:

      identify the pattern (repeating, continuing, or both)

      identify the numbers to complete the pattern

      read the pattern

money

- Ask your child to put the coins in the coin cup. Count the money in the coin cup together.
- Ask your child for another way to show that amount of money. Count these coins together to check the amount.

clock

- Set the clock to a five-minute interval.
- Ask the following:

      "It's (morning/afternoon/evening). What time is it?"

      time one hour ago

      time one hour from now

- Ask your child to write the digital time on the meeting strip.
- Record on the meeting strip the time an activity will occur.

number of the day

- Write three number sentences for the number of the day on the meeting strip.

fact practice

- Write three fact family numbers (e.g., 2, 7, 9) on the chalkboard.
- Allow time for your child to write the four fact family number sentences on the chalkboard.

ASSESSMENT

Written Assessment

"Today I would like to see what you remember from what we have been practicing."

- Give your child **Written Assessment #21**.
- Read the directions for each problem. Allow time for your child to complete that problem before continuing.
- Correct the paper, noting your child's mistakes on the **Individual Recording Form**. Review the errors with your child.

Oral Assessment

- Record your child's response(s) to the oral interview questions on the interview sheet.

THE LESSON

Covering The Same Design in Different Ways Using Tangram Pieces

"A few weeks ago you used tangram pieces to cover designs."

"Today you will learn how to cover the same design in different ways."

- Give your child **Master 2-110**.

"Each of these designs can be covered in more than one way."

"You will try to find two different ways of covering each design."

"Use your tangram pieces to cover shape A."

- Allow time for your child to do this.

"Now you will trace the tangram pieces you used."

- Assist your child as he/she removes one piece at a time and traces the edge of the remaining pieces.

  "Now find a different way to cover shape A."

- Allow time for your child to do this.

- Repeat with shapes B, C, and D.

CLASS PRACTICE

number fact practice

- Use the fact cards to practice all the subtraction facts with your child.

- Give your child **Fact Sheet S 5.4**.

- Time your child for one minute.

- Correct the fact sheet with your child and record the score.

- Allow time for your child to complete the unfinished facts.

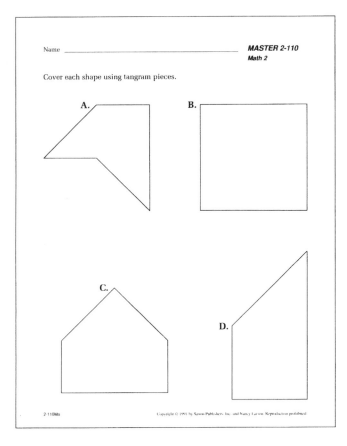

Teacher _____ **MATH 2 LESSON 110**
Date _____   **Oral Assessment # 11 Recording Form**

Materials: Demonstration clock Individual clock Students	• Show a time to a five-minute interval on the demonstration clock. *"It's morning." "What time is it?"*	• Give the child an individual clock. *"Show five forty-five on your clock."* (Vary the time used.)	• Reassess children who did not show mastery on previous assessment questions.

Name _____   ***ASSESSMENT 21***
Date _____   ***LESSON 110***
                                          Math 2

1. There are 22 children in Room 7. Twelve of these children are wearing sneakers. There are 24 children in Room 8. Fifteen of these children are wearing sneakers. Altogether, how many children are wearing sneakers?

   Number sentence ___12 + 15 = 27 children___   Answer _27 children_

2. Carla has 10 quarters. Draw the quarters.

   (Q) (Q) (Q) (Q) (Q) (Q) (Q) (Q) (Q) (Q)

   How much money is that? _$2.50_

3. Show how to share the balloons equally.

   12 balloons   9 balloons

   999999 | 999999 [extras]   9999 | 9999 [extras 9]

   One half of 12 is _6_   One half of 9 is _4 R1_

4. Draw a picture to show four hundred fifty-two. (Use ☐ for 100, ☐ for 10, and ☐ for 1.)

   ☐ ☐ ☐ ☐ ☐ ☐ ☐ ☐ ☐ ☐

   Write this number in expanded form. ___400 + 50 + 2___

   Write four hundred fifty-two using digits. _452_

5. Find the answers.

				3 9
4 × 10 = _40_	9 × 100 = _900_	7 8	9 2	1 6
7 × 10 = _70_	10 − 6 = _4_	+ 3 7	+ 4 9	+ 2 3
		1 1 5	1 4 1	7 8

Lesson 111

subtracting two-digit numbers (part 2)

---

lesson preparation

materials

scrap paper

Fact Sheet M 14.0

in the morning

• Write the following in the pattern box on the meeting strip:

$$\_\_\_, \_\_\_, \_\_\_, 12, 15, 18, \_\_\_, \_\_\_, \_\_\_$$

Answer: 3, 6, 9, 12, 15, 18, 21, 24, 27

• Write $1.02 on the meeting strip. Provide a cup of 6 quarters, a cup of 10 dimes, a cup of 10 nickels, and a cup of 20 pennies.

THE MEETING

calendar

• Ask your child to write the date on the calendar and meeting strip.

• Ask your child the following two or three times a week:

date ____ days ago, date ____ days from now

day of the week ____ days ago, day of the week ____ days from now

____th month, month before, month after

• Record on the meeting strip a special event and the number of days until it occurs.

weather graph

• Ask your child to read and graph today's temperature to the nearest two degrees.

• Count by 10's and 2's to check the temperature on the graph.

• Ask your child to connect the dot for yesterday's temperature to the dot for today's temperature and compare the temperatures.

counting

"Today you will learn how to count by 4's to 40."

"We will stand up as we learn how to do this."

"When we count by 4's, we will touch our toes, touch our knees, touch our waists, and put our hands on our heads."

"We will keep doing this until I say to stop."

"Let's try that."

"Toes, knees, waist, head; toes, knees, waist, head; toes, knees, waist, head."

- Stop after your child has learned the sequence.

"Now instead of saying toes, knees, waist, head, we will count by 1's as we repeat this pattern."

"We will whisper the numbers we say when we touch our toes, knees, and waists."

"We will say aloud the number we say when we touch our head."

"Watch as I do this."

- Repeat the movements as you whisper 1, 2, 3, and say 4 aloud when you touch your head. Whisper 5, 6, 7, and say 8 aloud when you touch your head. Repeat until you reach 40.

"Let's try that together."

- Do this with your child.

"The numbers we say aloud are the numbers we will say when we count by 4's."

- Count by 25's to 300 and backward from 300 by 25's.

- Count by 3's to 30 and backward from 30 by 3's.

- Do the following once a week:

  count by 10's to 400 and backward from 400 by 10's

  count by 5's to 100 and backward from 50 by 5's

  say the even numbers to 100 and backward from 50

  say the odd numbers to 49 and backward from 49

graph questions

- You and your child each ask a question about any of the graphs.

patterning

- Ask your child to do the following:

  identify the pattern (repeating, continuing, or both)

  identify the numbers to complete the pattern

read the pattern

money

- Ask your child to put the coins in the coin cup. Count the money in the coin cup together.

- Ask your child for another way to show that amount of money. Count these coins together to check the amount.

clock

- Set the clock to a five-minute interval.

- Ask the following:

  "It's (morning/afternoon/evening). What time is it?"

  time one hour ago

  time one hour from now

- Ask your child to write the digital time on the meeting strip.

- Record on the meeting strip the time an activity will occur.

number of the day

- Write three number sentences for the number of the day on the meeting strip.

fact practice

- Write three fact family numbers (e.g., 2, 7, 9) on the chalkboard.

- Allow time for your child to write the four fact family number sentences on the chalkboard.

THE LESSON

Subtracting Two-Digit Numbers (Part 2)

"Today you will continue to learn how to subtract two-digit numbers without using dimes and pennies."

- Give your child a piece of scrap paper.

- Write the following on the chalkboard:

$$\begin{array}{r} 94¢ \\ -\ 48¢ \\ \hline \end{array}$$

"Write this problem on your paper."

- Allow time for your child to write the problem.

"Make up a story for this problem."

- Ask your child to make up a some, some went away story for this problem.

  "How much money do we have to begin with?"   94¢

  "How much money are we giving away?"   48¢

  "What do we take away first?"   pennies

  "How many pennies do we have?"   4

  "How many pennies do we have to take away?"   8

  "Do we have enough pennies to do that?"   no

  "What should we do?"   trade 1 dime for 10 pennies

  "If we trade one of the dimes for ten pennies, how many dimes will we have?"   8

- Record the following on the chalkboard problem:

$$\begin{array}{r} {}^{8}\!\!\not{9}\quad 4¢ \\ -\ 4\quad 8¢ \\ \hline \end{array}$$

  "Show this on your paper."

  "Now we will show the ten pennies moving to the penny column."

  "How many pennies do we have now?"   14

- Write a small 1 in front of the 4. **Do not** have your child cross out the 4 and write "14" above.

$$\begin{array}{r} {}^{8}\!\!\not{9}\quad {}^{1}4¢ \\ -\ 4\quad 8¢ \\ \hline \end{array}$$

  "Can we take away eight pennies now?"   yes

  "How many pennies do we have left?"   6

  "What will we do next?"   take away 4 dimes

  "How many dimes do we have?"   8

  "How many dimes are we taking away?"   4

  "How many dimes are left?"   4

  "Show this on your paper."

  "Let's try another problem."

- Write the following on the chalkboard:

$$\begin{array}{r} 85¢ \\ -\ 61¢ \\ \hline \end{array}$$

  "Write this problem on your paper."

- Allow time for your child to write the problem.

  "Make up a story for this problem."

- Ask your child to make up a some, some went away story for this problem.

  "How much money do we have to begin with?"   85¢

  "How much money are we giving away?"   61¢

  "What do we take away first?"   pennies

  "How many pennies do we have?"   5

  "How many pennies do we have to take away?"   1

  "Do we have enough pennies?"   yes

  "How many pennies do we have left?"   4

  "What will we do next?"   take away 6 dimes

  "How many dimes do we have?"   3

  "How many dimes are we taking away?"   6

  "How many dimes are left?"   2

  "Show this on your paper."

  "Let's try another problem."

- Write the following on the chalkboard:

$$57¢$$
$$-19¢$$

  "Write this problem on your paper."

- Allow time for your child to write the problem.

  "Make up a story for this problem."

- Ask your child to make up a some, some went away story for this problem.

  "How much money do we have to begin with?"   57¢

  "How much money are we giving away?"   19¢

  "What do we take away first?"   pennies

  "How many pennies do we have?"   7

  "How many pennies do we have to take away?"   9

  "Do we have enough pennies?"   no

  "What should we do?"   trade 1 dime for 10 pennies

  "If we trade one of the dimes for ten pennies, how many dimes will we have?"   4

- Record the following on the chalkboard problem:

$$^4\cancel{5}\ 7¢$$
$$-1\ \ 9¢$$

  "Show this on your paper."

"Now we will show the ten pennies moving to the penny column."

"How many pennies do we have now?"   17

• Write a small 1 in front of the 7. **Do not** have your child cross out the 7 and write "17" above.

$$^4\cancel{5} \quad ^17¢$$
$$\underline{-\ 1 \quad 9¢}$$

"Can we take away nine pennies now?"   yes

"How many pennies do we have left?"   8

"What will we do next?"   take away 1 dime

"How many dimes do we have?"   4

"How many dimes are we taking away?"   1

"How many dimes are left?"   3

"Show this on your paper."

"There are three steps for finding the answer for a subtraction problem."

"First you must check to see if you have enough pennies to begin with."

"If you don't have enough pennies, the second step will be to trade a dime for ten pennies."

"The third and last step is to subtract the pennies and then the dimes."

"Let's try that with these problems."

• Write the following problem on the chalkboard:

$$55¢$$
$$\underline{-\ 27¢}$$

"Write this problem on your paper."

• Ask the following questions:

"Do we have enough pennies?"

• If the answer is no, ask the next question:

"What will we do?"   trade 1 dime for 10 pennies

"What is the last step?"   subtract the pennies and subtract the dimes

"Show how to find the answer for this problem."

• Repeat with the following problems:

$$63¢ \qquad\qquad 82¢$$
$$\underline{-\ 56¢} \qquad\quad \underline{-\ 12¢}$$

"Make up a subtraction problem where we will have to trade a dime for ten pennies."

• Write your child's problem on the chalkboard.

"How do you know that you will have to trade a dime for ten pennies?"

"Find the answer for this problem."

"Make up a subtraction problem where we will not have to trade a dime for ten pennies."

• Write your child's problem on the chalkboard.

"How do you know that you will not have to trade a dime for ten pennies?"

"Find the answer for this problem."

CLASS PRACTICE

number fact practice

• Use the yellow multiplication fact cards to practice the multiplying by five facts with your child.

• Give your child **Fact Sheet M 14.0.**

• Time your child for one minute.

• Correct the fact sheet with your child and record the score.

• Allow time for your child to complete the unfinished facts.

WRITTEN PRACTICE

• Complete **Worksheet 111A** with your child.

• Complete **Worksheet 111B** with your child later in the day.

Name _____
(Draw a 10 cm line segment.)

LESSON 111A
Math 2

Date _____
(Measure this line segment using centimeters. __4__ cm)

1. There are 16 markers in a package. Kristina will share them equally with her cousin Michael. Show how the children will share the markers.

   Kristina   Michael

   How many markers will each child have? __8__

2. Color $3\frac{3}{4}$ squares.

3. Write the fact family number sentences for 7, 16, and 9.

   $7 + 9 = 16$ $9 + 7 = 16$

   $16 - 9 = 7$ $16 - 7 = 9$

4. Find the answers.

   $\begin{array}{r} 6\,3\,¢ \\ -\,3\,7\,¢ \\ \hline 2\,6\,¢ \end{array}$ $\begin{array}{r} 4\,8\,¢ \\ -\,1\,6\,¢ \\ \hline 3\,2\,¢ \end{array}$ $\begin{array}{r} 5\,0\,¢ \\ -\,3\,4\,¢ \\ \hline 1\,6\,¢ \end{array}$

5. This is the time I get up in the morning.

   What time is it? __6:15 a.m.__

6. Find the answers.

   $6 \times 5 = $ __30__ $4 \times 10 = $ __40__ $\begin{array}{r} 2\,4 \\ 3\,6 \\ +\,5\,9 \\ \hline 1\,1\,9 \end{array}$ $\begin{array}{r} 4\,8 \\ 9\,3 \\ +\,1\,6 \\ \hline 1\,5\,7 \end{array}$

   $7 \times 1 = $ __7__ $9 \times 5 = $ __45__

   $3 \times 100 = $ __300__ $0 \times 1 = $ __0__

Name _____

LESSON 111B
Math 2

Date _____

1. There are 10 postcards in a package. Bruce will share them equally with his sister Sarah. Show how the children will share the postcards.

   Bruce   Sarah

   How many postcards will each child have? __5__

2. Color $4\frac{1}{4}$ squares.

3. Write the fact family number sentences for 12, 3, and 9.

   $3 + 9 = 12$ $9 + 3 = 12$

   $12 - 9 = 3$ $12 - 3 = 9$

4. Find the answers.

   $\begin{array}{r} 5\,1\,¢ \\ -\,1\,7\,¢ \\ \hline 3\,4\,¢ \end{array}$ $\begin{array}{r} 7\,0\,¢ \\ -\,5\,2\,¢ \\ \hline 1\,8\,¢ \end{array}$ $\begin{array}{r} 6\,7\,¢ \\ -\,4\,3\,¢ \\ \hline 2\,4\,¢ \end{array}$

5. This is the time I eat lunch.

   What time is it? __11:45 a.m.__

6. Find the answers.

   $4 \times 5 = $ __20__ $6 \times 10 = $ __60__ $\begin{array}{r} 1\,9 \\ 2\,3 \\ +\,8\,2 \\ \hline 1\,2\,4 \end{array}$ $\begin{array}{r} 2\,1 \\ 5\,3 \\ +\,6\,4 \\ \hline 1\,3\,8 \end{array}$

   $0 \times 10 = $ __0__ $7 \times 5 = $ __35__

   $7 \times 100 = $ __700__ $4 \times 1 = $ __4__

Lesson 112

measuring weight using pounds

lesson preparation

materials

Meeting Book

bathroom scale

2 food articles that weigh 1 pound (butter, sugar, etc.)

10-pound object (e.g., bag of potatoes)

3 heavy objects that will fit on a bathroom scale

Fact Sheet S 8.1

in the morning

• Write the following in the pattern box on the meeting strip:

☆, ☆, ☆, ☆, ——, ——, ——, ——, ——, ——

Answer: ☆, ☆, ☆, ☆, ☆, ☆, ☆, ☆, ☆, ☆

• Write $1.09 on the meeting strip. Provide a cup of 6 quarters, a cup of 10 dimes, a cup of 10 nickels, and a cup of 20 pennies.

THE MEETING

calendar

• Ask your child to write the date on the calendar and meeting strip.

"How many days are there in one week?"

"How many days are there in two weeks?"

"Let's count to check."

• Ask your child the following two or three times a week:

date _____ days ago, date _____ days from now

day of the week _____ days ago, day of the week _____ days from now

_____th month, month before, month after

• Record on the meeting strip a special event and the number of days until it occurs.

weather graph

- Ask your child to read and graph today's temperature to the nearest two degrees.

- Count by 10's and 2's to check the temperature on the graph.

- Ask your child to connect the dot for yesterday's temperature to the dot for today's temperature and compare the temperatures.

counting

"Yesterday you learned how to count by 4's to 40."

"How did we do that?" we counted by 4's as we touched our toes, touched our knees, touched our waists, and put our hands on our heads

"Remember, we will whisper the numbers we say when we touch our toes, knees, and waist."

"We will say aloud the number we say when we touch our head."

"Let's try that."

- Do this with your child.

"The numbers we say aloud are the numbers we will say when we count by 4's."

"Let's count by 4's to 40 together as I write the numbers on a counting strip in the Meeting Book."

- Record the numbers on the Meeting Book counting strip as you count by 4's to 40 with your child.

- Count by 25's to 300 and backward from 300 by 25's.

- Count by 3's to 30 and backward from 30 by 3's.

- Do the following once a week:

    count by 10's to 400 and backward from 400 by 10's

    count by 5's to 100 and backward from 50 by 5's

    say the even numbers to 100 and backward from 50

    say the odd numbers to 49 and backward from 49

graph questions

- You and your child each ask a question about any of the graphs.

patterning

- Ask your child to do the following:

    identify the pattern (repeating, continuing, or both)

    identify the shapes to complete the pattern

    read the pattern

money

- Ask your child to put the coins in the coin cup. Count the money in the coin cup together.
- Ask your child for another way to show that amount of money. Count these coins together to check the amount.

clock

- Set the clock to a five-minute interval.
- Ask the following:

  "It's (morning/afternoon/evening). What time is it?"

  time one hour ago

  time one hour from now
- Ask your child to write the digital time on the meeting strip.
- Record on the meeting strip the time an activity will occur.

number of the day

- Write three number sentences for the number of the day on the meeting strip.

fact practice

- Write three fact family numbers (e.g., 2, 7, 9) on the chalkboard.
- Allow time for your child to write the four fact family number sentences on the chalkboard.

THE LESSON

Measuring Weight Using Pounds

"A few weeks ago we measured our height in feet and inches."

"When you go to the doctor's office, what else does the nurse usually do before you see the doctor?" checks your weight

"How does the nurse do that?" using a scale

"When we measure our weight, do we tell people that we weigh 50 miles or 50 tons?"

"What unit of measure do we usually use?" pounds

"We can tell people how much we weigh in different ways."

"We can tell them how many pounds we weigh or we can tell them how many kilograms we weigh."

"Today you will learn how to measure weight using pounds."

"We can't see a pound."

"It is something we have to feel."

"Something small can weigh a pound or something large can weigh a pound."

- Show your child two food products that each weigh a pound, but are different sizes. (Use a pound of butter, a pound of sugar, or any other object that weighs a pound.)

"This (package of butter) weighs a pound."

"This (bag of potatoes) weighs ten pounds."

"Do you weigh more than a bag of potatoes?"

"How do you know?" I can pick up the potatoes

"Let's try to estimate how many pounds you weigh."

"If this bag of potatoes weighs ten pounds, how much do you think you weigh?"

- Write the estimate on the chalkboard.

"What can we use to find out how much something weighs?" a scale

"Where do we find scales?" in grocery stores, in doctors' offices, in homes

- Show your child the bathroom scale.

"Let's look at how our scale is marked."

- Allow time for your child to examine the scale.

"What will happen when you step on the scale?"

"Let's try it."

- Ask your child to step on the scale.

"Let's read the scale to find your weight."

"What ten did we pass?"

"How many more pounds do you weigh?"

"We can say that you weigh _____ pounds."

- Write the weight on the chalkboard.

"Do you think I weigh more or less than you?"

"Let's try to estimate how many pounds I weigh."

- Write the estimate on the chalkboard.

- Step on the scale.

"Let's read the scale to find my weight."

"What ten did we pass?"

"How many more pounds do I weigh?"

> *"We can say that I weigh _____ pounds."*

- Write the weight on the chalkboard.

- Use a heavy object that will fit on the scale.

  "Do you think that this _____ weighs more or less than you?"

  "Why do you think that?"   I can pick up the object

  "Do you think this weighs more than ten pounds?"

  "Let's use the bag of potatoes to check."

- Ask your child to hold the bag of potatoes and then the object.

  "Do you think it weighs more than ten pounds?"

  "Let's weigh it to see."

- Weigh the object.

- Repeat with two other objects.

CLASS PRACTICE

number fact practice

- Use the white fact cards to practice the last sixteen subtraction facts with your child.

- Give your child **Fact Sheet S 8.1.**

- Time your child for one minute.

- Correct the fact sheet with your child and record the score.

- Allow time for your child to complete the unfinished facts.

WRITTEN PRACTICE

- Complete **Worksheet 112A** with your child.

- Complete **Worksheet 112B** with your child later in the day.

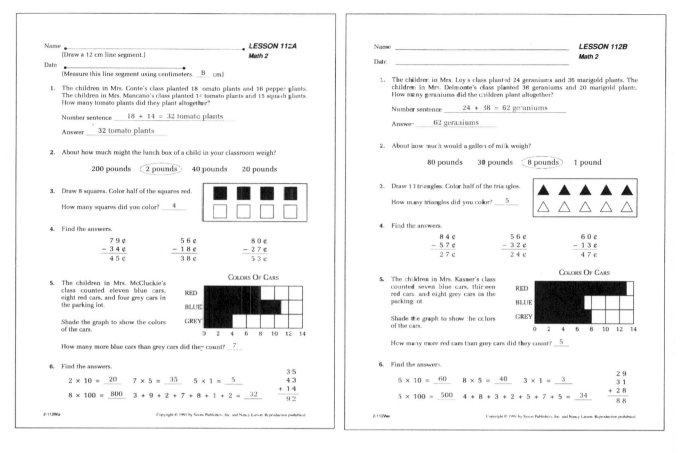

Name _____   **LESSON 112A**
(Draw a 12 cm line segment.)   **Math 2**

Date _____

(Measure this line segment using centimeters. __8__ cm)

1. The children in Mrs. Conte's class planted 18 tomato plants and 16 pepper plants. The children in Mrs. Mancano's class planted 14 tomato plants and 15 squash plants. How many tomato plants did they plant altogether?

   Number sentence ___18 + 14 = 32 tomato plants___

   Answer ___32 tomato plants___

2. About how much might the lunch box of a child in your classroom weigh?

   200 pounds   (2 pounds)   40 pounds   20 pounds

3. Draw 8 squares. Color half of the squares red.

   How many squares did you color? __4__

4. Find the answers.

7 9 ¢	5 6 ¢	8 0 ¢
− 3 4 ¢	− 1 8 ¢	− 2 7 ¢
4 5 ¢	3 8 ¢	5 3 ¢

5. The children in Mrs. McCluckie's class counted eleven blue cars, eight red cars, and four grey cars in the parking lot.

   Shade the graph to show the colors of the cars.

   COLORS OF CARS

   RED / BLUE / GREY   0 2 4 6 8 10 12 14

   How many more blue cars than grey cars did they count? __7__

6. Find the answers.

   2 × 10 = __20__   7 × 5 = __35__   5 × 1 = __5__

   8 × 100 = __800__   3 + 9 + 2 + 7 + 8 + 1 + 2 = __32__

3 5
4 3
+ 1 4
9 2

2-112Wa

Name _____   **LESSON 112B**
    Math 2

Date _____

1. The children in Mrs. Ley's class planted 24 geraniums and 36 marigold plants. The children in Mrs. Delmonte's class planted 38 geraniums and 20 marigold plants. How many geraniums did the children plant altogether?

   Number sentence ___24 + 38 = 62 geraniums___

   Answer ___62 geraniums___

2. About how much would a gallon of milk weigh?

   80 pounds   30 pounds   (8 pounds)   1 pound

3. Draw 10 triangles. Color half of the triangles.

   How many triangles did you color? __5__

4. Find the answers.

8 4 ¢	5 6 ¢	6 0 ¢
− 5 7 ¢	− 3 2 ¢	− 1 3 ¢
2 7 ¢	2 4 ¢	4 7 ¢

5. The children in Mrs. Kasner's class counted seven blue cars, thirteen red cars, and eight grey cars in the parking lot.

   Shade the graph to show the colors of the cars.

   COLORS OF CARS

   RED / BLUE / GREY   0 2 4 6 8 10 12 14

   How many more red cars than grey cars did they count? __5__

6. Find the answers.

   6 × 10 = __60__   8 × 5 = __40__   3 × 1 = __3__

   5 × 100 = __500__   4 + 8 + 3 + 2 + 5 + 7 + 5 = __34__

2 9
3 1
+ 2 8
8 8

2-112Wa

Lesson 113

finding perimeter

THE MEETING

calendar

• Ask your child to write the date on the calendar and meeting strip.

"How many days are there in one week?"

"How many days are there in two weeks?"

"Let's count to check."

• Ask your child the following two or three times a week:

  date _____ days ago, date _____ days from now

  day of the week _____ days ago, day of the week _____ days from now

  _____th month, month before, month after

• Record on the meeting strip a special event and the number of days until it occurs.

weather graph

- Ask your child to read and graph today's temperature to the nearest two degrees.
- Count by 10's and 2's to check the temperature on the graph.
- Ask your child to connect the dot for yesterday's temperature to the dot for today's temperature and compare the temperatures.

counting

"Let's count by 4's to 40."

"How did we do that yesterday?"   we counted by 4's as we touched our toes, touched our knees, touched our waists, and put our hands on our heads

"Remember, we will whisper the numbers we say when we touch our toes, knees, and waist."

"We will say aloud the number we say when we touch our head."

- Do this with your child.

"Now let's count by 4's to 40 and backward from 40 by 4's using our Meeting Book counting strip."

- Count by 25's to 300 and backward from 300 by 25's.
- Count by 3's to 30 and backward from 30 by 3's.
- Do the following once a week:

    count by 10's to 400 and backward from 400 by 10's

    count by 5's to 100 and backward from 50 by 5's

    say the even numbers to 100 and backward from 50

    say the odd numbers to 49 and backward from 49

graph questions

- You and your child each ask a question about any of the graphs.

patterning

- Ask your child to do the following:

    identify the pattern (repeating, continuing, or both)

    identify the numbers to complete the pattern

    read the pattern

money

- Ask your child to put the coins in the coin cup. Count the money in the coin cup together.

- Ask your child for another way to show that amount of money. Count these coins together to check the amount.

clock

- Set the clock to a five-minute interval.
- Ask the following:

  "It's (morning/afternoon/evening). What time is it?"

  time one hour ago

  time one hour from now

- Ask your child to write the digital time on the meeting strip.
- Record on the meeting strip the time an activity will occur.

number of the day

- Write three number sentences for the number of the day on the meeting strip.

fact practice

- Write three fact family numbers (e.g., 2, 7, 9) on the chalkboard.
- Allow time for your child to write the four fact family number sentences on the chalkboard.

THE LESSON

Finding Perimeter

"Today you will learn how to find the perimeter of a shape."

"To find the perimeter of a shape, we measure all the sides of a shape and add them together."

"Finding perimeter is just like a some, some more problem."

- Show your child the 20 cm × 30 cm construction paper rectangle.

"I would like to put a piece of string along the outside edge of this rectangle to make a border."

"Let's measure to see how much string I will need."

"How many sides will we measure?"   4

"Today we will measure using the centimeter side of the ruler."

"When we measure a side, we put the beginning of the ruler (the zero) below the endpoint of one side of the rectangle."

- Demonstrate.

"Look along the ruler until you come to the other endpoint."

"What is the length of this side?"

• Write the length next to the side of the rectangle using a dark marker.

"I will need _____ centimeters of string for this side."

"Now we will measure the next side."

"How many centimeters long do you think it is?"

• Turn the rectangle to measure the next side.

"How will we measure this side?"

• Ask your child to restate the steps.

• Write the length next to the side of the rectangle using a dark marker.

"I will need _____ centimeters of string for this side."

"Now we will measure the next side."

"How many centimeters long do you think it is?"

"How do you know?"   it is the same as the first side

• Turn the rectangle to measure the next side.

"Let's measure this side to check."

• Write the length next to the side of the rectangle.

"Now we will measure the last side."

"How many centimeters long do you think it is?"

"How do you know?"   it is the same as the opposite side

"Let's measure this side to check."

• Write the length next to the side of the rectangle.

"What do you notice about the lengths of the sides of the rectangle?"
opposite sides are the same length

"How will we find how much string I will need to go around the outside of the rectangle?"   add the sides

"How many sides does this shape have?"   4

"How many numbers will we add to find the perimeter?"   4

"What number sentence will we write to show how to find the perimeter of this rectangle?"   20 cm + 30 cm + 20 cm + 30 cm = 100 cm

• Write the number sentence on the chalkboard.

"What is the perimeter of the rectangle?"   100 cm

"Now you will have a chance to find the perimeter of some shapes."

• Give your child **Master 2-113** and a ruler.

"You will measure the sides of each of these shapes and find the perimeter."

"You will use the centimeter side of your ruler."

"How many sides does shape A have?" 3

"What do we call this shape?" triangle

"How many sides will you measure?" 3

"Point to the vertical line segment."

"Measure this line segment."

- Assist your child as he/she measures.

"How many centimeters long is the side?" 8 cm

"Write this number next to the side."

- Repeat with the oblique and horizontal line segments.

"How will you find the perimeter of the triangle?" add the lengths of the sides

"Write a number sentence to show how to find the perimeter of the triangle."

"What number sentence did you write?"

"What is the perimeter of the triangle?" 24 cm

"What do we call shape B?" square

"How many sides will you measure to find the perimeter?"

- Your child might say 4 sides because a square has 4 sides, 2 sides because opposite sides are equal, or 1 side because all the sides of a square have the same length.

"Point to the vertical line segment on the left."

"Measure this line segment."

"How many centimeters long is this side?" 6 cm

"Write this number next to the side."

"Point to the vertical line segment on the right."

"How long do you think this line segment is?"

"Why?"

"Measure this line segment."

"How many centimeters long is this side?" 6 cm

"Write this number next to the side."

- Repeat with the horizontal line segments.

"How will you find the perimeter of the square?" add the lengths of the sides

"Write a number sentence to show how to find the perimeter of the square."

"What number sentence did you write?"

"What is the perimeter of the square?"　24 cm

"What do we call shape C?"　rectangle

"How many sides will you measure to find the perimeter?"

- Your child might say 4 sides because the rectangle has 4 sides or 2 sides because opposite sides are equal.

"Point to the vertical line segment on the left."

"Measure this line segment."

"How many centimeters long is this side?"

"Write this number next to the side."

"Point to the vertical line segment on the right."

"How long do you think this line segment will be?"

"Why?"

"Measure this line segment."

"How many centimeters long is this side?"　5 cm

"Write this number next to the side."

- Repeat with the horizontal line segments.

"How will you find the perimeter of the rectangle?"　add the lengths of the sides

"Write a number sentence to show how to find the perimeter of the rectangle."

"What number sentence did you write?"

"What is the perimeter of the rectangle?"　24 cm

- Repeat with the last rectangle.

CLASS PRACTICE

number fact practice

- Give your child **Fact Sheet A 2-10C**.
- Time your child for five minutes.
- Correct the fact sheet with your child and record the score.
- Allow time for your child to complete the unfinished facts.

WRITTEN PRACTICE

- Complete **Worksheet 113A** with your child.
- Complete **Worksheet 113B** with your child later in the day.

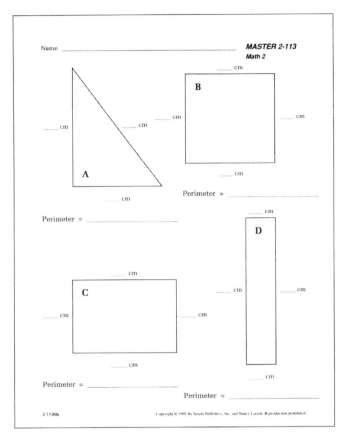

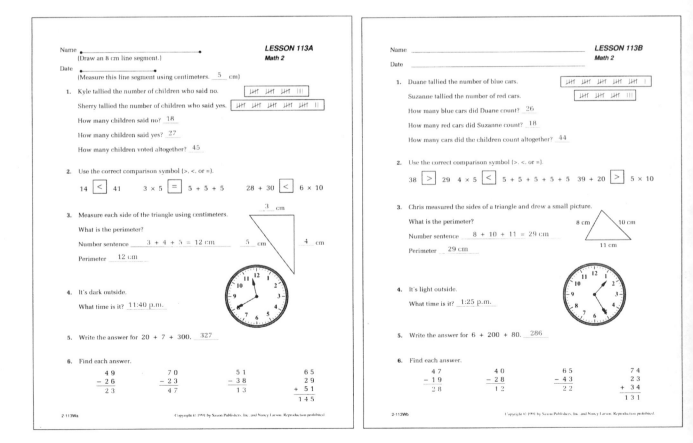

Name _____
(Draw an 8 cm line segment.)

LESSON 113A
Math 2

Date _____
(Measure this line segment using centimeters. __5__ cm)

1. Kyle tallied the number of children who said no. | LHT LHT LHT III |

   Sherry tallied the number of children who said yes. | LHT LHT LHT LHT LHT II |

   How many children said no? __18__

   How many children said yes? __27__

   How many children voted altogether? __45__

2. Use the correct comparison symbol (>. <. or =).

   14 | < | 41   3 × 5 | = | 5 + 5 + 5   28 + 30 | < | 6 × 10

3. Measure each side of the triangle using centimeters.

   What is the perimeter?

   Number sentence __3 + 4 + 5 = 12 cm__   __5__ cm   __4__ cm

   __3__ cm

   Perimeter __12 cm__

4. It's dark outside.

   What time is it? __11:40 p.m.__

5. Write the answer for 20 + 7 + 300. __327__

6. Find each answer.

49	70	51	65
− 26	− 23	− 38	29
23	47	13	+ 51
			145

2-113Wa   Copyright © 1991 by Saxon Publishers, Inc. and Nancy Larson. Reproduction prohibited.

Name _____

LESSON 113B
Math 2

Date _____

1. Duane tallied the number of blue cars. | LHT LHT LHT LHT LHT I |

   Suzanne tallied the number of red cars. | LHT LHT LHT III |

   How many blue cars did Duane count? __26__

   How many red cars did Suzanne count? __18__

   How many cars did the children count altogether? __44__

2. Use the correct comparison symbol (>. <. or =).

   38 | > | 29   4 × 5 | < | 5 + 5 + 5 + 5 + 5   39 + 20 | > | 5 × 10

3. Chris measured the sides of a triangle and drew a small picture.

   What is the perimeter?   8 cm   10 cm

   Number sentence __8 + 10 + 11 = 29 cm__   11 cm

   Perimeter __29 cm__

4. It's light outside.

   What time is it? __1:25 p.m.__

5. Write the answer for 6 + 200 + 80. __286__

6. Find each answer.

47	40	65	74
− 19	− 28	− 43	23
28	12	22	+ 34
			131

2-113Wb   Copyright © 1991 by Saxon Publishers, Inc. and Nancy Larson. Reproduction prohibited.

Lesson 114

writing observations from a graph

---

lesson preparation

materials

Meeting Book
scrap paper
Fact Sheet S 8.2

in the morning

• Write the following in the pattern box on the meeting strip:

$$5\tfrac{1}{2},\ 5,\ 4\tfrac{1}{2},\ 4,\ \underline{\quad},\ \underline{\quad},\ \underline{\quad},\ \underline{\quad},\ \underline{\quad},\ \underline{\quad},\ \underline{\quad}$$

Answer: $5\tfrac{1}{2}$, 5, $4\tfrac{1}{2}$, 4, $3\tfrac{1}{2}$, 3, $2\tfrac{1}{2}$, 2, $1\tfrac{1}{2}$, 1, $\tfrac{1}{2}$

• Write $1.11 on the meeting strip. Provide a cup of 6 quarters, a cup of 10 dimes, a cup of 10 nickels, and a cup of 20 pennies.

THE MEETING

calendar

• Ask your child to write the date on the calendar and meeting strip.

"How many days are there in one week?"

"How many days are there in two weeks?"

"Let's count to check."

• Ask your child the following two or three times a week:

date _____ days ago, date _____ days from now

day of the week _____ days ago, day of the week _____ days from now

_____th month, month before, month after

• Record on the meeting strip a special event and the number of days until it occurs.

weather graph

• Ask your child to read and graph today's temperature to the nearest two degrees.

- Count by 10's and 2's to check the temperature on the graph.
- Ask your child to connect the dot for yesterday's temperature to the dot for today's temperature and compare the temperatures.

counting

"Let's count by 4's to 40."

"How did we do that yesterday?" we counted by 4's as we touched our toes, touched our knees, touched our waists, and put our hands on our heads

"Remember, we will whisper the numbers we say when we touch our toes, knees, and waist."

"We will say aloud the number we say when we touch our head."

- Do this with your child.

"Now let's count by 4's to 40 and backward from 40 by 4's using our Meeting Book counting strip."

- Count by 25's to 300 and backward from 300 by 25's.
- Count by 3's to 30 and backward from 30 by 3's.
- Do the following once a week:

    count by 10's to 400 and backward from 400 by 10's

    count by 5's to 100 and backward from 50 by 5's

    say the even numbers to 100 and backward from 50

    say the odd numbers to 49 and backward from 49

graph questions

- You and your child each ask a question about any of the graphs.

patterning

- Ask your child to do the following:

    identify the pattern (repeating, continuing, or both)

    identify the numbers to complete the pattern

    read the pattern

money

- Ask your child to put the coins in the coin cup. Count the money in the coin cup together.
- Ask your child for another way to show that amount of money. Count these coins together to check the amount.

clock

- Set the clock to a five-minute interval.

- Ask the following:

  "It's (morning/afternoon/evening). What time is it?"

  time one hour ago

  time one hour from now

- Ask your child to write the digital time on the meeting strip.
- Record on the meeting strip the time an activity will occur.

number of the day

- Write three number sentences for the number of the day on the meeting strip.

fact practice

- Write three fact family numbers (e.g., 2, 7, 9) on the chalkboard.
- Allow time for your child to write the four fact family number sentences on the chalkboard.

THE LESSON

Writing Observations from a Graph

"Today you will learn how to write observations about a graph."

"Observations are the facts about the graph."

"They are the things you know when you look at the graph."

"Let's pretend we are going to have an ice cream party."

"Who could we invite to an ice cream party?"

- List the names on a piece of paper.

"I will buy three flavors of ice cream for the party."

"Before I buy the ice cream, I will need to know what ice cream flavors the people who come to the party like."

"I also will need to know how much of each flavor to buy."

"How can I find this information?"

- Ask your child for suggestions.

"What are some of the ice cream flavors you like?"

"What are some of the ice cream flavors other people like?"

- List the flavors on the chalkboard.

"Which three flavors should we use for our party?"

- Open the Meeting Book to page 35.

- Ask your child to color the ice cream scoops to show the flavors selected.

  "What flavor would you like?"

  "Write your initials in the box next to the name of the ice cream flavor you chose."

- Allow time for your child to do this.

  "I like _____."

  "Write my initials in the box next to _____."

  "Let's ask the other people on our list which flavor ice cream they would choose."

- Ask your child to call each person and record the person's initials in the correct box.

- When your child finishes the graph, continue the lesson.

  "Today, instead of talking about our graph, you will write all the things the graph tells us about the ice cream flavors we like."

- Give your child a piece of paper.

  "Write your observations about the graph on this paper."

  "Write as many facts as you can about the information on this graph."

- Provide time for your child to write his/her observations.

- Ask your child to read his/her observations.

- Optional: Have an ice cream party based on the results of the graph.

CLASS PRACTICE

number fact practice

- Use the white fact cards to practice the last sixteen subtraction facts with your child.

- Give your child **Fact Sheet S 8.2.**

- Time your child for one minute.

- Correct the fact sheet with your child and record the score.

- Allow time for your child to complete the unfinished facts.

WRITTEN PRACTICE

- Complete **Worksheet 114A** with your child.

- Complete **Worksheet 114B** with your child later in the day.

Name _____ **LESSON 114A**
(Draw a 10 cm line segment. • **Math 2**
Date _____

(Measure this line segment using centimeters. __9__ cm)

1. There are 124 children in grade 1, 147 children in grade 2, and 119 children in grade 3.

    Which grade has the most children? ___grade 2___

    Which grade has the fewest children? ___grade 3___

    Write the names of the grades in order from the one that has the most children to the one that has the fewest children.

    ___grade 2___   ___grade 1___   ___grade 3___

2. I have 4 dimes, 3 nickels, and 7 pennies. Draw the coins. How much money do I have? Write the amount two ways.

    (D)(D)(N) (N)(P)(P)(P)
    (D)(D)(N) (P)(P)(P)(P)

    ___62¢___   ___$0.62___

3. Measure each side of the rectangle in Problem 2 using centimeters. Find the perimeter.

    Number sentence __8 + 8 + 2 + 2 = 20 cm__   What is the perimeter? __20 cm__

4. Fill in a number to make each number sentence true. some answers may vary

    [ ] < 6   13 < [ ]   4 + 7 = 7 + [4]

5. The children in Miss Quinn's class made the following graph. Write something you know about the children.

    ___answers may vary___

    FAVORITE ICE CREAM FLAVORS
    chocolate
    vanilla
    strawberry

6. Find the answers.

    77 + 46 = __123__   9 + 88 = __97__

    8 6   9 8
    − 2 7   − 3 5
    5 9   6 3

2-114Wa

Name _____ **LESSON 114B**
  Math 2
Date _____

1. There are 195 children in grade 4, 215 children in grade 5, and 178 children in grade 6.

    Which grade has the most children? ___grade 5___

    Which grade has the fewest children? ___grade 6___

    Write the names of the grades in order from the one that has the most children to the one that has the fewest children.

    ___grade 5___   ___grade 4___   ___grade 6___

2. I have 5 dimes, 3 nickels, and 8 pennies. Draw the coins. How much money do I have? Write the amount two ways. 3 cm

    7 cm
    (D)(N)(P)(P)
    (D)(N)(P)(P)(P)
    (D)(N)(P)(P)

    ___73¢___   ___$0.73___

3. Someone measured each side of the rectangle in Problem 2 using centimeters. Find the perimeter.

    Number sentence __7 + 7 + 3 + 3 = 20 cm__   What is the perimeter? __20 cm__

4. Fill in a number to make each number sentence true. some answers may vary

    [ ] > 9   7 > [ ]   8 + [5] = 5 + 8

5. The children in Miss Padilla's class made the following graph. Write something you know about the children.

    ___answers may vary___

    FAVORITE ICE CREAM FLAVORS
    chocolate
    vanilla
    strawberry

6. Find the answers.

    86 + 35 = __121__   7 + 94 = __101__

    9 1   2 3
    − 3 5   − 1 1
    5 6   1 2

2-114Wb

Lesson 115

identifying parallel lines

---

lesson preparation

---

materials

Written Assessment #22

2 geoboards

8 geobands

Fact Sheet S 8.2

in the morning

• Write the following in the pattern box on the meeting strip:

| 70, 75, 80, 85, ____, ____, ____, ____, ____, ____ |

Answer: 70, 75, 80, 85, 90, 95, 100, 105, 110, 115

• Write $1.23 on the meeting strip. Provide a cup of 6 quarters, a cup of 10 dimes, a cup of 10 nickels, and a cup of 20 pennies.

THE MEETING

calendar

• Ask your child to write the date on the calendar and meeting strip.

"How many days are there in one week? ... two weeks?"

"How many days are there in three weeks?"

"Let's count to check."

• Ask your child the following two or three times a week:

date _____ days ago, date _____ days from now

day of the week _____ days ago, day of the week _____ days from now

_____th month, month before, month after

• Record on the meeting strip a special event and the number of days until it occurs.

---

weather graph

- Ask your child to read and graph today's temperature to the nearest two degrees.

- Count by 10's and 2's to check the temperature on the graph.

- Ask your child to connect the dot for yesterday's temperature to the dot for today's temperature and compare the temperatures.

counting

- Count by 4's to 40 and backward from 40 by 4's.

- Count by 25's to 300 and backward from 300 by 25's.

- Count by 3's to 30 and backward from 30 by 3's.

- Do the following once a week:

    count by 10's to 400 and backward from 400 by 10's

    count by 5's to 100 and backward from 50 by 5's

    say the even numbers to 100 and backward from 50

    say the odd numbers to 49 and backward from 49

graph questions

- You and your child each ask a question about any of the graphs.

patterning

- Ask your child to do the following:

    identify the pattern (repeating, continuing, or both)

    identify the numbers to complete the pattern

    read the pattern

money

- Ask your child to put the coins in the coin cup. Count the money in the coin cup together.

- Ask your child for another way to show that amount of money. Count these coins together to check the amount.

clock

- Set the clock to a five-minute interval.

- Ask the following:

    "It's (morning/afternoon/evening). What time is it?"

    time one hour ago

    time one hour from now

- Ask your child to write the digital time on the meeting strip.

• Record on the meeting strip the time an activity will occur.

number of the day

• Write three number sentences for the number of the day on the meeting strip.

fact practice

• Write three fact family numbers (e.g., 2, 7, 9) on the chalkboard.

• Allow time for your child to write the four fact family number sentences on the chalkboard.

ASSESSMENT

Written Assessment

"Today I would like to see what you remember from what we have been practicing."

• Give your child **Written Assessment #22**.

• Read the directions for each problem. Allow time for your child to complete each problem before continuing.

• Correct the paper, noting your child's mistakes on the **Individual Recording Form**. Review the errors with your child.

THE LESSON

Identifying Parallel Lines

"Today you will learn a new word that mathematicians use."

"The new word is 'parallel.' "

• Draw examples of parallel lines like the following on the chalkboard:

"What is the same about all of these pairs of lines?"

"Parallel lines will always stay the same distance apart."

"Parallel lines will never meet."

"Can you find an example of parallel lines in this room?"

• Allow time for your child to find several examples.

• Write the word "parallel" on the chalkboard.

"We can remember what the word parallel means by remembering the two L's in the middle of the word parallel."

"Now you will use a geoboard to make examples of parallel line segments."

• Give your child and yourself a geoboard and four geobands.

"Make a line segment on your geoboard using one geoband."

"Now make a parallel line segment on your geoboard."

• Allow time for your child to make the line segments.

"Take the geobands off the geoboard."

"Make two different parallel line segments on your geoboard."

• Allow time for your child to make the line segments.

"Take the geobands off the geoboard."

• Make the following shape on your geoboard. Do not show your child.

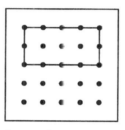

"I made a shape on my geoboard."

"I will give you some clues."

"Try to make a shape just like mine."

"Use a different geoband for each side of the shape."

"One side of my shape stretches from the top left peg of the geoboard to the top right peg of the geoboard."

• Allow time for your child to put the geoband on the geoboard.

"Hold up your geoboard for me to see."

• Repeat the clue, if necessary.

"Another side is parallel to that side."

"It is two pegs below that side."

"It is the same length as the first side."

"Show that side on your geoboard."

• Repeat the clues, if necessary.

"My shape has another set of parallel sides."

"Show the sides on your geoboard."

• Allow time for your child to put the geobands on the geoboard.

"What is the shape?"   a rectangle

• Show your child your geoboard.

"Does your shape look just like mine?"

"What do we call shapes that are exactly alike?"   congruent shapes

"Take your geobands off the geoboard."

• Make the following shape on your geoboard. Do not show your child.

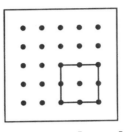

"I made a different shape on my geoboard."

"I will give you some clues."

"Try to make a shape just like mine."

"Use a different geoband for each side of the shape."

"One side of my shape stretches from the center peg on the geoboard to the peg two pegs to the right."

"Show that side on your geoboard."

• Allow time for your child to put the geoband on the geoboard.

"Hold up your geoboard for me to see."

• Repeat the clue, if necessary.

"Another side is parallel to that side."

"It is two pegs below that side."

"It is the same length as the first side."

"Show that side on your geoboard."

• Repeat the clue, if necessary.

"My shape has another set of parallel line segments."

"Show the other two sides on your geoboard."

• Allow time for your child to put the geobands on the geoboard.

"What is the shape?"   square

• Show your child your geoboard.

"Does your shape look just like mine?"

"Take your geobands off the geoboard."

"Make another shape that has a pair of parallel line segments."

• Allow time for your child to make the shape.

"Where are the parallel line segments?"

CLASS PRACTICE

number fact practice

- Use the white fact cards to practice the last sixteen subtraction facts with your child.

- Give your child **Fact Sheet S 8.2**.

- Time your child for one minute.

- Correct the fact sheet with your child and record the score.

- Allow time for your child to complete the unfinished facts.

WRITTEN PRACTICE

- Complete **Worksheet 115A** with your child.

- Complete **Worksheet 115B** with your child later in the day.

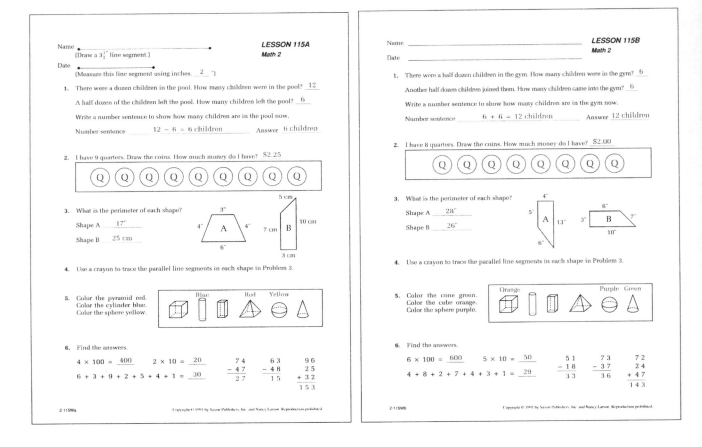

Name _____

ASSESSMENT 22

Date _____

LESSON 115

Math 2

1. Barbara has 314 pennies, Celina has 276 pennies, Amber has 358 pennies, and Megan has 298 pennies.

Who has the most pennies? __Amber__   Who has the fewest pennies? __Celina__

Write the names of the children in order from the one who has the most pennies to the one who has the fewest pennies.

__Amber__   __Barbara__   __Megan__   __Celina__

2. Color the cone yellow.
Color the sphere red.
Color the cylinder blue.
Color the cube green.

Green   Blue   Red   Yellow

3. Use the correct comparison symbol (>, <, or =).

16 [>] 9   3 + 7 [=] 10   8 + 1 [<] 8 + 2

4. Five children have red lunch boxes.

Twelve children have yellow lunch boxes.

Eleven children have blue lunch boxes.

Shade the graph to show the number of children with each color lunch box.

LUNCH BOX COLORS

red
yellow
blue

0 2 4 6 8 10 12 14

5. Find the answers.

$6 \times 100 =$ __600__ $7 \times 5 =$ __35__

$4 \times 10 =$ __40__ $4 \times 5 =$ __20__

$7 \times 100 =$ __700__ $8 \times 1 =$ __8__

$$\begin{array}{r} 6\,7 \\ 4\,3 \\ +\,2\,8 \\ \hline 1\,3\,8 \end{array}$$

$$\begin{array}{r} 7\,3 \\ 9 \\ +\,3\,6 \\ \hline 1\,1\,8 \end{array}$$

2-115Aa

Name _____
(Draw a $3\frac{1}{2}$" line segment.)

LESSON 115A

Math 2

Date _____
(Measure this line segment using inches. __2__ ")

1. There were a dozen children in the pool. How many children were in the pool? __12__

A half dozen of the children left the pool. How many children left the pool? __6__

Write a number sentence to show how many children are in the pool now.

Number sentence ___12 – 6 = 6 children___   Answer __6 children__

2. I have 9 quarters. Draw the coins. How much money do I have? __$2.25__

Q Q Q Q Q Q Q Q Q

3. What is the perimeter of each shape?

Shape A __17"__

Shape B __25 cm__

3"

4"   A   4"

6"

5 cm

7 cm   B   10 cm

3 cm

4. Use a crayon to trace the parallel line segments in each shape in Problem 3.

5. Color the pyramid red.
Color the cylinder blue.
Color the sphere yellow.

Blue   Red   Yellow

6. Find the answers.

$4 \times 100 =$ __400__ $2 \times 10 =$ __20__

$6 + 3 + 9 + 2 + 5 + 4 + 1 =$ __30__

$$\begin{array}{r} 7\,4 \\ -\,4\,7 \\ \hline 2\,7 \end{array}$$

$$\begin{array}{r} 6\,3 \\ -\,4\,8 \\ \hline 1\,5 \end{array}$$

$$\begin{array}{r} 9\,6 \\ 2\,5 \\ +\,3\,2 \\ \hline 1\,5\,3 \end{array}$$

2-115Wa

Name _____

LESSON 115B

Math 2

Date _____

1. There were a half dozen children in the gym. How many children were in the gym? __6__

Another half dozen children joined them. How many children came into the gym? __6__

Write a number sentence to show how many children are in the gym now.

Number sentence ___6 + 6 = 12 children___   Answer __12 children__

2. I have 8 quarters. Draw the coins. How much money do I have? __$2.00__

Q Q Q Q Q Q Q Q

3. What is the perimeter of each shape?

Shape A __28"__

Shape B __26"__

4"

5"   A   13"

6"

6"

3"   B   7"

10"

4. Use a crayon to trace the parallel line segments in each shape in Problem 3.

5. Color the cone green.
Color the cube orange.
Color the sphere purple.

Orange   Purple   Green

6. Find the answers.

$6 \times 100 =$ __600__ $5 \times 10 =$ __50__

$4 + 8 + 2 + 7 + 4 + 3 + 1 =$ __29__

$$\begin{array}{r} 5\,1 \\ -\,1\,8 \\ \hline 3\,3 \end{array}$$

$$\begin{array}{r} 7\,3 \\ -\,3\,7 \\ \hline 3\,6 \end{array}$$

$$\begin{array}{r} 7\,2 \\ 2\,4 \\ +\,4\,7 \\ \hline 1\,4\,3 \end{array}$$

2-115Wb

esson 116

multiplying by two
acting out equal groups stories

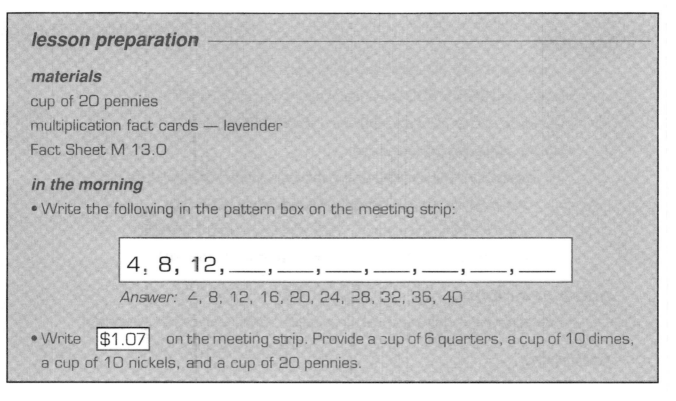

lesson preparation ———————————————————————————

materials

cup of 20 pennies

multiplication fact cards — lavender

Fact Sheet M 13.0

in the morning

• Write the following in the pattern box on the meeting strip:

> 4, 8, 12, ___, ___, ___, ___, ___, ___, ___

> *Answer:* 4, 8, 12, 16, 20, 24, 28, 32, 36, 40

• Write $1.07 on the meeting strip. Provide a cup of 6 quarters, a cup of 10 dimes, a cup of 10 nickels, and a cup of 20 pennies.

THE MEETING

calendar

• Ask your child to write the date on the calendar and meeting strip.

"How many days are there in one week? ... two weeks?"

"How many days are there in three weeks?"

"Let's count to check."

• Ask your child the following two or three times a week:

date _____ days ago, date _____ days from now

day of the week _____ days ago, day of the week _____ days from now

_____th month, month before, month after

• Record on the meeting strip a special event and the number of days until it occurs.

weather graph

- Ask your child to read and graph today's temperature to the nearest two degrees.

- Count by 10's and 2's to check the temperature on the graph.

- Ask your child to connect the dot for yesterday's temperature to the dot for today's temperature and compare the temperatures.

counting

- Count by 4's to 40 and backward from 40 by 4's.

- Count by 25's to 300 and backward from 300 by 25's.

- Count by 3's to 30 and backward from 30 by 3's.

- Do the following once a week:

    count by 10's to 400 and backward from 400 by 10's

    count by 5's to 100 and backward from 50 by 5's

    say the even numbers to 100 and backward from 50

    say the odd numbers to 49 and backward from 49

graph questions

- You and your child each ask a question about any of the graphs.

patterning

- Ask your child to do the following:

    identify the pattern (repeating, continuing, or both)

    identify the numbers to complete the pattern

    read the pattern

money

- Ask your child to put the coins in the coin cup. Count the money in the coin cup together.

- Ask your child for another way to show that amount of money. Count these coins together to check the amount.

clock

- Set the clock to a five-minute interval.

- Ask the following:

    "It's (morning/afternoon/evening). What time is it?"

    time one hour ago

    time one hour from now

- Ask your child to write the digital time on the meeting strip.

- Record on the meeting strip the time an activity will occur.

number of the day

- Write three number sentences for the number of the day on the meeting strip.

fact practice

- Write three fact family numbers (e.g., 2, 7, 9) on the chalkboard.

- Allow time for your child to write the four fact family number sentences on the chalkboard.

THE LESSON

Multiplying by Two
Acting Out Equal Groups Stories

"Today you will learn how to multiply by two."

"You also will learn how to act out equal groups stories."

- Give your child a cup of 20 pennies.

"Show one group of two pennies."

"How many pennies did you use?"   2

- Write "1 group of 2 is 2" on the chalkboard.

"Show three groups of two pennies."

- Allow time for your child to do this

"How many pennies did you use?"   6

- Write "3 groups of 2 is 6" on the chalkboard.

"Show five groups of two pennies."

- Allow time for your child to do this

"How many pennies did you use?"   10

- Write "5 groups of 2 is 10" on the chalkboard.

"Show ten groups of two pennies."

- Allow time for your child to do this.

"How many pennies did you use?"   20

- Write "10 groups of 2 is 20" on the chalkboard.

- Point to "5 groups of 2 is 10" on the chalkboard.

"What is a shorter way of writing 'five groups of two is ten'?"

- Write "5 × 2 = 10" next to the corresponding problem.

"Writing the × is a short way of writing the words 'groups of.' "

"Some people call this the times symbol."

"They say 'five times two is ten.' "

"How many times did you put two pennies on the table for this problem?" 5

"Let's write all of the other problems the short way."

"How will we do that?"

- Ask your child to write each problem on the chalkboard.

- Write the following on the chalkboard:

$$0 \times 2 = \qquad 6 \times 2 =$$
$$1 \times 2 = \qquad 7 \times 2 =$$
$$2 \times 2 = \qquad 8 \times 2 =$$
$$3 \times 2 = \qquad 9 \times 2 =$$
$$4 \times 2 = \qquad 10 \times 2 =$$
$$5 \times 2 =$$

"Let's fill in these answers."

"How many pennies are in one group of two?" 2

"How many pennies are in two groups of two?" 4

- Repeat with three through ten groups of two. Record each answer next to the problem.

"If we didn't have any groups of two, how many pennies would we have?" zero

"Do you see a pattern in the answers?" they are the even numbers

"We can write these problems another way."

- Write the following on the chalkboard:

$$\begin{array}{ccccccccccc} 2 & 2 & 2 & 2 & 2 & 2 & 2 & 2 & 2 & 2 & 2 \\ \underline{\times\,0} & \underline{\times\,1} & \underline{\times\,2} & \underline{\times\,3} & \underline{\times\,4} & \underline{\times\,5} & \underline{\times\,6} & \underline{\times\,7} & \underline{\times\,8} & \underline{\times\,9} & \underline{\times\,10} \end{array}$$

"Let's read these problems together as we say the answers."

- Read the problems as "Zero groups of two equal zero," "One group of two equals two," etc.

"Which number tells us the number of groups now?" bottom number

"Show four groups of two pennies."

- Allow time for your child to show this.

"How many pennies did you use?" 8

"We can write the number sentence to show what you did like this."

- Write "4 × 2 = 8" on the chalkboard.

  "This means four groups of two."

  "Leave this on the table."

  "Now show two groups of four pennies."

- Allow time for your child to show this.

  "How many pennies did you use to show two groups of four?" *8*

  "We can write the number sentence to show what you did like this."

- Write "2 × 4 = 8" on the chalkboard.

  "This means two groups of four."

  "What do you notice about these number sentences?" *the 2 and 4 changed places; the total number of pennies used in each problem is 8*

- Write 2 below 4 .

$$\times\ 4 \qquad \times\ 2$$

  "We call these problems 'switcharounds.' "

  "We can switch the numbers around when we multiply and the answers will be the same."

- Write 2 and 6 × 2 = on the chalkboard.

$$\times\ 9$$

  "Write a switcharound problem for each of these problems."

- Ask your child to write the switcharound problems on the chalkboard.

  "What are the answers?" *18, 12*

  "Now you will use the pennies to act out equal groups stories."

- Write the following on the chalkboard.

  There are 7 children at the party.

  Each child ate two pieces of pizza.

- Ask your child to read the story.

  "Pretend that your pennies are pieces of pizza."

  "Use the pennies to show what happened in this story."

- Allow time for your child to act out the story using pennies.

  "How did you show what happened in this story?" *made 7 groups of 2 pennies*

  "How many pieces of pizza did the seven children eat?" *14*

  "What number sentence can we write for this story?" *7 × 2 = 14*

- Write "7 × 2 pieces of pizza = 14 pieces of pizza" on the chalkboard. Read the number sentence as "7 groups of 2 pieces of pizza is 14 pieces of pizza."

- Write the following on the chalkboard:

> Two children went to the store.
>
> Each child bought 8 pencils.

- Ask your child to read the story.

  "Pretend that your pennies are pencils."

  "Use the pennies to show what happened in this story."

- Allow time for your child to act out the story using pennies.

  "How did you show what happened in this story?"　*made 2 groups of 8 pennies*

  "How many pencils did the 2 children buy?"　*16*

  "What number sentence can we write for this story?"　*2 × 8 = 16*

- Write "2 × 8 pencils = 16 pencils" on the chalkboard. Read the number sentence as "2 groups of 8 pencils is 16 pencils."

- Write the following on the chalkboard:

> Nine children are in a play.
>
> Each child will need 2 rabbit ears.

- Ask your child to read the story.

  "Pretend that your pennies are rabbit ears."

  "Use the pennies to show what happened in this story."

- Allow time for your child to act out the story using pennies.

  "How did you show what happened in this story?"　*made 9 groups of 2 pennies*

  "How many rabbit ears will the nine children need?"　*18*

  "What number sentence can we write for this story?"　*9 × 2 = 18*

- Write "9 × 2 ears = 18 ears" on the chalkboard. Read the number sentence as "9 groups of 2 ears is 18 ears."

- Write the following on the chalkboard:

> The children will make two cakes.
>
> Each cake needs 3 eggs.

- Ask your child to read the story.

  "Pretend that your pennies are eggs."

  "Use the pennies to show what happened in this story."

- Allow time for your child to act out the story using pennies.

  "How did you show what happened in this story?"　*made 2 groups of 3 pennies*

  "How many eggs will the children use to make the 2 cakes?"　*6*

  "What number sentence can we write for this story?"　*2 × 3 = 6*

- Write "2 × 3 eggs = 6 eggs" on the chalkboard. Read the number sentence as "2 groups of 3 eggs is 6 eggs."

  "Put your pennies in the cup."

  "I will give you the multiplying by two fact cards."

- Give your child the lavender multiplication fact cards.

  "Match the switcharound facts."

  "When you finish, practice saying the answers to yourself."

CLASS PRACTICE

number fact practice

- Give your child **Fact Sheet M 13.0**.

- Time your child for one minute.

- Correct the fact sheet with your child and record the score.

- Allow time for your child to complete the unfinished facts.

WRITTEN PRACTICE

- Complete **Worksheet 116A** with your child.

- Complete **Worksheet 116B** with your child later in the day.

Name _____ **LESSON 116A**
(Draw a 12 cm line segment.) **Math 2**
Date _____
(Measure this line segment using centimeters. ___10___ cm)

1. Dolores has 3 dimes and 6 pennies. Francis has 7 pennies and 2 dimes.

   How much money does Dolores have? __36¢__

   How much money does Francis have? __27¢__

   How much money do they have altogether?

   Number sentence ___36¢ + 27¢ = 63¢___   Answer __63¢__

2. Put these numbers in order from least to greatest.

176 284 373 181 279	176	181	279	284	373
	least				greatest

3. Circle the letters that have parallel line segments.

   A (E) L (M) V (Z)

4. Find the answers.

2	2	2	2	2	2	2	2	2	2	2
×4	×8	×3	×7	×5	×9	×1	×6	×2	×0	×10
8	16	6	14	10	18	2	12	4	0	20

5. Calvin has four markers. Draw the markers. One marker is red. Color that marker.

   What fractional part of the markers is red? $\frac{1}{4}$

6. Find the answers.

52 + 39 + 48	7 + 36	31 − 14	74 − 43
139	43	17	31

2-116Wa   Copyright © 1991 by Saxon Publishers, Inc. and Nancy Larson. Reproduction prohibited.

Name _____ **LESSON 116B**
Math 2
Date _____

1. Juan has 6 pennies and 4 dimes. Gary has 3 dimes and 9 pennies.

   How much money does Juan have? __46¢__

   How much money does Gary have? __39¢__

   How much money do they have altogether?

   Number sentence ___46¢ + 39¢ = 85¢___   Answer __85¢__

2. Put these numbers in order from least to greatest.

192 284 194 487 291	192	194	284	291	487
	least				greatest

3. Circle the letters that have parallel line segments.

   (H) Y (I) (N) T X

4. Fill in the missing numbers.

2	2	2	2	2	2	2	2
× 5	× 2	× 8	× 1	× 9	× 6	× 3	× 7
10	4	16	2	18	12	6	14

5. Steve has three markers. Draw the markers. One marker is green. Color that marker.

   What fractional part of the markers is green? $\frac{1}{3}$

6. Find the answers.

31 + 49 + 24	6 + 58	43 − 29	82 − 51
104	64	14	31

2-116Wb   Copyright © 1991 by Saxon Publishers, Inc. and Nancy Larson. Reproduction prohibited.

Lesson 117

counting quarters, dimes, nickels, and pennies showing money amounts using quarters, dimes, nickels, and pennies

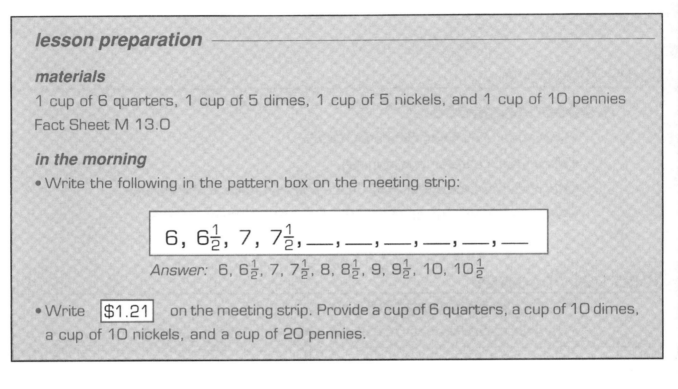

lesson preparation

materials

1 cup of 6 quarters, 1 cup of 5 dimes, 1 cup of 5 nickels, and 1 cup of 10 pennies
Fact Sheet M 13.0

in the morning

• Write the following in the pattern box on the meeting strip:

$$6, 6\tfrac{1}{2}, 7, 7\tfrac{1}{2}, \underline{\quad}, \underline{\quad}, \underline{\quad}, \underline{\quad}, \underline{\quad}, \underline{\quad}$$

Answer: $6, 6\tfrac{1}{2}, 7, 7\tfrac{1}{2}, 8, 8\tfrac{1}{2}, 9, 9\tfrac{1}{2}, 10, 10\tfrac{1}{2}$

• Write $\boxed{\$1.21}$ on the meeting strip. Provide a cup of 6 quarters, a cup of 10 dimes, a cup of 10 nickels, and a cup of 20 pennies.

THE MEETING

calendar

• Ask your child to write the date on the calendar and meeting strip.

"How many days are there in one week? ... two weeks?"

"How many days are there in three weeks?"

"Let's count to check."

• Ask your child the following two or three times a week:

    date _____ days ago, date _____ days from now

    day of the week _____ days ago, day of the week _____ days from now

    _____th month, month before, month after

• Record on the meeting strip a special event and the number of days until it occurs.

weather graph

- Ask your child to read and graph today's temperature to the nearest two degrees.
- Count by 10's and 2's to check the temperature on the graph.
- Ask your child to connect the dot for yesterday's temperature to the dot for today's temperature and compare the temperatures.

counting

- Count by 4's to 40 and backward from 40 by 4's.
- Count by 25's to 300 and backward from 300 by 25's.
- Count by 3's to 30 and backward from 30 by 3's.
- Do the following once a week:

     count by 10's to 400 and backward from 400 by 10's

     count by 5's to 100 and backward from 50 by 5's

     say the even numbers to 100 and backward from 50

     say the odd numbers to 49 and backward from 49

graph questions

- You and your child each ask a question about any of the graphs.

patterning

- Ask your child to do the following:

     identify the pattern (repeating, continuing, or both)

     identify the numbers to complete the pattern

     read the pattern

money

- Ask your child to put the coins in the coin cup. Count the money in the coin cup together.
- Ask your child for another way to show that amount of money. Count these coins together to check the amount.

clock

- Set the clock to a five-minute interval.
- Ask the following:

     "It's (morning/afternoon/evening). What time is it?"

     time one hour ago

     time one hour from now

- Ask your child to write the digital time on the meeting strip.

• Record on the meeting strip the time an activity will occur.

number of the day

• Write three number sentences for the number of the day on the meeting strip.

fact practice

• Write three fact family numbers (e.g., 2, 7, 9) on the chalkboard.

• Allow time for your child to write the four fact family number sentences on the chalkboard.

THE LESSON

Counting Quarters, Dimes, Nickels, and Pennies
Showing Money Amounts Using Quarters, Dimes, Nickels, and Pennies

"Today you will learn how to count quarters, dimes, nickels, and pennies."

"You also will learn how to show money amounts using these coins."

"Each morning you have been counting the coins in the coin cup."

"When we count quarters, dimes, nickels, and pennies, which coin do we count first?"   quarter

"Which coin do we count next?"   dime

"Which coin do we count after that?"   nickel

"Which coin do we count last?"   penny

• Give your child a cup of 6 quarters, a cup of 5 dimes, a cup of 5 nickels, and a cup of 10 pennies.

"Put two quarters, one dime, three nickels, and four pennies on the table."

• Allow time for your child to do this.

"Let's put the coins in the coin cup. Count the money."

"Which coin will we count first?"   quarter

"When we count quarters, what do we count by?"   25's

"Let's count by 25's as you slide the quarters."

• Count with your child.

"Which coin will we count next?"   dime

"When we count dimes, what do we count by?"   10's

"Fifty and ten more is sixty."

"Which coin will we count next?" nickel

"When we count nickels, what do we count by?" 5's

"We will count by 5's from sixty as you slide the nickels."

"Sixty, sixty-five, seventy, seventy-five."

"Now you have four more pennies."

"We will count by 1's from seventy-five."

- Count with your child.

"How much money do you have?" seventy-nine cents

"Can you trade any of these coins for a coin that is worth more?" 1 dime and 3 nickels for a quarter

"Trade one dime and three nickels for a quarter."

"What coins do you have now?" 3 quarters, 4 pennies

"Let's put the coins in the coin cup. Count the money now."

- Ask your child to put the coins in the coin cup. Count the money with your child.

"How much money do you have now?" 79¢

- Repeat, using the following combinations:

1 quarter, 3 dimes 2 nickels, and 6 pennies	71¢
3 quarters, 1 dime 1 nickel, and 5 pennies	95¢
2 quarters, 4 dimes, 3 nickels, and 1 penny	$1.06
5 quarters, 1 dime 4 nickels, and 6 pennies	$1.61

"Now I will tell you a certain amount of money."

"Show that amount of money using quarters, dimes, nickels, and pennies."

"Show 37¢."

- Allow time for your child to do this.

"What coins did you use?"

"Let's count this money as I point to each coin."

- Repeat with 45¢, 68¢, 81¢, $1.20, $1.16, and $1.43.

CLASS PRACTICE

number fact practice

- Use the lavender multiplication fact cards to practice the multiplying by two facts with your child.

• Give your child **Fact Sheet M 13.0.**

• Time your child for one minute.

• Correct the fact sheet with your child and record the score.

• Allow time for your child to complete the unfinished facts.

WRITTEN PRACTICE

• Complete **Worksheet 117A** with your child.

• Complete **Worksheet 117B** with your child later in the day.

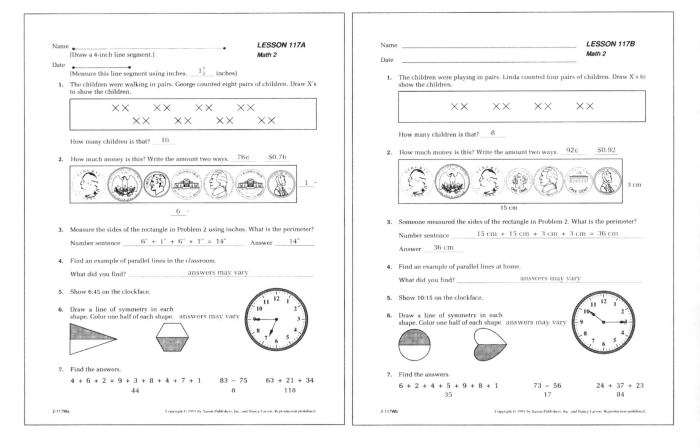

Name _____
[Draw a 4-inch line segment.]

Date _____
(Measure this line segment using inches. $1\frac{1}{2}$ inches)

LESSON 117A
Math 2

1. The children were walking in pairs. George counted eight pairs of children. Draw X's to show the children.

XX   XX   XX   XX
XX   XX   XX   XX

How many children is that? __16__

2. How much money is this? Write the amount two ways. 76¢ $0.76

1"

6"

3. Measure the sides of the rectangle in Problem 2 using inches. What is the perimeter?

Number sentence ___6" + 1" + 6" + 1" = 14"___   Answer ___14"___

4. Find an example of parallel lines in the classroom.

What did you find? _____ answers may vary _____

5. Show 6:45 on the clockface.

6. Draw a line of symmetry in each shape. Color one half of each shape. answers may vary

7. Find the answers.

4 + 6 + 2 + 9 + 3 + 8 + 4 + 7 + 1   83 − 75   63 + 21 + 34
44   8   118

2-117Wa

Name _____

Date _____

LESSON 117B
Math 2

1. The children were playing in pairs. Linda counted four pairs of children. Draw X's to show the children.

XX   XX   XX   XX

How many children is that? __8__

2. How much money is this? Write the amount two ways. 92¢ $0.92

3 cm

15 cm

3. Someone measured the sides of the rectangle in Problem 2. What is the perimeter?

Number sentence ___15 cm + 15 cm + 3 cm + 3 cm = 36 cm___

Answer ___36 cm___

4. Find an example of parallel lines at home.

What did you find? _____ answers may vary _____

5. Show 10:15 on the clockface.

6. Draw a line of symmetry in each shape. Color one half of each shape. answers may vary

7. Find the answers.

6 + 2 + 4 + 5 + 9 + 8 + 1   73 − 56   24 + 37 + 23
35   17   84

2-117Wb

Lesson 118

rounding to the nearest ten

---

lesson preparation

materials

Meeting Book

11 pink 2" × 3" construction paper tags

38 yellow 2" × 3" construction paper tags

Fact Sheet S 5.4

the night before

• Write the following numbers on the yellow and pink construction paper tags. (Colors other than pink and yellow can be used.)

Pink   0, 10, 20, 30, 40, 50, 60, 70, 80, 90, 100

Yellow   1, 2, 5, 7, 9, 11, 14, 18, 22, 23, 25, 27, 31, 34, 36, 39, 43, 45, 47, 52, 53, 57, 59, 64, 65, 67, 71, 72, 75, 76, 78, 82, 84, 88, 92, 93, 95, 97

in the morning

• Write the following in the pattern box on the meeting strip:

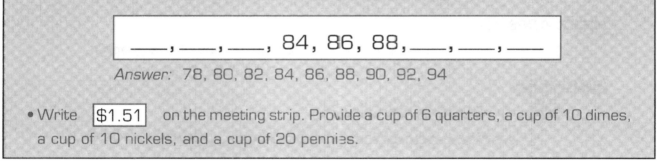

_____, _____, _____, 84, 86, 88, _____, _____, _____

Answer: 78, 80, 82, 84, 86, 88, 90, 92, 94

• Write $1.51   on the meeting strip. Provide a cup of 6 quarters, a cup of 10 dimes, a cup of 10 nickels, and a cup of 20 pennies.

THE MEETING

calendar

• Ask your child to write the date on the calendar and meeting strip.

"How many days are there in one week? ... two weeks?"

"How many days are there in three weeks?"

"Let's count to check."

• Ask your child the following two or three times a week:

date _____ days ago, date _____ days from now

---

day of the week ____ days ago, day of the week ____ days from now

____th month, month before, month after

- Record on the meeting strip a special event and the number of days until it occurs.

weather graph

- Ask your child to read and graph today's temperature to the nearest two degrees.

- Count by 10's and 2's to check the temperature on the graph.

- Ask your child to connect the dot for yesterday's temperature to the dot for today's temperature and compare the temperatures.

counting

- Count by 4's to 40 and backward from 40 by 4's.

- Count by 25's to 300 and backward from 300 by 25's.

- Count by 3's to 30 and backward from 30 by 3's.

- Do the following once a week:

  count by 10's to 400 and backward from 400 by 10's

  count by 5's to 100 and backward from 50 by 5's

  say the even numbers to 100 and backward from 50

  say the odd numbers to 49 and backward from 49

graph questions

- You and your child each ask a question about any of the graphs.

patterning

- Ask your child to do the following:

  identify the pattern (repeating, continuing, or both)

  identify the numbers to complete the pattern

  read the pattern

money

- Ask your child to put the coins in the coin cup. Count the money in the coin cup together.

- Ask your child for another way to show that amount of money. Count these coins together to check the amount.

clock

- Set the clock to a five-minute interval.

- Ask the following:

    "It's (morning/afternoon/evening). What time is it?"

    time one hour ago

    time one hour from now

- Ask your child to write the digital time on the meeting strip.
- Record on the meeting strip the time an activity will occur.

number of the day

- Write three number sentences for the number of the day on the meeting strip.

fact practice

- Write three fact family numbers (e.g., 2, 7, 9) on the chalkboard.
- Allow time for your child to write the four fact family number sentences on the chalkboard.

THE LESSON

Rounding to the Nearest Ten

"Each day you have been recording the temperature on a weather graph."

- Open the Meeting Book to this month's weather graph.

    "What was today's temperature?"

    "Was the temperature closer to _____ or _____?" (Use the 10's above and below the temperature.)

    "How can you tell?"

    "What was the temperature on _____?"

    "Was the temperature closer to _____ or _____?" (Use the 10's above and below the temperature.)

    "How can you tell?"

- Repeat with one or two more temperatures.
- Mix the pink number cards. Show your child the cards.

    "I have some cards with numbers on them."

    "What is the same about all of these numbers?" they are the numbers we say when we count by 10's

    "Put these numbers in order on the floor so they look like a number line."

    "Put each number card two feet apart."

    "How can we check to make sure that the cards are two feet apart?" use rulers

"We will use two one-foot rulers to space the number cards."

- Mix the yellow number cards.

"We are going to put these number cards on our number line."

"Choose a number card."

"What is the number?"

"Between which two 10's is _____?"

"Which ten is _____ closer to?"

- Note: All numbers ending in 5 are rounded to the next highest 10.

"Where will you put this number on the number line?"

- Ask your child to place the number card on the number line.

"Choose another number card."

"What is the number?"

"Between which two 10's is _____?"

"Which ten is it closer to?"

"Where will we put this number on the number line?"

- Ask your child to place the number card on the number line.

- Repeat two more times.

"When we round a number to the nearest ten, we name the ten the number is closest to."

"What numbers are in the middle between the 10's?"   *5, 15, 25, 35, 45, 55, 65, 75, 85, 95*

- Write these numbers on the chalkboard.

"How can we tell when a number will be in the middle between two 10's?"   *it ends with a 5*

"When a number is halfway between the tens, mathematicians round the number to the next highest ten."

"We round 45 to 50, 75 to 80, and 5 to 10."

- Hand your child the rest of the number cards, one at a time. Ask the following questions:

"Where will we put this card?"

"How do you know?"

"Which ten is it closest to?"

"Now you will have a chance to practice rounding some numbers to the nearest ten."

- Write "42" on the chalkboard.

"Write the 10's that are on each side of 42."

- Ask your child to write the numbers on the chalkboard:

40   42   50

"Circle the 10 that 42 is closest to on the number line."

• Allow time for your child to do this.

"Forty-two rounded to the nearest ten is forty."

• Repeat, using 67, 81, 35, 11, and 93.

CLASS PRACTICE

number fact practice

• Use the fact cards to practice all the subtraction facts with your child.

• Give your child **Fact Sheet S 5.4**.

• Time your child for one minute.

• Correct the fact sheet with your child and record the score.

• Allow time for your child to complete the unfinished facts.

WRITTEN PRACTICE

• Complete **Worksheet 118A** with your child.

• Complete **Worksheet 118B** with your child later in the day.

LESSON 118A
Math 2

Name _____
(Draw a 10 cm line segment.)
Date _____
(Measure this line segment using centimeters. __2__ cm)

1. Quinton has 9 baseball cards. Curtis has twice as many baseball cards as Quinton.
   How many baseball cards does Curtis have? __18__

2. There are five children in line. Draw the children. You are fourth. Circle yourself.
   How many children are before you? __3__
   How many children are after you? __1__

3. Each number is between what two tens?
   __10__ , 17, (20)   (40) , 43. 50   __30__ , 35, (40)
   Circle the 10 each number is closest to.

4. Gina has 2 quarters, 3 dimes, 2 nickels, and a penny. Draw the coins. How much money does she have?
   Q Q L D D N N P
   Write the answer two different ways. __91¢__ __$0.91__

5. Fill in the missing numbers in the number pattern.
   80, 85, 90, __95__ , __100__ , __105__ , __110__ , __115__ , __120__
   __16__ , __26__ , __36__ , 46, 56, 66, __76__ , __86__ , __96__

6. Find the answers.
   6 × 2 = __12__   8 × 2 = __16__   4 × 5 = __20__
   18 + 6 + 52 = __76__   63 − 49 = __4__   38 + 42 = __80__

2-118Wa   Copyright © 1991 by Saxon Publishers, Inc. and Nancy Larson. Reproduction prohibited.

LESSON 118B
Math 2

Name _____
Date _____

1. Ariana has 7 stuffed animals. Her sister has twice as many stuffed animals.
   How many stuffed animals does Ariana's sister have? __14__

2. There are six children in line. Draw the children. You are third. Circle yourself.
   How many children are before you? __2__
   How many children are after you? __3__

3. Each number is between what two tens?
   (20) , 24, __30__   __50__ , 58, (60)   __20__ , 25, (30)
   Circle the 10 each number is closest to.

4. Daniel has 1 quarter, 2 dimes, 4 nickels, and 2 pennies. Draw the coins. How much money does he have?
   Q D D N N N N P P
   Write the answer two different ways. __67¢__ __$0.67__

5. Fill in the missing numbers in the number patterns.
   92, 94, 96, __98__ , __100__ , __102__ , __104__ , __106__ , __108__
   __17__ , __27__ , __37__ , 47, 57, 67, __77__ , __87__ , __97__

6. Find the answers.
   7 × 2 = __14__   9 × 2 = __18__   6 × 5 = __30__
   6 + 5 − 54 = __75__   72 − 48 = __24__   49 + 31 = __80__

2-118Wb   Copyright © 1991 by Saxon Publishers, Inc. and Nancy Larson. Reproduction prohibited.

Lesson 119

acting out equal groups stories
drawing pictures to show equal groups

---

lesson preparation

materials

cup of 20 pennies

Master 2-119

8 pieces of 3" × 4" construction paper

crayons

Fact Sheet S 8.2

in the morning

• Write the following in the pattern box on the meeting strip:

⌞ , ⌜ , ⌝ , ⌟ , ⌞ , ___ , ___ , ___ , ___ , ___ , ___

Answer: ⌞ , ⌜ , ⌝ , ⌟ , ⌞ , ⌜ , ⌝ , ⌟ , ⌞ , ⌜ , ⌝

• Write $2.07 on the meeting strip. Provide a cup of 10 quarters, a cup of 10 dimes, a cup of 10 nickels, and a cup of 20 pennies.

THE MEETING

calendar

• Ask your child to write the date on the calendar and meeting strip.

"How many days are there in one week? ... two weeks?"

"How many days are there in three weeks?"

"Let's count to check."

• Ask your child the following two or three times a week:

date _____ days ago, date _____ days from now

day of the week _____ days ago, day of the week _____ days from now

_____th month, month before, month after

• Record on the meeting strip a special event and the number of days until it occurs.

weather graph

- Ask your child to read and graph today's temperature to the nearest two degrees.
- Count by 10's and 2's to check the temperature on the graph.
- Ask your child to connect the dot for yesterday's temperature to the dot for today's temperature and compare the temperatures.

counting

- Count by 4's to 40 and backward from 40 by 4's.
- Count by 25's to 300 and backward from 300 by 25's.
- Count by 3's to 30 and backward from 30 by 3's.
- Do the following once a week:

  count by 10's to 400 and backward from 400 by 10's

  count by 5's to 100 and backward from 50 by 5's

  say the even numbers to 100 and backward from 50

  say the odd numbers to 49 and backward from 49

graph questions

- You and your child each ask a question about any of the graphs.

patterning

- Ask your child to do the following:

  identify the pattern (repeating, continuing, or both)

  identify the shapes to complete the pattern

  read the pattern

money

- Ask your child to put the coins in the coin cup. Count the money in the coin cup together.
- Ask your child for another way to show that amount of money. Count these coins together to check the amount.

clock

- Set the clock to a five-minute interval.
- Ask the following:

  "It's (morning/afternoon/evening). What time is it?"

  time one hour ago

  time one hour from now

- Ask your child to write the digital time or the meeting strip.

• Record on the meeting strip the time an activity will occur.

number of the day

• Write three number sentences for the number of the day on the meeting strip.

fact practice

• Write three fact family numbers (e.g., 2, 7, 9) on the chalkboard.

• Allow time for your child to write the four fact family number sentences on the chalkboard.

THE LESSON

Acting Out Equal Groups Stories
Drawing Pictures to Show Equal Groups

"A few days ago you acted out equal groups stories."

"Today you will learn how to draw pictures for equal groups stories."

• Give your child a cup of 20 pennies.

"We will use pennies to act out equal groups stories."

"Then we will draw pictures of the stories we acted out."

"Today we will pretend that pennies are wheels, buttons, raisins, and cookies."

• Give your child five 3" × 4" pieces of construction paper.

"Let's pretend that these are five cars."

"How many wheels does a car have?"   4

"Use your pennies to show the wheels on the cars."

• Allow time for your child to do this.

"How many cars are there?"   5

"How many wheels are on each car?"   4

"How many wheels did you use altogether?"   20

• Give your child **Master 2-119**.

"Now we will draw a picture of this equal groups story."

"We can draw a rectangle to show each car."

"How many rectangles will we draw for this story?"   5

• Draw 5 rectangles on the chalkboard.

"Use a pencil to draw five rectangles in the first box on your paper."

• Allow time for your child to do this.

"We can draw circles to show the wheels."

"How many wheels will we draw on each car?"   4

• Draw the following on the chalkboard:

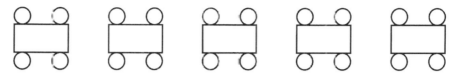

"Use a crayon to draw the wheels."

• Allow time for your child to do this.

"Put the pennies back in the cup."

• Take away two of the construction paper pieces.

"Now we will pretend that the pieces of paper are shirts."

"Pretend that the pennies are buttons."

"Put five buttons on each shirt."

• Allow time for your child to do this.

"How many shirts are there?"   3

"How many buttons are on each shirt?"   5

"How many buttons did you use altogether?"   15

"Now we will draw a picture of this equal groups story."

"We can draw a rectangle to show a shirt."

"How many rectangles will we draw for this story?"   3

• Draw 3 rectangles on the chalkboard.

"Use a pencil to draw this in the second box on your paper."

• Allow time for your child to do this.

"We will draw circles to show the buttons."

"How many buttons will we draw on each shirt?"   5

• Draw the following on the chalkboard:

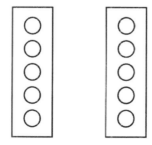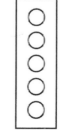

"Use a crayon to draw the buttons."

• Allow time for your child to do this.

"Put the pennies back in the cup."

• Give your child 3 more construction paper pieces.

"Now we will pretend that the pennies are raisins."

"We will pretend that the pieces of paper are cupcakes."

"Each cupcake has three raisins on top."

"There are six cupcakes."

"How will you show that using the paper and the pennies?"   *6 pieces of paper with 3 pennies on each*

"Use the pennies to show the raisins on the cupcakes."

- Allow time for your child to do this.

"How many cupcakes are there?"   6

"How many raisins are on each cupcake?"   3

"How many raisins did you use altogether?"   18

"Now you will draw a picture of our equal groups story."

"How will you do that?"

- Ask your child to describe what to draw.

"Use a pencil to draw the six cupcakes in the third box on your paper."

- Allow time for your child to do this.

"Use a crayon to draw the raisins on the cupcakes."

- Allow time for your child to do this.

"Put the pennies back in the cup."

- Give your child 2 more construction paper pieces.

"Now you will pretend that the pennies are bicycle wheels."

"We will pretend that the pieces of paper are bicycles."

"How many wheels does a bicycle have?"   2

"There are eight bicycles."

"How will you show that using the paper and the pennies?"   *8 pieces of paper with 2 pennies on each*

"Use the pennies to show the wheels on the bicycle."

- Allow time for your child to do this.

"How many bicycles are there?"   8

"How many wheels are on each bicycle?"   2

"How many bicycle wheels is that altogether?"   16

"Now we will quickly draw a picture of our equal groups story."

"How can we draw this quickly?"

- Demonstrate drawing a bicycle quickly.

"Use a pencil to draw the bicycles in the fourth box on your paper."

"Use a crayon to draw the wheels."

- Allow time for your child to do this.

"Now we will pretend that the pennies are cookies."

"We will pretend that the pieces of paper are children."

"Each child has two cookies."

"There are four children."

"How will you show that using the paper and the pennies?"   *4 pieces of paper with 2 pennies on each*

"Show this using the pieces of paper and the pennies."

- Allow time for your child to do this.

"How many children are there?"   *4*

"How many cookies will each child have?"   *2*

"How many cookies is this altogether?"   *8*

"Now we will quickly draw a picture of our equal groups story."

"We can draw a circle to show each child's face."

"Use a pencil to draw the faces in the fifth box on your paper."

- Allow time for your child to do this.

"Use a crayon to draw the small circles next to each face to show the cookies."

- Allow time for your child to do this.

"You will use this paper again when you learn how to write number sentences for equal groups stories."

- Save **Master 2-119** for use in Lesson 123.

Class Practice

number fact practice

- Use the white fact cards to practice the last sixteen subtraction facts with your child.
- Give your child **Fact Sheet S 8.2**.
- Time your child for one minute.
- Correct the fact sheet with your child and record the score.
- Allow time for your child to complete the unfinished facts.

Written Practice

- Complete **Worksheet 119A** with your child.
- Complete **Worksheet 119B** with your child later in the day.

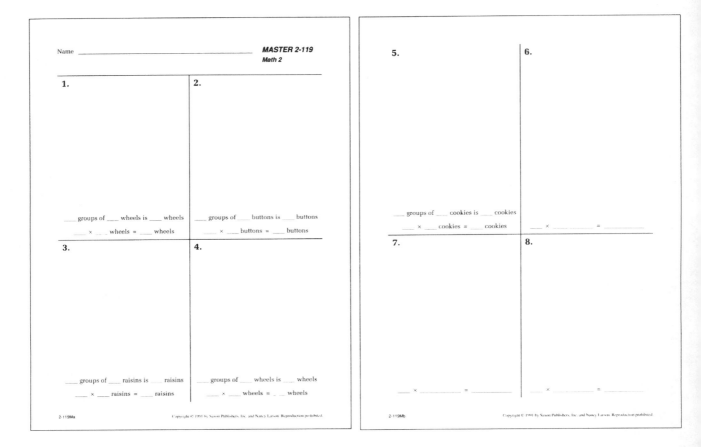

1.

2.

____ groups of ___ wheels is ____ wheels

____ × ___ wheels = ____ wheels

____ groups of ___ buttons is ____ buttons

____ × ___ buttons = ____ buttons

3.

4.

____ groups of ___ raisins is ____ raisins

____ × ___ raisins = ____ raisins

____ groups of ___ wheels is ____ wheels

____ × ___ wheels = ___ wheels

5.

6.

____ groups of ___ cookies is ____ cookies

____ × ___ cookies = ____ cookies

____ × ____ = ____

7.

8.

____ × ____ = ____

____ × ____ = ____

2-119Ma · Copyright © 1991 by Saxon Publishers, Inc. and Nancy Larson. Reproduction prohibited.

2-119Mb · Copyright © 1991 by Saxon Publishers, Inc. and Nancy Larson. Reproduction prohibited.

Name •_____ **LESSON 119A**
(Draw a 4½" line segment.)   **Math 2**
Date •_____
(Measure this line segment using inches. __4__ ")

1. There are five desks in the room. T.J. put three books on each desk. Draw the books on the desks.

   [boxes with books]

   How many books did you draw altogether? __15__

2. Round each number to the nearest 10.

   27 __30__   52 __50__   45 __50__

3. Measure the vertical line segment on the left using centimeters.   __7__ cm

   Measure the vertical line segment on the right using centimeters.   __4__ cm

   Measure the horizontal line segment using centimeters.   __4__ cm

   Measure the oblique line segment using centimeters.   __5__ cm

   What is the perimeter of the shape?   __20__ cm

4. Trace the parallel line segments in Problem 3 using a red crayon.

5. Find an example of a cylinder in the classroom. What did you find? __answers__

   Find an example of a sphere in the classroom. What did you find? __may vary__

6. Write six hundred seventeen using digits. __617__

   Write this number in expanded form. __600 + 10 + 7__

7. Find the answers.

   9 × 5 = __45__   7 × 2 = __14__

   7 × 10 = __70__   8 × 100 = __800__

   $$\begin{array}{r} 45 \\ +26 \\ \hline 71 \end{array}\qquad \begin{array}{r} 71 \\ -53 \\ \hline 18 \end{array}\qquad \begin{array}{r} 85 \\ -32 \\ \hline 53 \end{array}$$

2-119Wa · Copyright © 1991 by Saxon Publishers, Inc. and Nancy Larson. Reproduction prohibited.

---

1. There are two tables in the room. Brenden put six books on each table. Draw the books on the tables.

   [boxes with books]

   How many books did you draw altogether? __12__

2. Round each number to the nearest 10.

   19 __20__   63 __60__   75 __80__

3. How long is the vertical line segment on the right? __8__ cm

   How long is the vertical line segment on the left? __4__ cm

   How long is the oblique line segment? __5__ cm

   How long is the horizontal line segment? __3__ cm

   What is the perimeter of the shape? __20__ cm

   5 cm

   8 cm

   4 cm

   3 cm

4. Trace the parallel line segments in Problem 3 using a red crayon.

5. Find an example of a cylinder at home. What did you find? __answers__

   Find an example of a sphere at home. What did you find? __may vary__

6. Write two hundred thirty-seven using digits. __237__

   Write this number in expanded form. __200 + 30 + 7__

7. Find the answers.

   3 × 5 = __15__   6 × 2 = __12__

   4 × 10 = __40__   5 × 100 = __500__

   $$\begin{array}{r} 71 \\ +69 \\ \hline 140 \end{array}\qquad \begin{array}{r} 24 \\ -16 \\ \hline 8 \end{array}\qquad \begin{array}{r} 58 \\ -17 \\ \hline 41 \end{array}$$

2-119Wb · Copyright © 1991 by Saxon Publishers, Inc. and Nancy Larson. Reproduction prohibited.

L esson 120

choosing a survey question and choices
representing data using a graph

---

lesson preparation

materials

Written Assessment #23

Oral Assessment #12

2 envelopes

20 tags

writing paper

markers or crayons

Masters 2-120A and 2-120B

tape

10 dimes

20 pennies

scrap paper

Fact Sheet M 13.0

in the morning

- Write the following in the pattern box on the meeting strip:

$$Z, \measuredangle, \sqcap, N, Z, \measuredangle, \_\_, \_\_, \_\_, \_\_, \_\_, \_\_$$

Answer: Z, ∑, И, N, Z, ∑, И, N, Z, ∑, И, N

- Write $1.82 on the meeting strip. Provide a cup of 10 quarters, a cup of 10 dimes, a cup of 10 nickels, and a cup of 20 pennies.

THE MEETING

calendar

- Ask your child to write the date on the calendar and meeting strip.

- Ask your child to identify the number of days in 1 week, 2 weeks, and 3 weeks

- Ask your child the following two or three times a week:

    date _____ days ago, date _____ days from now

    day of the week _____ days ago, day of the week _____ days from now

    _____th month, month before, month after

- Record on the meeting strip a special event and the number of days until it occurs.

weather graph

- Ask your child to read and graph today's temperature to the nearest two degrees.

- Count by 10's and 2's to check the temperature on the graph.

- Ask your child to connect the dot for yesterday's temperature to the dot for today's temperature and compare the temperatures.

counting

- Count by 4's to 40 and backward from 40 by 4's.

- Count by 25's to 300 and backward from 300 by 25's.

- Count by 3's to 30 and backward from 30 by 3's.

- Do the following once a week:

    count by 10's to 400 and backward from 400 by 10's

    count by 5's to 100 and backward from 50 by 5's

    say the even numbers to 100 and backward from 50

    say the odd numbers to 49 and backward from 49

graph questions

- You and your child each ask a question about any of the graphs.

patterning

- Ask your child to do the following:

    identify the pattern (repeating, continuing, or both)

    identify the shapes to complete the pattern

    read the pattern

money

- Ask your child to put the coins in the coin cup. Count the money in the coin cup together.

- Ask your child for another way to show that amount of money. Count these coins together to check the amount.

clock

- Set the clock to a five-minute interval.
- Ask the following:

    "It's (morning/afternoon/evening). What time is it?"

    time one hour ago

    time one hour from now

- Ask your child to write the digital time on the meeting strip.
- Record on the meeting strip the time an activity will occur.

number of the day

- Write three number sentences for the number of the day on the meeting strip.

fact practice

- Write three fact family numbers (e.g., 2, 7, 9) on the chalkboard.
- Allow time for your child to write the four fact family number sentences on the chalkboard.

ASSESSMENT

Written Assessment

"Today I would like to see what you remember from what we have been practicing."

- Give your child **Written Assessment #23**.
- Read the directions for each problem. Allow time for your child to complete each problem before continuing.
- Correct the paper, noting your child's mistakes on the **Individual Recording Form**. Review the errors with your child.

Oral Assessment

- Record your child's response(s) to the oral interview questions on the interview sheet.

THE LESSON

Choosing a Survey Question and Choices
Representing Data Using a Graph

"What are some of the graphs we have made this year?"

"Today you will learn how to ask your own question and how to make a graph to show the answers."

- Give your child **Master 2-120A.**

"First you will choose a question to ask."

"Then you will make up four choices for your question."

"When you have made up your question and the choices, you will ask 20 people to answer the question."

"Then you will make a graph to show their answers."

"What kinds of questions could you ask?"

- List possible questions on the chalkboard. Possible topics for a graph might be:

> favorite vacation spot
>
> favorite animal
>
> favorite flower
>
> favorite season
>
> favorite food
>
> favorite cookie

"What question would you like to ask?"

- Write the question on the chalkboard.

"Write the question on Master 2-120A."

"Now you will make up four choices for answers."

"Try to make the choices something people might like."

"For example, if my question was 'What is your favorite food?', why would liver, beets, spinach, and cookies not be good choices?" most people would probably choose cookies

"What choices would you like to use for your question?"

- Write the choices on the chalkboard.

"Write the choices on your paper."

"Now you will survey 20 people."

"You will ask 20 different people to answer your question."

"Each person you ask will choose one of the four choices."

"You will need 20 small pieces of paper so that each person can vote for his/her choice."

- Give your child an envelope of 20 tags. Give your child another empty envelope labeled "votes."

- After each person has written his/her choice on a tag, put it in the "votes" envelope.

- Assist your child in tallying the results.

  "Now you will make a graph to show the results of your survey."

- Give your child a copy of **Master 2-120B**.

  "Write the title at the top of the graph."

  "Write the choices below each column."

  "If more than seven people chose the same answer, you will need to number your graph by 2's."

- If necessary, assist your child as he/she numbers the graph by 2's.

  "Color the graph to show how many people chose each choice."

- Allow time for your child to do this.

  "Now you will write your observations about the graph on a piece of writing paper."

  "Observations are all the things you know about the graph."

- Allow time for your child to write observations.

- Tape the observations to the graph.

- Encourage your child to share the graph and observations with the people surveyed.

CLASS PRACTICE

number fact practice

- Use the lavender multiplication fact cards to practice the multiplying by two facts with your child.

- Give your child **Fact Sheet M 13.0**.

- Time your child for one minute.

- Correct the fact sheet with your child and record the score.

- Allow time for your child to complete the unfinished facts.

MASTER 2-120A
Math 2

Name _____

Question _____
_____

Choices   1) _____

              2) _____

              3) _____

              4) _____

The title of the graph will be:
_____

Choices	Tally of votes
1) _____	[ ]
2) _____	[ ]
3) _____	[ ]
4) _____	[ ]

If any choice receives more than 7 votes, number your graph by 2's

2-120AMa   Copyright © 1991 by Saxon Publishers, Inc. and Nancy Larson. Reproduction prohibited.

MASTER 2-120B
Math 2

Name _____

Title _____

Class surveyed: Grade _____   Teacher _____

2-120BMa   Copyright © 1991 by Saxon Publishers, Inc. and Nancy Larson. Reproduction prohibited.

Teacher _____
Date _____

MATH 2   LESSON 120
Oral Assessment # 12 Recording Form

Materials:
10 dimes
20 pennies
scrap paper

Students

	"Put 32¢ on the paper." *"Give me 15¢ of that money."* *"How much money do you have left?"*				• Write 52¢ - 15¢ on a piece of scrap paper *"Show how to find the answer for this example."* *"Explain each step."*		
	Recognizes that Trading is Necessary	Trades a Dime for 10 Pennies	Gives Away 15¢	Counts Money	Trades Correctly	Subtracts Correctly	Explains Steps

2-120La   Copyright © 1991 by Saxon Publishers, Inc. and Nancy Larson. Reproduction prohibited.

ASSESSMENT 23
LESSON 120
Math 2

Name _____

Date _____

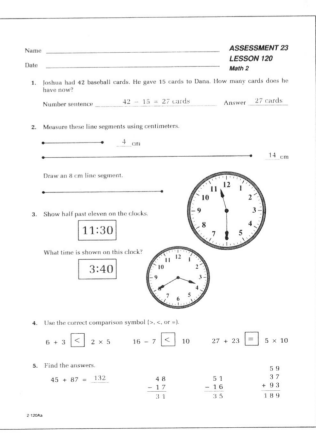

1. Joshua had 42 baseball cards. He gave 15 cards to Dana. How many cards does he have now?

   Number sentence ___42 – 15 = 27 cards___   Answer ___27 cards___

2. Measure these line segments using centimeters.

   •————————•   ___4___ cm

   •————————————————————————•   ___14___ cm

   Draw an 8 cm line segment.

   •————————————————————•

3. Show half past eleven on the clocks.

   [11:30]

   What time is shown on this clock?

   [3:40]

4. Use the correct comparison symbol (>, <, or =).

   6 + 3 [<] 2 × 5   16 – 7 [<] 10   27 + 23 [=] 5 × 10

5. Find the answers.

   45 + 87 = ___132___

   $\begin{array}{r} 48 \\ -17 \\ \hline 31 \end{array}$ $\begin{array}{r} 51 \\ -16 \\ \hline 35 \end{array}$ $\begin{array}{r} 59 \\ 37 \\ +93 \\ \hline 189 \end{array}$

2-120Aa

Lesson 121

making and labeling an array

lesson preparation

materials

20 color tiles (all one color)

Master 2-121

black, red, orange, and yellow crayons

Fact Sheet A 2-100

in the morning

• Write the following in the pattern box on the meeting strip:

$$\frac{1}{3}, \frac{2}{3}, \frac{3}{3}, \frac{4}{3}, \frac{5}{3}, \underline{\quad}, \underline{\quad}, \underline{\quad}, \underline{\quad}, \underline{\quad}, \underline{\quad}$$

Answer: $\frac{1}{3}, \frac{2}{3}, \frac{3}{3}, \frac{4}{3}, \frac{5}{3}, \frac{6}{3}, \frac{7}{3}, \frac{8}{3}, \frac{9}{3}, \frac{10}{3}, \frac{11}{3}$

• Write $\boxed{\$1.65}$ on the meeting strip. Provide a cup of 10 quarters, a cup of 10 dimes, a cup of 10 nickels, and a cup of 20 pennies.

THE MEETING

calendar

• Ask your child to write the date on the calendar and meeting strip.

• Ask your child to identify the number of days in 1 week, 2 weeks, and 3 weeks.

• Ask your child the following two or three times a week:

    date _____ days ago, date _____ days from now

    day of the week _____ days ago, day of the week _____ days from now

    _____th month, month before, month after

• Record on the meeting strip a special event and the number of days until it occurs.

weather graph

• Ask your child to read and graph today's temperature to the nearest two degrees.

• Count by 10's and 2's to check the temperature on the graph.

• Ask your child to connect the dot for yesterday's temperature to the dot for today's temperature and compare the temperatures.

counting

• Count by 4's to 40 and backward from 40 by 4's.

• Count by 25's to 300 and backward from 300 by 25's.

• Count by 3's to 30 and backward from 30 by 3's.

• Do the following once a week:

count by 10's to 400 and backward from 400 by 10's

count by 5's to 100 and backward from 50 by 5's

say the even numbers to 100 and backward from 50

say the odd numbers to 49 and backward from 49

graph questions

• You and your child each ask a question about any of the graphs.

patterning

• Ask your child to do the following:

identify the pattern (repeating, continuing, or both)

identify the numbers to complete the pattern

read the pattern

money

• Ask your child to put the coins in the coin cup. Count the money in the coin cup together.

• Ask your child for another way to show that amount of money. Count these coins together to check the amount.

clock

• Set the clock to a five-minute interval.

• Ask the following:

"It's (morning/afternoon/evening). What time is it?"

time one hour ago

time one hour from now

• Ask your child to write the digital time on the meeting strip.

• Record on the meeting strip the time an activity will occur.

number of the day

• Write three number sentences for the number of the day on the meeting strip.

fact practice

- Write three fact family numbers (e.g., 2, 7, 9) on the chalkboard.
- Allow time for your child to write the four fact family number sentences on the chalkboard.

THE LESSON

Making and Labeling an Array

"Today you will learn how to make an array."

"You also will learn how to label an array."

"Have you seen a band march in a parade?"

"What did you notice about how they marched?"

"When people march in a parade, they march in rows."

"We call this an array"

"The date squares on a calendar make an array."

"They are arranged in rows of seven days."

"Today you will use color tiles to make an array."

- Give your child a bag of 20 color tiles.

"Make a row of four tiles."

"Rows are horizontal."

"Now make another row of four tiles below the first row."

- The array should look like the following:

"How many rows of tiles do you have now?" 2

"How many tiles are in each row?" 4

"You have two groups of four tiles."

"This is a two by four array."

"How many tiles is that altogether?" 8

"Now make another row of four tiles below the first row."

- The rows of tiles should look like the following:

"How many rows of tiles do you have now?" 3

"How many tiles are in each row?" 4

"You have three rows of four tiles."

"This is a three by four array."

- Write "3 by 4 array" on the chalkboard.

"How many tiles is that altogether?" 12

"Now you will make a different array."

"Make a row of eight tiles."

- If necessary, tell your child that rows are horizontal.

"Now make another row of eight tiles below the first row."

- The array should look like the following:

"How many rows of tiles do you have now?" 2

"How many tiles are in each row?" 8

"You have two rows of eight tiles."

"This is a two by eight array."

- Write "2 by 8 array" on the chalkboard.

"What does the first number mean?" number of rows

"What does the second number mean?" number in each row

"How many tiles is that altogether?" 16

"Now you will make a different array."

"Make a row of two tiles."

"Now make another row of two tiles below the first row."

"How many rows of tiles do you have now?" 2

"Now make another row of two tiles below the other rows."

- The array should look like the following:

"How many rows of tiles do you have now?" 3

"How many tiles are in each row?" 2

"You have three rows of two tiles."

"You have made a three by two array."

"Now make another row of two tiles."

"How many rows of tiles do you have now?" 4

"You have four rows of two tiles."

"What will we call this array?" 4 by 2 array

"Now make another row of two tiles."

• The array should look like the following:

"How many rows of tiles do you have now?"   5

"You have five rows of two tiles."

"What will we call this array?"   5 by 2 array

• Write "5 by 2 array" on the chalkboard.

"What does the first number mean?"   number of rows

"What does the second number mean?"   number in each row

"How many tiles is that altogether?"   10

"Now you will make a different array."

"You will make a two by three array."

"You will make two rows of three tiles each."

"Make one row of three tiles."

"Now make another row of three tiles below the first row."

• The array should look like the following:

"What will we call this array?"   2 by 3 array

"How many tiles is that altogether?"   6

• Give your child **Master 2-121.**

"Now you will draw a picture of a two by three array."

"You will use the squares in part A."

"You will make the picture look just like the tiles."

"How do you think you will do that?"   color one square for each tile

"You will color one square for each tile."

"Use a red crayon to color the squares to show a two by three array."

"Start in the upper left-hand corner of your paper."

• When your child finishes, continue.

"Label the array like this."

• Write the following on the chalkboard:

"Write the number of tiles you used inside of the array using a black crayon."

- Write "2 by 3 array" on the chalkboard.

"Write 'two by three array' next to the letter A."

"Now you will make a four by two array."

- Write "4 by 2 array" on the chalkboard.

"What does the first number mean?"   *4 rows*

"What does the second number mean?"   *2 in each row*

"You will show four rows of two tiles."

"Make one row of two tiles."

"Now make another row of two tiles."

"Make a third row of two tiles."

"Now make a fourth row of two tiles."

"Do you have four rows of two tiles?"

"You have made a four by two array."

"How many tiles are in your four by two array?"   *8*

"Now you will draw a picture of the four by two array."

"You will use the squares in part B."

"How many rows will you have?"   *4*

"How many tiles will you have in each row?"   *2*

"Make your picture look just like the tiles."

"Use an orange crayon to draw a picture of your four by two array."

- When your child finishes, continue.

"Label the array like this."

- Write the following on the chalkboard:

"Write the number of tiles you used inside of the array using a black crayon."

"Write 'four by two array' next to the letter B."

"Now you will make a three by five array."

- Write "3 by 5 array" on the chalkboard.

"What does the first number mean?"   *3 rows*

"What does the second number mean?"   *5 in each row*

"How many rows will you make?"   *3*

"How many tiles will be in each row?"   5

"Make one row of five tiles."

"Now make two more rows of five tiles."

"How many tiles are in our three by five array?"   15

"Now you will draw a picture of the three by five array."

"You will use the squares in part C."

"How many rows will you have?"   3

"How many tiles will you have in each row?"   5

"Make your picture look just like the tiles."

"Use a yellow crayon to draw a picture of your three by five array."

- When your child finishes, continue.

  "Label the array like this."

- Write the following on the chalkboard:

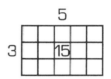

"Write the number of tiles you used inside of the array using a black crayon."

"Write 'three by five array' next to the letter C."

CLASS PRACTICE

number fact practice

- Give your child **Fact Sheet A 2-100**.

- Time your child for five minutes.

- Correct the fact sheet with your child and record the score.

- Allow time for your child to complete the unfinished facts.

WRITTEN PRACTICE

- Complete **Worksheet 121A** with your child.

- Complete **Worksheet 121B** with your child later in the day.

Name _____ **MASTER 2-121**
Math 2

A.

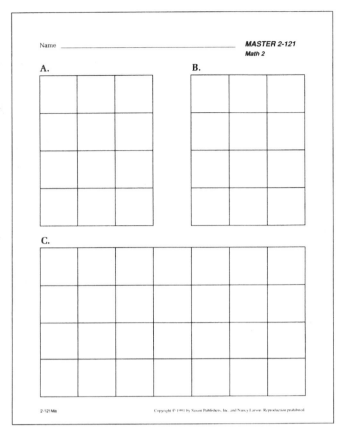

B.

C.

2-121Ma   Copyright © 1991 by Saxon Publishers, Inc. and Nancy Larson. Reproduction prohibited.

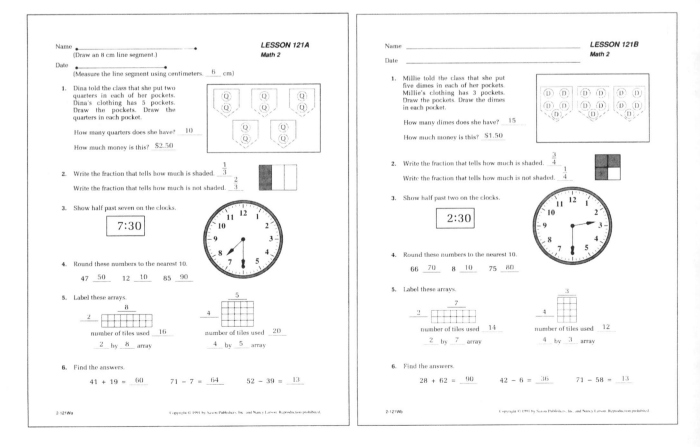

Name _____   **LESSON 121A**
(Draw an 8 cm line segment.)   **Math 2**
Date _____
(Measure the line segment using centimeters. __6__ cm)

1. Dina told the class that she put two
    quarters in each of her pockets.
    Dina's clothing has 5 pockets.
    Draw the pockets. Draw the
    quarters in each pocket.

    How many quarters does she have? __10__

    How much money is this? __$2.50__

2. Write the fraction that tells how much is shaded. __$\frac{1}{3}$__

    Write the fraction that tells how much is not shaded. __$\frac{2}{3}$__

3. Show half past seven on the clocks.

    7:30

4. Round these numbers to the nearest 10.

    47 __50__   12 __10__   85 __90__

5. Label these arrays.

    8
    2 [ ] number of tiles used __16__
    __2__ by __8__ array

    5
    4 [ ] number of tiles used __20__
    __4__ by __5__ array

6. Find the answers.

    41 + 19 = __60__   71 − 7 = __64__   52 − 39 = __13__

2-121Wa   Copyright © 1991 by Saxon Publishers, Inc. and Nancy Larson. Reproduction prohibited.

Name _____   **LESSON 121B**
Math 2
Date _____

1. Millie told the class that she put
    five dimes in each of her pockets.
    Millie's clothing has 3 pockets.
    Draw the pockets. Draw the dimes
    in each pocket.

    How many dimes does she have? __15__

    How much money is this? __$1.50__

2. Write the fraction that tells how much is shaded. __$\frac{3}{4}$__

    Write the fraction that tells how much is not shaded. __$\frac{1}{4}$__

3. Show half past two on the clocks.

    2:30

4. Round these numbers to the nearest 10.

    66 __70__   8 __10__   75 __80__

5. Label these arrays.

    7
    3 [ ] number of tiles used __14__
    __2__ by __7__ array

    3
    4 [ ] number of tiles used __12__
    __4__ by __3__ array

6. Find the answers.

    28 + 62 = __90__   42 − 6 = __36__   71 − 58 = __13__

2-121Wb   Copyright © 1991 by Saxon Publishers, Inc. and Nancy Larson. Reproduction prohibited.

Lesson 122

identifying right angles

---

lesson preparation

materials

1 rectangular piece of paper

1 geoboard

1 geoband

Master 2-122

Fact Sheet M 14.1

in the morning

• Write the following in the pattern box on the meeting strip:

> 0, ___, 10, ___, 20, ___, 30, ___, 40, ___, 50

Answer: 0, 5, 10, 15, 20, 25, 30, 35, 40, 45, 50

• Write | $1.17 | on the meeting strip. Provide a cup of 10 quarters, a cup of 10 dimes, a cup of 10 nickels, and a cup of 20 pennies.

THE MEETING

calendar

• Ask your child to write the date on the calendar and meeting strip.

• Ask your child to identify the number of days in 1 week, 2 weeks, and 3 weeks.

• Ask your child the following two or three times a week:

date _____ days ago, date _____ days from now

day of the week _____ days ago, day of the week _____ days from now

_____th month, month before, month after

• Record on the meeting strip a special event and the number of days until it occurs.

weather graph

• Ask your child to read and graph today's temperature to the nearest two degrees.

- Count by 10's and 2's to check the temperature on the graph.
- Ask your child to connect the dot for yesterday's temperature to the dot for today's temperature and compare the temperatures.

counting

- Count by 4's to 40 and backward from 40 by 4's.
- Count by 25's to 300 and backward from 300 by 25's.
- Count by 3's to 30 and backward from 30 by 3's.
- Do the following once a week:

    count by 10's to 400 and backward from 400 by 10's

    count by 5's to 100 and backward from 50 by 5's

    say the even numbers to 100 and backward from 50

    say the odd numbers to 49 and backward from 49

graph questions

- You and your child each ask a question about any of the graphs.

patterning

- Ask your child to do the following:

    identify the pattern (repeating, continuing, or both)

    identify the numbers to complete the pattern

    read the pattern

money

- Ask your child to put the coins in the coin cup. Count the money in the coin cup together.
- Ask your child for another way to show that amount of money. Count these coins together to check the amount.

clock

- Set the clock to a five-minute interval.
- Ask the following:

    "It's (morning/afternoon/evening). What time is it?"

    time one hour ago

    time one hour from now
- Ask your child to write the digital time on the meeting strip.
- Record on the meeting strip the time an activity will occur.

number of the day

- Write three number sentences for the number of the day on the meeting strip.

fact practice

- Write three fact family numbers (e.g., 2, 7, 9) on the chalkboard.
- Allow time for your child to write the four fact family number sentences on the chalkboard.

THE LESSON

Identifying Right Angles

"Today you will learn how to identify right angles."

- Hold up a rectangular piece of paper.

"What is the shape of this piece of paper?" rectangle

"How many angles does a rectangle have?" 4

"Mathematicians have a special name for angles that look like the angles on this piece of paper."

"They call them right angles."

"We can turn these angles in any direction and they are still called right angles."

"If we trace along the edge of the angle, we have a capital L."

- Trace along the edge of the paper to show a capital L on the chalkboard.

"The L can be turned in any direction."

- Turn the paper and trace along the paper in the following ways:

"Mathematicians use a special symbol to show that an angle is a right angle."

"They draw a small square in the corner of the angle to show that it is a right angle."

- Draw a small square in the corner of each chalkboard L. Draw small squares in each corner of the paper rectangle.

"Let's try to find examples of right angles in this room."

"We can check to see if an angle is a right angle by putting a corner of a piece of paper inside the angle."

"If the corner of the paper fits exactly, we have a right angle."

• Allow 1–2 minutes for your child to suggest as many right angles as possible. Check the angles using the corner of a piece of paper.

"Now you will have a chance to make some shapes that have right angles."

"We will use the corner of a piece of paper to check to see if you have a right angle."

• Give your child a geoboard, a geoband, and a copy of **Master 2-122**.

"Make a square on your geoboard."

• When your child finishes, continue.

"Put the corner of your paper in one of the angles of your square."

"Does it fit exactly?"

"This is a right angle."

"Try another angle."

"Does the corner of the paper fit exactly?"

"How many right angles does a square have?" 4

"Draw a picture of your square on the first small geoboard on Master 2-122."

• Allow time for your child to do this.

"Now you will draw a small square in each right angle."

"You will do that like this."

• Demonstrate on the chalkboard.

"Make a rectangle on your geoboard."

• When your child finishes, continue.

"How many right angles do you think a rectangle has?"

"Put the corner of your paper in one of the angles of your rectangle."

"Does it fit exactly?"

"This is a right angle."

"Try the other three angles."

"Does the corner of the paper fit exactly?"

• Allow time for your child to check the angles.

"How many right angles does a rectangle have?" 4

"Draw a picture of your rectangle on the second small geoboard on Master 2-122."

- Allow time for your child to do this.

   "Draw a small square to show each right angle."

- Allow time for your child to do this.

   "Lift the band off one peg of the geoboard."

   "What shape do you have now?"   triangle

   "Check the angles of the triangle with the corner of your paper."

   "Do you have any right angles?"

   "How many?"   1

   "A triangle with a right angle is called a right triangle."

   "Draw a picture of your right triangle on the third small geoboard on Master 2-122."

- Allow time for your child to do this.

   "Draw a small square to show the right angle."

- Allow time for your child to do this.

   "Make a design on your geoboard that has at least one right angle."

   "Check the angle using the corner of your piece of paper."

   "When you have made a shape that has a right angle, draw a picture of it on the fourth small geoboard on Master 2-122."

   "Draw a small square in the corner of each right angle."

- Allow time for your child to make and draw the design.

CLASS PRACTICE

number fact practice

- Use all the multiplication fact cards to practice the multiplication facts with your child.

- Give your child **Fact Sheet M 14.1.**

- Time your child for one minute.

- Correct the fact sheet with your child and record the score.

- Allow time for your child to complete the unfinished facts.

WRITTEN PRACTICE

- Complete **Worksheet 122A** with your child.

- Complete **Worksheet 122B** with your child later in the day.

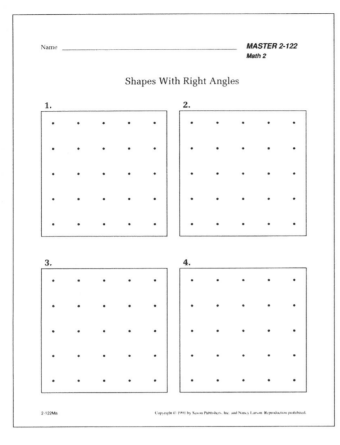

Name _____ **MASTER 2-122**
                                        Math 2

Shapes With Right Angles

1.

2.

3.

4.

2-122Ma   Copyright © 1991 by Saxon Publishers, Inc. and Nancy Larson. Reproduction prohibited.

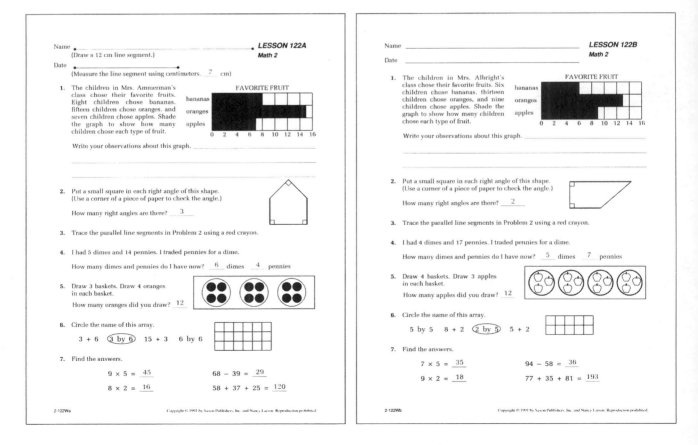

LESSON 122A
Math 2

Name _____
(Draw a 12 cm line segment.)

Date _____
(Measure the line segment using centimeters. __7__ cm)

1. The children in Mrs. Ammerman's class chose their favorite fruits. Eight children chose bananas, fifteen children chose oranges, and seven children chose apples. Shade the graph to show how many children chose each type of fruit.

    FAVORITE FRUIT

    Write your observations about this graph. _____

2. Put a small square in each right angle of this shape. (Use a corner of a piece of paper to check the angle.)

    How many right angles are there? __3__

3. Trace the parallel line segments in Problem 2 using a red crayon.

4. I had 5 dimes and 14 pennies. I traded pennies for a dime.

    How many dimes and pennies do I have now? __6__ dimes __4__ pennies

5. Draw 3 baskets. Draw 4 oranges in each basket.

    How many oranges did you draw? __12__

6. Circle the name of this array.

    3 + 6   ③ by 6   15 + 3   6 by 6

7. Find the answers.

    9 × 5 = __45__   68 − 39 = __29__

    8 × 2 = __16__   58 + 37 + 25 = __120__

2-122Wa   Copyright © 1991 by Saxon Publishers, Inc. and Nancy Larson. Reproduction prohibited.

LESSON 122B
Math 2

Name _____

Date _____

1. The children in Mrs. Albright's class chose their favorite fruits. Six children chose bananas, thirteen children chose oranges, and nine children chose apples. Shade the graph to show how many children chose each type of fruit.

    FAVORITE FRUIT

    Write your observations about this graph. _____

2. Put a small square in each right angle of this shape. (Use a corner of a piece of paper to check the angle.)

    How many right angles are there? __2__

3. Trace the parallel line segments in Problem 2 using a red crayon.

4. I had 4 dimes and 17 pennies. I traded pennies for a dime.

    How many dimes and pennies do I have now? __5__ dimes __7__ pennies

5. Draw 4 baskets. Draw 3 apples in each basket.

    How many apples did you draw? __12__

6. Circle the name of this array.

    5 by 5   8 + 2   ② by 5   5 + 2

7. Find the answers.

    7 × 5 = __35__   94 − 58 = __36__

    9 × 2 = __18__   77 + 35 + 81 = __193__

2-122Wb   Copyright © 1991 by Saxon Publishers, Inc. and Nancy Larson. Reproduction prohibited.

esson 123

writing number sentences for equal groups stories

THE MEETING

calendar

• Ask your child to write the date on the calendar and meeting strip.

• Ask your child to identify the number of days in 1 week, 2 weeks, and 3 weeks.

• Ask your child the following two or three times a week:

> date _____ days ago, date _____ days from now
>
> day of the week _____ days ago, day of the week _____ days from now
>
> _____th month, month before, month after

• Record on the meeting strip a special event and the number of days until it occurs.

weather graph

• Ask your child to read and graph today's temperature to the nearest two degrees.

• Count by 10's and 2's to check the temperature on the graph.

• Ask your child to connect the dot for yesterday's temperature to the dot for today's temperature and compare the temperatures.

counting

- Count by 4's to 40 and backward from 40 by 4's.
- Count by 25's to 300 and backward from 300 by 25's.
- Count by 3's to 30 and backward from 30 by 3's.
- Do the following once a week:

    count by 10's to 400 and backward from 400 by 10's

    count by 5's to 100 and backward from 50 by 5's

    say the even numbers to 100 and backward from 50

    say the odd numbers to 49 and backward from 49

graph questions

- You and your child each ask a question about any of the graphs.

patterning

- Ask your child to do the following:

    identify the pattern (repeating, continuing, or both)

    identify the numbers to complete the pattern

    read the pattern

money

- Ask your child to put the coins in the coin cup. Count the money in the coin cup together.
- Ask your child for another way to show that amount of money. Count these coins together to check the amount.

clock

- Set the clock to a five-minute interval.
- Ask the following:

    "It's (morning/afternoon/evening). What time is it?"

    time one hour ago

    time one hour from now

- Ask your child to write the digital time on the meeting strip.
- Record on the meeting strip the time an activity will occur.

number of the day

- Write three number sentences for the number of the day on the meeting strip.

fact practice

- Write three fact family numbers (e.g., 2, 7, 9) on the chalkboard.
- Allow time for your child to write the four fact family number sentences on the chalkboard.

THE LESSON

Writing Number Sentences for Equal Groups Stories

"We have been drawing pictures for equal groups stories."

"Today you will learn how to write number sentences for equal groups stories."

- Give your child **Master 2-119** from Lesson 119.

"Let's fill in the missing numbers below each picture."

"What will we write below the first picture?"   *5 groups of 4 wheels is 20 wheels*

"Fill in the numbers on your paper."

- Allow time for your child to do this.

"What will we write below the second picture?"   *3 groups of 5 buttons is 15 buttons*

"Fill in the numbers on your paper."

- Repeat with the next three pictures.

"Now I will write an equal groups story on the chalkboard."

- Write the following story on the chalkboard:

    Cheryl has 5 pieces of paper.

    She put 2 stickers on each piece of paper.

- Ask your child to read the story.

"How will you draw a picture to show this?"

"Draw a picture for this story in the sixth box on your paper."

- Allow time for your child to do this.

"How will we write a number sentence for this story?" *5 × 2 stickers = 10 stickers*

- Write "5 × 2 stickers = 10 stickers" on the chalkboard.

"The first number in our equal groups number sentence will always tell us the number of groups."

"The next number will tell us how many are in each group."

"The last number tells us the total."

"We will label our answer so we know if we are talking about stickers or donuts or chocolate chips."

"Write this number sentence below your picture."

- Write the following story on the chalkboard:

> Mrs. Rawden has 3 boxes of donuts.
>
> Each box has 6 donuts.

- Ask your child to read the story.

"How will you draw a picture to show this?"

"Draw a picture for this story in the seventh box on your paper."

- Allow time for your child to do this.

"How will we write a number sentence for this story?" *3 × 6 donuts = 18 donuts*

- Write "3 × 6 donuts = 18 donuts" on the chalkboard.

"What does the first number in our equal groups number sentence tell us?" *the number of groups*

"What does the next number tell us?" *how many are in each group*

"What does the last number tell us?" *the total*

"We will label our answer so we know if we are talking about stickers or donuts or chocolate chips."

"Write this number sentence below your picture."

- Write the following story on the chalkboard:

> Each chocolate chip cookie has 5 chocolate chips.
>
> There are seven cookies.

- Ask your child to read the story.

"How will you draw a picture to show this?"

"Draw a picture for this story in the eighth box on your paper."

- Allow time for your child to do this.

"How will we write a number sentence for this story?" *7 × 5 chocolate chips = 35 chocolate chips*

- Write "7 × 5 chocolate chips = 35 chocolate chips" on the chalkboard.

"What does the first number in our equal groups number sentence tell us?" *the number of groups*

"What does the next number tell us?" *how many are in each group*

"What does the last number tell us?" *the total*

"We will label our answer so we know if we are talking about stickers or donuts or chocolate chips."

"Write this number sentence below your picture."

CLASS PRACTICE

number fact practice

- Give your child **Fact Sheet S-100**.

- Time your child for five minutes.

- Correct the fact sheet with your child and record the score.

- Allow time for your child to complete the unfinished facts.

WRITTEN PRACTICE

- Complete **Worksheet 123A** with your child.

- Complete **Worksheet 123B** with your child later in the day.

Name _____ **LESSON 123A**
(Draw a 9 cm line segment.) **Math 2**

Date _____
(Measure the line segment using centimeters. __11__ cm)

1. There were 6 children at the party. Aunt Angie put 5 strawberries on each child's dish of ice cream.

   What type of story is this? _____ equal groups
   Draw a picture to show the strawberries on the dishes of ice cream.

   How many strawberries did Aunt Angie use altogether?

   Number sentence __6 × 5 = 30 strawberries__ Answer __30 strawberries__

2. How much money is this? Write the amount two ways.

   __46¢__ __$0.46__

3. Measure the sides of the rectangle in Problem 2 using centimeters. What is the perimeter?

   Number sentence __3 cm + 10 cm + 3 cm + 10 cm = 26 cm__

   Answer __26 centimeters__

4. Draw a triangle with a right angle.

5. Label this array. __10__

   __3__

   number of tiles used __30__

   __3__ by __10__ array

6. Write two hundred seven using digits. __207__

   Write two hundred seven in expanded form. __200 + 7__

7. Find the answers.

   7 × 2 = __14__   8 × 5 = __40__   9 × 100 = __900__

2-123Wa

Name _____ **LESSON 123B**

Date _____ **Math 2**

1. There were 8 children. Miss Natiello put 2 stickers on each child's paper.

   What type of story is this? _____ equal groups
   Draw a picture to show the stickers on the papers.

   How many stickers did Miss Natiello use altogether?

   Number sentence __8 × 2 stickers = 16 stickers__ Answer __16 stickers__

2. How much money is this? Write the amount two ways. __61¢__ __$0.61__

3. Find the perimeter of the rectangle in Problem 2.

   Number sentence __3 cm + 11 cm + 3 cm + 11 cm = 28 cm__

   Answer __28 centimeters__

4. Draw a shape with 4 right angles.

5. Label this array. __9__

   __2__

   answers may vary

   number of tiles used __18__

   __2__ by __9__ array

6. Write four hundred sixteen using digits. __416__

   Write four hundred sixteen in expanded form. __400 + 10 + 6__

7. Find the answers.

   3 × 5 = __15__   9 × 2 = __18__   4 × 100 = __400__

2-123Wb

Lesson 124

multiplying by three

---

lesson preparation ———————————————————

materials

bag of 30 color tiles

multiplication fact cards — pink

Fact Sheet M 17.0

the night before

• Separate the pink multiplication fact cards.

in the morning

• Write the following in the pattern box on the meeting strip:

∩, ⌣⌣, ⌢⌢, ⌣⌣⌣, ⌢⌢⌢⌢, _____, _____, _____

Answer: ∩, ⌣⌣, ⌢⌢, ⌣⌣⌣, ⌢⌢⌢⌢, ⌣⌣⌣⌣⌣, ⌢⌢⌢⌢⌢⌢, ⌣⌣⌣⌣⌣⌣

• Write $1.47 on the meeting strip. Provide a cup of 10 quarters, a cup of 10 dimes, a cup of 10 nickels, and a cup of 20 pennies.

THE MEETING

calendar

• Ask your child to write the date on the calendar and meeting strip.

• Ask your child to identify the number of days in 1 week, 2 weeks, and 3 weeks.

• Ask your child the following two or three times a week:

  date _____ days ago, date _____ days from now

  day of the week _____ days ago, day of the week _____ days from now

  _____th month, month before, month after

• Record on the meeting strip a special event and the number of days until it occurs.

weather graph

• Ask your child to read and graph today's temperature to the nearest two degrees.

- Count by 10's and 2's to check the temperature on the graph.
- Ask your child to connect the dot for yesterday's temperature to the dot for today's temperature and compare the temperatures.

counting

- Count by 4's to 40 and backward from 40 by 4's.
- Count by 25's to 300 and backward from 300 by 25's.
- Count by 3's to 30 and backward from 30 by 3's.
- Do the following once a week:

    count by 10's to 400 and backward from 400 by 10's

    count by 5's to 100 and backward from 50 by 5's

    say the even numbers to 100 and backward from 50

    say the odd numbers to 49 and backward from 49

graph questions

- You and your child each ask a question about any of the graphs.

patterning

- Ask your child to do the following:

    identify the pattern (repeating, continuing, or both)

    identify the shapes to complete the pattern

    read the pattern

money

- Ask your child to put the coins in the coin cup. Count the money in the coin cup together.
- Ask your child for another way to show that amount of money. Count these coins together to check the amount.

clock

- Set the clock to a five-minute interval.
- Ask the following:

    "It's (morning/afternoon/evening). What time is it?"

    time one hour ago

    time one hour from now
- Ask your child to write the digital time on the meeting strip.
- Record on the meeting strip the time an activity will occur.

number of the day

- Write three number sentences for the number of the day on the meeting strip.

fact practice

- Write three fact family numbers (e.g., 2, 7, 9) on the chalkboard.
- Allow time for your child to write the four fact family number sentences on the chalkboard.

THE LESSON

Multiplying by Three

"Today you will learn how to multiply by three."

- Give your child a bag of 30 color tiles.

"Put the color tiles in stacks of three."

- Allow time for your child to stack the tiles.

"How many groups of three do you have?"   *10*

"We can write that like this."

- Write the following on the chalkboard:

$$10 \text{ groups of } 3 = 30$$
$$10 \times 3 = 30$$

"How many tiles are in one group?"   *3*

"Let's count by 3's to find out how many tiles are in five groups."   *15*

"Let's count by 3's to find out how many tiles are in eight groups."   *24*

- Write the following on the chalkboard:

$4 \times 3 =$ $6 \times 3 =$ $8 \times 3 =$ $7 \times 3 =$ $9 \times 3 =$ $0 \times 3 =$

"The first problem is read 'four groups of three.' "

- Write "four groups of three" under 4×3.

"Read the second problem."

- Repeat for each problem.

"We will count the groups of tiles by 3's to find each answer."

"How many groups of tiles will we count for the first problem?"   *4 groups*

"Count by 3's as you point to each group of tiles."

"How many tiles are in four groups of three?"   *12*

- Record "$4 \times 3 = 12$" on the chalkboard.

- Repeat with all of the problems.
- Write the following on the chalkboard:

0 × 3 =	6 × 3 =
1 × 3 =	7 × 3 =
2 × 3 =	8 × 3 =
3 × 3 =	9 × 3 =
4 × 3 =	10 × 3 =
5 × 3 =	

"Let's fill in the answers."

"How many tiles are in zero groups of three?" *0*

"How many tiles are in one group of three?" *3*

"How many tiles are in two groups of three?" *6*

- Repeat with each problem.

"We can write the multiplication facts for three a different way."

- Write the following on the chalkboard:

$$6 \times 3 = \qquad \begin{array}{r} 3 \\ \times\ 6 \\ \hline \end{array}$$

"These are two different ways of writing the same problem."

- Write the following next to the previous problem:

$$9 \times 3 = \qquad \begin{array}{r} 3 \\ \times\ 9 \\ \hline \end{array}$$

"We read these problems as 'six groups of three' and 'nine groups of three.' "

"What switcharound problems can we write for these two problems?"

$$3 \times 6 = \qquad \begin{array}{r} 6 \\ \times\ 3 \\ \hline \end{array} \qquad\qquad 3 \times 9 = \qquad \begin{array}{r} 9 \\ \times\ 3 \\ \hline \end{array}$$

- Ask your child to write the problems on the chalkboard.

"I will give you the multiplying by three fact cards."

- Give your child the pink multiplication fact cards.

"Match the switcharound problems."

"When you finish, practice saying the answers for these problems."

CLASS PRACTICE

number fact practice

- Give your child **Fact Sheet M 17.0.**
- Time your child for one minute.
- Correct the fact sheet with your child and record the score.
- Allow time for your child to complete the unfinished facts.

WRITTEN PRACTICE

- Complete **Worksheet 124A** with your child.
- Complete **Worksheet 124B** with your child later in the day.

Name _____   **LESSON 124A**
(Draw a 12 cm line segment.)   Math 2

Date _____
(Measure the line segment using centimeters. __10__ cm)

1. Pencils are sold in packages of 3. Mrs. Doster bought 7 packages of pencils.

   What type of story is this? ___equal groups___

   Draw a picture to show the packages of pencils.

   How many pencils did she buy?

   Number sentence ___7 × 3 pencils = 21 pencils___   Answer __21 pencils__

2. Write the fraction that tells how much is shaded. __$\frac{3}{5}$__

   Write the fraction that tells how much is not shaded. __$\frac{2}{5}$__

3. Davina has 4 quarters, a nickel, and 7 pennies. Draw the coins.

   How much money is this? __$1.12__

4. Find the answers.

3	3	3	3	3	3	3	3	3	3	3
×7	×3	×9	×1	×4	×10	×2	×6	×8	×0	×5
21	9	27	3	12	30	6	18	24	0	15

5. Draw a small square to show the right angle.

6. Find the answers.

   58   85   94
   −34   +35   −25
   ‾‾24 ‾‾120 ‾‾69

   6 + 2 + 3 + 7 + 9 + 1 = __28__

2-124Wa   Copyright © 1991 by Saxon Publishers, Inc. and Nancy Larson. Reproduction prohibited.

Name _____   **LESSON 124B**
   Math 2

Date _____

1. Pencils are sold in packages of 10. Mrs. Campion bought 9 packages.

   What type of story is this? ___equal groups___

   Draw a picture to show the packages of pencils.

   How many pencils did she buy?

   Number sentence ___9 × 10 pencils = 90 pencils___   Answer __90 pencils__

2. Write the fraction that tells how much is shaded. __$\frac{2}{6}$__

   Write the fraction that tells how much is not shaded. __$\frac{4}{6}$__

3. Karen has 5 quarters, a nickel, and 2 pennies. Draw the coins.

   How much money is this? __$1.32__

4. Fill in the missing numbers.

3	3	3	3	3	3	3	3	3
×3	×8	×5	×1	×9	×4	×0	×6	×2	×7
9	24	15	3	27	12	0	18	6	21

5. Draw a small square to show the right angle.

6. Find the answers.

   85   91   62
   −43   +47   −57
   ‾‾42 ‾‾138 ‾‾5

   7 + 3 + 2 + 9 + 8 + 4 + 1 = __34__

2-124Wb   Copyright © 1991 by Saxon Publishers, Inc. and Nancy Larson. Reproduction prohibited.

L esson 125

identifying intersecting lines
identifying perpendicular lines

---

lesson preparation

materials

Written Assessment #24

1 geoboard

2 geobands

Master 2-125

Fact Sheet M 17.0

in the morning

• Write the following in the pattern box on the meeting strip:

> 450, 460, 470, ____, ____, ____, ____, ____, ____

Answer: 450, 460, 470, 480, 490, 500, 510, 520, 530

• Write $2.76 on the meeting strip. Provide a cup of 10 quarters, a cup of 10 dimes, a cup of 10 nickels, and a cup of 20 pennies.

THE MEETING

calendar

• Ask your child to write the date on the calendar and meeting strip.

• Ask your child to identify the number of days in 1 week, 2 weeks, and 3 weeks.

• Ask your child the following two or three times a week:

   date _____ days ago, date _____ days from now

   day of the week _____ days ago, day of the week _____ days from now

   _____th month, month before, month after

• Record on the meeting strip a special event and the number of days until it occurs.

---

weather graph

- Ask your child to read and graph today's temperature to the nearest two degrees.
- Count by 10's and 2's to check the temperature on the graph.
- Ask your child to connect the dot for yesterday's temperature to the dot for today's temperature and compare the temperatures.

counting

- Count by 4's to 40 and backward from 40 by 4's.
- Count by 25's to 300 and backward from 300 by 25's.
- Count by 3's to 30 and backward from 30 by 3's.
- Do the following once a week:

    count by 10's to 400 and backward from 400 by 10's

    count by 5's to 100 and backward from 50 by 5's

    say the even numbers to 100 and backward from 50

    say the odd numbers to 49 and backward from 49

graph questions

- You and your child each ask a question about any of the graphs.

patterning

- Ask your child to do the following:

    identify the pattern (repeating, continuing, or both)

    identify the numbers to complete the pattern

    read the pattern

money

- Ask your child to put the coins in the coin cup. Count the money in the coin cup together.
- Ask your child for another way to show that amount of money. Count these coins together to check the amount.

clock

- Set the clock to a five-minute interval.
- Ask the following:

    "It's (morning/afternoon/evening). What time is it?"

    time one hour ago

    time one hour from now
- Ask your child to write the digital time on the meeting strip.

• Record on the meeting strip the time an activity will occur.

number of the day

• Write three number sentences for the number cf the day on the meeting strip.

fact practice

• Write three fact family numbers (e.g., 2, 7, 9) on the chalkboard.

• Allow time for your child to write the four fact family number sentences on the chalkboard.

ASSESSMENT

Written Assessment

"Today I would like to see what you remember from what we have been practicing."

• Give your child **Written Assessment #24.**

• Read the directions for each problem. Allow time for your child to complete each problem before continuing.

• Correct the paper, noting your child's mistakes on the **Individual Recording Form.** Review the errors with your child.

THE LESSON

Identifying Intersecting Lines
Identifying Perpendicular Lines

"We have been talking about parallel lines and line segments."

"Where do you see parallel lines or line segments in this room?"

• Give your child a geoboard and two geobands.

"Make parallel line segments on your geoboard."

• Allow time for your child to do this.

"What do we know about parallel lines?"   they never meet; they are an equal distance apart

"Take the geobands off the geoboard."

"Today you will learn about line segments that do meet."

"You also will learn about lines and line segments that meet in a special way."

"Make two line segments that meet on your geoboard."

- Allow time for your child to do this.

  "Put your finger on the point where the line segments meet."

  "Mathematicians call this the intersection of the line segments."

  "These are intersecting line segments."

  "This is why we say that when two streets meet, we have an intersection."

  "There are special types of intersecting line segments."

- Draw the following on the chalkboard:

  "When two line segments intersect like these line segments do, we call them perpendicular line segments."

  "What do you notice about perpendicular line segments?"

- Allow time for your child to offer observations.

  "Perpendicular lines and line segments have at least one right angle."

  "We can use the corner of a piece of paper to check for perpendicular lines and line segments."

- Demonstrate on the chalkboard examples.

  "Where do you see an example of perpendicular lines or line segments in this room?"

- Allow time for your child to locate as many right angles as possible.

  "Make perpendicular line segments on your geoboard."

- Allow time for your child to do this.

  "Let's check to see if the line segments are perpendicular."

  "How can we do this?"   use the corner of a piece of paper

- Give your child **Master 2-125**.

  "Use the corner of this paper to make sure that you have at least one right angle."

  "Copy your perpendicular line segments on the first small geoboard picture."

  "Draw a small square in the corner of the right angle."

- Allow time for your child to do this.

  "Make a different pair of perpendicular line segments on your geoboard."

- Allow time for your child to do this.

  "Check to see if they are perpendicular."

"Copy your perpendicular line segments on the second small geoboard picture."

"Draw a small square in the corner of the right angle."

- Allow time for your child to do this.

"Make two more different examples of perpendicular line segments."

"Copy your perpendicular line segments on the third and fourth small geoboard pictures."

- Allow time for your child to do this.

CLASS PRACTICE

number fact practice

- Use the pink fact cards to practice the multiplying by three facts with your child.
- Give your child **Fact Sheet M 17.0.**
- Time your child for one minute.
- Correct the fact sheet with your child and record the score.
- Allow time for your child to complete the unfinished facts.

WRITTEN PRACTICE

- Complete **Worksheet 125A** with your child.
- Complete **Worksheet 125B** with your child later in the day.

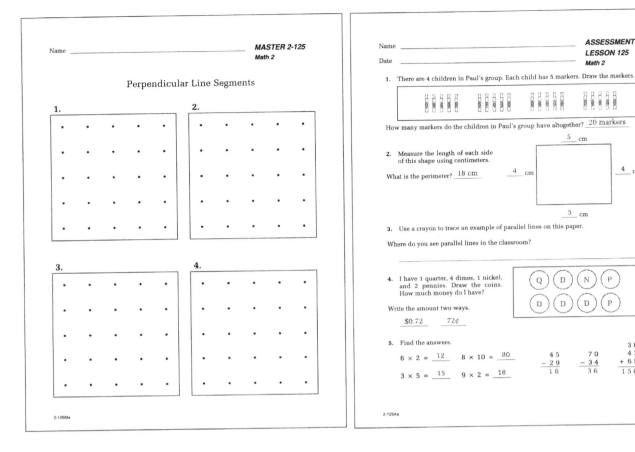

Name _____ **MASTER 2-125**
Math 2

Perpendicular Line Segments

1.

2.

3.

4.

2-125Ma

Name _____ **ASSESSMENT 24**
Date _____ **LESSON 125**
Math 2

1. There are 4 children in Paul's group. Each child has 5 markers. Draw the markers.

How many markers do the children in Paul's group have altogether? __20 markers__

2. Measure the length of each side of this shape using centimeters.

What is the perimeter? __18 cm__

5 cm
4 cm
4 cm
5 cm

3. Use a crayon to trace an example of parallel lines on this paper.

Where do you see parallel lines in the classroom?

_____

4. I have 1 quarter, 4 dimes, 1 nickel, and 2 pennies. Draw the coins. How much money do I have?

Write the amount two ways.

__$0.72__   __72¢__

Q D N P
D D D P

5. Find the answers.

$6 \times 2 =$ __12__ $8 \times 10 =$ __80__

$3 \times 5 =$ __15__ $9 \times 2 =$ __18__

$\begin{array}{r} 45 \\ -29 \\ \hline 16 \end{array}$ $\begin{array}{r} 70 \\ -34 \\ \hline 36 \end{array}$ $\begin{array}{r} 38 \\ 47 \\ +65 \\ \hline 150 \end{array}$

2-125Aa

Name •————————————• **LESSON 125A**
(Draw a 3" line segment.) Math 2
Date _____
(Measure this line segment using inches. __4__ ")

1. Twenty-six children were in the gym. Seventeen children from another class joined them.

What type of story is this? ____some, some more____

How many children are in the gym now?

Number sentence ___26 + 17 = 43 children___   Answer __43 children__

2. Circle the perpendicular line segments.

3. About how much might a 7-year-old child weigh?

200 pounds   (60 pounds)   15 pounds   2 pounds

4. Round each number to the nearest 10.

78 __80__   13 __10__   25 __30__

5. Circle all the geometric solids that have at least one point.

(pyramid)   cylinder   (cone)   sphere   (cube)

6. Find the answers.

$\begin{array}{r} 62 \\ -38 \\ \hline 24 \end{array}$ $\begin{array}{r} 68 \\ 37 \\ +25 \\ \hline 130 \end{array}$ $2 \times 3 =$ __6__ $8 \times 10 =$ __80__

$7 \times 3 =$ __21__ $3 \times 100 =$ __300__

$9 \times 3 =$ __27__ $5 \times 3 =$ __15__

2-125Wa

Name _____ **LESSON 125B**
Date _____ Math 2

1. There were forty-three children in the gym. Fifteen children went back to class.

What type of story is this? ____some, some went away____

How many children are in the gym now?

Number sentence ___43 − 15 = 28 children___   Answer __28 children__

2. Circle the perpendicular line segments.

3. About how much might a 10-year-old child weigh?

25 pounds   300 pounds   (90 pounds)   4 pounds

4. Round each number to the nearest 10.

31 __30__   9 __10__   15 __20__

5. Circle all the geometric solids that will roll.

pyramid   (cylinder)   (cone)   (sphere)

6. Find the answers.

$\begin{array}{r} 71 \\ -25 \\ \hline 46 \end{array}$ $\begin{array}{r} 79 \\ 26 \\ +35 \\ \hline 140 \end{array}$ $4 \times 3 =$ __12__ $7 \times 10 =$ __70__

$6 \times 3 =$ __18__ $2 \times 100 =$ __200__

$3 \times 3 =$ __9__ $8 \times 3 =$ __24__

2-125Wb

698

Lesson 126

writing number sentences for arrays

THE MEETING

calendar

- Ask your child to write the date on the calendar and meeting strip.

- Ask your child to identify the number of days in 1 week, 2 weeks, and 3 weeks.

- Ask your child the following two or three times a week:

    date _____ days ago, date _____ days from now

    day of the week _____ days ago, day of the week _____ days from now

    _____th month, month before, month after

- Record on the meeting strip a special event and the number of days until it occurs.

weather graph

- Ask your child to read and graph today's temperature to the nearest two degrees.

- Count by 10's and 2's to check the temperature on the graph.

- Ask your child to connect the dot for yesterday's temperature to the dot for today's temperature and compare the temperatures.

counting

- Count by 4's to 40 and backward from 40 by 4's.
- Count by 25's to 300 and backward from 300 by 25's.
- Count by 3's to 30 and backward from 30 by 3's.
- Do the following once a week:

  count by 10's to 400 and backward from 400 by 10's

  count by 5's to 100 and backward from 50 by 5's

  say the even numbers to 100 and backward from 50

  say the odd numbers to 49 and backward from 49

graph questions

- You and your child each ask a question about any of the graphs.

patterning

- Ask your child to do the following:

  identify the pattern (repeating, continuing, or both)

  identify the numbers to complete the pattern

  read the pattern

money

- Ask your child to put the coins in the coin cup. Count the money in the coin cup together.
- Ask your child for another way to show that amount of money. Count these coins together to check the amount.

clock

- Set the clock to a five-minute interval.
- Ask the following:

  "It's (morning/afternoon/evening). What time is it?"

  time one hour ago

  time one hour from now
- Ask your child to write the digital time on the meeting strip.
- Record on the meeting strip the time an activity will occur.

number of the day

- Write three number sentences for the number of the day on the meeting strip.

fact practice

- Write three fact family numbers (e.g., 2, 7, 9) on the chalkboard.

- Allow time for your child to write the four fact family number sentences on the chalkboard.

THE LESSON

Writing Number Sentences for Arrays

"Today you will learn how to write a number sentence for an array."

- Draw the following array on the chalkboard:

"How will we label this array?"

- Ask your child to label the chalkboard array.

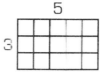

"We say that this is a three by five array."

"How many tiles are in a three by five array?"   15

"We can write that like this."

- Write the following on the chalkboard below the array:

$$3 \times 5 = 15$$

- Repeat with the following arrays one at a time: $4 \times 2 = 8$; $5 \times 6 = 30$; $2 \times 7 = 14$

 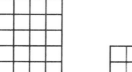

"Now you will have a chance to write some number sentences for arrays."

- Draw the following on the chalkboard:

"Write a number sentence for this array." $2 \times 6 = 12$

- Ask your child to write the number sentence on the chalkboard.

- Repeat with the following arrays one at a time: $4 \times 3 = 12$; $5 \times 3 = 15$; $3 \times 7 = 21$

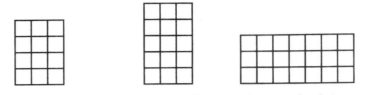

"Sometimes people draw shapes or letters instead of squares to make an array."

- Draw the following on the chalkboard one at a time:

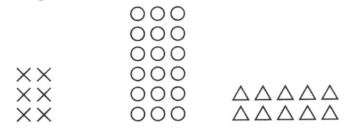

- Ask your child to write the number sentence for each array on the chalkboard. [$3 \times 2 = 6$; $6 \times 3 = 18$; $2 \times 5 = 10$]

CLASS PRACTICE

number fact practice

- Use the fact cards to practice the multiplication facts with your child.
- Give your child **Fact Sheet M 14.1.**
- Time your child for one minute.
- Correct the fact sheet with your child and record the score.
- Allow time for your child to complete the unfinished facts.

WRITTEN PRACTICE

- Complete **Worksheet 126A** with your child.
- Complete **Worksheet 126B** with your child later in the day.

Name _____ •———————————• **LESSON 126A**
(Draw a $2\frac{1}{2}$" line segment.) **Math 2**
Date _____
•———————————•
(Measure this line segment using inches. __4__ ")

1. Three children can sit at each table in Room 7. There are ten tables in the room. Draw a picture to show the tables and chairs in Room 7.

How many children can sit in Room 7?

Number sentence _____ $10 \times 3 = 30$ children _____   Answer 30 children

2. Find an example of perpendicular line segments on this paper. Trace them with a crayon.

3. Round each number to the nearest 10 and add the rounded numbers.

63 + 29   31 + 48
60 + 30 = 90   30 + 50 = 80

4. Label this array. Write a number sentence for the array.

_____ $3 \times 2 = 6$ _____   2

3 ▦

5. Use the correct comparison symbol (>, <, or =).

35 [<] 53   8 + 42 [=] 5×10 16 − 7 [>] 14 − 6

6. Find the answers.

59 + 87 = __146__   74 − 28 = __46__   7 + 63 + 51 = __121__

2-126Wa

Name _____ **LESSON 126B**
Math 2
Date _____

1. Mrs. Wagoner has 5 games for the children to use during recess. Four children can play each game. Draw a picture to show the games and children.

How many children can play the games?

Number sentence _____ $5 \times 4 = 20$ children _____   Answer 20 children

2. Find an example of perpendicular line segments on this paper. Trace them with a crayon.

3. Round each number to the nearest 10 and add the rounded numbers.

28 + 49   52 + 39
30 + 50 = 80   50 + 40 = 90

4. Label this array. Write a number sentence for the array.   4

$2 \times 4 = 8$ _____   2 ▦

5. Use the correct comparison symbol (>, <, or =).

68 [<] 86   3 + 27 [=] 5×6 17 − 9 [<] 13 − 4

6. Find the answers.

65 + 86 = __151__   85 − 27 = __58__   59 + 61 + 8 = __128__

2-126Wb

703

Lesson 127

writing the date using digits

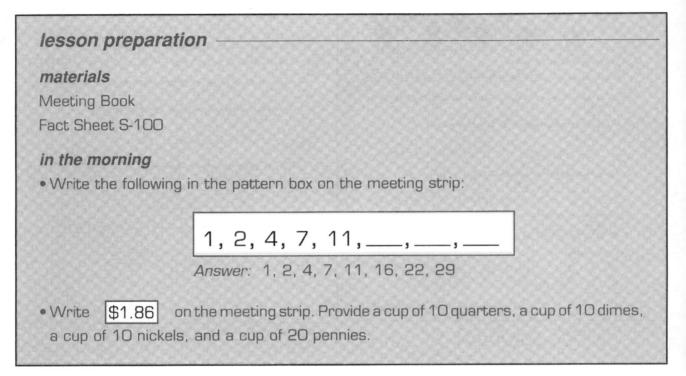

lesson preparation

materials

Meeting Book

Fact Sheet S-100

in the morning

• Write the following in the pattern box on the meeting strip:

> 1, 2, 4, 7, 11, ___, ___, ___
>
> *Answer:* 1, 2, 4, 7, 11, 16, 22, 29

• Write $1.86 on the meeting strip. Provide a cup of 10 quarters, a cup of 10 dimes, a cup of 10 nickels, and a cup of 20 pennies.

THE MEETING

calendar

• Ask your child to write the date on the calendar and meeting strip.

• Ask your child to identify the number of days in 1 week, 2 weeks, and 3 weeks.

• Ask your child the following two or three times a week:

  date _____ days ago, date _____ days from now

  day of the week _____ days ago, day of the week _____ days from now

  _____th month, month before, month after

• Record on the meeting strip a special event and the number of days until it occurs.

weather graph

• Ask your child to read and graph today's temperature to the nearest two degrees.

• Count by 10's and 2's to check the temperature on the graph.

- Ask your child to connect the dot for yesterday's temperature to the dot for today's temperature and compare the temperatures.

counting

- Count by 4's to 40 and backward from 40 by 4's.

- Count by 25's to 300 and backward from 300 by 25's.

- Count by 3's to 30 and backward from 30 by 3's.

- Do the following once a week:

  count by 10's to 400 and backward from 400 by 10's

  count by 5's to 100 and backward from 50 by 5's

  say the even numbers to 100 and backward from 50

  say the odd numbers to 49 and backward from 49

graph questions

- You and your child each ask a question about any of the graphs.

patterning

- Ask your child to do the following:

  identify the pattern (repeating, continuing, or both)

  identify the numbers to complete the pattern

  read the pattern

money

- Ask your child to put the coins in the coin cup. Count the money in the coin cup together.

- Ask your child for another way to show that amount of money. Count these coins together to check the amount.

clock

- Set the clock to a five-minute interval.

- Ask the following:

  "It's (morning/afternoon/evening). What time is it?"

  time one hour ago

  time one hour from now

- Ask your child to write the digital time on the meeting strip.

- Record on the meeting strip the time an activity will occur.

number of the day

- Write three number sentences for the number of the day on the meeting strip.

fact practice

- Write three fact family numbers (e.g., 2, 7, 9) on the chalkboard.
- Allow time for your child to write the four fact family number sentences on the chalkboard.

THE LESSON

Writing the Date Using Digits

"Today you will learn how to write the date a different way."

"You will write the date using only digits."

"When we write today's date using digits, we write it like this."

- Write the date using digits on the chalkboard in the following way:

5/6/92   (Use today's date.)

"How do you think I knew which digits to write?"

"The first digit tells us the month."

"Why did I write the number (five)?"   because (May) is the (5th) month of the year

"Let's write the digits we will use for the months on the chart of the names of the months in the Meeting Book."

- Open the Meeting Book to this month's calendar.

"Which digit will we use for the month of January?"   1

"Why?"   it is the first month of the year

- Write the digit on the chart.
- Repeat with each month.

"The digits between the slanted lines tell us the date."

"They are the same as the digits you write on the date tag."

"The last two digits tell us the year."

"How will we know what digits to write?"   the last two digits of the year

"How will we write tomorrow's date using digits?"

- Write the date using digits on the chalkboard.
- Write the following on the chalkboard:

January 14, 1982

"How will we write this date using digits?"

- Write the date using digits on the chalkboard as your child describes the digits.

- Repeat with the following dates:

  March 1, 1997

  December 25, 1971

"Now I will write a date using digits."

- Write the following on the chalkboard:

  2/14/95

"What is the full date?"   *February 14, 1995*

- Write the full date on the chalkboard.

- Repeat with the following dates:

  6/25/63

  10/9/04

"Now you will have a chance to practice writing dates using digits."

- Write the following dates on the chalkboard:

  August 13, 1950

  April 15, 1902

  November 1, 1911

"Write these dates using digits."

- Allow time for your child to write the dates using digits on the chalkboard.

"Now I will write three dates using digits."

"You will write the full dates."

- Write the following dates on the chalkboard:

  3/27/98

  9/23/40

  7/2/09

- Allow time for your child to write the dates on the chalkboard.

"Write a date using only 3's as digits."

"What is the full date?"   *March 3, 1933*

"Write a date using only 2's as digits."

"What is the full date?"   *February 2, 1922 or February 22, 1922*

"Write a date using only 9's as digits."

"What is the full date?"   *September 9, 1999*

- Repeat with the other digits, if desired.

"Write April 9, 1994 using digits."

"Now write December 8, 1921 using digits."

"What do you notice about the digits in each of these dates?"   *they are the same read forward as they are read backward*

> *"Numbers that are the same read forward or backward are called palindromes."*

> *"Can you write another date where the digits are a palindrome?"*

- There are many possibilities. Possible dates include January 3, 1931; March 4, 1943; November 22, 1911; etc.

- Encourage your child to find as many palindromic dates as possible. Write the dates in the Meeting Book, if desired.

CLASS PRACTICE

number fact practice

- Give your child **Fact Sheet S-100**.

- Time your child for five minutes.

- Correct the fact sheet with your child and record the score.

- Allow time for your child to complete the unfinished facts.

WRITTEN PRACTICE

- Complete **Worksheet 127A** with your child.

- Complete **Worksheet 127B** with your child later in the day.

Name _____ LESSON 127A
(Draw an 11 cm line segment.) Math 2
Date _____
(Measure the line segment using inches. __3__ " Write the date using digits.)

1. Nine children in Mrs. O'Neill's class assembled books. The children used three staples for each book. Draw a picture to show the staples in the books.

   How many staples did they use for the books?

   Number sentence __9 × 3 = 27 staples__   Answer __27 staples__

2. Write number sentences for these arrays.

   × × × × × × ×
   × × × × × × ×
   __2 × 7 = 14__

   ○ ○ ○
   ○ ○ ○
   ○ ○ ○
   ○ ○ ○
   __4 × 3 = 12__

3. Measure each side of this shape using centimeters.

   How long is the vertical line segment? __3 cm__

   How long is the oblique line segment? __5 cm__

   How long is the horizontal line segment? __4 cm__

   What is the perimeter? __12 cm__

4. Trace the perpendicular line segments in Problem 3 using a crayon.

5. I have 2 quarters, 2 dimes, 3 nickels, and 7 pennies. Draw the coins.

   How much money do I have? __92¢__

   Ⓠ Ⓓ Ⓝ Ⓝ Ⓟ Ⓟ Ⓟ
   Ⓠ Ⓓ Ⓝ Ⓟ Ⓟ Ⓟ Ⓟ

6. Find the answers.

   63 + 94 = __157__   94 − 77 = __17__

2-127Wa   Copyright © 1991 by Saxon Publishers, Inc. and Nancy Larson. Reproduction prohibited.

Name _____ LESSON 127B
  Math 2
Date _____

1. Four children in Mrs. Sheehan's class made picture books. Each book had 10 pictures. Draw a picture to show the pictures in the books.

   How many pictures are in the books?

   Number sentence __4 × 10 pictures = 40 pictures__   Answer __40 pictures__

2. Write number sentences for these arrays.

   △ △ △ △ △
   △ △ △ △ △
   __2 × 5 = 10__

   ○ ○ ○ ○
   ○ ○ ○ ○
   ○ ○ ○ ○
   __3 × 4 = 12__

3. Someone measured each side of this shape using inches.

   How long is the vertical line segment? __5"__

   How long is the oblique line segment? __13"__

   How long is the horizontal line segment? __12"__

   What is the perimeter? __30"__

4. Trace the perpendicular line segments in Problem 3 using a crayon.

5. I have 1 quarter, 3 dimes, 1 nickel, and 9 pennies. Draw the coins.

   How much money do I have? __69¢__

   Ⓠ Ⓓ Ⓝ Ⓟ Ⓟ Ⓟ Ⓟ
   Ⓓ Ⓓ Ⓟ Ⓟ Ⓟ Ⓟ Ⓟ

6. Find the answers.

   75 + 37 = __112__   81 − 43 = __38__

2-127Wb   Copyright © 1991 by Saxon Publishers, Inc. and Nancy Larson. Reproduction prohibited.

Lesson 128

locating points on a coordinate graph

lesson preparation

materials

2 geoboards

2 geobands

ruler

Fact Sheet M 17.0

the night before

• If the geoboards are not labeled in the following way, use small pieces of masking tape and a marker to label them.

```
4 • • • • •
3 • • • • •
2 • • • • •
1 • • • • •
0 • • • • •
  0 1 2 3 4
```

in the morning

• Write the following in the pattern box on the meeting strip:

③ , ⑥ , ⑨ , ⑫ , ⑮ , ⑱ , ___ , ___ , ___ , ___

Answer: ③ , ⑥ , ⑨ , ⑫ , ⑮ , ⑱ , ㉑ , ㉔ , ㉗ , ㉚

• Write ⟨$2.37⟩ on the meeting strip. Provide a cup of 10 quarters, a cup of 10 dimes, a cup of 10 nickels, and a cup of 20 pennies.

THE MEETING

calendar

• Ask your child to write the date on the calendar and meeting strip.

• Ask your child to identify the number of days in 1 week, 2 weeks, and 3 weeks.

• Ask your child the following two or three times a week:

date ____ days ago   date ____ days from now

day of the week _____ days ago, day of the week _____ days from now

_____th month, month before, month after

- Record on the meeting strip a special event and the number of days until it occurs.

weather graph

- Ask your child to read and graph today's temperature to the nearest two degrees.
- Count by 10's and 2's to check the temperature on the graph.
- Ask your child to connect the dot for yesterday's temperature to the dot for today's temperature and compare the temperatures.

counting

- Count by 4's to 40 and backward from 40 by 4's.
- Count by 25's to 300 and backward from 300 by 25's.
- Count by 3's to 30 and backward from 30 by 3's.
- Do the following once a week:

    count by 10's to 400 and backward from 400 by 10's

    count by 5's to 100 and backward from 50 by 5's

    say the even numbers to 100 and backward from 50

    say the odd numbers to 49 and backward from 49

graph questions

- You and your child each ask a question about any of the graphs.

patterning

- Ask your child to do the following:

    identify the pattern (repeating, continuing, or both)

    identify the numbers and shapes to complete the pattern

    read the pattern

money

- Ask your child to put the coins in the coin cup. Count the money in the coin cup together.
- Ask your child for another way to show that amount of money. Count these coins together to check the amount.

clock

- Set the clock to a five-minute interval.

- Ask the following:

    "It's (morning/afternoon/evening). What time is it?"

    time one hour ago

    time one hour from now
- Ask your child to write the digital time on the meeting strip.
- Record on the meeting strip the time an activity will occur.

number of the day

- Write three number sentences for the number of the day on the meeting strip.

fact practice

- Write three fact family numbers (e.g., 2, 7, 9) on the chalkboard.
- Allow time for your child to write the four fact family number sentences on the chalkboard.

THE LESSON

Locating Points on a Coordinate Graph

"Today you will learn how to locate points on a different type of graph."

"Mathematicians call this a coordinate graph."

"You will use a geoboard to learn how to do this."

"We played a game using the geoboard when you were learning about parallel lines and line segments."

"I told you to start at a peg and stretch the band to another peg."

"It takes a lot of words to describe where a peg is."

"There is an easier way to do this."

"We will name the pegs using two numbers."

"This is how we will do that."

- Draw the following on the chalkboard:

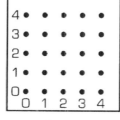

"We will always number our geoboard the same way."

"We will start at the peg at the bottom left-hand corner."

"This is called zero, zero."

"Mathematicians call it the origin."

"We name the other pegs by saying how far right to move and then how far up to move."

"We will go to the right first and then up."

"Let's try that."

"The point is four to the right and one up."

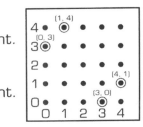

• Point to the point (4, 1). Write (4, 1) next to the point.

"The point is one to the right and four up."

• Point to the point (1, 4). Write (1, 4) next to the point.

"The point is three to the right and zero up."

• Point to the point (3, 0). Write (3, 0) next to the point.

"The point is zero to the right and three up."

• Point to the point (0, 3). Write (0, 3) next to the point.

"Now you will have a chance to find some points on the geoboard."

• Give your child and yourself a geoboard and a geoband.

"Hold your geoboard so it looks like my geoboard."

"Put your finger on the zero, zero peg."

• Check to make sure your child is pointing to the correct peg.

"Put your finger on the four, one peg."

• Check to make sure your child is pointing to the correct peg.

• Repeat with the one, four peg; the three, zero peg; and the zero, three peg.

"Point to the two, four peg."

"Put one end of your geoband over that peg."

"Point to the two, zero peg."

"Stretch your geoband to that peg."

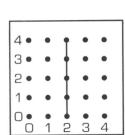

• Allow time for your child to do this.

"What type of line segment did you make?" vertical

"Stretch your geoband from the four, three peg to the zero, three peg."

• Allow time for your child to do this.

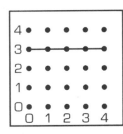

"What type of line segment did you make?" horizontal

"Stretch your geoband from the four, two peg to the one, four peg."

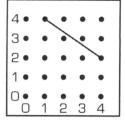

• Allow time for your child to do this.

"What type of line segment did you make?" oblique

"Let's play a game."

"I will make a line segment on my geoboard."

"See if you can make the same line segment without looking at my geoboard."

"I will write the location of the endpoints of my line segment on the chalkboard."

"Try to find my endpoints and make a line segment just like mine."

• Write the following on the chalkboard:

$$(1, 3) \qquad (4, 0)$$

"Remember, the first number tells us to go across and the second number tells us to go up."

"Use your geoband to show what you think my line segment looks like."

• Allow time for your child to do this.

• Put your geoboard next to your child's geoboard.

"Do our line segments look alike?"

• Repeat, using the following points:

$$(0, 3) \text{ and } (2, 1)$$
$$(4, 3) \text{ and } (1, 2)$$
$$(4, 0) \text{ and } (3, 4)$$

"Now make the longest line segment possible on your geoboard."

• Allow time for your child to do this.

"We can use a ruler to measure the line segment."

"Tell me the endpoints of the line segment you think is the longest possible line segment."

• Write the endpoints on the chalkboard.

"Let's use a ruler to check."

• Measure the geoboard line segment using a ruler. The longest line segments will be from (0, 0) to (4, 4) and from (0, 4) to (4, 0).

CLASS PRACTICE

number fact practice

- Use the pink fact cards to practice the multiplying by three facts with your child.

- Give your child **Fact Sheet M 17.0**.

- Time your child for one minute.

- Correct the fact sheet with your child and record the score.

- Allow time for your child to complete the unfinished facts.

WRITTEN PRACTICE

- Complete **Worksheet 128A** with your child.

- Complete **Worksheet 128B** with your child later in the day.

Name _____ **LESSON 128A**
(Draw an 11 cm line segment.) **Math 2**

Date
(Measure this line segment using centimeters. 4 cm)

1. The children in Mrs. Cambias's class read 37 books, the children in Mrs. Roberts's class read 28 books, and the children in Mrs. Ambrose's class read 25 books. How many books did the children in the three classes read altogether?

   Number sentence __37 + 28 + 25 = 90 books__ Answer __90 books__

2. How can you check to see if a shape has a right angle?

   Draw an example
   of a right angle.

   _____

   _____

3. Round each number to the nearest 10.

   13 __10__   45 __50__   76 __80__

4. Put a red dot at (3, 0).

   Put a blue dot at (2, 4).

   ```
   4 • • ●B • •
   3 • • • • •
   2 • • • • •
   1 • • • • •
   0 • • • ●R •
     0 1 2 3 4
   ```

5. Write the full date for 3/14/95.

   __March 14, 1995__

6. Find the answers.

   5 × 3 = __15__   6 × 2 = __12__   9 × 100 = __900__

   70 − 42 = __28__   63 − 18 = __45__

   ```
    4 2      2 9
    3 8      5 9
   + 1 7    + 3 2
    9 7     1 2 0
   ```

2-128Wa

Name _____ **LESSON 128B**

Date _____ **Math 2**

1. Craig has 36 markers, Eric has 24 markers, and Ryan has 48 markers. How many markers do the three boys have altogether?

   Number sentence __36 + 24 + 48 = 108 markers__ Answer __108 markers__

2. Use the corner of this piece of paper to find four examples of right angles at home. What did you find?

   _____   _____

   _____   _____

3. Round each number to the nearest 10.

   18 __20__   74 __70__   85 __90__

4. Put a red dot at (3, 4).

   Put a blue dot at (1, 2).

   ```
   4 • • • ●R •
   3 • • • • •
   2 • ●B • • •
   1 • • • • •
   0 • • • • •
     0 1 2 3 4
   ```

5. Write the full date for 7/9/97.

   __July 9, 1997__

6. Find the answers.

   9 × 3 = __27__   7 × 2 = __14__   3 × 100 = __300__

   18 + 54 = __72__   52 − 17 = __35__

   ```
    3 8      1 6
    5 2      5 7
   + 7 6    + 6 2
   1 6 6    1 3 5
   ```

2-128Wb

L esson 129

multiplying by four

lesson preparation

materials

bag of 40 color tiles

multiplication fact cards — blue

Fact Sheet M 15.0

the night before

• Separate the multiplying by four fact cards.

in the morning

• Write the following in the pattern box on the meeting strip:

> A, 5, B, 15, A, 25, B, 35, A, ___, ___, ___, ___, ___, ___

Answer: A, 5, B, 15, A, 25, B, 35, A, 45, B, 55, A, 65, B

• Write $1.88 on the meeting strip. Provide a cup of 10 quarters, a cup of 10 dimes, a cup of 10 nickels, and a cup of 20 pennies.

THE MEETING

calendar

• Ask your child to write the date on the calendar and meeting strip.

• Ask your child to identify the number of days in 1 week, 2 weeks, and 3 weeks.

• Ask your child the following two or three times a week:

> date _____ days ago, date _____ days from now

> day of the week _____ days ago, day of the week _____ days from now

> _____th month, month before, month after

• Record on the meeting strip a special event and the number of days until it occurs.

weather graph

- Ask your child to read and graph today's temperature to the nearest two degrees.
- Count by 10's and 2's to check the temperature on the graph.
- Ask your child to connect the dot for yesterday's temperature to the dot for today's temperature and compare the temperatures.

counting

- Count by 4's to 40 and backward from 40 by 4's.
- Count by 25's to 300 and backward from 300 by 25's.
- Count by 3's to 30 and backward from 30 by 3's.
- Do the following once a week:

    count by 10's to 400 and backward from 400 by 10's

    count by 5's to 100 and backward from 50 by 5's

    say the even numbers to 100 and backward from 50

    say the odd numbers to 49 and backward from 49

graph questions

- You and your child each ask a question about any of the graphs.

patterning

- Ask your child to do the following:

    identify the pattern (repeating, continuing, or both)

    identify the numbers and letters to complete the pattern

    read the pattern

money

- Ask your child to put the coins in the coin cup. Count the money in the coin cup together.
- Ask your child for another way to show that amount of money. Count these coins together to check the amount.

clock

- Set the clock to a five-minute interval.
- Ask the following:

    "It's (morning/afternoon/evening). What time is it?"

    time one hour ago

    time one hour from now

- Ask your child to write the digital time on the meeting strip.

• Record on the meeting strip the time an activity will occur.

number of the day

• Write three number sentences for the number of the day on the meeting strip.

fact practice

• Write three fact family numbers (e.g., 2, 7, 9) on the chalkboard.
• Allow time for your child to write the four fact family number sentences on the chalkboard.

THE LESSON

Multiplying by Four

"Today you will learn how to multiply by four."

• Give your child a bag of 40 color tiles.

"Put the color tiles in piles of four."

• Allow time for your child to pile the tiles.

"How many groups of four do you have?" **10**

"We can write that like this."

• Write the following on the chalkboard:

$$10 \text{ groups of } 4 = 40$$
$$10 \times 4 = 40$$

"How many tiles are in one group?" **4**

"Let's count by 4's to find out how many tiles are in five groups." **20**

"Let's count by 4's to find out how many tiles are in eight groups." **32**

• Write the following on the chalkboard:

$3 \times 4 = \quad 6 \times 4 = \quad 8 \times 4 = \quad 4 \times 4 = \quad 7 \times 4 = \quad 9 \times 4 = \quad 0 \times 4 =$

"The first problem is read 'three groups of four.' "

• Write "3 groups of 4" under 3×4.

"Read the second problem."

• Repeat for each problem.

"We will count the groups of tiles by 4's to find each answer."

"How many groups of tiles will we count for the first problem?" **3 groups**

"Count by 4's as you point to each group of tiles."

"How many tiles are in three groups of four?"   12

- Record "3 × 4 = 12" on the chalkboard.

- Repeat with all of the problems.

- Write the following on the chalkboard:

0 × 4 =	6 × 4 =
1 × 4 =	7 × 4 =
2 × 4 =	8 × 4 =
3 × 4 =	9 × 4 =
4 × 4 =	10 × 4 =
5 × 4 =	

"Let's fill in the answers."

"How many tiles are in zero groups of four?"   0

"How many tiles are in one group of four?"   4

"How many tiles are in two groups of four?"   8

- Repeat with each problem.

"We can write the multiplication facts for four a different way."

- Write the following on the chalkboard:

$$6 \times 4 = \qquad \begin{array}{r} 4 \\ \times\ 6 \\ \hline \end{array}$$

"These are two different ways of writing the same problem."

- Write the following next to the previous problem:

$$9 \times 4 = \qquad \begin{array}{r} 4 \\ \times\ 9 \\ \hline \end{array}$$

"We read these problems as 'six groups of four' and 'nine groups of four.' "

"What switcharound problems can we write for these two problems?"

$$4 \times 6 = \qquad \begin{array}{r} 6 \\ \times\ 4 \\ \hline \end{array} \qquad\qquad 4 \times 9 = \qquad \begin{array}{r} 9 \\ \times\ 4 \\ \hline \end{array}$$

- Ask your child to write the problems on the chalkboard.

"I will give you fact cards for the multiplying by four facts."

- Give your child the blue multiplication fact cards.

"Match the switcharound problems."

"When you finish, practice saying the answers for these problems."

Class Practice

number fact practice

- Give your child **Fact Sheet M 15.0**.
- Time your child for one minute.
- Correct the fact sheet with your child and record the score.
- Allow time for your child to complete the unfinished facts.

Written Practice

- Complete **Worksheet 129A** with your child.
- Complete **Worksheet 129B** with your child later in the day.

Name _____ **LESSON 129A**
(Draw an 8 cm line segment.) **Math 2**
Date _____
(Measure this line segment using centimeters. 12 cm)

1. The boys in Mrs. Topalanchik's class collected 74 buttons. The girls collected 17 buttons. The boys gave the girls 25 of their buttons. How many buttons do the boys have now?

   Number sentence   74 − 25 = 49 buttons   Answer 49 buttons

2. Write December 31, 1965 using digits.   12/31/65

3. Write a number sentence for this array.
   3 × 6 = 18

4. Measure the sides of the rectangle in Problem 3 using centimeters.
   What is the perimeter of the rectangle? 18 cm

5. Color half of the small squares in Problem 3 red. Color the other half of the small squares in Problem 3 yellow.
   How many squares are red? 9
   How many squares are yellow? 9

   4 • • • • •
   3 • ●R • • •
   2 • • • • •
   1 • • • • •
   0 • • ●B • •
     0 1 2 3 4

6. Put a red dot at (1, 3).
   Put a blue dot at (2, 0).

7. Find the answers.

4	4	4	4	4	4	4	4	4
×3	×6	×0	×9	×4	×1	×7	×5	×2
12	24	0	36	16	4	28	20	8

   2-129Wa

Name _____ **LESSON 129B**
Date _____ **Math 2**

1. Harvey collected 27 brown stones and 16 white stones. Andy collected 36 brown stones and 18 white stones. How many brown stones do the two boys have altogether?

   Number sentence   27 + 36 = 63 stones   Answer 63 stones

2. Write October 17, 1958 using digits.   10/17/58

3. Write a number sentence for this array.
   2 × 8 = 16
   (8 cm, 2 cm)

4. Someone measured the sides of the rectangle in Problem 3 using centimeters.
   What is the perimeter of the rectangle? 20 cm

5. Color half of the small squares in Problem 3 red. Color the other half of the small squares in Problem 3 yellow.
   How many squares are red? 8
   How many squares are yellow? 8

   4 ●R • • • •
   3 • • • • ●B
   2 • • • • •
   1 • • • • •
   0 • • • • •
     0 1 2 3 4

6. Put a red dot at (0, 4).
   Put a blue dot at (4, 3).

7. Fill in the missing numbers.

4	4	4	4	4	4	4	4	4
×4	×7	×2	×9	×3	×0	×5	×8	×1	×6
16	28	8	36	12	0	20	32	4	24

   2-129Wb

Lesson 130

creating two graphs using dominoes

---

lesson preparation

materials

Written Assessment #25

Oral Assessment #13

1 set of double six dominoes in a small brown paper bag

Masters 2-130A (optional), 2-130B, and 2-130C

crayons

5 quarters

10 dimes

10 nickels

20 pennies

Fact Sheet M 15.0

the night before

- If you do not have a set of dominoes, cut apart the dominoes on Master 2-130A and put them in a small brown paper bag. (The master can be glued to tagboard before cutting to make a sturdier set of dominoes.)

in the morning

- Write the following in the pattern box on the meeting strip:

> 25, 50, 75, 100, ____, ____, ____, ____, ____, ____

Answer: 25, 50, 75, 100, 125, 150, 175, 200, 225, 250

- Write $3.28 on the meeting strip. Provide a cup of 10 quarters, a cup of 10 dimes, a cup of 10 nickels, and a cup of 20 pennies.

THE MEETING

calendar

- Ask your child to write the date on the calendar and meeting strip.

- Ask your child to identify the number of days in 1 week, 2 weeks, and 3 weeks.

- Ask your child the following two or three times a week:

  date _____ days ago, date _____ days from now

  day of the week _____ days ago, day of the week _____ days from now

  _____th month, month before, month after

- Record on the meeting strip a special event and the number of days until it occurs.

weather graph

- Ask your child to read and graph today's temperature to the nearest two degrees.

- Count by 10's and 2's to check the temperature on the graph.

- Ask your child to connect the dot for yesterday's temperature to the dot for today's temperature and compare the temperatures.

counting

- Count by 4's to 40 and backward from 40 by 4's.

- Count by 25's to 300 and backward from 300 by 25's.

- Count by 3's to 30 and backward from 30 by 3's.

- Do the following once a week:

  count by 10's to 400 and backward from 400 by 10's

  count by 5's to 100 and backward from 50 by 5's

  say the even numbers to 100 and backward from 50

  say the odd numbers to 49 and backward from 49

graph questions

- You and your child each ask a question about any of the graphs.

patterning

- Ask your child to do the following:

  identify the pattern (repeating, continuing, or both)

  identify the numbers to complete the pattern

  read the pattern

money

- Ask your child to put the coins in the coin cup. Count the money in the coin cup together.

- Ask your child for another way to show that amount of money. Count these coins together to check the amount.

clock

- Set the clock to a five-minute interval.
- Ask the following:

  "It's (morning/afternoon/evening). What time is it?"

  time one hour ago

  time one hour from now

- Ask your child to write the digital time on the meeting strip.
- Record on the meeting strip the time an activity will occur.

number of the day

- Write three number sentences for the number of the day on the meeting strip.

fact practice

- Write three fact family numbers (e.g., 2, 7, 9) on the chalkboard.
- Allow time for your child to write the four fact family number sentences on the chalkboard.

ASSESSMENT

Written Assessment

"Today I would like to see what you remember from what we have been practicing."

- Give your child **Written Assessment #25**.
- Read the directions for each problem. Allow time for your child to complete each problem before continuing.
- Correct the papers, noting your child's mistakes on the **Individual Recording Form**. Review the errors with your child.

Oral Assessment

- Record your child's response(s) to the oral interview questions on the interview sheet.

THE LESSON

Creating Two Graphs Using Dominoes

"Today you will use a set of double six dominoes to make two graphs."

- Put the dominoes face up on the table.

"What do you notice about the dominoes?" there are two halves on each domino; there are zero to six dots per half

- Give your child **Master 2-130B.**

  "You will make a graph to show how many dominoes have zero dots, one dot, two dots, and so forth."

  "Instead of coloring the graph, you will write the number fact that shows how many dots are on each half of the domino."

  "We will call this the 'Domino Sums Graph.' "

- Point to one domino.

  "What number fact will you write for this domino?"

  "In which column will you write that?"

- Allow time for your child to write the number fact in the appropriate column.

- Give your child a small bag.

  "Put this domino in the bag."

  "Now choose another domino."

  "Write the number fact in the correct column on your graph."

- Allow time for your child to do this.

  "Put this domino in the bag."

  "Keep working until you have used all of the dominoes."

- When your child finishes, continue.

  "What do you notice about the 'Domino Sums Graph?' "

- Allow time for your child to offer as many observations as possible.

  "Now you will make a different graph."

- Give your child **Master 2-130C.**

  "We will call this graph the 'Domino Mix and Pick Graph.' "

  "This time you will take a domino out of the bag without looking in the bag."

  "You will use a crayon to color a box to show the total number of dots on the domino."

  "Then you will put the domino back into the bag and gently shake the bag to mix the dominoes."

  "Then you will take another domino from the bag."

  "You will color a box on the graph that shows that number of dots."

  "Keep doing this until I tell you to stop."

- Allow time for your child to record at least 40 dominoes.

  "What do you notice about the 'Domino Mix and Pick Graph?' " it looks like the Domino Sums Graph

"Why do you think that happened?" *there are more dominoes with the sums of 5, 6, 7, and 8*

CLASS PRACTICE

number fact practice

- Use the blue fact cards to practice the multiplying by four facts with your child.

- Give your child **Fact Sheet M 15.0.**

- Time your child for one minute.

- Correct the fact sheet with your child and record the score.

- Allow time for your child to complete the unfinished facts.

Teacher _____

Date _____

MATH 2   LESSON 130
Oral Assessment #13 Recording Form

Materials:
5 quarters
10 dimes
10 nickels
20 pennies
Students

	• Hand the child a selection of coins including at least one of each coin with a total value less than $1.00. *"Count the money."*	*"Show 82¢."* (Vary the amounts for different children.) *"Show 82¢ using different coins."*

2-130La

Name _____

Date _____

ASSESSMENT 25
LESSON 130
Math 2

1. There are 5 tables in Room 7. There are 3 books on each table. Draw a picture of the books on the tables.

How many books are on the tables altogether?

Number sentence _____ 5 × 3 books = 15 books _____   Answer _15 books_

2. Circle the shapes that have a right angle.

3. Write a number sentence for this array.

_____ 3 × 5 = 15 _____

4. Round each number to the nearest 10.

23 _20_   35 _40_   87 _90_

5. Use the graph to answer the question.

How many children have cats? _9_

Write two facts about this graph.

_____

CHILDREN'S PETS

dogs
cats
birds

0 2 4 6 8 10 12 14

6. Find the answers.

```
  7 9        6 1      16 + 54 = _70_      84 − 57 = _27_
+ 5 3      − 2 5
-----      -----
1 3 2        3 6
```

2-130Aa

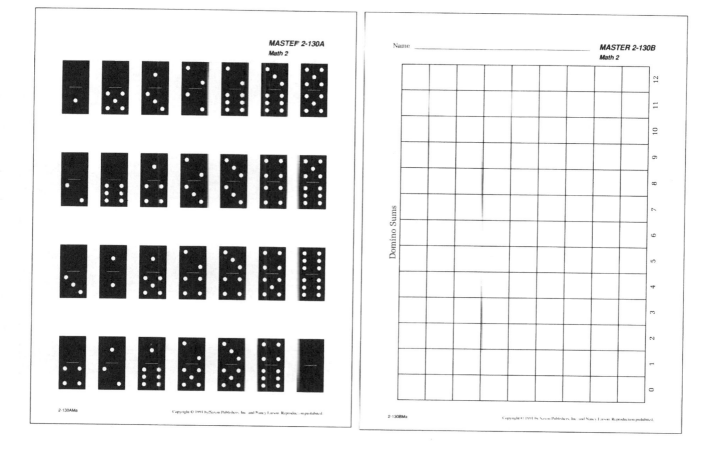

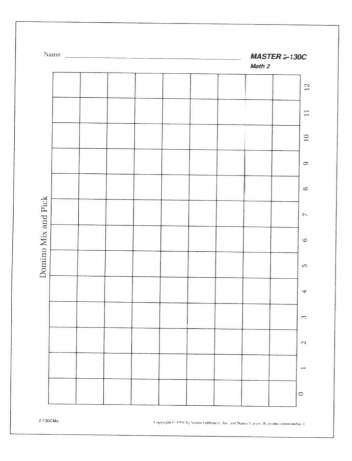

Lesson 131

doubling a number

THE MEETING

calendar

• Ask your child to write the date on the calendar and meeting strip.

• Ask your child to identify the number of days in 1 week, 2 weeks, and 3 weeks.

• Ask your child the following two or three times a week:

date _____ days ago, date _____ days from now

day of the week _____ days ago, day of the week _____ days from now

_____th month, month before, month after

• Record on the meeting strip a special event and the number of days until it occurs.

weather graph

• Ask your child to read and graph today's temperature to the nearest two degrees.

• Count by 10's and 2's to check the temperature on the graph.

• Ask your child to connect the dot for yesterday's temperature to the dot for today's temperature and compare the temperatures.

counting

- Count by 4's to 40 and backward from 40 by 4's.
- Count by 25's to 300 and backward from 300 by 25's.
- Count by 3's to 30 and backward from 30 by 3's.
- Do the following once a week:

    count by 10's to 400 and backward from 400 by 10's

    count by 5's to 100 and backward from 50 by 5's

    say the even numbers to 100 and backward from 50

    say the odd numbers to 49 and backward from 49

graph questions

- You and your child each ask a question about any of the graphs.

patterning

- Ask your child to do the following:

    identify the pattern (repeating, continuing, or both)

    identify the numbers to complete the pattern

    read the pattern

money

- Ask your child to put the coins in the coin cup. Count the money in the coin cup together.
- Ask your child for another way to show that amount of money. Count these coins together to check the amount.

clock

- Set the clock to a five-minute interval.
- Ask the following:

    "It's (morning/afternoon/evening). What time is it?"

    time one hour ago

    time one hour from now
- Ask your child to write the digital time on the meeting strip.
- Record on the meeting strip the time an activity will occur.

number of the day

- Write three number sentences for the number of the day on the meeting strip.

fact practice

- Write three fact family numbers (e.g., 2, 7, 9) on the chalkboard.

• Allow time for your child to write the four fact family number sentences on the chalkboard.

THE LESSON

Doubling a Number

"Today you will learn how to double a number."

• Give your child one coupon.

"What is this?"

"What do people do with it?"

"How do people use it?"

"This coupon has a picture of _____ and it has _____ ¢ printed on it."

"Does this mean that _____ will cost only _____ ¢?"   no

"What does that mean?"   that is the amount of money subtracted from the cost of the item

• Give your child another coupon.

"What item is on this coupon and how much is the coupon worth?"

• Repeat with one more coupon.

"Look carefully at your coupons."

"Find things that are the same about all of your coupons."

• Allow time for your child to list as many similarities as possible.

"Most coupons have an expiration date."

"This means that the coupon cannot be used after this date."

"This date is usually written in digits."

"Find the expiration date on your coupons."

"What is the expiration date on each coupon?"

• Ask your child to read the full expiration date on each coupon.

• Write the following on the chalkboard:

25¢   40¢   15¢

"Let's pretend that I have coupons for these amounts."

"Some stores advertise that they will double the value of the coupon."

"What does that mean?   it is worth twice as much

"Doubling a coupon means that it is worth twice as much or two times as much."

"If I have a coupon worth 25¢, how much is it worth in a store that will double its value?"   50¢

"Let's double the value of my other coupons."

"What is 40¢ doubled?" 80¢

"What is 15¢ doubled?" 30¢

"Doubling is like adding the number twice or multiplying by two."

"I will write the original value of my coupons in one column and the double value in another column."

- Write the following on the chalkboard:

original value	doubled value
25¢	50¢
40¢	80¢
15¢	30¢

"Let's double the value of each of your coupons."

- Draw another chalkboard chart next to the first chart.

"Write the original value of your coupons in the first column of the chalkboard chart and write the doubled value in the second column."

- When your child finishes, continue.

"How can we find out how much money I will save if I use all my coupons at a store that doesn't double the value of the coupon?" add the first column

"Let's add the amounts on my coupons."

- Do the addition with your child.

"How can we find out how much money I will save if I use all my coupons at a store that doubles the value of the coupons?" add the second column

- Do the addition with your child.

"What do you notice about the answer?" it is double the amount of the first answer

"Now add to find how much money you will save without doubling your coupons."

"Then add to find how much money you will save if you double your coupons."

- Check your child's addition.

CLASS PRACTICE

number fact practice

- Use the fact cards to practice the multiplication facts with your child.

• Give your child **Fact Sheet M 14.1.**

• Time your child for one minute.

• Correct the fact sheet with your child and record the score.

• Allow time for your child to complete the unfinished facts.

WRITTEN PRACTICE

• Complete **Worksheet 131A** with your child.

• Complete **Worksheet 131B** with your child later in the day.

Name _____

(Draw a 9 cm line segment.)

Date _____

(Measure this line segment using centimeters. __5__ cm)

LESSON 131A
Math 2

1. Each child in Mrs. Velardi's class has 3 pencils. There are seven children in the class. Draw a picture to show the children's pencils.

How many pencils do the children have altogether?

Number sentence ____7 × 3 pencils = 21 pencils____ Answer _21 pencils_

2. Use the Venn diagram to answer the questions.

Which letters on the graph have only parallel line segments? __N, M, Z__

Which letters have perpendicular line segments but not parallel line segments? __T, L__

Which letters have both parallel and perpendicular line segments? __E, F, I, H__

LETTERS
PARALLEL LINE SEGMENTS   PERPENDICULAR LINE SEGMENTS

3. Round each number to the nearest 10 and add the rounded numbers.

57 + 12
_60_ + _10_ = _70_

43 + 21
_40_ + _20_ = _60_

4. Double the value of these coupons.

| 30¢ | 60¢ | | 25¢ | 50¢ |

5. 8 dimes and 16 pennies = 9 dimes and __6__ pennies

6. Find the answers.

```
  4 9 ¢        5 7 ¢
+ 3 3 ¢      − 2 9 ¢
-------      -------
  8 2 ¢        2 8 ¢
```

68¢ + 52¢ = _$1.20_   73¢ − 41¢ = _32¢_

2-131Wa   Copyright © 1991 by Saxon Publishers, Inc. and Nancy Larson. Reproduction prohibited.

Name _____

Date _____

LESSON 131B
Math 2

1. Eddie counted 6 cars. Each car has 4 wheels. Draw a picture to show the wheels on the cars.

How many wheels do 6 cars have?

Number sentence ____6 × 4 wheels = 24 wheels____ Answer _24 wheels_

2. Use the Venn diagram to answer the questions.

Which letters on the graph have perpendicular line segments but not parallel line segments? __T, L__

Which letters have parallel line segments? __N, M, Z, E, F, H, I__

Which letters have both parallel and perpendicular line segments? __E, F, H, I__

LETTERS
PARALLEL LINE SEGMENTS   PERPENDICULAR LINE SEGMENTS

3. Round each number to the nearest 10 and add the rounded numbers.

21 + 58
_20_ + _60_ = _80_

38 + 18
_40_ + _20_ = _60_

4. Double the value of these coupons.

| 50¢ | $1.00 | | 15¢ | 30¢ |

5. 5 dimes and 18 pennies = 6 dimes and __8__ pennies

6. Find the answers.

```
  5 4 ¢        8 5 ¢
+ 4 5 ¢      − 7 3 ¢
-------      -------
  9 9 ¢        1 2 ¢
```

47¢ + 28¢ = _75¢_   91¢ − 36¢ = _55¢_

2-131Wb   Copyright © 1991 by Saxon Publishers, Inc. and Nancy Larson. Reproduction prohibited.

730

Lesson 132

dividing by two

---

lesson preparation

materials

bag of 20 color tiles

2 small plates

Fact Sheet S-100

in the morning

• Write the following in the pattern box on the meeting strip:

> 43, 48, 53, 58, ____, ____, ____, ____, ____, ____

Answer: 43, 48, 53, 58, 63, 68, 73, 78, 83, 88

• Write $1.59 on the meeting strip. Provide a cup of 10 quarters, a cup of 10 dimes, a cup of 10 nickels, and a cup of 20 pennies.

THE MEETING

calendar

- Ask your child to write the date on the calendar and meeting strip.

- Ask your child to identify the number of days in 1 week, 2 weeks, and 3 weeks.

- Ask your child the following two or three times a week:

  date _____ days ago, date _____ days from now

  day of the week _____ days ago, day of the week _____ days from now

  _____th month, month before, month after

- Record on the meeting strip a special event and the number of days until it occurs.

weather graph

- Ask your child to read and graph today's temperature to the nearest two degrees.

- Count by 10's and 2's to check the temperature on the graph.

- Ask your child to connect the dot for yesterday's temperature to the dot for today's temperature and compare the temperatures.

counting

- Count by 4's to 40 and backward from 40 by 4's.
- Count by 25's to 300 and backward from 300 by 25's.
- Count by 3's to 30 and backward from 30 by 3's.
- Do the following once a week:

    count by 10's to 400 and backward from 400 by 10's

    count by 5's to 100 and backward from 50 by 5's

    say the even numbers to 100 and backward from 50

    say the odd numbers to 49 and backward from 49

graph questions

- You and your child each ask a question about any of the graphs.

patterning

- Ask your child to do the following:

    identify the pattern (repeating, continuing, or both)

    identify the numbers to complete the pattern

    read the pattern

money

- Ask your child to put the coins in the coin cup. Count the money in the coin cup together.
- Ask your child for another way to show that amount of money. Count these coins together to check the amount.

clock

- Set the clock to a five-minute interval.
- Ask the following:

    "It's (morning/afternoon/evening). What time is it?"

    time one hour ago

    time one hour from now

- Ask your child to write the digital time on the meeting strip.
- Record on the meeting strip the time an activity will occur.

number of the day

- Write three number sentences for the number of the day on the meeting strip.

fact practice

- Write three fact family numbers (e.g., 2, 7, 9) on the chalkboard.
- Allow time for your child to write the four fact family number sentences on the chalkboard.

THE LESSON

Dividing by Two

"Today you will learn how to divide by two."

"Dividing by two is the same as finding half of a group of objects."

- Put 14 color tiles in a pile on the table.

"Let's pretend that these are hard candies."

"I would like to share these 14 candies with you."

"When two people share something, they divide it into two equal groups."

"I will use two plates to show the two groups."

"How many candies do you think will be on each plate when we divide the candies into two equal groups?"

"Divide the candies into two equal groups."

- Allow time for your child to do this.

"Do both groups have the same number?"

"How many candies are on each plate?" 7

"We can write what you did like this."

- Write the following on the chalkboard:

  14 candies ÷ 2 groups = 7 candies in each group

- Put the tiles in a pile. Remove 4 tiles.

"Let's divide ten candies into two equal groups."

"How many candies do you think will be on each plate?"

"Divide the candies into two equal groups."

- Allow time for your child to do this.

"Do both groups have the same number?"

"How many candies are on each plate?" 5

"We can write what you did like this."

- Write the following on the chalkboard:

  10 candies ÷ 2 groups = 5 candies in each group

> *"Now you will have a chance to practice dividing by two."*

- Give your child 6 more color tiles.
- Write the following on the chalkboard:

$$20 \div 2 =$$

$19 \div 2 =$	$9 \div 2 =$
$18 \div 2 =$	$8 \div 2 =$
$17 \div 2 =$	$7 \div 2 =$
$16 \div 2 =$	$6 \div 2 =$
$15 \div 2 =$	$5 \div 2 =$
$14 \div 2 =$	$4 \div 2 =$
$13 \div 2 =$	$3 \div 2 =$
$12 \div 2 =$	$2 \div 2 =$
$11 \div 2 =$	$1 \div 2 =$
$10 \div 2 =$	$0 \div 2 =$

> *"When you divided 14 candies into two groups, how many candies were in each group?"* 7

> *"The first number tells us the number of candies."*

> *"The two tells us the number of groups or plates we will use."*

> *"When you divided ten candies into two groups, how many candies were in each group?"* 5

> *"Do you know another one of these answers?"*

- Record your child's suggested answer on the chalkboard.

> *"Let's check the answer to see if you are correct."*

> *"Divide _____ candies equally between the plates."*

- Allow time for your child to divide the tiles.

> *"How many candies are on each plate?"*

> *"Do you have any extra candies?"*

- Record the answer. If there is an extra tile, record the remainder as R1.
- Repeat until all the answers are listed.

> *"Which numbers of candies can we divide into two groups without a remainder?"* even numbers

> *"Do you see a pattern in the answers?"*

> *"What is it?"*

> *"I will erase all the answers that have remainders."*

- Erase the odd numbers divided by two.

"What do you notice about the numbers being divided?"   they are the even numbers

"Division is the opposite of multiplication."

"How can we use multiplication to help us find these answers?"   we can think what number times two will give us the number being divided; we can multiply backward to check our answer

"Now I will say a problem."

"See how fast you can say the answer."

- Allow your child to refer to the answers on the chalkboard.
- Name problems at random.
- Erase the answers and repeat.

CLASS PRACTICE

number fact practice

- Give your child **Fact Sheet S-100**.
- Time your child for five minutes.
- Correct the fact sheet with your child and record the score.
- Allow time for your child to complete the unfinished facts.

WRITTEN PRACTICE

- Complete **Worksheet 132A** with your child.
- Complete **Worksheet 132B** with your child later in the day.

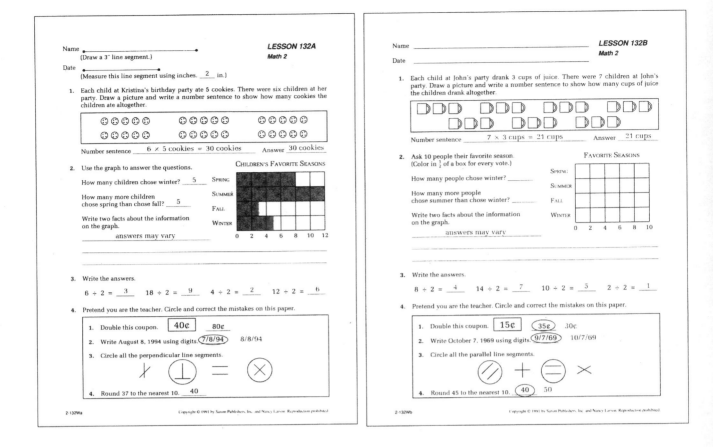

LESSON 132A
Math 2

Name ●————————————————●
(Draw a 3" line segment.)

Date ————————————————
(Measure this line segment using inches. __2__ in.)

1. Each child at Kristina's birthday party ate 5 cookies. There were six children at her party. Draw a picture and write a number sentence to show how many cookies the children ate altogether.

Number sentence __6 × 5 cookies = 30 cookies__   Answer __30 cookies__

2. Use the graph to answer the questions.

CHILDREN'S FAVORITE SEASONS

How many children chose winter? __5__

How many more children chose spring than chose fall? __5__

Write two facts about the information on the graph.
_____ answers may vary _____

3. Write the answers.

$6 ÷ 2 =$ __3__ $18 ÷ 2 =$ __9__ $4 ÷ 2 =$ __2__ $12 ÷ 2 =$ __6__

4. Pretend you are the teacher. Circle and correct the mistakes on this paper.

1. Double this coupon. ☐40¢   80¢
2. Write August 8, 1994 using digits. (7/8/94)   8/8/94
3. Circle all the perpendicular line segments.
4. Round 37 to the nearest 10. __40__

2-132Wa   Copyright © 1991 by Saxon Publishers, Inc. and Nancy Larson. Reproduction prohibited.

LESSON 132B
Math 2

Name ————————————————

Date ————————————————

1. Each child at John's party drank 3 cups of juice. There were 7 children at John's party. Draw a picture and write a number sentence to show how many cups of juice the children drank altogether.

Number sentence __7 × 3 cups = 21 cups__   Answer __21 cups__

2. Ask 10 people their favorite season. (Color in ½ of a box for every vote.)

FAVORITE SEASONS

How many people chose winter? _____

How many more people chose summer than chose winter? _____

Write two facts about the information on the graph.
_____ answers may vary _____

3. Write the answers.

$8 ÷ 2 =$ __4__ $14 ÷ 2 =$ __7__ $10 ÷ 2 =$ __5__ $2 ÷ 2 =$ __1__

4. Pretend you are the teacher. Circle and correct the mistakes on this paper.

1. Double this coupon. ☐15¢   (35¢)   30¢
2. Write October 7, 1969 using digits. (9/7/69)   10/7/69
3. Circle all the parallel line segments.
4. Round 45 to the nearest 10. (40)   50

2-132Wb   Copyright © 1991 by Saxon Publishers, Inc. and Nancy Larson. Reproduction prohibited.